D1100485

Dow's Dictionary of Railway Quotations

Dow's Dictionary of Railway Quotations

Compiled and Edited by Andrew Dow

The Johns Hopkins University Press

Baltimore

© 2006 The Johns Hopkins University Press

All rights reserved. Published 2006

Printed in the United States of

America on acid-free paper

1 2 3 4 5 6 7 8 9

The Johns Hopkins University Press

2715 North Charles Street

Baltimore, Maryland 21218-4363

www.press.jhu.edu

Library of Congress Cataloging-in-Publication Data

Dow, Andrew.

Dow's dictionary of railway quotations / compiled and edited by Andrew Dow.

p. cm.

Includes index.

ISBN 0-8018-8292-3 (hardcover : alk. paper)

1. Railroads—Quotations, maxims, etc. I. Title: Dictionary of railway quotations. II. Title.

TF9.D68 2006

385′.03—dc22   2005013812

A catalog record for this book is available

from the British Library.

385

Pages 353–54 are an extension

of the copyright page.

To my darling wife, Stephanie,
without whose love and encouragement
I would not have written this book

At the back of my mind, in writing of railway matters,
is the desire that railwaymen may be better understood and
sympathised with by a public which is usually too busy to heed.

James Scott, Railway Romance and Other Essays, *1913*

Contents

Foreword

In his letter to me enclosing the draft manuscript of this book, the compiler commented that his work "may be regarded by some as unusual, but I have established that the specialist dictionary of quotations is much accepted these days . . . So maybe it is due time for the railways to have one."

Time indeed. In 2004, as I write this, the railway is to an increasing number of people a near-forgotten anachronism, though to others it remains the artery that encompasses and facilitates their lives. In Britain, it is some four hundred years since the railway emerged in distinctive tramroad form, two hundred years since mechanical power first displaced muscle—or gravity—as traction for rail transport, and one hundred years since the speed of transport by rail first topped, reputedly, one hundred miles per hour.

But the celebration of the *origins* of this form of transport is much less important than the celebration of its unparalleled social and economic *impact,* for the railway defined and enabled an era. In the 150 years after 1804 it reigned with little challenge. It pushed the boundaries of individual experience for whole populations far beyond local horizons. Throughout our planet, lives were increasingly sustained by a growing complexity of trade on a national and international scale. The railway, unheralded, enabled whole countries to function far more effectively as entities, whatever their size. Nowhere was this more true than in the nascent United States, as the expanding bounds of settlement pushed ever westward and shining twin ribbons of steel linked frontiers to eastern homelands in journeys of hours rather than weeks.

The bones of such change have long been grist to the economic, social, and technical historians' mill, but too often those bones are dry indeed and may come to life only through the unwittingly vivid words of the contemporary observer. For me, the strict localism of the pre-railway world is nowhere better expressed than through the pen of Thomas Hardy in *Far from the Madding Crowd,* first published in 1874 but presenting a gloriously evocative picture of earlier country innocence and isolation in the days before the railway penetrated the rural fastnesses of southern England. The young shepherd boy Cainy Bell, visiting Bath from Dorset for the first time—a distance of some forty-five miles—explodes in naive wonder, "I've seed the world at last." For me, the first impact of the speed of mechanical traction on a world previously constrained by human or animal muscle is most famously recalled in a letter written by the actress Fanny Kemble, a beautiful girl who, given a ride on the primitive footplate of the *Rocket* by the susceptible George

Stephenson in August 1830, before the formal opening of the pioneer Liverpool & Manchester Railway, breathlessly exclaimed, "You cannot conceive what that sensation of cutting the air was . . . When I closed my eyes this sensation of flying was quite delightful, and strange beyond description." At thirty miles per hour indeed!

Images like this random pair bring life to history in a way so many other forms of illustration fail to. But they occur in scattered places, often tucked away in unrelated and obscure forms of literature, most of a non-railway nature, and rarely emerge in a casual library browse. Even the best read of railway historians, such as Jack Simmons, Michael Robbins, or John White, can touch only the margins of this treasury in their writings. In this unique book, Andrew Dow seeks to bring together more of these images than has any previous author and succeeds in what has been a truly Herculean labor of discovery and compilation, however absorbing and enjoyable.

The scale of his achievement is truly staggering. This volume includes more than three thousand entries covering over seven hundred themes. Andrew Dow's choices have been truly eclectic: there is a grave danger of casually picking up this volume and being lost beyond time in its treasures. His patience, his scholarship, and his range are equally outstanding. Read, for example, the appendix "The Engine Driver's Epitaph." The compiler unearthed— almost literally—four versions of this epitaph, three carved on British graves between 1840 and 1842 and one in the United States from 1853, and all four were visited and their provenance, wording, and spelling checked.

Andrew Dow, son of the renowned British railway manager and historian George Dow, has proved the ideal candidate for such a historic task. From a senior business position in the United States, he became the Head of Britain's National Railway Museum at York. He later returned to industry, but to Britain's privatized railway, and he now enjoys a highly active retirement. He is a prolific railway author in his own right, with a stirring record of achievement in railway preservation. In this volume he releases that breadth of enthusiasm and knowledge for a wider audience. For me, however, the keynote of his achievement is not the scale of this work, impressive though it is, but rather the sheer bubbling enthusiasm it generates and shares. I would love to have finished this foreword with an apposite quotation, but every one that came to mind was already here!

Allan Patmore, CBE

Acknowledgments

Books such as this always take far longer to write than one wishes, and it is sobering to have to record that three of the people from whom I received earliest support are no longer with us. My predecessor as Head of the National Railway Museum, Dr John Coiley, was, as ever, unfailingly courteous and helpful, and pointed me to much of the poetry that has been written about railways. Professor Jack Simmons spent a lot of time searching for items and recommending sources, and was a remarkably generous host whenever I visited him to discuss this project in its early days. Michael Robbins, a railwayman and railway historian of distinction, gave a lot of thought to important sayings and quotations of obscure origin.

Another distinguished railway historian was of great assistance to me: Jack White, who was helpful with sources and references in America. Richard Hardy, engineman and railwayman extraordinary, was generous with his files and memories, bringing the railway and its people so much to life. Peter Gilliver at the *Oxford English Dictionary* was particularly helpful and patient with a variety of queries on words, meanings, and origins, and George Drury, former librarian at Kalmbach, the publisher of *Trains Magazine,* provided valuable leads. Roger Ford of *Modern Railways* dug quickly and willingly into the recesses of his memory more often than I can remember, and Nigel Harris of *Rail* was very helpful with good advice and contributions. Adrian Ettlinger, a fellow member of the Railway & Locomotive Historical Society, most helpfully supplied entries that I would not have otherwise found, while Cathie Lamere of Hayner Public Library in Alton, Illinois, diligently followed leads and unearthed the story of the American use of the famous epitaph described in appendix A. Terry Worrall was very helpful in ensuring that I correctly recorded the story of the headline by which he was famously misquoted, and Karen Douglas of the *Evening Standard* helped me to find out who wrote the headline. Michael Max, a bibliophile, fellow admirer of railways, and good friend, generously made his collection available to me and gave me several leads that I would otherwise surely have missed. At the National Railway Museum in York, Beverley Cole helped to identify and date legends on posters and opened collections and files for me, while librarian Philip Atkins pointed unerringly to much else. Recently retired from NRM, Richard Gibbon threw himself, with his customary enthusiasm, into helping me to fill some vital holes in the collection. Bill Withuhn at the Smithsonian Institution was also of material assistance. The staff at the British Library and the Newspaper Library were always helpful, and Rohn Roaché, head of book service at the

Library of Congress, made particular efforts to ensure that a research session there was very fruitful.

I am also indebted to have received help of all kinds from Ian Allan, Mark Allatt, Chris Austin, George Behrend, Gordon Biddle, the late John Blyth, Anthony Bulleid, Anthony Burton, Malcolm Crawley, Dick Dawson, David Elliot, Chris Ford, Allan Garraway, Andrew Gillitt, Dr Terry Gourvish, Peter Handford, Jo Hornsby, Dr Geoff Hughes, Ed King, Dr Michael Lewis, Steve McColl, Ian McKellar, Donald F Morrison, Graham Nicholas, Philip Pacey, the late Sir Peter Parker, Sir Christopher Pinsent, Paul Ramsey, Sir Bob Reid, Dr Peter Rodgers, the late Ted Rose, Gregg Ryan, Dr Richard Saunders Jr, Hermann Schmidtendorf, Roy Thomson, Peter Trewin, Roger Waller, Bob Webber, Grant Woodruff, and Kyle Wyatt. Several fellow members of the Railway & Locomotive Historical Society were helpful, especially Bob Guhr, Rob MacKavanagh, and Tommy Meehan, who helped me to find original material concerning William Vanderbilt's infamous condemnation of the public. My sister Margaret Moore kindly provided some references.

Professor Allan Patmore, who has been a good friend and colleague for a dozen or more years, kindly interrupted his busy days to write a foreword for which I am truly grateful. My fine friend Richard Hall was of assistance more often than he knew; his wife Jackie kindly translated two passages from French.

It has been a joy to work with the staff of the Johns Hopkins University Press to bring this book to life. Henry Tom and Claire McCabe in the acquisitions department took it on with assurance and enthusiasm; Susan Lantz edited it in the face of all my intrusions; Thomas Lovett, Jennifer Gray, and Kathy Alexander in the marketing department dealt expertly with the needs of establishing the book on both sides of the Atlantic, and many of their colleagues, unknown and unseen to me, helped to bring this book to market. I am indebted to them all.

Indirectly, but influentially, I received much assistance from my father, for although he died many years ago, his catholic appreciation (and, it appeared to me, all-embracing knowledge) of railways, coupled with his love of the English language, seemed always to be there to help me to identify passages of value, significance, interest, and amusement. My son James found some entries but more vitally rescued me many times from ignorance of my computer, which is able to detect stress in its user and to react proportionately with fresh challenges; my daughter Alyx helped me with proofreading. Above all I thank my dear wife Stephanie, who all too often set aside her own more demanding doctoral studies to help me in so many ways. In particular she was an invaluable adviser in the more intellectual aspects of research and presentation of material.

The help and generosity of all these good people is very much appreciated, but then, of the railway fraternity one could expect no less.

Introduction

Countless millions of words have been written about railways over the many generations since they started to shape the development of the modern world. The creation, operation, and, in too many cases, abandonment of railways have generated words ranging from the dry office memorandum to highly emotive language; from Acts of Parliament to poetry, song, and obituary; from loathing and bitter criticism to love and admiration; and from occasions of great gravity to moments of exquisite comedy.

This book is intended to draw together, perhaps for the first time, a wide selection from this treasure-trove. In doing so, few boundaries have been observed. In particular, this is not simply a book of literary quotations. Railways are, after all, agents of commerce and industry, and therefore relatively few railway writings are to be found in literature. To have restricted the collection thus would have excluded many entries that tell us so much about attitudes to railways. Their exclusion would not have allowed the collection to reflect the vast range of subjects and emotions that railways inspire. Nor would it have allowed this collection to perform a wider service for the reader, summed up thus: first, to gather together, in one authoritative record, well-known words spoken and written about railways; second, to draw attention to many more words that, while less well known, are no less worthy of record; and third, to present the whole in a form that allows readers—especially authors and commentators—to find an apt expression, opinion, or statement on a particular railway subject with which to inform and illustrate their own work on railways.

The collection has therefore been arranged by subject matter rather than by author. This has made easy the comparison of many interesting points of view and modes of expression. It has also imposed upon the compiler an obligation to find diverse views while resisting the temptation to include particular passages simply to serve that purpose. The inclusion of an entry regardless of merit has not been the aim. Rather, this book is intended as a repository of the memorable.

Particular care has been taken to obtain precise quotations from original sources and to use spelling and punctuation as they appeared there. Attention has not been drawn to unusual words or spellings by means of the device [*sic*]; the reader may take it that the unexpected is faithfully reproduced from the original. Occasionally authors, in quoting others, interrupt the quoted passage with narrative words such as *he said*. These have been omitted to restore the quotation to its original spoken state, in preference to duplicating the form in which it appeared in print.

In many cases, entries have been found as quotations in other works that have not cited their source. In these cases the citation in this book uses words such as "quoted by —— in ——."

It is a reflection of the vast scope and scale of railway writing that more than thirty-four hundred entries have been found to give what the compiler hopes is adequate weight and balance to the book. Sources include not only published material but also private documents and conversations. Here again, a good balance has been sought, made possible only by the generous involvement of many people who responded to requests for suggested entries. Such sources may not customarily be used for a work of this kind but are included because documents such as internal office memoranda and personal letters often contain unfettered views or even undiplomatic truths. Conversations and informal addresses may yield unexpected *bon mots* or truisms. These reflect railways, their impact, and their history just as much as do carefully drafted reports, historical accounts, or speeches.

A few quotations have achieved fame, or infamy, well beyond the railway fraternity. In America, railroadmen were for generations tainted quite wrongly by the reported words of William H Vanderbilt: "The public be damned." Accounts of the origin and context of these words vary, as will be learned from their entry in this book (542.2) and from appendix B. Either way, it seems, Vanderbilt was in part misquoted or, at best, taken out of context. It is worrying for railwaymen to have to acknowledge that three other quotations that have haunted them over the years are also examples of misquotes, willful or otherwise. One is "Charge what the traffic will bear" (565.2). This does not mean "charge the shipper all you can," as is often believed, but the expression, used as a term of art on both sides of the Atlantic starting more than one hundred years ago, resulted in many who heard it thinking that railways and railroads were rapacious and grasping. In fact, the expression arose principally because the nature of railway operation makes calculation of the unit cost of production almost impossible; therefore, custom on both sides of the Atlantic has always dictated that railways do not charge prices derived directly from the cost of providing service. Next, "What a way to run a railroad," which today is taken to mean puzzled criticism of an institution by a third party, started life as a caption to a cartoon in which the statement was made by the person responsible for the chaos, who was quite oblivious to what he had done (412.8). The third expression is "The wrong kind of snow," which was not stated by the railwayman to whom it has been attributed (with the implication that incompetent engineers had been taken unawares by a natural phenomenon that should have been anticipated) but was a newspaper headline writer's shorthand (753.12). Indeed, investigation after the event showed that the meteorological conditions were highly unusual and had caused problems in many countries across Europe. Although these three quotes may have since taken on lives and meanings of their own, they all started as misunderstandings, misquotations, or simply inadequate reporting.

This book is intended to be a true dictionary of quotations and not in any sense an anthology. Several collections of longer railway writings have been published, and this book offers no competition to them. Rather, true to the established form of dictionaries of quotations, this book concentrates on the self-contained comment, the pithy summation, the single view on the single subject. Some writers, however, are not always pithy. A point has to be developed or put into context, and in all cases the compiler has tried to ensure that a passage quoted has lost nothing of its meaning or intent or impact by being taken out of its original context. This desire has often justified a longer passage where to have selected a shorter entry would have destroyed the impact that the writer or speaker had desired.

Throughout, the intention has been to record remarkable, memorable words—words, perhaps, that compelled the speaker to share their quality with friends and acquaintances, knowing that they would be appreciated and would strike a chord. Thus, many of the quotations concern commonplace matters. And while memorable words on commonplace matters have been used to compile this book, ordinary words on extraordinary facts have not: it has never been the intention that this would be a record of remarkable facts and events in the history of railways. Nor is it intended to be a repository of anecdotes and funny stories. This book is more about attitudes and opinions, memorably expressed, whether explicit or implicit, than about simple cold facts. This rule, however well founded, has tempted me sorely. For example, I would have loved to include Albert Speer's brief account of Hermann Goering's attempt to have him produce steam locomotives from concrete because of the shortage of steel in wartime Germany: "Of course the concrete locomotives would not last as long as the steel ones, he said; but to make up for that we would simply have to produce more of them." Apart from the fact that this story was not originally told in the English language, it is more a remarkable fact than a memorable quotation, and the distinction has to be observed.

Of the very many statements of very many kinds included here, a few may not be wholly accurate in their dealing with this or that aspect of railways: surely Margaret Thatcher was wrong in her criticism of British Railways managers (412.11); the *Chattanooga Choo-choo* could not have left Track 29 because Pennsylvania Station did not have that many tracks (712.16); it is not strictly true that the common rail spike was derived in America because it was used elsewhere beforehand (628.1). And surely David P Morgan exaggerated the strategic importance of the Pennsylvania's electrified line between New York and Washington (746.29). But that is not the point; this book serves to record, not to judge. With only very few exceptions, it has not been thought proper or necessary to comment upon the accuracy of statements or sentiments, but rather it has been for the compiler to decide which words were of import, and, for those that were selected, to ensure that the words themselves were accurately recorded.

There remains the question of what exactly constitutes a railway quotation. All of the entries in this book have been chosen because they tell us something about railways. They do this directly, or they show the writer's bias toward railways, or they show the railwayman's view of matters connected with the railway. The mere fact of the railway appearing in a description or narrative does not ensure inclusion if the railway is incidental or if no opinion is given of the railway. Certainly the metaphorical use of the railway and its components, once very common in literature, has been sparingly illustrated here. There are, of course, exceptions to these exclusions if the circumstances are significant or if, as in the case of the extraordinarily intense writing of Thomas Wolfe, it is absolutely irresistible. I do not regret any inconsistency here: early in the preparation of this book the late Professor Jack Simmons gave sound advice: "By all means have some rules, but be prepared to break them!"

The research for this book has certainly demonstrated, never to the compiler's surprise, the seemingly boundless wealth of railway writing. It would be possible for the compilation of a book such as this to absorb a lifetime. That temptation has had to be resisted, even though the price may be letters from readers wondering why the writings of particular authors are not better represented, or why there is no entry on a particular subject, or why the book does not contain a fondly remembered passage. The answers are simple: some authors, for all their proliferation of works, have not appealed to the compiler as producing worthy material for inclusion, and, unfortunately, some subjects have not, it appears, attracted authors who can produce the memorable turn of phrase that might warrant inclusion. For example, there is a huge contrast between the writings of O S Nock, who first and last was a technical author and whose writings reflect the mechanical approach he had to mechanical things, and David P Morgan, whose ability to express thoughts and feelings about all manner of experiences of railways, great and small, was without equal on either side of the Atlantic.

These considerations, and many others, point not only to some of the few limitations of the subject but more particularly to the limitations of the compiler. The line had to be drawn somewhere and the book published, if only to discover from readers what it lacks. Suggestions for inclusion in an expanded edition of this work will be gratefully received, provided that they are accompanied wherever possible by details of the source from which they come: author, title of work, page number, and date of publication. This also applies to quotations within other published works. A photocopy of the words, whether published previously or not, will be particularly welcome, as accuracy is of paramount importance. Some indication of the extent of the proposed extract is desirable; this is vital as this is a book of short quotations, not an anthology of longer passages. The writer or speaker quoted has to be seen to be making a single point, not giving a general description of many matters or a narrative of several events. In the case of speech for which no written record exists, the compiler will need to know precisely the circumstances in which the words were spoken and, if possible, whether or not the speaker can verify the account. The rule applied to spoken quotations is the same as for written quotations: they must be from the original source if at all possible. Of the spoken word a simple question must be asked: would the speaker easily deny the accuracy of the quoted words? Suggestions will be gratefully received by the compiler by email at railquote@btinternet.com.

Quotations in this book are arranged according to the precise subject the writer or speaker addresses. For example, comments about stations in general appear under "Stations," but comments about particular stations appear under the station names. A particular effort has been made to identify the core subject a writer or speaker is discussing. Sometimes this is not easy: the song "I Thought about You" has been entered under "Absence" because that seems to be the point the songwriter is making. But some readers might have expected to find it entered under "Travel." In the particular cases of verse and song, in fact, while many ideas are often expressed in one song or poem, in most cases it has not been thought desirable to spread them around the book under various headings, but instead to keep the text together. The principal exceptions occur when separate ideas are expressed in separate verses.

For these reasons there are two indexes. The first lists writers and speakers, and the second deals with subjects within entries containing key words and thoughts other than the heading under which the entry appears. An example of this is the song *Chattanooga Choo-choo,* in which the songwriter touches upon platforms, shoeshine service at stations, fares, Pennsylvania Station, timetables, reading on a train, eating on a train, whistles, firing locomotives, and arrival, all under the general theme of travel. It is under the last heading that the song will be found. To find some familiar words, therefore, it may be necessary to look first under the main headings in the dictionary itself and then in the second index under a key word, in case the desired quotation has been placed elsewhere. To find the words of a particular person, look in the index of writers and speakers. To find a quotation to suit a particular subject it might be quicker to start in the index of subjects and key words. All entries under each heading are arranged as closely as possible to the date order in which they were originally published or spoken, but where they have been found quoted in a later work without the original date or source given, they are arranged by the later publication date.

Some other comments on songs are necessary. It is commonplace for individual singers to render their own versions of others' songs. In some cases this is understandable, for example, when a song written for a man to sing is sung by a woman. In other cases, it seems that singers simply take liberties with the words. In such instances, should this book contain the popular, familiar version, or should it show the song as written, even if never performed that way? As a general rule, the original version has been used when found, and has been attributed to the writer. In the case of some American folk songs, many versions exist and even the most assiduous research by historians has failed to determine which came first. It is hoped that the reader will agree that this book contains acceptable versions. Usually, as with poetry, song quotations are limited to the first verse, and occasionally the second. Most of the poems are undated, and I accept full responsibility for this omission. Most poems were found in books that did not themselves give the date of first publication of each work, and in my indolence I decided to look for more entries rather than spend time searching for first publication dates. Of course, to give the publication date of the books in which I found the poems would be misleading.

The entries in this book reveal something about railway writing. The first point that has to be made is that recent and current general dictionaries of quotations are all remarkably reticent on the matter of railways. Even the trusty old *Bartlett's Familiar Quotations,* in its 1946 edition, at a time when one would have assumed a far greater presence of railways in the public mind, displays a remarkable lack of awareness of railway writing. Today, any reader seeking significant railway subject matter in such a general work will simply fail. Only half a dozen entries will be found, under the essential headings of "Railways," "Trains," "Stations," and perhaps "Locomotives."

This sobering fact illustrates the next, namely that when railways were closer to the heart of the life of nations, seventy-five and more years ago, and were more used by their peoples than they are today, the railway figured more often in literature and comment. An astonishingly broad selection of people, from all parts of society, had something to say about railways in the nineteenth century and, to a lesser extent, in the beginning of the twentieth. Since the motor car became the most popular form of personal transport, the railway and the great variety of things that one can do on a train have receded from the forefront of people's lives. This coincided, of course, with all the circumstances that made the rise of the motor car possible, including not only its apparent affordability and the network of roads built at public expense but also the destruction of much of the railway system by the same governments. Those who were sufficiently interested in railways to write about them began to look increasingly inward at the railways themselves. As Jack Simmons noticed, people today too often regard railways more as me-

chanical contrivances than as instruments of social change. This is reflected
in the nature of much contemporary railway writing. Books from the 1960s
and thereafter have poured out multitudes of facts and minutiae but very little analysis and critical comment: a lot of *What* without much *Why*. Whether
we compare modern railway writing with that of Charles Dickens, who appreciated railways but was no enthusiast, or with that of E L Ahrons, who was
never afraid to judge but invariably did so with good nature, or with the
aesthetic appreciations of Hamilton Ellis or Lucius Beebe, we find that modern railway writing seldom has as much to offer as we can find in the first 175
years of the railways' public existence. Perhaps the railway has indeed gained a
new status "as a minority and persecuted transport interest" (52.2). Certainly
the abject failure of politicians in Great Britain to get to grips with their
responsibilities, from the Modernisation Plan of 1955, through the Beeching
era, to the agonies of Privatisation and its many aftermaths, or in America,
where the vast Interstate Highway System was constructed and the airline
industry given significant infrastructure support, without due thought to the
consequences for the railroad industry or the environment, has not helped
the railways to provide the breadth of service that we are aware, from a
knowledge of history, can be offered. By this process of alienation of the
railway from the affections of whole nations, and its removal from the heart
of nations' affairs, railway writing has been affected. In both the United States
and Great Britain, the much-reduced exposure of the population to the experience of railway travel resulted, inevitably, in fewer people with the breadth
and depth of experience of railways to write about them. Indeed, in America
it has been observed that, for many people today, their only contact with
railroads is the delay occasioned by waiting for a mile-long freight to pass the
grade crossing, and there are poems to illustrate that point. While many
American universities are home to serious and excellent historians who have
added many significant works of railway writing to our bookshelves, this has
not occurred on anything like the same scale in Great Britain. And while
many newspapers and magazines in America have on their staffs commentators who add much to our understanding of the modern railroad, even if only
by prompting contrary thinking, it seems that in Great Britain interest in
railways has become so marginalized that few come forward to write. Most
specialist magazines are written by a tiny cadre of knowledgeable journalists,
and few national newspapers have transport specialists on their staffs. Some
of those who do write seem to believe it is necessary to excuse their showing
an interest in railways—what was once lightly referred to as *anoraksia nervosa*.
Worse still, instead of treating railways as a serious and vital industry, many
modern media commentators seem unable to resist the temptation to ignore
important events on railways, or to report them in heavily clichéd terms of
toy trains, socially immature enthusiasts, and incompetent railway manage-

ment. Much of the writing by well-informed and technically articulate commentators has as a result appeared to be inward-looking. The same trend of introspection can be seen in the choice of names for locomotives in Great Britain in the last several years, for it is seen by many merely as a marketing function. The tendency is for much modern railway writing to be split crudely between those on the inside, who write for like-minded people and with great technical knowledge but with rarely expressed appreciation of the place of the railway in the grand scheme of things, and those on the outside, who tend to be slightly condescending, apologetic to their peers for having shown an interest in an unfashionable subject, and who tend to associate the railway only with the steam age, not paying nearly enough attention to the modern railway and its potential for far greater things than it offers at present.

There are other differences in railway writing on the two sides of the Atlantic. For historical reasons the railway has a strong place in the affections of the American and Canadian people to this day. The railroad was the means by which much of America was built and by which the government could manage such a vast nation; it was also the means by which the Canadian nation was unified. It was commemorated in many American folk songs concerning life, love, and death, and it lives on in the country and western genre of music. References to "hopping a midnight freight" seem to have survived the facility to do so on today's more security-conscious railroads. There is no equivalent to this music in Great Britain, and one is hard put to think of any British popular song written with a railway theme. On the other hand, Great Britain, which gave railways to the world (559.22), seems to have produced more people ready to compose poetry about railways, and there is a good tradition of this throughout railway history.

Some lessons may therefore be drawn from the entries in this book. Comparisons of views and modes of expression old and new show that, while some attitudes have changed, few problems are wholly new, and a reference to matters past is, as ever, instructive in our conduct of matters present and our intentions for the future. It is sincerely hoped that the selection of entries is balanced and presents a fair view of railways as they have been seen through the years. If this collection is of use to writers, commentators, students, or any person who simply wants to gain an insight into the extraordinary breadth and depth of the railway, then this book will have served its purpose.

A Note on Clichés, Catchphrases, and Familiar Words

In compiling this dictionary I hoped to unearth the origins of many familiar expressions with roots in railways. As this is not a dictionary of technical terminology, I have been selective and only rarely have included any expression not in general use or any confined to professional railwaymen. Those few well-known expressions with which I have had success are duly included, including William Vanderbilt's "the public be damned," the widely used "what a way to run a railroad" (sometimes spoken as "not a way to run . . ."), and the more recent "the wrong kind of snow."

With too many, I have not been successful. I have not found expressions such as "a full head of steam" or "full steam ahead" in early railway writing, and it is as likely that they originated in maritime usage as that of the railway. "Painting the Forth Bridge," meaning a seemingly endless task, was first used at the time of the completion of the great bridge in 1890, but I regret that I have not found who introduced it into more general use. As with many expressions, it may well have been in conversational use long before it was committed to writing. It appears, from a reference in Lucius Beebe's book *Hear the Train Blow!*, that the circumstances giving rise to "the wrong side of the tracks" go back to the days of Wyatt Earp, who cleaned up Abilene, Kansas, by relocating saloons and brothels to one side of town, using the tracks as a dividing line. But I have not discovered precisely who first used the expression. Similarly it was Beebe who stated, in *The Trains We Rode,* that the expression "roll out the red carpet" comes from this practice for passengers on the *Twentieth Century Limited,* but I have not found proof of it. An expression that I believe was first used in the 1960s to describe the economic benefits to a community from the electrification of its railway service, "the sparks effect," has similarly defied discovery of its first use. To "get off at Wigan," although it seems to hint at the pleasures of travel on the West Coast Main Line in comparison with the charms of Wigan, has nothing to do with railways.

Then there are expressions that do not mention railways but that may first have been used in a railway context. "The difference between men and boys is the price of their toys" (218.26) does not seem to appear in dictionaries of catchphrases, and so its use in 1976 by the wife of an owner of an operating main line steam locomotive is recorded here on the off chance (and, frankly, hope) that this was its origin. Similarly, "profit is not a four-letter word" (535.4) may well be original.

On these, and other railway-related expressions, the commonly available dictionaries of catchphrases and clichés are generally silent, but at least the process of research was educative: there are some expressions that I had hoped to find were inspired by railways, but that, it turned out, were not. Robert Louis Stevenson's belief that it is "better to travel hopefully than to arrive," which I had at one time innocently thought referred to the joys of railway travel, turned out to be all about Life. Horace Greeley's urging to "Go west, young man!" was said long before one could do so by railroad. And there are others.

With the help of Peter Gilliver of the *Oxford English Dictionary,* I was able to discover the earliest acknowledged use of the word *gricer,* but I did not find *anorak* where I expected it, referring to trainspotters. *OED* does not acknowledge that its origins are peculiar to the railway, even though the particular choice of word points toward an outdoor activity, unlike many other activities to which the word is now applied. There are many other words peculiar to railways, but I have not attempted to find their first use, for this is not an etymological dictionary nor a repository of railway terminology and slang. Technical terms used by railwaymen to describe the components and activities of the railway, such as *traxcavate, push-and-pull,* and *getting one's buffers locked,* belong more in a dictionary of technical terms than in this book, although the origins of *decelerate* and *derail* appeared in the course of my research and seemed worthy of inclusion. The language of railway enthusiasts, who often seem deliberately to eschew correct technical terms or to invent their own terminology and nicknames not used by railwaymen, has similarly been given no special treatment. Railway company nicknames appear only very sparingly, when they seem to be of some merit in describing the railway accurately rather than being—as they too often are—simply derogatory words selected solely to fit the initials. I am, anyway, reluctant to look too far into this aspect of railway language, as doing so would oblige me to track down the origin of *Festering Nignog Railway,* and that would be unkind.

Words used to describe those who are passionately interested in railways do, however, deserve some attention, particularly as many of them will, I hope, buy this book. Identified by the compiler thus far, with considerable assistance from fellow members of the Railway & Locomotive Historical Society and other friends are the following:

anorak. British. Probably 1970s. Originally from the use of the anorak as a highly practical garment, with many large pockets (for notebooks and food), worn while watching trains in inclement British weather, from the platform-end. Since, more widely used to describe any person with an extensive and deep interest in systems and devices, from railways and accounting to computers and buses. The habit of treating railway enthusiasts with disdain led to the use of the expression *anoraksia nervosa,* to describe the condition of those who looked down on enthusiasts or who treated them as a lower form of life, in *Steam World* magazine a few years ago. Current.

buffer kisser. German. Train chasers and photographers. Probably mildly critical. Current. *Contributed by Hermann Schmidtendorf.*

enthusiast. Universal. Those with any interest in railways at any level, and therefore the least discriminating of these terms. Few geographical boundaries, if any. Used in a derogatory sense by those who should know better, including in an infamous after-dinner speech at which many academics who were enthusiasts were present, and thereby gravely insulted. Current.

ferroequinologist. First noted by Emmett Gatewood in a letter to *Trains Magazine,* April 1947. He was an early member of the Central Coast Railway Club and later explained that in college, a friend had told him "I know what you are" and had written the word *ferroequinologist* on a piece of paper and passed it to Emmett, who asked him to explain. "Ferro is iron. Equin is horse. Ologist is a student, so you are a student of the iron horse." This may have been the earliest use of the word, and apparently the conversation resulted in its adoption by the Central Coast Railway Club (now Central Coast Chapter, National Railway Historical Society) in the San Francisco Bay Area. They used it as a name for the club's newsletter, first published in March 1952, and also on a former Southern Pacific observation car, number 2901. Later the word was used by Lucius Beebe, and it probably crossed the Atlantic in the early 1950s. Also attributed to Rogers E M Whittaker, a member of the editorial board of the *New Yorker* magazine who wrote about railroads under the name E M Frimbo, but apparently later than its emergence in California. Earliest known use in England was in the mid-1950s, in the introduction of a catalogue of the Doncaster Grammar School Railway Museum. Rarely heard these days, but undoubtedly a most superior form of identification. Rarer still is the variation *ferroequinarchaeologist,* probably because it is almost beyond pronunciation. *Contributed by Jack Gibson via Kyle Wyatt, and by others.*

foamer. American. A railfan who foams at the mouth at the sight of a train; a railfan with a superficial level of interest, held in disdain by serious scholars or other participants in the hobby. Also known as *foamo erectus:* "reported by nauseated enginemen everywhere on the continent. The species is usually seen with excessive amounts of photographic equipment, radio scanner, incredibly tasteless clothes, and bits of foam running down its chin." Current. *Contributed by Adrian Ettlinger and Dick Dawson.*

*f***ing railroad nut or FRN.* American. Usually used by professional railroaders to refer to a person taking unnecessary risks in the pursuit of his hobby. *Railroad nut* is common, and needs no explanation, but the addition of the expletive is apparently confined to the purveyor of dangerous

practices. This, one assumes, is an indication of the frustration professional railroaders feel when encountering such a person. Current. *Contributed by Dick Dawson.*

gerf. American. Acronym for *Glassy-Eyed Rail Fan.* Generally applied to
those who chase trains but never support the excursions by buying a
ticket. Coined circa 1985 by Carl Jensen, then manager of the Norfolk
Southern Steam Program. Probably no longer current. *Contributed by
Frank Corley.*

gricer. British, mid or late 1960s. Those with interests in steam, diesel, or
electric locomotives. Known to the editors of the *Oxford English Dictionary.* Believed by some to originate from *gracer*—a railwayman's expression
meaning "grace your property with their presence," and almost certainly
in sarcasm. Adopted by gricers themselves as a self-description, and by
others as a slightly demeaning term from those higher in the pecking order
of enthusiasts. Current.

locospotter. Very British. Coined by Ian Allan for his *Locospotters' Club* of the
late 1940s to 1960s, which eventually ran to 290,000 members. Current;
only occasionally heard, and usually in adverse criticism.

nut counter. German equivalent of *rivet counter.* Current. *Contributed by
Hermann Schmidtendorf.*

power slave. American. Probably no longer current. *Contributed by Dick
Dawson.*

puffer nutter. Derisive term used by some organs of the cynical press. Current.

railbuff or *railroad buff.* American, widespread. Current.

railcamerist. American. Amateur railroad photographer. Used often by *Railroad Magazine,* 1930s and 1940s.

railfan. American, universal in application. Used as part of a magazine title.
Current.

railfanette. American. The female of the species, frequently used by *Railroad
Magazine* in the 1930s and 1940s.

railgeek. American, not very widespread, possibly 1970s. No longer current.
Contributed by Dick Dawson.

railroadian. American. Used by the group Railroadians of America, founded
in 1939, and by Lucius Beebe in the first edition of *Trains Magazine,*
November 1940. The root word, presumably, of railroadiana, meaning
railroad collectibles. (The English equivalent is railwayana, used by Ian
Wright of Sheffield Railwayana Auctions.) Probably obsolescent.

railroad nut. American, widespread. Current.

railwayac. British. Noted in 1922 as "popular," and thus certainly older; an
early effort to identify those with an interest in railways. Archaic.

railwayist. British. Contemporary with, and in rivalry with, *railwayac.*
Archaic.

rivet counter. British. Used primarily to refer to railway model-makers, but later to anyone with a great interest in small details. Current.

steamer. American. Probably no longer current. *Contributed by Dick Dawson.*

train-spotter (sometimes *train spotter* or *trainspotter*). British. First use noted by the *Oxford English Dictionary* as 1958. Originally, one who liked to watch trains and take notes of numbers of locomotives seen, and more or less synonymous with *Locospotter.* More recently a term of general and often indiscriminate use by the tabloid press, to pigeonhole anyone with an above average interest in railways. Not often applied to model engineers or model-makers. Usually used in a condescending manner. Current.

trolley jolly. Person keen on streetcars and wishing to retain them in the face of bus competition. The term was in use in the 1960s in Pittsburgh, if not earlier, but at that time intentionally derisively to stigmatize those who were opposed to conversion of the streetcar ("trolley" in eastern dialect) system and its replacement by an elevated, rubber-tired device sponsored by Westinghouse. The pro-rail faction invested in lapel buttons reading "Trolley Jolly," and wore them with pride and attitude. They neither supported the preservation of steam nor cared about diesel power. Later broadened to refer to anyone with an interest in streetcars. Probably no longer used. *Contributed by Tom Matoff and Henry H. Deutch.*

vaporazzi. coined by Albi Glatt. Those who follow, ride on, and climb trees and fences to photograph steam excursion trains. A beautiful word. Current, but not seen often enough. *Contributed by Roger Waller.*

Dow's Dictionary of Railway Quotations

1. A1 CLASS LOCOMOTIVES
(Great Britain)

1.1 A big engine in a small loading gauge.
Heiner Vogel, Swiss steam engineer, conversation with the compiler, November 2002.

2. A4 CLASS LOCOMOTIVES
(Great Britain)

2.1 In a sheltered bay a dozen drake shoveller were cruising in line ahead, looking, with their necks drawn in and their bills slanting down their breasts, like the streamlined locomotives of the LNER.
Anon., A Country Diary, Manchester Guardian, late 1930s.

2.2 Locomotive history probably has no parallel to a feat such as this.
Cecil J Allen, writing on the fact that the first A4 had run the new Silver Jubilee *service without a reserve locomotive available, for its first two weeks.* The Railway Magazine, *November 1935.*

2.3 Sitting in the fireman's seat, looking ahead through the strip of safety-glass at the side of the cab, the ballast streaming by beneath, and the wind singing its song in my ears, I tuned in my senses on this matter of the smooth drive, with its absence of threshing weights and unbalanced couples. I felt myself coming into harmony with this beautiful machine, as if I were part of it, feeling for it and with it in the work it was doing so nobly, and a glow began to creep over me. Surely this was the poetry of locomotive motion, beyond anything I had yet experienced.
Edward H Livesay, The Locomotive Magazine, *15 November 1938. His use of the expression 'poetry of motion' is a neat reply to Kenneth Graham's use of the expression, referring to the motor car, in* The Wind in the Willows, *1908.*

2.4 We have nothing to touch this engine.
Western Region Locomotive Inspector, Paddington station, during locomotive trials, 1948. Quoted in RCTS Locomotives of the LNER, *part 2A.*

3. ABANDONMENT

3.1 It's like cutting down a sturdy old oak tree to make room for a hot dog stand.
Anon., caption of photograph of removing the track of a narrow gauge railroad, Railroad Magazine, *August 1941.*

4. ABERDEEN

4.1 The Silver City by the Sea.
Joint LMS/LNER poster, c. 1935.

5. ABSENCE

5.1 I took a trip on a train,
and I thought about you.
Johnny Mercer, song I Thought about You, *1939.*

5.2 And every stop that we made
Oh, I thought about you.
And when I pulled down the shade
Then I really felt blue.
I peeked through the crack,
and looked at the track,
the one leading back to you,
And what did I do?
I thought about you.
Ibid.

6. ACCIDENTS and WRECKS
See also Derailment, Safety

6.1 I have met my death!
William Huskisson, MP, immediately after being run over by Rocket *at the opening of the Liverpool & Manchester Railway, 15 September 1830, as recorded by Charles Francis Adams,* Notes on Railway Accidents, *1879.*

6.2 That is very fine; but it is impossible to make the men perfect; the men will always remain the same as they are now; and no legislation will make a man have more presence of mind, or, I believe, make him more cautious: and besides that, the next time such an accident occurs, the circumstances will be so different, that the instructions given to the men, in consequence of the former accident, will not apply.
I K Brunel, Evidence before the Select Committee on Railways, para. 567, Parliamentary Papers, 22 March 1841.

6.3 But the real source of the danger, and the only one of which there is any hope of removing, is in a complication of imperfections in a great number of the mechanical parts of the system. We have gradually discovered, that the wheels had better be a little wider in gauge than we made them at first; that that they had better not be quite so wide in the external gauge; that they should be about half an inch wider on the tire; that the guard rails had better not touch them that increased care should be

given to the gauge of the rail; and that the tail-lamps should be put in a position in which they shall be less likely to be obscured.
Ibid.

6.4 Ye who're not weary of dear life, forswear
The locomotive, and ye lovely fair,
Who teeth and noses would preserve, to you
This counsel I address,—the train eschew.
Old, young, or neither, fat, thin, short or tall
Travel by turnpike road or not at all;
Or, if you *must* by railway travel still
Think what may happen and—*first make your will.*
Basil Montague, Railroad Eclogues, *1846.*

6.5 The Govt Inspector could not find out the cause, altho' a bad rail made it plain enough to me. It was not our policy to enlighten the public and it was put down as one of those cases which no fellow understands.
Sir Daniel Gooch, of the derailment of a train on the Great Western Railway at Ealing, 24 February 1853; Memoirs and Diary, *1972.*

6.6 A railway accident! Ah, poets! how much of poetry could you find in that, were you so minded. Odes and ballads, sapphics, alcaics and dactylics, strophes, chorusses and semi-chorusses might be sung—rugged poems, rough as the rocky numbers of Ossian, soothing poems, "soft pity to infuse," running "softly sweet in Lydian measure" upon the woes of railway accidents, the widowhoods and orphanages that have been made by the carelessness of a driver, a faulty engine, an unturned "point", a mistaken signal.
George A Sala, Poetry on Railways, Household Words, *2 June 1855.*

6.7 No doubt the cost of this accident will be equal to the sum necessary to protect the whole road from such accidents in future.
Comment in Engineer, *4 October 1860, about an accident to a Pennsylvania Central express train near Pittsburgh caused by a bull on the track, and leading to many injuries. Quoted by Jack White Jr in* American Locomotives, *1997.*

6.8 When the rigors of the law are so sure that the mere sight of a doubtful-looking rail or an ancient sleeper appalls the switchman and the laborer with the fear that if it splits or breaks or rots their individual pockets will suffer, and when the jerking, jumping plunge of the train along the rickety track brings the heart of the director and stockholder into his mouth, not with the fear of losing his life, but his dividend, we may be very sure that whatever human skill and foresight can do to avoid accidents will be done.
Editorial Comment, Harper's Weekly, *10 December 1864, quoted by P Harvey Middleton in* Railways and Public Opinion, *1941.*

6.9 Anything but collisions! If it is semaphores by the dozen or flagmen by the hundred!
Robert Harris, Chicago Burlington & Quincy Railroad, manuscript note on inter-office memo, 16 August 1870. Quoted by Steven W Usselman in Regulating Railroad Innovation, *2002.*

6.10 The Queen must again bring most seriously & earnestly before Mr Gladstone & the Cabinet the vy alarming & serious state of the railways. Every day almost something occurs & every body trembles for their friends & for every one's life.
Queen Victoria, letter to the Prime Minister, 3 October 1873.

6.11 If some people were punished for manslaughter who neglect their duties—or if a Director was bound to go with the Trains we shld soon see a different state of things!
Ibid.

6.12 ... experience has shown that to bring about any considerable reform, railroad disasters have, as it were, to be emphasized by loss of life.
Charles Francis Adams, Notes on Railway Accidents, *1879.*

6.13 Writers and orators seem always to forget that, next to the immediate sufferers and their families, the unfortunate officials concerned are the greatest losers by railroad accidents. For them, not only reputation but bread is involved. A railroad employé implicated in the occurrence of an accident lives under a stigma. And yet, from the tenor of public comment it might fairly be supposed that these officials are in the custom of plotting to bring disasters about, and take a fiendish delight in them.
Ibid.

6.14 It is no cause for surprise that such an event makes the community in which it happens catch its breath; neither is it unnatural that people should think more of the few who are killed, of whom they hear so much, than of the myriads who are carried in safety and of whom they hear nothing.
Ibid.

6.15 I am first going in for repairs, and then for damages.
Anon., a person injured in a railway accident, when asked what he was going to do. Quoted in Railway Adventures and Anecdotes, *Richard Pike, 1888.*

6.16 Collisions four
Or five she bore,
The signals were in vain;
Till old and rusted,
Her boiler *busted*,
And smashed the excursion train.
Anon., of an old locomotive at Crewe, The Great Western Railway Magazine, *August 1892, but attributed to* Punch *by John Aye in* Railway Humour, *1931.*

6.17 Oh the Empire State must learn to wait,
And the Cannonball go hang,
When the westbound's ditched, and the toolcar's
    hitched,
And it's 'way for the Breakdown Gang (Ta-rara!),
'Way for the Breakdown Gang!
    Rudyard Kipling, .007, Scribner's Magazine, August
    1897.

6.18 On a cold frosty morning in the month of September
When the clouds were hanging low,
Ninety-seven pulled out of the Washington station
Like an arrow shot from a bow.
    David Graves George, verse one, Wreck of the Old 97,
    1903 or 1906. The authorship of this song was the subject of
    a long and complex court case, as related by Norm Cohen
    in Long Steel Rail, *1981.*

6.19 He was going down hill at ninety miles an hour
When the whistle broke into a scream—
He was found in the wreck with his hand on the throttle
And scalded to death by steam.
    Ibid., verse six.

6.20 Wrecks are a reflection of administration.
    Charles Delano Hine, Letters from an Old Railway
    Official, 1904.

6.21 Superintindint wuz Flannigan;
Boss av th' siction wuz Finnigan.
Whiniver th' cyars got off th' thrack
An' muddle up things t' th' divvle an' back,
Finnigin writ it t' Flannigan,
Afther th' wrick wuz all on agin;
That is, this Finnigin
Repoorted t' Flannigan.
    Strickland Gillilan, on reporting after derailments;
    poem, Finnigan to Flannigan, *1908.*

6.22 He wuz shantyin' thin, wuz Finnigin,
As minny a railroader's been agin,
An' 'is shmoky ol' lamp wuz burning bright
In Finnigin' shanty all that night—
Bilin' down 's report, wuz Finnigin,
An' he writed this here: "Musther Flannigan—
Off agin, on agin
Gone agin.—Finnigan."
    Ibid., last verse, after Finnigan had been asked to write
    more succinct reports.

6.23 It is as well for its peace of mind that the travelling
public takes railroads so much for granted. The only
men who are incurably nervous about railway travel are
the railroad operatives. A railroad man never forgets that
the next run may be his turn.
    Willa Cather, The Song of the Lark, *1915.*

6.24 I've been lucky. My engines have only killed two
persons.

Jack Rigney, New York Central locomotive engineer, of
grade crossing accidents, American Magazine, *September*
1921.

6.25 He should patiently investigate in person the causes
leading up to every accident and apply the preventive
remedy without fear or favor. He is the best judge of the
discipline suited to each particular case. He must take
care to get past appearances down to realities; and in
dealing with his men he must keep in mind the difference, the very searching distinction, between reputation
and character.
    L F Loree, of the Division Superintendent, Railroad
    Freight Transportation, *1922.*

6.26 On a dark and stormy night, the rain was fallin' fast,
The two crack trains on the Southern road with a scream
    and a whistle blast
Were speedin' down the line, for home and Christmas
    day,
On the *Royal Palm* and the *Ponce de Leon*, was laughter
    bright and gay.
    Rev. Andrew Jenkins, song, The Wreck of the *Royal*
    Palm Express, *c. 1927.*

6.27 Then comin' 'round the curve at forty miles an hour
The *Royal Palm* was makin' time amid the drenchin'
    shower;
There come a mighty crash, the two great engines met,
And in the minds of those who live is a scene they can't
    forget.
    Ibid.

6.28 Accidents do not happen by accident.
    Sir Herbert Walker, after the Sevenoaks derailment,
    24 August 1927, quoted by Sir John Elliot, On and Off the
    Rails, *1982.*

6.29 The Trans-Siberian Express sprawled foolishly down
the embankment. The mail-van and the dining-car,
which had been in front, lay on their sides at the bottom.
Behind them the five sleeping cars, headed by my own,
were disposed in attitudes which became less and less
grotesque until you got to the last, which had remained,
primly, on the rails. Fifty yards down the line the engine,
which had parted company with the train, was dug in,
snorting, on top of the embankment. It had a truculent
and naughty look; it was defiantly conscious of
indiscretion.
    Peter Fleming, One's Company, *1934.*

6.30 There she lay, in the middle of a wide green plain: the
crack train, the Trans-Siberian Luxury Express. For more
than a week she had bullied us. She had knocked us
about when we tried to clean our teeth, she had jogged
our elbows when we wrote, and when we read she had
made the print dance tiresomely before our eyes. Her
whistle had arbitrarily curtailed our frenzied excursions

on the wayside platforms. Her windows we might not open on account of the dust, and when closed they had provided perpetual attraction to small, sabotaging boys with stones. She had annoyed us in a hundred little ways: by spilling tea in our laps, by running out of butter, by regulating her life in accordance with Moscow time, now six hours behind the sun. She had been our prison, our Little Ease. We had not liked her.
Ibid.

6.31 Now she was down and out. We left her lying there, a broken, buckled toy, a thick black worm without a head, awkwardly twisted: a thing of no use, above which larks sang in an empty plain.
If I know Russia, she is lying there still.
Ibid.

6.32 Not much news; it's going to rain, and, oh, another wreck on the New Haven!
Clarence Day, Life with Father, *1935.*

6.33 Now gather round, you sheddies, and I'll spin you a yarn.
It's got a little moral that'll do you no harm.
It's all about a driver from the Crewe North Shed;
He forgot to read his orders and he ended up dead.
Attributed to Geoff Timperley and Colin Pearson, poem, The Ballad of Arthur Swabey; *six verses parodying* The Ballad of Casey Jones *but telling the story of an accident in October 1945 at Bourne End, Hertfordshire; c. 1946.*

6.34 There is no railroad in world, according to its employees and executives, where a wreck or minor disaster is looked forward to with eager and happy anticipation by so many people as along the right of way of the Frankfort and Cincinnati.
Lucius Beebe, Mixed Train Daily, *1947. The Frankfort and Cincinnati served a distillery at Elsinore.*

6.35 The mob demanded scalps and, as usual, the politicians obliged them.
Editorial comment, Railway Age, *about accusations levied at the trustees (in bankruptcy) of the Long Island Railroad, after two accidents. Quoted in* Trains Magazine, *November 1951.*

6.36 The debris of the wrecked train was seasonably but incongruously littered with Christmas puddings . . .
George Dow, description of aftermath of accident at Aylesbury, 23 December 1904. Great Central, *vol. III. 1965.*

6.37 It always takes an eternity to stop after you've hit something.
Ed King, Trains Magazine, *January 1996.*

6.38 The main body of American railroad songs is an unremittingly morbid litany of fatal collisions, exploding boilers, failed brakes, death-dealing blizzards, catastrophic floods, washed-out bridges, collapsing trestles, split switches and misread signals, all accompanied by tragic loss of life, usually on the part of helpless women and children.
Edward A Fagen, The Engine's Moans, *2001.*

7. ACCOUNTING PRACTICE

7.1 The accountants appear to me to treat the value of the railways and stock as if the concern were about to be broken up and sold for what it would fetch.
Robert Stephenson, on the report of a committee of investigation into the affairs of the Midland Railway, 1849.

7.2 It is a fact of common knowledge that so-called cost accounting has been applied to every important branch of industry except to railway transportation.
Balthasar H Meyer, Member, ICC, paper presented to the American Economic Association, 29 December 1913. Quoted in The Railway Library, *1913, 1914.*

7.3 Bookkeeping is to the average operating officer an esoteric mystery with which he is not in the way of becoming familiar unless he deliberately acquires a new and highly technical and difficult art. On the other hand, the accounting staff has neither the foundation of technical training nor experience in transportation movements that supply a base on which it can build up a body of knowledge. Much attention will have to be given to broadening the field of knowledge of both parties.
L F Loree, Railroad Freight Transportation, *1922.*

7.4 Calculating machines are as much a part of Southern Pacific's business as its much better known cab-forward articulateds.
Anon., caption to photograph, Trains Magazine, *December 1941.*

7.5 The refinements of depreciation, fixed charges, contingent interest, and so forth mean nothing to a real short line. It is simply how much cash comes in and how much goes out. Perhaps this is one reason why many an independent short line keeps going while comparable big railroad branches are abandoned. The fine points of standard railroad accounting would have placed most of the independents in the obituary column years ago.
Anon., Trains Magazine, *March 1944.*

7.6 In other words, there is considerable uncertainty about measures of cost—uncertainty which seems in some cases to impede action. There is obviously a general awareness of the need for action, but between the awareness and action there seems to stand the bookkeeper.
Dwight R Ladd, Cost Data for the Management of Railroad Passenger Service, *1957.*

7.7 The orientation of the railroad accounting and statistical systems is not downward but upward. It strives to submerge a host of details into broad statements and

reports which will be coherent to a sort of Orwellian supermanager.
Ibid.

8. ACCURACY

8.1 At Horwich they had gone all scientific and talked in 'thous' although apparently some of their work was to the nearest half inch. At Crewe they just didn't care so long as their engines could roar and rattle along with a good paying load, which they usually did.
D W Sanford, A Modern Locomotive History, *Proceedings of the Institution of Locomotive Engineers, No 190, 1946.*

ACRONYMS
See Mnemonics and Acronyms

9. ACTS OF PARLIAMENT

9.1 Acts of Parliament do not, as a rule, afford entertainment to the general reader, especially those relating to such prosaic and sternly business-like undertakings such as railways and other works of public utility.
Ammon Beasley, How Parliament Harassed Early Railways, *within* The Railway Magazine, vol. XXIII, *1908.*

10. ADLESTROP

10.1 Yes. I remember Adlestrop—
The name, because one afternoon
Of heat the express-train drew up there
Unwontedly. It was late June.
The steam hissed. Someone cleared his throat.
No one left and no one came
On the bare platform. What I saw
Was Adlestrop—only the name.
Edward Thomas, poem, Adlestrop.

11. ADVERTISING
See also Posters

11.1 It will thus be seen that advertising plays a powerful part in the drama of the railway; that the railway uses advertising as a sword's point with which to best its rivals.
Comment, Printers' Ink, *3 February 1892.*

11.2 The successful railroad advertisement is the one which most distinctly and indelibly associates in the public mind the name of the road and the territory or cities it reaches. All talk about superior service has become to a great extent airy persiflage.
Frank Presbrey, Printers' Ink, *14 January 1903.*

11.3 Do you mean to tell me that you make the public purchase your advertising?
Anon. American railroad president, commenting upon

the jigsaw puzzles sold by the Great Western Railway; recorded by Sir Felix Pole in His Book, 1954.

11.4 Advertisements on the LNER compel attention.
London & North Eastern Railway poster, date not known.

11.5 Come, Hygiene, goddess of the growing boy,
I here salute thee in Sanatogen!
Anaemic girls need Virol, but for me
Be Scott's Emulsion, rusks, and Mellin's Food,
Cod-liver oil and malt, and for my neck
Wright's Coal Tar Soap, Euthymol for my teeth.
Come friends of Hygiene, Electricity
And those young twins, Free Thought and clean Fresh
     Air:
Attend the long express from Waterloo
That takes us down to Cornwall.
John Betjeman, of advertisements commonly seen at railway stations; Summoned by Bells, *1960.*

12. AIR COMPETITION

12.1 You cannot compete in time with airlines on transcontinental runs, but you can outstrip them in comfort, safety, dependability of service, and also show the passenger the countryside. This, we believe, is a permanent market.
Edward G Budd Jr, speech before the American Association of Passenger Traffic Officers, Chicago, 24 April 1957, excerpted in Trains Magazine, *August 1957.*

13. AIR QUALITY

13.1 Finest Air in the Country
Great Western Railway advertisement for Manor House Hotel, Moretonhampstead, 1939.

14. ALASKA RAILROAD

14.1 Historically, the role of the Alaska Railroad has been subject to considerable confusion. At various times it has been a frontier development tool, a part of the national defense system, a vehicle for implementing Federal policies, a resource recovery mechanism, a repository for surplus Federal material, a means of social service delivery, and a marketplace railroad.
Comment, Biven & Associates Inc. consulting study, July 1981, quoted in Trains Magazine, *November 1983.*

15. ALCOHOL
See also Beer, Temperance

15.1 There are some mistakes that you don't make on a railroad.
Anon. Boston & Maine official, c. 1914, when leniency had been pleaded for an intoxicated engineer who had coupled clumsily to his train. His fireman had claimed that the

engineer "had made a mistake". Contributed by Donald Morrison.

16. ALDERSGATE STREET STATION

16.1 Snow falls in the buffet of Aldersgate station,
Soot hangs in the tunnel in clouds of steam.
City of London! before the next desecration
Let your steepled forest of churches be my theme.
John Betjeman, who did not need to wait for the destruction of the Euston propyleum ("arch") to stir him on the preservation of fine old buildings; poem Monody on the Death of Aldersgate Street Station, *within* Punch, *16 March 1955.*

17. IAN ALLAN

17.1 ... the patron saint of trainspotting ...
Nicholas Whittaker, Platform Souls, *1995.*

18. CECIL J ALLEN

18.1 The Pope of Shepperton.
Anon. Epithet applied while Allen was writing for Ian Allan publications (of Shepperton) such as Railway World, *late in his life.*

19. JAMES ALLPORT

19.1 He was the Midland personified, the Bismarck of railway policy, the Nunquam Dormio of the line.
W J Gordon, Our Home Railways, *1910.*

AMALGAMATION

See Grouping of Railways and Mergers

20. AMERICAN LOCOMOTIVE COMPANY (ALCO)

20.1 The Mark of Modern Locomotion.
Advertising slogan, mid-1940s.

21. AMERICANS

21.1 Americans take to this little contrivance, the railroad, as if it were the cradle in which they were born.
Attributed to Ralph Waldo Emerson by Peter K Gifford and Robert D Ilisevich, editors, European Capital, British Iron, and an American Dream, *by William Reynolds, 2002.*

21.2 Railroads continue to be my laboratory to learn about who we are.
Andy Warren, Railway & Locomotive Historical Society Newsgroup, *27 May 2004.*

22. AMTRAK

22.1 To cook up a future for the American passenger train, blackmail 20 railroads into the frying pan, congeal with Woodrow Wilson-era work-rules and Truman-second-term-era diesels and coaches, season with red jackets and a pointless arrow, bring to a boil over Congressional oratory, dilute with one cupful of ICC, add a pinch of NARP, spice with Turbo, employ non-rail cooks, and pour into airline plastic. Serves 43 states and D.C.
David P Morgan, Trains Magazine, *November 1971.*

22.2 If we had been permitted to cut service to the level that Amtrak has cut it—we never would have joined Amtrak.
W T Rice, Chairman of Seaboard Coast Line, Richmond News Leader, *quoted in* Trains Magazine, *May 1972.*

22.3 We're making the trains worth traveling again.
Advertising slogan, early 1970s.

22.4 I saw the old ovals and diamonds and shields and wings of private ownership come down and the pointless arrow of a quasi-public corporation go up.
David P Morgan, Trains Magazine, *January 1974.*

22.5 This ain't no railroad, it's Amtrak.
Anon. ex-Pullman attendant, to Ralph Carrington, recorded in Those Pullman Blues, *David D Perata, 1996.*

22.6 Apart from the question whether it has any excuse for existence, Amtrak seems to me to have all the antique charm of a Travelodge and a Denny's—and to be equally worthy of enthusiasm.
George W Hilton, Trains Magazine, *September 1984.*

22.7 If Washington, D.C., is the brain of Amtrak, then Chicago is the heart.
Mike Shafer, caption to photograph, Trains Magazine, *June 1991.*

22.8 You guys have given Amtrak a Ford Trimotor and a few bucks for gas and told us to go out and compete with the 747s.
Alan Boyd, of the paucity of U.S. government funding, 1970s, quoted in Trains Magazine, *October 2001.*

22.9 What a difference the Train makes.
Advertising slogan, 2003.

23. ANGST

23.1 Angoisse des Gares: A particularly violent form of Angst. Bad when we meet someone at the station, much worse when we are seeing them off; not present when departing oneself, but unbearable when arriving in London, if only from a day in Brighton.
Cyril Connolly, The Unquiet Grave, *1948.*

24. APPROVAL

24.1 Now you're railroading!
Anon., given by Eric Partridge in Dictionary of Catchphrases *as an expression of approval and admiration, described as a U.S. railroad catchphrase, twentieth century.*

25. ARCHITECTURE

25.1 Will a single traveller be willing to pay an increased fare on the South Western, because the columns of the terminus are decorated with patterns from Nineveh?
John Ruskin, The Seven Lamps of Architecture, *1849.*

25.2 Railroad architecture has, or would have, a dignity of its own if it were only left to its work. You would not put rings on the fingers of a smith at his anvil.
Ibid.

25.3 Euston station reminds us of an architect's house, where a magnificent portico and hall leads to dungeon-like dining-room, and mean drawing-room. Why are our architects so inferior to our engineers?
Samuel Sidney, Rides on Railways, *1851.*

25.4 To be consistent, the directors should not confine their expressions of artistic feeling to station buildings only; all their porters might be dressed as javelin men, their guards as beefeaters, and their station-masters might assume the picturesque attire of Garter-king-at-arms; their carriages might be copied from the Lord Mayor's show, and even their large locomotive wheels might imitate the Gothic window near their terminus at York. These things, however, will eventually come; the water tank is moulded in the Gothic style.
John T Emmet, The State of English Architecture, *April 1872, quoted in* Culture and Society in Britain 1850–1890, *1986.*

25.5 Make no little plans. They have no magic to stir men's blood.
Daniel H Burnham, architect of Union Station, Washington D.C., c. 1905. Quoted in Trains Magazine, *May 1989.*

25.6 No one visits railway stations to study architecture!
James Scott, Railway Romance and other Essays, *1913.*

25.7 To enter some terminals you walk directly in from the street, a simple task for even a low-grade moron. But in others it's a matter of going up or down, and in one case you actually go up *and* down, and then you begin to wonder if the architect was not a high-grade moron.
Frank P Donovan Jr, of stations in Chicago. Trains Magazine, *August 1948.*

25.8 As so often happens, the modern idea seems to lack the solid permanence of early railway architecture.
C A Johns, One Hundred Years at Kings Cross, *within* The Railway Magazine, *vol. 98, 1952.*

25.9 It would be instructive if considered opinions about—let us say—Waterloo Station, which is just old enough to be moving into a period of adverse criticism, could be recorded at ten-year intervals for a century. But the upshot might teach us a good deal more about architectural taste than about railway architecture itself.
Quoted from The Times *in* The Railway Magazine, *vol. 100, 1954.*

25.10 The first importance of the Italian stations is that they are great architecture: but, secondarily, they again witness that faith in railways. When a nation can build such things it is easy to believe in the survival of her railways. When Italy builds them, it is even possible to conceive of a Railway Renaissance.
Bryan Morgan, The End of the Line, *1955.*

25.11 Proletarian-Baroque.
Anon., of the style of stations on the Moscow underground railway, built in the 1930s; noted by Hamilton Ellis in The Pictorial Encyclopedia of Railways, *1968.*

25.12 Since its establishment in 1947 British Rail has acquired for itself an all too deserved reputation as the biggest corporate vandal and iconoclast Britain has seen since the Tudor dissolution of the monasteries.
Marcus Binney and David Pearce, Railway Architecture, *1979.*

25.13 We do our best, but we know it can't be good enough. But let me make it plain, we have a creative concern and for two reasons. First, the heritage justifies itself, and, secondly, if we can do it justice, it's an added attraction to our customer service. Our paramount concern is our customer, and his confidence in us. That means the look of the railway matters.
Sir Peter Parker, of British Railways' awareness of its responsibility for preserving railway architecture; launch of BR exhibition train, August 1977. Quoted by Binney and Pearce in Railway Architecture, *1979.*

26. ARMORED TRAINS

26.1 An armoured train! The very name seems strange; a locomotive disguised as a knight-errant; the agent of civilisation in the habiliments of chivalry.
Winston Churchill, quoted by Violet Bonham-Carter in Churchill As I Knew Him, *1965.*

27. ARMY (British)

27.1 If only the Army could operate with as few written instructions as the Southern Railway does!
Anon. general, during the Second World War, quoted by Bernard Darwin in War on the Line, *1946.*

28. ARRIVAL
See also Departure, Stations, Whistles

28.1 The train calls at stations in the woods, where the wild impossibility of anybody having the smallest reason to get out, is only to be equalled by the apparently desperate hopelessness of there being anybody to get in.
Charles Dickens, of a journey on the Boston & Lowell Railroad, American Notes, *1842.*

28.2 Behold, smoke-panoplied, the wond'rous car,
Strong and impetuous, but obedient still;
Behold it comes, loud panting, from afar,
As if it lived, and of its own free will
Ran a free race with wild winds blowing shrill!
Fire-bowell'd, iron ribb'd, of giant length,
Snake-like it comes exulting its strength,
The pride of art—the paragon of skill,
Triumph of mind! what hand thy bound shall mark?
Lo! through the curtain of the country time,
Seem looming palpably 'mid cloud and dark,
Yet other triumphs, more than this sublime,
Rise numerous on the far seening ken
Of those who watch, and hope the good of men.

> *Charles Mackay, poem,* The Arriving Train, *from* Legends of the Isles and Other Poems, *1845, but reproduced as found in* The Euston Arch, *1968. In the thirteenth line "seening" may be a misprint for "seeing".*

28.3 No sound is heard in the cold air but the hissing of a pilot engine, which, like a restless spirit advancing and retrograding is stealing along the intermediate rails, waiting to carry off the next down-train; its course being marked by white steam meandering above it and by red hot coals of different sizes which are continually falling from beneath it. In this obscure scene the Company's interminable lines of gaslights (there are 232 at the Euston Station) economically screwed down to the minimum of existence, are feebly illuminating the damp varnished panels of the line of carriages in waiting, the brass doorhandles of the cabs, the shining haims, brass browbands and other ornaments on the drooping heads and motionless backs of the cab-horses; and while the blood-red signal lamp is glaring near the tunnel to deter unauthorised intrusion, the stars of heaven cast a faint silvery light though the long strips of plateglass in the roof above the platform. On a sudden is heard—the stranger hardly knows whence—the mysterious moan of compressed air, followed by the ringing of a bell. That instant every gaslight on and above a curve of 900 feet burst into full power. The carriages, cabs, &c. appear, comparatively speaking, in broad daylight, and the beautiful iron reticulation which sustains the glazed roof appears like fairy work.

> *Sir Francis Head,* Quarterly Review, *December 1848.*

28.4 A jerk—a stop. A gruff shout. 'BROMBRO!' A great fuss to get the window on the other side from us open; calling to the conductor, having the door unlocked; squeezing through the ladies' knees, and dragging our packs over their laps—all borne with a composure that shows them to be used to it, and that they take it as a necessary evil of railroad travelling. The preparations for rain are just completed as we emerge upon a platform, and now it comes down in a torrent. We rush, with a quantity of flying muslin, white ankles, and thin shoes, under an arch. With a sharp whistle and hoarse puffing the train rumbles onward; grooms pick up the lap-dog and baskets; flaunting white skirts are moved again across the track; another rush, in which a diminutive French sun-shade is assisted by a New York umbrella to protect a new English bonnet; a graceful bow in return, with lifting eyebrows, as if in enquiry; and we are altogether crowded in the station-house.

> *Frederick Law Olmstead,* Walks and Talks of an American Farmer in England, *1852. The station is assumed to be Bromborough, in the Wirral peninsula, not far from Birkenhead, where Olmstead started his journey.*

28.5 The seizure of the station with a fit of trembling, gradually deepening to a complaint of the heart, announced the train. Fire and steam, and smoke, and a red light; a hiss, a crash, a bell, and a shriek; Louisa put into one carriage, Mrs Sparsit into another, the little station a desert speck in the thunderstorm.

> *Charles Dickens,* Hard Times, *1854.*

28.6 Gentlemen, here's Julesburg!

> *Anon., brakeman, to the waiting crowd as he arrived with a trainload of prefabricated wooden buildings in Julesburg, Colorado, during the construction of the Union Pacific, late 1867. Possibly apocryphal, related by Dee Brown in* Hear that Lonesome Whistle Blow, *1977.*

28.7 Isn't this invigorating?
No, sir, it's Croydon.

> *Exchange in early morning between arriving passenger and porter, recorded in* Railway Adventures and Anecdotes, *Richard Pike, 1888.*

28.8 Just then a distant whistle sounded, and there was a shuffling of feet on the platform. A number of lanky boys of all ages appeared as suddenly and slimily as eels wakened by the crack of thunder; some came from the waiting-room, where they had been warming themselves by the red stove, or half asleep on the slat benches; others uncoiled themselves from baggage trucks or slid out of express wagons. Two clambered down from the driver's seat of a hearse that stood backed up against the siding. They straightened their stooping shoulders and lifted their heads, and a flash of momentary animation kindled their dull eyes at that cold, vibrant scream, the world-wide call for men. It stirred them like the note of a trumpet; just as it had often stirred the man who was coming home tonight, in his boyhood.

> *Willa Cather,* The Troll Garden, *1905.*

28.9 Only three people got out of the 11.54. The first was a countrywoman with two baskety boxes full of live chickens who stuck their russet heads out anxiously through the wicker bars; the second was Miss Peckitt, the grocer's wife's cousin, with a tin box and three brown-paper parcels; and the third—

'Oh! My Daddy, my daddy!' That scream went like a knife to the heart of everyone in the train and people put their heads out of the window to see a tall pale man with lips set in a thin close line, and a little girl clinging to him with arms and legs, while his arms went tightly round her.
Edith Nesbit, The Railway Children, *1906.*

28.10 Came a whistle from the distance. The breast of an engine was descried, and a long train curving after it, under a flight of smoke.
Max Beerbohm, Zuleika Dobson, *1911.*

28.11 The trains are never called at the little junction towns; everyone knows when they come in.
Willa Cather, My Antonia, *1918.*

28.12 In a few minutes the train was running through the disgrace of outspread London. Everybody in the carriage was on the alert, waiting to escape. At last they were under the huge arch of the station, in the tremendous shadow of the town.
D H Lawrence, of an arrival at St Pancras, Women in Love, *1920.*

28.13 Smoothly, proudly, with all that vast dignity which had surrounded its exit from London, the express moved along the platform. It was the entrance into a gorgeous drawing room of a man that was sure of everything.
Stephen Crane, The Scotch Express, *within* Men, Women and Boats, *1921.*

28.14 However, at that moment the engines, looking very big, rocked with screaming whistles into view around the bend. Behind them followed ten box-waggons, crowded with rifle-muzzles at the windows and doors; and in the little sand-bag nests on the roofs Turks precariously held on, to shoot at us. I had not thought of two engines, and on the moment decided to fire the charge under the second, so that however little the mine's effect, the uninjured engine should not be able to uncouple and drag the carriages away.
T E Lawrence, The Seven Pillars of Wisdom, *1926.*

28.15 Accordingly when the front 'driver' of the second engine was on the bridge, I raised my hand to Salem. There followed a terrific roar, and the line vanished from sight behind a spouting column of black dust and smoke a hundred feet high and wide. Out of the darkness came shattering crashes and long, loud metallic clangings of ripped steel, with many lumps of iron and plate; while one entire wheel of a locomotive whirled up suddenly out of the cloud against the sky, and sailed musically over our heads, to fall slowly and heavily into the desert behind. Except for the flight of these, there succeeded a deathly silence, with no cry of men or rifle-shot, as now the grey mist of the explosion drifted from the line towards us, over our ridge until it was lost in the hills.
Ibid.

28.16 The little platform received her with indifference.
Francis Brett Young, Portrait of Clare, *1927.*

28.17 And a moment later, like a spent runner who finishes the course half-humorously as a matter of form, the little local came puffing in.
Ibid.

28.18 Up and down the train, porters were in the vestibules with their little footstools. One or two doors were already open. Half a dozen passengers, hatted and furred, were in the aisles, waiting for the train to stop. Green berth curtains swayed outward and inward with the motion of the train. Yard lights winked past. The long line of eastbound sleepers snaked through the yards, slouched slowly over switches, past slow-breathing engines not in motion, across the beams of locomotive headlights. Wreathed and blanketed in snow, the train steamed into the furry glow of the platform lights, past redcaps in little groups, steel handtrucks laden with baggage.
Frederick Nebel, Sleepers East, *1933.*

28.19 They passed long strings of silent, darkened railway compartments, and as they neared the station, several suburban trains steamed past them, loaded with people going home. Some of the trains were the queer little double-deckers that one sees in France: Eugene felt like laughing every time he saw them and yet, with their loads of Frenchmen going home, they too were like something he had always known. As the train came into the station, and slowed down to its halt, he could see a boat-train ready for departure on another track. Sleek as a panther, groomed, opulent, ready, purring softly as a cat, the train waited there like a luxurious projectile, evoking perfectly, and at once, the whole structure of the world of power and wealth and pleasure that had created it.
Thomas Wolfe, Of Time and the River, *1935.*

28.20 But now the train was coming. Down the powerful shining tracks a half a mile away, the huge black snout of the locomotive swung slowly round the magnificent bend and flare of the rails that went into the railway yards of Altamont two miles away, and with short explosive thunders of its squat funnel came barging slowly forward.
Ibid.

28.21 The striking form of the Shredded Wheat Factory at Welwyn, white by day and flooded by coloured lamps at night, tells the passenger from the north that he is within half an hour of King's Cross.
Anon., LNER Magazine, *February 1936.*

28.22 Many important foreign missions have arrived at Victoria, but never before in this manner.
Official Report of the crash of a German aircraft on the roof of Victoria station, London, late 1940. Quoted by Evan John in Time Table for Victory, *c. 1946.*

28.23 This lost generation has missed more. It has missed the sight of an awaited train coming around a bend and slowing for the station. It has missed the grinding sound of brakes and the hiss of escaping steam as the engine passes, and the casual, friendly glance of the engineer. It has missed the quick look into the open door of the baggage car , and the half-opened door of the car carrying mail, no doubt with armed guards. It has missed also the diner, slowing as if in invitation, and that moment when the conductor swings down and beckons important passengers aboard. It has missed the one true and great moment of any trip: the "All Aboard," the brisk notes of the bell as the train starts slowly off again. With no experience of that, it is a lost generation indeed.

Anon. The Lost Generation, *of children who have never been on a train, in the* New York Times, *reprinted in* Trains Magazine, *October 1949.*

28.24 It was six o'clock of a March morning, and still dark. The long train came sidling through the scattered lights of the yard, clicking gently over the points. Into the glow of the signal cabin and out again. Under the solitary emerald among the rubies of the signal bridge. On towards the grey waste of platform that waited under the arcs.

The London mail was at the end of its journey.
Josephine Tey, The Singing Sands, *1952.*

28.25 Gaily into Ruislip Gardens,
Runs the red electric train,
With a thousand Ta's and Pardon's
Daintily alights Elaine.
John Betjeman, poem Middlesex, *within* A Few Late Chrysanthemums, *1954.*

28.26 The train drew under the boat-shed at Newport News. The terrific locomotive, as beautiful as any ship, breathed with unlaborious fatigue at the rail-head. There, by lapping water, she came to rest, like a completed destiny.
Thomas Wolfe, Look Homeward, Angel, *1957.*

28.27 Inside the coaches the lamps came on early on winter nights; the shadows leaned as we creaked around the curves—left and back and right and back—and a spot of hot metal glowed in the stove at the end of the aisle and the stale air grew—brassy, is the only word I can think of, and everyone sank farther in his seat, head lolling, not resting, just numb. And then after we pulled off the last hill below Telluride onto the willow flats, speed quickened. Everyone stirred, the click of the rails deepened as the door opened, the conductor bawled his invariable wit, "To Hell You Ride!" and we knew we were home.
David Lavender, quoted by Lucius Beebe in Narrow Gauge in the Rockies, *1958.*

28.28 Moments later I heard the signal wires whisper along the track and I knew that their train was coming.
Otto Kuhler, My Iron Journey, *1967.*

28.29 Or when, at one of the great termini,
One stood the main arrival platform by,
Did all Elysium contain a grace
More sanctified than such a resting-place?
Henry Maxwell, poem, Ave Atque Vale, *verse nineteen, in* A Railway Rubaiyyat, *1968.*

28.30 Like a shot grouse from a very great height, a short passenger train came tumbling down the 1 in 54 from the Waterworks and with a shriek of tortured brakes stopped at the water column in Girvan station.
Introduction, Steam in Scotland, *1972.*

28.31 ... et expecto Hiawatham ...
George Drury, caption to photograph of nuns awaiting the Hiawatha at New Lisbon, Wisconsin, paraphrasing a Latin Credo "Et expecto resurrectionem mortuorum"; Trains Magazine, *November 1977.*

28.32 In the hours that had elapsed since the Panamerican had arrived at the station the atmosphere had changed. The train had brought squalor to the station and had transformed it into a muck-heap. There were orange peels and banana skins under every window—the station was too respectable to have pigs nearby to eat them; and water poured from beneath the coaches, and there were heaps of shit under each toilet pipe. The sun had grown stronger, and flies collected around the coaches. This express train, so dramatic when it was on the move, became foul when it was stationary.
Paul Theroux, at Humahuaca; The Old Patagonian Express, *1979.*

28.33 There was something very disconcerting about leaving a train in the middle of nowhere. It was all activity and warm upholstery and then the clang of the carriage door and the train pulled out and left me in a sort of pine-scented silence. Lairg Station was two miles from Lairg, but even Lairg was nowhere.
Paul Theroux, The Kingdom by the Sea, *1983.*

28.34 At Bexhill I got out. It wasn't easy. The train didn't stop there.
Spike Milligan, BBC television, An Evening with Spike Milligan, *1996.*

29. ASPIRATES

29.1 The late Sir Daniel Gooch put a stop to the exasperating misuse of the aspirate in an admirable manner. 'Hacton! Hacton!' roared the man. 'Dear me!' said the chairman of the Great Western from the railway carriage; 'let that fellow be transferred to Hanwell; he can't give us too much *h* there.'

W J Gordon, of an incident at Acton, Everyday Life on the Railroad, *1898.*

30. ATCHISON TOPEKA & SANTA FE

30.1 Extra Fast. Extra Fine. Extra Fare.
Advertising slogan, 1912. See also 93.1

30.2 The Chief is still chief.
Advertising slogan, referring to its premier train, between Los Angeles and Chicago, 1929. See also 30.4

30.3 Santa Fe All the Way
Advertising slogan, painted on boxcars, late 1930s.

30.4 Too many Chiefs and not enough passengers.
Anon., editorial comment, referring to the Santa Fe's keeping many passenger trains running after passengers had deserted the railroad for air and road; quoted by Keith L Bryant Jr, in History of the Atchison Topeka and Santa Fe Railway, *1974.*

31. ATLANTIC COAST LINE

31.1 The Standard Railroad of the South.
Advertising slogan, 1930s.

32. ATLANTIC LOCOMOTIVES

32.1 ... it is evident that Mr Ivatt has brought out his splendid ten-wheeled giant No 990 not a day too soon.
Comment in The Railway Magazine vol. III, *1898.*

33. ATMOSPHERIC RAILWAYS

33.1 Please your Royal Highness, I think it is a Noomboog.
George Hudson, to Prince Albert, 1845, as related by Edward Lear (who was present) in a letter to Lord Carlingford, 21 December 1884. The meaning of Noomboog is not recorded elsewhere, but George Stephenson is quoted as referring to the atmospheric railway as " a humbug" in A History of the English Railway *by John Francis. It is pos-* *sible that Hudson was saying "an 'umbug", and his flat York accent was thus interpreted by Edward Lear.*

33.2 These difficulties were not only such as had not been anticipated, but such as no one was justified in anticipating.
I K Brunel, of the atmospheric railway in South Devon, Life of Isambard Kingdom Brunel, *1870.*

33.3 I don't think this rope of wind will answer so well as the rope of wire would.
George Stephenson, comparing the atmospheric railway with the outdated rope-hauled railway. Quoted by Samuel Smiles in Life of George Stephenson, *1881.*

33.4 Mr Brunel introduced it on the South Devon Line, where, after spending £300,000 in experiments, it was found that the conveyance of one hundred pounds worth of passengers cost about £108.
Sam Fay, A Royal Road, *1883.*

34. AUSTERITY LOCOMOTIVES
(in the Second World War)

34.1 A "Q1" after the war will perhaps remind us that we should persevere in our efforts to obtain a better world!
Editorial comment on a singularly inelegant wartime design, The Railway Magazine, *Jan/Feb 1943.*

AUTOMOBILES
See Cars and Trains

35. JOHN AXON

35.1 John Axon was a railwayman, to steam trains born and bred,
He was an engine driver at Edgeley loco shed;
For 40 years he travelled and served the iron way,
He lost his life upon the track one February day.
Ewen McColl, song, The Ballad of John Axon, *broadcast 2 July 1958.*

36. BACKYARDS

36.1 People's back yards are much more interesting than their front gardens, and houses that back on to railways are public benefactors.

John Betjeman, quoted in The Observer, *13 March 1983.*

BAGGAGE

See Luggage

37. BALDWIN LOCOMOTIVE WORKS

37.1 Baldwin serves the Nation which the railroads helped to build.

Anon., advertising slogan, mid-1940s.

37.2 Every 15 hours for 111 years.

Anon., advertising slogan referring to Baldwin's production of locomotives.

38. BALTIMORE & OHIO RAILROAD

38.1 Hail the Baltimore & Ohio, 'tis the road of service fine
Hail the men whose toil has made it a mighty transportation line
For a hundred years it has served, and its spirit n'er can fail,
Hail the Baltimore & Ohio, she's the queen of the rail.

Margaret Talbott Stevens, chorus of The Baltimore & Ohio Railroad Song, *1927.*

38.2 As long as was practicable in a degenerating society and as long as how you traveled was who you were, the Baltimore & Ohio was a carrier with style in every aspect of its varnish trains.

Lucius Beebe, The Trains We Rode, *1965.*

38.3 The New York Central was fancy, and the Pennsylvania was awesome, but the B&O was good.

David M Vrooman, Daniel Willard and Progressive Management on the Baltimore & Ohio Railroad, *1991.*

39. BARNSLEY STATION
(L&Y AND SY JOINT, England)

39.1 In fact, the station arrangements violated the most ordinary requirements of decency.

Frederick Williams, The Midland Railway, *1878.*

40. BASTARDY

40.1 Shunter: Come on, bang the bastards up!
W D Phillips: Those waggons may have no father, but I am their rich uncle.

Peter Cheeseman and others, play, The Knotty, *1966. W D Phillips was the General Manager of the North Staffordshire Railway, which was known locally as the Knotty.*

41. BATH—DOMESTIC

41.1 Barf, guv, barf, ain't got a bleedin' barf.

Anonymous fireman, when allowing water tank of his locomotive to be overfilled, and upon being asked if he let his bath at home overflow in a similar manner. Quoted by Richard Hardy in Steam in the Blood, *1971.*

42. H P M BEAMES

42.1 Beames, the CME manqué at Crewe, sat surlily in that citadel of past glories and viewed with a sour gaze all that went on at Derby, and with a still more jaundiced eye any instructions appertaining to locomotive design which arrived from that despised headquarters.

E S Cox, A Locomotive Panorama, *1965.*

43. LUCIUS BEEBE

43.1 Riding on trains is to him an essentially esthetic satisfaction. He sleeps better in a Pullman than at home. He admires inordinately the ingenuity of their washroom appliances and all the internal economy of the car builders' devising. He is soothed by the cornfields of Iowa or alarmed by the Colorado Rockies when viewed from the lounge of the *Forty-niner* or the *Exposition Flyer*. He eats prodigiously in the restaurant car of the *Super-Chief* and he is gentled and delighted by brews, vintages and strong waters on anything from the *Congressional Limited* to the *Daylight*. These are the softer and sissier aspects of his devotion.

Lucius Beebe, of himself, Trains Magazine, *November 1940.*

43.2 He cheerfully acknowledged, and made capital of, the simple circumstance that eluded his detractors—that railroading is to be enjoyed.

David P Morgan, obituary of Beebe, Trains Magazine, *April 1966.*

43.3 He was also aware by instinct that his subject defied monosyllabic grasp, hence he sensibly employed big words to describe big things.

Ibid.

43.4 As he approached middle age, however, his interests turned increasingly to trains of the steam locomotive era, to gourmet cooking, and to rare wines. The first enthusiasm led to the production (and that is the exact word) of 19 railroad books in 25 years, while a passion for the other two eventually helped to cut him down at the age of 63.

Carl W Condit, Railway History No 142, 1980.

44. BEECHING ERA

44.1 The Report is somewhat technical in character and the reading of certain passages requires a little patience.

Lord Morrison of Lambeth, Lords debate on the Beeching proposals, Hansard vol. 249, 1 May 1963.

44.2 I don't think that we shall antagonise the main users of inter city services by the closure of branch lines.

Dr Richard Beeching, press conference, 1963, BBC Sound Archive.

44.3 In general there's a great emotional upsurge every time we intend to cancel a service, and the week after the service is discontinued the whole thing dies away.

Ibid.

44.4 You cannot really believe, although you sometimes pretend to, that I joined the railways to destroy them. That would be such an odious task that I would never have accepted the job at all if I had seen it in that light.

Dr Richard Beeching, manuscript of address to National Union of Railwaymen, 24 May 1964.

44.5 We don't want to be prevented from developing rail services which are basically sound and desirable merely because of the artificiality of competitive road costs.

Ibid.

44.6 Closure is beginning to look less like a negative facet of overall policy, to be carried out only where the situation is absolutely irremediable, and more like a positive programme to be pushed through in obedience to a formula, regardless of its side-effects.

Editorial comment, Modern Railways, October 1964.

44.7 Beeching's Way.

New name applied to the station approach at Alford, Lincolnshire, after the line was closed.

44.8 No more will I go to Blandford Forum and Mortehoe,
On the slow train from Midsomer Norton and Mumby Road,
No churns, no porter, no cat on a seat,
At Chorlton-cum-Hardy or Chester-le-Street,

We won't be meeting again on the slow train.

I'll travel no more from Littleton Badsey to Openshaw,
At Long Stanton I'll stand well clear of the doors no more,
No whitewashed pebbles, no Up and no Down,
From Formby, Four Crosses to Dunstable Town,
I won't be going again on the slow train.

Song, Slow Train, Michael Flanders, 1964.

44.9 The sleepers sleep at Audlem and Ambergate.

Ibid.

44.10 Cockermouth for Buttermere,
on the slow train,
Armley Moor, Arram,
on the slow train,
Pye Hill and Somercotes
on the slow train,
Windmill End.

Ibid.

44.11 People in fields at Chipping Norton,
How will they know the time of day,
Should the Twelve Twenty cease to run
Eight minutes late down Didcot way?
And oh! the anguish at the deep heart's core,
The milk train does not stop here any more!

Caryl Brahms, That Was The Week That Was, c. 1964.

44.12 It is no longer dangerous to be a railwayman ahead of one's time.

Editorial comment, Modern Railways, February 1965.

44.13 To a great extent the future of British Railways depends on what the travelling public and industry think of us.

Dr Richard Beeching, introduction to BR's new corporate style manual, 1965.

44.14 Lord Beeching is praised by many for his ruthlessness. He was given a job of making the railways pay, and to this task he applied precisely the same tools that he would use to make a cosmetics company pay its way. His recipe was to cut out all uneconomic lines. Simple, but utterly naive and in callous disregard of the convenience of millions. He could make the hospitals or the schools pay by similar methods.

Bernard Hollowood, Punch, 23 August 1967.

44.15 The Great and Good Doctor.

Gerry Fiennes, I Tried to Run a Railway, 1967.

44.16 It did not pay, it seemed, to innovate, to suggest the abandonment of practices that many had for long found comfortable and profitable. The old traditions died hard, the old methods had continued to be used long after they had proved uneconomic and even ruinous, and those who attempted to break into established markets by the use of new methods were apt to find themselves treated as though they had introduced bubonic plague rather than rational economics.

Bernard Levin, of the reception and subsequent watering down of Richard Beeching's proposals; The Pound in Your Pocket, *within* Pendulum Years, *1970.*

44.17 But, of course, a railway of half a million employees cannot be managed by an investment approval veto power alone. Nor could the appointment of a super-manager, aided by a large number of very temporary marketing and finance managers, provide lasting solutions when they had not even identified the primary problems.

Stewart Joy, The Train that Ran Away, *1973.*

44.18 En route to utopia, Beeching's Reshaping Plan was compromised by politics and tinkering. The English are masters of both.

David P Morgan, Trains Magazine, *July 1975.*

44.19 The concentration on obvious loss-makers, cutting away the dead wood, was meant to leave a healthier, well-pruned, more fruitful tree. In fact, that is not what happened. The analogy would be more accurate if one saw the process as cutting off the tributaries and drying up the main flow of the river—or perhaps it is Ibsen's onion in *Peer Gynt*, where layer after layer is stripped and the heart never found.

Peter Parker (later Sir), Haldane Memorial Lecture, *23 February 1978.*

44.20 The Beeching era continued to be less than favourable to engineers. It was a time when dis-investment appeared to take priority over investment. The new marketing aggression, which had grown from regulatory freedom, made it fashionable to assert that a railway needed a market before it needed rails or rolling stock. The early railway builders concentrated on the engineering problems of supply rather than the commercial questions of demand, which never flagged. It was only natural that, as a result of road competition, when supply exceeded demand, the positions should be reversed. There was a general feeling that the railway should be reduced in size before further investment could be contemplated.

John Johnson and Robert A Long, British Railways Engineering, 1948–1980, *1981.*

44.21 In the mid-sixties the railway to Largo had closed. It was the worst thing that had ever happened in this part of Fife. The end of the railway was the end of the village.

Paul Theroux, The Kingdom by the Sea, *1983.*

44.22 The massacre of country stations which accompanied the dismemberment of the British railway network in the 1960s permanently diminished the lives of the villages these stations had served. Nothing equivalent took their place. The bus services which replaced the train service did not long survive. Instead the plan-

ners condemned many villages to long, slow, lingering decline. Like so many of the decisions of sixties planners, this cruel decision has brought both spiritual and physical desolation in its wake.

J Richards and J M MacKenzie, The Railway Station, *1986.*

44.23 The Plan reads now like Overconfidence on a Monument sighing at the grief-stricken, mis-spent old age of the Railways. Its final sentences would fit an anthology of famous last words: it concluded that much of the railways' deficit would be eliminated by 1970.

Sir Peter Parker, For Starters, *1989.*

44.24 He did not act as a real Chairman, working from within the organisation, but as a high-level Consultant working from the outside. You and I know that the first thing one has to do with consultants is to educate them. Later on their brain may yield results. But Beeching seemed forced to produce results first, and there was little time left to educate him.

Michael Bonavia, letter to Richard Hardy, 5 October 1989.

44.25 The best chairman we ever had.

Lance Ibbotson, discussion with Richard Hardy (researching a book about Beeching), 13 October 1986.

45. BEER

See also Alcohol, Temperance

45.1 Beer.
1. Ascertain from passenger what kind of Beer is desired.
2. Arrange set-up on bar tray in buffet: one cold bottle of Beer, which has been wiped, standing upright; glass (No. 11), ⅔ full of finely chopped ice (for chilling purpose—making it a distinctive service); glass (No. 12); bottle opener; and paper cocktail napkin. Attendant should carry clean glass towel on his arm with fold pointing towards his hand while rendering this service.
3. Proceed to passenger with above set-up.
4. Place bar tray with set-up on table (or etc.).
5. Place paper cocktail napkin on table in front of passenger.
6. Present bottle of Beer to passenger displaying label and cap. Return bottle to bar tray.
7. Pour ice from chilled glass (No. 11) into glass (No.12).
8. Open bottle of Beer with bottle opener in presence of passenger (holding bottle at an angle), pointing neck of bottle away from passenger; wipe top of bottle with clean glass towel.
9. Pour Beer into glass (No. 11) by placing top of bottle into glass, and slide the beer down the side until beer reaches about two inches from top—then put a collar on the beer by dropping a little in the glass which now should be upright.

10. Place glass containing Beer on paper cocktail napkin.
11. Place bottle containing remainder of Beer on table before passenger, with label facing him.
12. Remove bar tray with equipment not needed by passenger and return to buffet.
Pullman Company, instructions to staff within service manual. Quoted in Trains Magazine, *November 1969.*

46. BEHAVING WELL (or otherwise)

46.1 A four-foot whistle on a two-foot boiler.
Attributed to Mark Twain by Edward Fagen in The Engine's Moan, *2001. Although Mark Twain wrote more about steamboats than railroads, the reference to physical attachment of whistle to boiler suggests that Twain had locomotives in mind when he wrote this description of a character.*

46.2 She is not a woman, she is a locomotive—with a Pullman car attached.
Henry James, of a powerful American lady, rich and eternally busy. Quoted by Jack Simmons, The Victorian Railway, *1991.*

46.3 As tight as a Pullman window.
Anon., of meanness, referring to a characteristic of Pullman car windows in the days before air conditioning when opening the window was often desirable. Recalled by Thomas Streeter, RLHS, 2004.

46.4 I am sure the railways of America owe a great deal to your blunt speech and your forceful character, Mr Diensen, but kindly remember that I am not an American Railway.
Noel Coward; words spoken by Serena, in act I scene II, Quadrille, *1952.*

46.5 Serena: Would it be a healthy sign if I laughed at that too?
Axel: Perhaps not a guffaw, ma'am, just a graceful smile.
Serena: Have all American railway magnates so light a touch?
Axel: Every man jack of them.
Noel Coward, exchange between Serena and Axel Diensen, act II scene II, Quadrille, *1952.*

46.6 He reached for a blue packet of Gauloises, fixed one in the corner of his mouth and talked through the blue clouds of smoke that puffed continuously out through his lips, as if somewhere inside him there was a small steam-engine.
Ian Fleming, On Her Majesty's Secret Service, *1963.*

47. ROBERT BELL

47.1 Professor of Organisation. Externally nearly as rugged as his native mountains, but internally kindly and understanding. Holds world's record for swift personal interviews; is saying a polite Goodbye to the unwanted

visitor before the latter has settled in his chair. Also holds Hoo College records for attendances (he has not missed a day since his arrival) and for walking.
"Keyhole Charlie" (possibly Kenelm Kerr), of the Assistant General Manager, Staff, of the LNER. Who's Who at the Hoo (wartime headquarters of the LNER), Ballyhoo Review, *19 December 1939.*

48. BELLS

48.1 Jacking up the old locomotive bell and rolling a new locomotive beneath it.
Anon. Expression from immediate post-Civil War days, when a tax was levied on new locomotives but not on repairs to old machines, no matter how extensive. Quoted by Charles Fisher, RLHS Bulletin No 102, April 1960.

49. CHARLES BEYER

49.1 A German who never designed an ugly engine.
George Dow, of the co-founder of Beyer, Peacock, locomotive builders, British Steam Horses, *1950.*

50. BEYER-GARRATT LOCOMOTIVES

50.1 The bucking engines beneath the tanks heel and sway in opposite directions but the fat boiler slung between them—like a comatose Roman patrician in his litter—scarcely cants.
Brian Fawcett, Trains Magazine, *June 1955.*

50.2 Someday, perhaps, someone will do a sociological study of the Garratt. It won't contain a line on pivot joints or cylinder stroke or adhesion rates. The book will dwell on the economic effort of the creature, noting that one Garratt could work 500 tons up 2.5 per cent on the same 60 pound rail over which displaced 4-8-0's could only haul 230 tons. And the conclusion drawn will explain more about the why and wherefore of Africa today than a hundred TV reports from the UN.
David P Morgan, Trains Magazine, *March 1966.*

51. BIG ENGINE POLICY

51.1 One engine, one train.
Old rule on the Great Northern Railway, quoted in Modern Engines of the Great Northern Railway, *within* The Railway Magazine, *vol. VXII, 1905.*

52. BIRMINGHAM NEW STREET (England)

52.1 Still, the station is more imposing than the services that issue from its roof . . .
E Foxwell and T C Farrer, Express Trains English and Foreign, *1889.*

52.2 Catching the train had been turned into a furtive, subterranean act, appropriate, perhaps to the new status of railways as a minority and persecuted transport interest.

J Richards and J M MacKenzie, of the new station early 1960s, The Railway Station, 1986.

52.3 . . . like an enormous underground corporation lavatory.

Anon., contemporary railwaymen's description of the station after its unsympathetic reconstruction beneath a large modern building, and extensively decorated with white-tiled walls. C. 1963.

53. BIRTH

53.1 He was as nearly born on the South Western as is consonant with correctly ordered life. He was a fortnight late in being born, and, advised by a wise family doctor, his prospective mother walked a mile through summer meadows to a level crossing. She watched a South Western train thunder between the gates, and then walked home to bed. The baby arrived some hours later, very early on the hundred-and-twenty-eighth anniversary of the birth of George Stephenson.

C Hamilton Ellis, of his own birth, Preface to The South Western Railway, 1956.

53.2 But W Averell Harriman, who died July 19, 1986, at age 94, was born with a gold spike in his hand.

Obituary comment on the son of E H Harriman, Trains Magazine, October 1986.

54. THE BLACK DIAMOND EXPRESS

54.1 The Handsomest Train in the World

Advertising slogan, Lehigh Valley Railroad, 1908.

55. BLACKOUT

55.1 We were alone, I hailed the fellow blindly,
"Excuse me, sir, I live at Wavertree.
Is it the next but one?" She answered kindly
"It was the last but three."

A A Milne (last verse of poem) Punch, c. 1940, quoted by Bernard Darwin, in War on the Line.

56. BLACKPOOL

56.1 The Zone of Ozone.

London Midland & Scottish Railway poster, c. 1930.

57. BLAST PIPES

57.1 The fire burns much better when the steam goes up the chimney than what it do when the engine is Idle.

Richard Trevithick, acknowledging the importance of the blast pipe, letter to Davies Giddy, 20 February 1804.

58. BLEA MOOR TUNNEL

58.1 But what is a mile and a half of Styx in seventy miles of the Delectable Mountains?

C Hamilton Ellis, The Midland Railway, 1953.

59. THE BLUE TRAIN (South Africa)

59.1 However, the Blue Train is not transportation. It is escape.

George H Drury, Railroad History 189, 2004.

60. BOARD OF TRADE

60.1 Resolved that Mr Bourne be authorised to run the trains to Pan Lane—but to be careful to avoid coming into collision with the Board of Trade.

Board Minute, Isle of Wight (Newport Junction) Railway, 1875. Quoted by Michael Robbins in The Railway Magazine, 1959.

61. BOILERS

61.1 Too much spaghetti.

Anon., railroad engineer, of the water tube boiler on Baldwin's experimental 4-10-2 No. 60000. Trains Magazine, April 1954.

61.2 Your Number Two her boiler blew, you'll pick her up in baskets;
So send another engine crew, and send a couple of caskets.
For when she blew, her engine crew took off for climates hotter;
They musta watched the gauge for steam, but plumb forgot the water.

Charles D Dulin, poem, first four lines, Trains Magazine, April 1995.

62. BOOKING CLERKS

62.1 Many things must be learned outright, and of other things one must know where to find at once the source of information. The self-confidence and self-control which accompany thorough familiarity in knowledge are among the best qualities of the ticket seller.

B C Burt, Railway Station Service, 1911.

62.2 The courtesy and patience necessary to the ticket seller are proverbial. Many persons traveling are ill informed, suspicious and nervous, scarcely knowing what they wish or what they do; unable to see, and unwilling to acknowledge, mistakes of their own making, even inclined to quarrel at the ticket window. A single person of this type may on a busy day task the peculiar virtues of the ticket seller to the utmost.

Ibid.

62.3 It's widely known all o'er the line,
That jolly joke and quip
Of the girl who asked for "two to Looe"
And the clerk who said "pip-pip."

Robert F Thurtle, poem Single, Aberystwith, in Great Western Railway Magazine, July 1923.

62.4 The booking clerk is a man apart. Secure in his little room, and only to be glimpsed by peeping through the little wicket, he is a sort of railway hermit, but, sad to say, a hermit of little feeling, who only displays emotion— and that of sombre satisfaction—when he is able to say that there are no cheap fares to the place to which you wish to travel.

John Aye, Railway Humour, _1931_.

62.5 The average railway traveller knows nothing of the activities of the Booking Clerk.

L T C Rolt, Railway Adventure, _1953_.

62.6 Contact with his public is restricted to a hatchway of the smallest possible dimensions, and even this he often seems reluctant to open until the last possible moment. Because of this excessively retiring disposition little is known about him except that, as a species, he obviously needs plenty of warmth. For on bitter winter mornings after we have sat shivering before the cold grate in the station waiting-room, how tantalising it is to glimpse through his hatchway that heaped fire of glowing coals which blazes perpetually in his sanctum!

Ibid.

63. BOOKS ON RAILWAYS

See also Dedication, History—Railway, Writing about Railways

63.1 The British drink 37 per cent more beer per capita than we do, and issue books on railways in apparently greater ratio.

George W Hilton, review of book on brewery railways, Railroad History No 154, _1986_.

64. HENRY BOOTH

64.1 He was more than a competent administrator, the tireless and efficient secretary of the Liverpool & Manchester company. He was a man of ideas.

Jack Simmons, The Railway in Town and Country, _1830–1914, 1986_.

65. BOSTON, Massachusetts

65.1 The best thing about Boston is the five o'clock train that takes you away from there to New York.

Attributed to Arnold Bennett by Lucius Beebe, referring to The Merchants _train of the New York, New Haven and Hartford Railroad;_ The Trains We Rode, _1965_.

66. _BRADSHAW'S GUIDE_

See also The Official Guide, _Timetables_

66.1 Do not buy a Bradshaw, unless you want a headache, but buy the guide of the particular line upon which you are going to travel. . . .

R S Surtees, Hints to Railway Travellers and Country Visitors to London, _in the_ New Monthly Magazine, _1851_.

66.2 . . . I had to content myself with a perusal of Bradshaw, with wondering whether anybody ever went

to Ambergate, Flotton Episcopi, or Bolton-le-Moors, and what they did when they got there, and with musing upon Heal's bedsteads, which, according to the advertisement, could be sent free by post, and upon the dismayed gentleman who, in the woodcut, cannot put up his umbrella, and is envious of the symphonia'd individual who finds "comfort in a storm."

Edmund Yates, A Fearful Night, _within_ Household Words, _17 May 1856_.

66.3 At what hour shall I get to Glasgow? I cannot learn without an amount of continued study of Bradshaw for which I have neither strength nor mental ability.

Anthony Trollope, Reprinted Pieces, _1869_.

66.4 "Bradshaw", to me, has never been the sealed book that it is to some, and I can, as a rule, find my way in it better than in the substitutes sometimes offered.

H Stanley Tayler, Some Joys of Railway Travelling, _within_ The Railway Magazine, _vol. VIII, 1901_.

66.5 I've seen you with Bradshaw. It takes Anglo-Saxon blood.

Henry James, The Golden Bowl, _1904_.

66.6 We always intervene between Bradshaw and anyone whom we see consulting him. "Madamoiselle will permit me to find that which she seeks?" asked Melisande. "Be quiet," said Zuleika. We always repulse, at first, anyone who intervenes between us and Bradshaw.

We always end by accepting the intervention. "See if it possible to go direct from here to Cambridge," said Zuleika, handing the book on. "If it isn't, then—well, see how one _does_ get there."

Max Beerbohm, Zuleika Dobson, _1911_.

66.7 "Stop!" she said suddenly. "I have a much better idea. Go down very early to the station. See the stationmaster. Order me a special train. For ten o'clock, say."

Ibid.

66.8 Harry set himself to one of his favourite studies— Bradshaw. He always handled Bradshaw like a master, accomplishing feats of interpretation that amazed his wife.

Arnold Bennett, These Twain, _1916_.

66.9 They take up their tickets, pay their fare
they're booked right through to everywhere
To lead a life of hopeless worry
With Bradshaw, Baedecker and Murray.

J Ashby Sterry, poem, At Charing Cross.

66.10 Did you ever?—No, you never—dreamt of such absurdities—
Enough to make your noddle ache—it is, upon my word it is;
A handy thing for travelling I've pretty often heard it is,
And so I've been investing in a 'Bradshaw's Guide'.

But with all the figures to be hunting up or diving at,
And what with all your efforts to discover what they're
   driving at—
The stations you are leaving, or the places you're arriving
   at,
I'm hanged if you can fathom in a 'Bradshaw's Guide'.
   Henry S Leigh, poem, Bradshaw's Guide.

66.11 Come, Angela, let us read together in a book more
moving than the Koran, more eloquent than Shakespeare, the book of books, the crown of all literature—
Bradshaw's Railway Guide.
   P G Wodehouse, Meet Mr Mulliner, 1927.

66.12 Its main drawback was that many people found it
difficult or impossible to understand.
   Jack Simmons, The Victorian Railway, 1991.

BRAKEMEN
See Guards and Conductors

67. BRAKES
See also Accidents, Derailment, Safety

67.1 I believe that if self-acting brakes were put upon
every carriage, scarcely any accident could take place.
   George Stephenson, quoted by Samuel Smiles in Life of
   George Stephenson, *1881.*

67.2 Do you pretend to tell me that you could stop trains
with wind? I'll give you to understand, young man, that I
am too busy to have any time taken up in talking to a
damned fool.
   *Cornelius Vanderbilt, on air brakes, proposed to him by
   George Westinghouse, c. 1869. Quoted by Professor John
   Stover, American Railroads, 1968.*

67.3 Simplicity as regards the application of railway
brakes is not obtained by the system now more commonly employed of brake handles to be turned by different men in different parts of the train; but it is obtained
when, by more complicated construction an engine
driver is able easily in an instant to apply brake-power
at pleasure with more or less force to every wheel of his
train; is obtained when, every time an engine driver
starts, or attempts to start his train, the brake itself
informs him if it is out of order; and is still more
obtained when, on the occasion of an accident and the
separation of a coupling, the brakes will unfailingly
apply themselves on every wheel of the train without the
action of the engine-driver or guards, and before even
they have the time to realise the necessity for it.
   *Sir Henry Tyler, on the principles and benefits of the
   continuous automatic brake, quoted by Charles Francis
   Adams, Notes on Railway Accidents, 1879.*

67.4 Stopping a train was found to be fully as important as
starting it.

*Alfred W Bruce, of early experience leading to the
development of the air brake. The Steam Locomotive in
America, 1952.*

67.5 Never before was there such degeneration in so short
a time. Shorn of his usefulness by the air brake, he has
been left a victim of evolution, an organ without a
purpose.
   *Anon., of brakemen, Harper's Weekly, 15 July, 1893,
   quoted by Richard Reinhardt in* Workin' on the Railroad,
   1970.

68. BRANCH LINES
See also Shortlines

68.1 Experience shows that where branches or side lines
are operated under the same local management as that
for the through, or main line, the inevitable tendency is
to do things on the branches in the same way as on the
main line, which is often a more expensive way than is
really necessary for a branch.
   Charles E Perkins, President, CB&Q, memo Organization of Railroads, *within Alfred Chandler,* The Railroads,
   The Nation's First Big Business, *1965.*

68.2 No railway ever sells branch lines to another railway.
   *Attributed to James J Hill by D B Hanna, Trains of Rec-
   ollection, 1924.*

68.3 It is axiomatic in railway practice that most branch
lines, of themselves, do not pay, and that, without the
branch lines that do not pay, the main lines can't pay—
at least their prosperity would be reduced.
   D B Hanna, Trains of Recollection, 1924.

69. RICHARD BRANSON

69.1 I'll be in my grave before that fucker gets his logo on
my trains.
   *John Welsby, inadvertently over an office intercom,
   October 1991, as recounted by Richard Branson, Losing My
   Virginity, 2002.*

70. BRECON & MERTHYR RAILWAY

70.1 . . . the Brecon & Merthyr was a railway upon which it
was always more amusing to travel than to arrive.
   Derek Barrie, The Brecon & Merthyr Railway, 1957.

70.2 . . . the puny, penurious and pushful little Brecon &
Merthyr Railway . . .
   *Derek Barrie, A Regional History of the Railways of
   Great Britain, vol. 12, 1980.*

71. BRIDGES

71.1 Brought here was I—and by this stream was seen,
The day when your Victoria bloom'd eighteen—
And in the annals of Britannia's page,

Attained the period of her legal age:
And ready then was I, on even lay'r,
This massy bridge, but lightsome arch to bear:
This first proud work, that 'cross the forky Lea,
Shall lead with Heaven's high will to Yarmouth Sea,
And join anon in one continuous chain,
The Western Ocean to the Eastern Main.

*Henry Bosanquet, Chairman of the Eastern Counties
Railway. The first of four verses inscribed on a foundation
stone of that company's bridge over the River Lea near
Stratford, July 1837. The full poem is quoted in* The Railway
Magazine, *vol. 100, 1954.*

71.2 To future ages these lines will tell
Who built this structure o'er the dell.
Gilkes Wilson, with his eighty men,
Raised Belah's Viaduct o'er the glen.

Quoted in The Railway Magazine, *vol. XLVI, 1920.*

71.3 See! now Beelah's beauteous sights begin!
Whose curling stream shall ever flow within,
And underneath this splendid monster Bridge,
Shall floods henceforth descend from every Ridge;
Westmoreland's honour form'd by the skill of man
Shall ever o'er thy spacious landscape span,
And thousands wonder at the glorious sight,
When trains will run aloft both day and night;
For ages past, no human tongue could tell
Of such a structure o'er thy monster gill.
Time will roll on, and mortals may increase
When those who see it now, we hope will rest in peace.

Charles Davis, verse, of Belah viaduct, 1859.

71.4 It is not at all improbable that the traffic crossing this
bridge may, at some future time, be even greater than
that passing up and down the river.

*Abraham Lincoln, acting as attorney for the Illinois
Central, when river interests sought to prevent the construction of the first bridge across the Mississippi. A land-
mark case, which Lincoln won. His words have long been
held as an example of forethought. Quoted by H U Mudge,
President, Chicago Rock Island & Pacific; speech at Commerce Club of Topeka, 11 April 1911;* The Railway Library,
1910, 1911.

71.5 Build a bridge. . . . where the trains as they pass will
catch the spray of the falling Zambesi.

Attributed to Cecil Rhodes, and recorded thus, Trains
Magazine, *December 1961.*

71.6 I have seen the most remarkable structure human
eyes ever rested upon. That man Haupt has built a bridge
across Potomac Creek 400 feet long and nearly 100 feet
high, over which loaded trains are running every hour
and, upon my word, gentlemen, there is nothing in it but
beanpoles and cornstalks.

Abraham Lincoln, to the War Committee, of the very

*many replacement bridges built by Herman Haupt during
the Civil War. Late May 1862. Quoted by George Edgar
Turner in* Victory Rode the Rails, *1953.*

71.7 The Yankees can build bridges faster than the Rebs
can burn them down.

Saying quoted by Herman Haupt, Reminiscences of
General Herman Haupt, *1901.*

71.8 At the station pertaining to the railway suspension-
bridge, you see in mid-air, beyond an interval of murky
confusion produced at once by the farther bridge, the
smoke of the trains, and the thickened atmosphere of the
peopled bank, a huge far-flashing sheet which glares
through the distance as a monstrous absorbent and irra-
diant of light. And here, in the interest of the picturesque, let me note that this obstructive bridge tends in a
way to enhance the first glimpse of the cataract. Its long
black span, falling dead along the shining brow of the
Falls, seems shivered and smitten by their fierce efful-
gence, and trembles across the field of vision like some
enormous mote in a light too brilliant.

Henry James, of Niagara, The American Scene, *1907.*

72. BRISTOL & EXETER RAILWAY

72.1 Here lies, from malediction free,
The niggard, grasping, B. and E.;
High fares and bad accommodation
Made it renowned throughout the nation;
In life its customers it bled,
And o'er its grave no tears are shed,
Such save as kind folks will be venting,
When their foes die without repenting.

Anon., poem, The Bristol & Exeter's Epitaph, *signed by
JUBILATE,* Great Western Railway Magazine, *May 1936.*

73. BRISTOL TEMPLE MEADS STATION

73.1 This is what stations were like before they were
messed-up by designers in jazzy bow-ties.

Nicholas Whittaker, Platform Souls, *1995.*

74. BRITISH RAILWAYS and BRITISH RAIL

74.1 Nobody quite believes in the title "British Railways,"
because the old names were so much more poetically
apposite. LNER, pronounced as a word, has a thin,
sneering, east-wind sound, suggestive of those long, cold
stretches by the North Sea. GWR is the sort of noise
made by a large, round-eyed but essentially domesticated
dog, worrying playfully through Mendip tunnels.

Paul Jennings, The Observer, *30 October 1949.*

74.2 See Britain by Train.

Advertising slogan, 1951–1980.

74.3 Fastest through crowded Britain.
Advertising slogan, c. 1960.

74.4 The train was slowing down. They slid past sidings full of empty freight cars bearing names from all over the States—'Lackawanna', 'Chesapeake and Ohio', 'Lehigh Valley', 'Seaboard Fruit Express', and the lilting 'Atchison, Topeka and Santa Fe'—names that held all the romance of the American railroads.
'British Railways?' thought Bond. He sighed and turned his thoughts back to the present adventure.
Ian Fleming, Live and Let Die, *1955.*

74.5 Now plum-and-spilt-milk's had its day and fades from vulgar view,
Will people rush to travel in our Flame Red, Grey and Blue?
Now we're desperately "with it" will our rightful dues prevail—
Will we transport lots more traffic now we're known as "British Rail"?
Whilst Stewart's super-salesmen strain their sinews round the clock,
Will Mr Hoyle keep chopping up the vital rolling stock?
Could anything have forfeited more public goodwill than
The trials and tribulations of the National Sundries Plan?
No more the poetry of names like Rhu and Bethnal Green,
And goods bound for the Scillies are consigned to B.18;
When the faithful railway-minded wish to go from Y to Z,
We'll have to tell them "Try the P.M.T. or Midland Red."
Whilst financial ills beset us is the best that we can do
Just to prune the system further of its hope of revenue?
But whilst we lack a policy then haply for the nonce
We'll pursue our apt new symbol and we'll go all ways at once.
Anon., Thoughts On Our New Image, or, The Outpourings of a Demented Railway Servant. *Found in the papers of George Dow, and thought to have been composed by a member of the Stoke on Trent Divisional staff, c. 1964.*

74.6 Inter-City makes the going easy (and the coming back).
Advertising slogan, 1970.

74.7 Let the train take the strain.
Advertising slogan, 1980s.

74.8 The Overground
Advertising slogan used in conjunction with maps designed in the style of the London Underground map, promoting Inter-City services, 1970.

74.9 It is accompanied by a new form of lettering known as *Rail* alphabet, the noble Gill Sans being unceremoniously interred, and the logotype British Rail in the upper and lower case of the new alphabet. These are the really questionable elements. Their protagonists would have one believe that the new alphabet is more legible than Gill and that the senseless emasculation *British Rail* is 'more forceful, simple to read and comprehend'. If this official blurb has a single grain of truth in it, then indeed the British have become a race of morons.
George Dow, of the new BR corporate style, 1965, Railway Heraldry, *1973.*

74.10 In fact, both exemplify a case of change for the sake of change by people who were content to swallow the gimmicks served up by their consultants.
Ibid.

74.11 Every day is an Open Day.
Sir Peter Parker, noted in Modern Railways, *July 1977.*

74.12 This is the Age of the Train.
Allen, Brady & Marsh, advertising slogan, 1980.

74.13 We're getting there.
Advertising slogan, presumably intended to be a contraction of We are in the business of getting you there, *but received with ridicule, mid-1980s.*

74.14 There always seemed to be a wisp of steam hanging around the safety valves of co-operation whenever Paddington or Swindon was involved and there was little doubt that, given the chance, the Western Region would have seceded from the new transport commonwealth.
John Johnson and Robert A Long, British Railways Engineering, 1948–1980, *1981.*

74.15 *This is the Age of the Train*, the British Rail posters said, showing a celebrity who was noted for his work on behalf of handicapped people and incurables. He had been hired to promote British Rail.
Paul Theroux, The Kingdom by the Sea, *1983.*

74.16 Railtrack has succeeded where generations of British Rail managers failed: it has made BR popular.
Christian Wolmar, Independent on Sunday, *1 October 1995.*

75. BRITISH TRANSPORT COMMISSION

75.1 The most important viceroyalty in the gift of the government.
Winston S Churchill, quoted by T R Gourvish in British Railways 1948–1973, *1986.*

76. BROAD GAUGE
See also Gauge; Gauge, Broad and Standard

76.1 Looking to the speeds which I contemplated would be adopted on railways, and the masses to be moved, it seemed to me that the whole machine was too small for the work to be done, and that it required that the parts should be on a scale commensurate with the mass and the velocity to be attained.

I K Brunel, Evidence to the Gauge Commissioners, 25 October 1845. Parliamentary Papers.

76.2 If the Great Western still existed, its superior comfort and speed would be enjoyed despairingly, not hopefully; and there can be no doubt that the historian of the next Railway generation would have to record the final extinction of the Mammoth class of engines. One or two might survive at Paddington, fossil-like, among the herds of smaller elephants of the narrow-gauge, to perpetuate the memory of the frustrated genius of Brunel: but the great race would be no more.

Henry Lushington, anticipating the demise of the broad gauge, The Broad and the Narrow Gauge, or Remarks on the Report of the Gauge Commission, *1846.*

76.3 A serious evil, a commercial evil, of the first magnitude.

Anon., conclusion of a meeting of Birmingham manufacturers, 1845, discussing the Break of Gauge at Gloucester and railway traffic to Bristol; reported by Samuel Sidney in A Brief History of the Gauge Question, *1846.*

76.4 What a result! England, which has given railways to the world, would see France, Belgium, Germany, Italy and the United States advancing in railway enterprise on a uniform plan—the gauge which England furnished to them; and would stand alone in the anomalous position of having (because one man of great genius disdained to pursue the path pursued by others, and because Parliament, being careless and indifferent to the subject, allowed one powerful company to deviate from the general plan) engraved on her railway a duplication, a complexity and a ruinous expense, of which I am satisfied it would be said that could they have been foreseen, they would never have been tolerated. Why, then, my Lord, should we pursue a policy which is gradually destroying the capital now invested in railways and why should we lessen the safety of railway travelling? Why was the Gauge Commission appointed, and why are its warnings disregarded?

Joseph Locke, pamphlet Railway gauge, a letter to the right Honourable Lord John Russell MP on the best mode of avoiding the evils of mixed gauge railways and the break of gauge, *1848.*

76.5 We are of the opinion that the continued existence of the double gauge is a national evil. We think it worthy of consideration whether it may not be desirable to require the Broad Gauge to be put an end to; and, as the evil has arisen to some extent from the proceedings of Parliament, whether a loan of public money should not be granted for that purpose, on the principle we have suggested for advances to Irish Railway Companies.

Quoted by E T MacDermot in History of the Great Western Railway, *vol. II, 1931,* Report of the Royal Commission on Railways, *May 1867.*

76.6 The exceptional gauge.

Anon. Noted as the epithet applied to the broad gauge by the directors of the London & South Western Railway, by Sam Fay in A Royal Road, *1883.*

76.7 Having come to the conclusion that the entire discontinuance of the broad gauge is unavoidable, and that the conversion of the railways of the company in the West of England from the broad to the narrow gauge cannot be longer postponed without detriment to the interests of the company, we have determined to carry out the conversion in the spring of 1892. Although the disappearance of the broad gauge will undoubtedly be regretted by many travellers in the West of England, we are satisfied that the facilities for the conduct of traffic which a uniform gauge will afford, and the gradual extension of double narrow-gauge lines in districts where at present single broad-gauge lines only exist, will be appreciated by the public, and that the economies which must result from the maintenance of one system of permanent-way and of rolling-stock will be attended with advantage to the proprietors.

Announcement made by the board of the Great Western Railway, 1891, quoted in Our Railways, *J Pendleton, 1896.*

76.8 So! I shall never see you more,
You mighty lord of railway-roar;
The splendid stroke of driving-wheel,
The burnished brass, the shining steel,
Triumphant pride of him who drives
From Paddington to far St Ives.
Another year, and then your place
Knows you no more; a pigmy race
Usurps the glory of the road,
And trails along a lesser load.
Drive on then, engine, drive amain,
Wrap me, like love, yet once again
A follower in your fiery train.

Horatio Brown, Drift, *1900.*

76.9 It was inconvenient, it was costly, but a journey in the old "Flying Dutchman" was a very enjoyable experience.

H Stanley Tayler, Some Joys of Railway Travelling, *within* The Railway Magazine, *vol. VIII, 1901.*

76.10 Not a whistle was heard, not a brass-bell note,
As his corse o'er the sleepers we hurried,
Not a fog-signal wailed from a husky throat
O'er the grave where our "Broad-Gauge" we buried.

We buried him darkly, at dead of night,
The sod with our pickaxes turning,
By the danger-signals ruddy light,
And our oil-lamps dimly burning.

E J Milliken, poem, The Burial of the Broad Gauge,

verses 1 and 2, after The Burial of Sir John Moore at Corunna, *by Charles Wolfe;* Punch, *4 June 1892; reproduced in* Great Western Railway Magazine, *August 1912.*

76.11　But the battle is ended, our task is done;
After forty year's fight he's retiring;
This hour sees thy triumph O Stephenson,
Old "Broad Gauge" no more will need firing.
　Ibid., verse 6.

76.12　The real incentive for this step went much further back, to the stubborn stand taken by I K Brunel and the other promoters of the Great Western Railway of England, for a broad-gauge track. They had builded their important line at seven-foot gauge and this extreme figure was adhered to on that railway for many years (until 1892). In the meantime such railroads of the United States as had originally adopted the broad gauge, and which in consequence were extremely embarrassed in their traffic relations with their fellow lines, were compelled gradually to return to the standard gauge—four feet eight and one-half inches—at no small cost to themselves.
　Edward Hungerford, The Story of the Baltimore & Ohio Railroad, vol. 1, *1927.*

77. BROAD STREET STATION (London)

77.1　The North London Railway brought you into Broad Street, that elevated chunk of Victorian-French-Baroque with the great sprawl of Liverpool Street under its grimy wing.
　C Hamilton Ellis, The Trains We Loved, *1947.*

78. BROAD STREET STATION
(Philadelphia)

78.1　Among the cloudy memories of early childhood it stands solidly, a home of thunders and shouting, of gigantic engines with their fiery droppings of coal and sudden jets of steam.
　Christopher Morley, Broad Street Station, *within* Travels in Philadelphia, *1920.*

78.2　Nowadays when I ramble about the station its enchantment is enhanced by the recollection of those early adventures. And as most people, when passing through a station, are severely intent upon their own problems and little conscious of scrutiny, it is the best of places to study the great human show.
　Ibid.

78.3　A sense of baffling excitement and motion keeps the mind alert as one wanders about the station. In the dim, dusky twilight of the trainshed this is all the more impressive. A gray-silver haze hangs in the great arches. Against the brightness of the western opening the locomotives come gliding in with a restful relaxation of

effort, black indistinguishable profiles. The locomotives are the only restful things in the scene—they and the red-capped porters, who have the priestly dignity of oracles who have laid aside all earthly passions.
　Ibid.

78.4　Cathedral of catarrh.
　Christopher Morley, within poem, Elegy in a railroad station, *about Broad Street station, Philadelphia. Published in* The Saturday Review *and reprinted in* Trains Magazine, *April 1955.*

78.5　As the stronghold of the Pennsylvania Railroad's administrative bureaucracy, the ten-story Gothic-Mooresque office building that rose behind its vaulted train shed symbolized not only the railroad but the character of Philadelphia itself, aloof, disdainfully possessive, conservative and faintly arrogant.
　Lucius Beebe, The Trains We Rode, *1965.*

78.6　It was the first thing millions saw on arriving at Philadelphia and the last thing they saw on departing and it made a lasting impression. It suggested that it had been built in the era of silk-hatted railroad presidents with gold-headed walking sticks who ate luncheons of six courses including terrapin at the Philadelphia Club. It had.
　Ibid.

79. THE BROADWAY LIMITED

79.1　There is leisure here, of body and spirit, leisure and charm and quiet thoughtful comfort, against the beautiful background of those soft gray-green cars.
　Katherine Woods, The Broadway Limited, *1927.*

79.2　The Broadway Limited between New York and Chicago isn't a Wing-jet, a Jumpjet, a Speedjet or a Jetjet.
It's called a train.
The last time you took it you probably called it a choo-choo train.
It doesn't go at the speed of sound. It goes at the speed of a train.
It takes the Broadway Limited a whole night to get to either Chicago or New York. But that's not its only advantage.
It offers the convenience of private rooms for sleeping, washing, working or contemplation of the great American countryside.
A separate dining car comes complete with real tables, real plates, and a wide variety of fresh foods to choose from.
There are two club cars, the perfect places to win friends and influence people.
And the Broadway Limited always operates in rain, fog, mist, sleet, snow, or anything that makes birds walk.

So it's sometimes the fastest means of transportation
available. As well as the pleasantest.
Advertisement, Pennsylvania Railroad, 1966.

80. BROCKENHURST

80.1 . . . a convenient railway junction, because it is in the
middle of nowhere . . .
Paul Theroux, The Kingdom by the Sea, *1983.*

81. ISAMBARD KINGDOM BRUNEL

81.1 . . . though an able and ingenious man, has himself
had no experience in railways and seems to hold in slight
regard the judgment of those who have.
Editorial, The Railway Times, *1838, quoted in* Railway
World, *vol. XXI, 1960.*

81.2 We do not take him for either a rogue or a fool but
an enthusiast, blinded by the light of his own genius, an
engineering knight-errant, always on the lookout for
magic caves to be penetrated and enchanted rivers to be
crossed, never so happy as when engaged 'regardless of
cost' in conquering some, to ordinary mortals,
impossibility.
Editorial Comment, The Railway Times, *c. 1845, quoted*
by Adrian Vaughan in Isambard Kingdom Brunel, *1991.*

81.3 Mr Brunel has learnt to shave on the chin of the
Great Western Proprietors.
"£.s.d.", author of The Broad Gauge. The Bane of the
Great Western Railway, *1846, on the title page, where it*
states "A barbe de fol, on apprend à raire, (which being
translated for the benefit of Country Gentlemen, means)
Mr Brunel has learnt . . . . etc." *The French is thought to*
contain a typographical error ("raire" instead of "rire"); it
probably means "One learns to laugh under the nose of stu-
pidity", in other words, "faced with stupidity one has to
smile to oneself." In its application here, "shave" is thought
to be used in the sense of depriving a person of their money,
in this case "under the noses" of the Great Western propri-
etors. Brunel certainly had the reputation of being less than
circumspect in his expenditure of company money.

81.4 . . . the engineers on the broad gauge lines appeared
to regard themselves less as the officers of the company
than as the channel of the will of Mr Brunel.
F R Conder, Personal Recollections of English Engineers and of the Introduction of the Railway System in
the United Kingdom, *1868.*

81.5 Pecuniary expense was a consideration to which
Brunel was indifferent.
J C Jeaffreson, quoted by N W Webster, Joseph Locke,
1970.

81.6 By his death the greatest of England's engineers was
lost, the man of the greatest originality of thought and

power of execution, bold in his plans but right. The commercial world thought him extravagant, but although he
was so, great things are not done by those who sit down
and count the cost of every thought and act.
Sir Daniel Gooch, Diaries, *1892.*

81.7 He made some errors in judgment but never in
craftsmanship.
Frank P Donovan Jr, Trains Magazine, *September 1948.*

81.8 . . . an Achilles in the camp of broad gauge thinkers.
Lucius Beebe, Narrow Gauge in the Rockies, *1958.*

82. BUCHANAN STREET STATION
(Glasgow)

82.1 The luckless men who are compelled to spend their
days and nights within its precincts have my heartfelt
sympathy. A sad-faced, melancholy band they are; they
never smile; they die young; and more sad-faced melancholy men come along to fill their places in this living
sepulchre.
Anon., St Mungo, 23 April 1897, quoted by Jack Sim-
mons in The Victorian Railway, *1991.*

83. BUFFERS

83.1 Bovril brightens up old buffers.
Advertisement for beef tea commonly seen on railway
stations, 1930s.

84. OLIVER BULLEID

84.1 It got so nobody thought he could do any wrong,
particularly himself. I think we would have given him
square wheels if he had asked for them.
Anon. Southern Railway man, quoted by Sean Day-
Lewis, Bulleid: Last Giant of Steam, *1964.*

84.2 A very clever man, Bulleid was rather eccentric and
he had some strange ideas. Indeed, he seemed to have a
new one every week, and of these, one a year would be
brilliant.
Eric Bannister, Trained by Sir Nigel Gresley, *1984.*

85. BURLINGTON NORTHERN

85.1 Burlington Northern is in the railroad business to
stay, and we're doing it by being a low-cost carrier.
Richard C Grayson, BN President, comments to North-
west Shippers Advisory Board, 4 November 1981, quoted in
Trains Magazine, *May 1982.*

86. BURTON ON TRENT

86.1 I exchanged the chemical exhalations of St. Peter's,
Bristol, for the balmy atmosphere of Burton, fragrant of
malt and hops, and was again on the pay list of the Midland Railway.

Robert Weatherburn, Leaves from the Log of a Locomotive Engineer, No XIX, *within* The Railway Magazine, vol. XXXIV, *1914.*

86.2 Some of the oldest Midland shunting engines are kept there to attend to the beer.

E L Ahrons, The Railway Magazine, *August 1919.*

86.3 For one whose highest value was the 0-4-0ST, Burton must have been a paradise beyond Baltimore!

George W Hilton, Railroad History, No 154, *1986.*

87. BUSES

87.1 These pleasure and comfort with safety combine,
They will neither blow up nor explode like a mine;
Those who ride on the railroad might half die with fear,
You can come to no harm in the safe Shillibeer.

Anon. One verse of a long song written to promote the horse-drawn omnibuses of George Shillibeer, operated in London, in this case in competition with the London & Greenwich Railway, 1834. The reference to These *in the first line refers to Shillibeer's "elegant omnis" in the preceding verse.*

87.2 I said, "Buses aren't the answer."
Mr Winch was looking at the oncoming train. He said, "Buses aren't even a good question."

Paul Theroux, The Kingdom by the Sea, *1983.*

87.3 A bus is for getting you to the station when there isn't a tram.

John Poyntz, conversation with the compiler, 16 March 2004.

BUSINESS CARS

See Inspection Saloons

88. CABLE CARS

88.1 Oh the rain is slanting sharply, and the norther's
blowing cold;
When the cable strands are loosened she is nasty hard to
hold;
There's little time for sitting down, and little chance for
gab,
For the bumper guards the crossing, and you'd best be
keeping tab,
Two-and-twenty "let-go's" every double trip—
It takes a bit of doing on the Hyde Street Grip!
Gelett Burgess, poem, Ballad of the Hyde Street Grip,
in Railroad Magazine, *July 1940.*

88.2 Throw her off at Powell Street, let her go at Post,
Watch her well at Geary and at Sutter when you coast!
Easy at the Power House, have a care at Clay,
Sacramento, Washington, Jackson—all the way!
Drop your rope at Union—never make a slip—
The lever keeps you busy on the Hyde Street Grip!
Ibid., verse two.

89. CABOOSE

89.1 We are jolly American railroad boys and braking is
our trade,
We're always on the go both day and night;
Throwing switches, makin' flagstops, along the line we go,
And to see that all the train is made up right.
You bet we're always ready when called upon to go,
No matter whether sunshine or in rain,
And a jolly crew you'd find us if you will come and see
In the little red caboose behind the train.
Anon. song, The Little Red Caboose, *first published c.*
1882.

89.2 O, the brake-wheel's old and rusty, the shoes are thin
and worn,
And she's loaded down with link and pin and chain,
And there's danger all around us as we try to pound our
ear
In the little red caboose behind the train.
Anon., old folksong quoted by William Knapke in The
Railroad Caboose, *1968. It is not one of the many versions
quoted by Norm Cohen in* Long Steel Rail.

89.3 Their motor was out of order, and there was no
passenger-train at an early hour. They went down by
freight-train, after the weighty and conversational business of leaving Hugh with Aunt Bessie. Carol was exultant over this irregular jaunting. It was the first unusual
thing, except the glance of Bresnahan, that had happened
since the weaning of Hugh. They rode in the caboose, the
small red cupola-topped car jerked along at the end of
the train. It was a roving shanty, the cabin of a land
schooner, with black oilcloth seats along the side, and for
desk, a pine board to be let down on hinges. Kennicott
played seven-up with the conductor and two brakemen.
Carol liked the blue silk kerchiefs around the brakemen's
throats; she liked their welcome to her, and their air of
friendly independence. Since there were no sweating passengers crammed in beside her, she reveled in the train's
slowness. She was part of these lakes and tawny wheat-
fields. She liked the smell of hot earth and clean grease
and the leisurely chug-a-chug, chug-a-chug of the trucks
was a song of contentment in the sun.
Sinclair Lewis, Main Street, *1922.*

89.4 Sentenced to roll at the end of the train
Trying ever with might and main
To keep abreast of the speeding load
But tagging behind on the roaring road.
Tardy at sidings, the last to leave,
Yet I may not complain or grieve,
Minding my own, the last of the train,
Pounding out but the one refrain.
C Milligan, first verse of poem, The Old Caboose, *first
printed in* Chicago Tribune *but reprinted in* Railroad
Magazine *December 1941.*

89.5 I must follow where engines lead,
Bending my will to an iron steed,
Every day till my frames are loose,
That is the lot of an old caboose.
Ibid.

89.6 Just as every store, factory or other place of business
must have an office, so must the freight train have an
office to transact its business. The freight train really
does a big business. It handles large quantities of merchandise every day. It is true, the train does not buy and
sell merchandise like a store, but it produces and sells
transportation—transportation of merchandise of every
sort—and it must keep a complete record of all transportation produced and sold.
Association of American Railroads, A Study of Railway
Transportation, *a teacher's manual for primary and intermediate grades, 1942.*

89.7　I roll along upon the rails
　　Hooked to the lengthy freights,
　　No matter what the day avails,
　　With long delays and waits.
　　The crew that rides me o'er the line
　　Through sunshine, snow and rain,
　　Consists of willing men as fine
　　As ever ran a train.

　　I jog along o'er highest grades,
　　I take the sharpest curves;
　　And when the daylight softly fades
　　A tail light always serves.
　　Sometimes the long train hits a spot
　　That makes me roughly ride;
　　But when a journal box runs hot
　　It greatly stings my pride.
　　　A W Munkittrick, poem Song of the Caboose, *within* A
　　　Study of Railway Transportation, vol. 2, *Association of*
　　　American Railroads, 1942.

89.8　The caboose is the Bohemian of railroad rolling stock
society.
　　　Lucius Beebe, Highball: a Pageant of Trains, *1945.*

89.9　In the gaudy lexicon of railroad jargon it has more
names than any other property in the economy of the
high iron, even more than there were for engines and
engine drivers. It is caboose, crummy, way car, van, cage,
doghouse, drone house, bouncer, bedhouse, buggy,
chariot, shelter house, glory wagon, go-cart, hack, hut,
monkey wagon, pavilion, palace, parlor, brainbox, zoo,
diner, kitchen perambulator, parlor, cabin car, and
shanty. There are probably others in a variety only
bounded by the limitations of human imagining and
the vocabulary of profane and uninhibited men.
　　　Lucuis Beebe, Highball, *1945.*

89.10　Oh, Lord, the whole of life seems newly washed,
seen from the open door of a caboose.
　　　Noel Coward, Quadrille, *1952.*

89.11　The tail end of a freight train, the last car of all, a
small shaky cabin with a twisted iron ladder climbing to
the roof, that is the home of the brakeman. There he sits,
hour in hour out, watching the trees marching along and
the cinders and earth and sands of America slipping
away beneath the wheels. He can watch the sun set over
the gentle farmlands of Wisconsin and rise over the
interminable prairies of Nebraska and Illinois and
Kansas. Those flat, flat lands bring the sky so low that on
clear nights you can almost feel that you are rattling
along through the stars. It is rougher going in the mountains where there are sharp curves and steep gradients
and the locomotive strains and gasps and fills the air with
steam and sparks; tunnels so close round you, infernos of

noise and sulphurous smoke, then suddenly you are in
the open and can breathe again and there are snow-
covered peaks towering above you and pine forests and
the sound of waterfalls. Over it all and through it all, the
familiar, reassuring noise of the train; a steady beat on
the level stretches when the wheels click over the joints in
the rails but changing into wilder rhythms when you
clatter over bridges and crossings and intersections.
　　　Ibid.

89.12　On a caboose you hear the engineer's parentage discussed, not to mention the trainmaster's, and on a
caboose you can drink the best coffee and attempt to
swallow the worst.
　　　David P Morgan, Introduction to The Railroad
　　　Caboose, *William Knapke, 1968.*

89.13　What is more graceful than a conductor catching
the rear step of a caboose moving past the yard office
at, say, 10 to 15 m.p.h.? Nothing, absolutely nothing.
　　　Ibid.

89.14　A caboose is a vehicle for carrying redundant rail-
road employees.
　　　Anon., from a senior railroad official recorded in Trains
　　　Magazine, *c. 1985.*

90. CAJON PASS

90.1　Five a.m. is a ridiculous time to wake up, but if that's
what it takes to enjoy Cajon Pass from a train, then so
be it.
　　　Kevin Keefe, Trains Magazine, *January 1992.*

91. CALEDONIAN RAILWAY

91.1　The country is greatly in a state of derangement, the
harvest, with its black potato fields, no great things, and
all roads and lanes overrun with drunken navvies; for
our great Caledonian Railway passes in this direction,
and all the world here, as elsewhere, calculates on getting
to heaven by steam!
　　　Thomas Carlyle, letter to Gavan Duffy, 29 August 1846.

91.2　The Caledonian Railway Company, the work neither
of lawyers, nor of old women, nor spendthrifts, but of
shrewd middle-aged mercantile men, is just such a tangle
as one might dream of after supping on lobster salad and
champagne.
　　　Comment, The Times, *30 September 1850.*

92. CALIFORNIA

92.1　California was, and remains, an artifact of the railroad. Few places have been so thoroughly created by one
industry, and fewer still have had such a profound influence on the nation and the world.
　　　John P Hankey, Trains Magazine, *June 1999.*

93. CALIFORNIA ZEPHYR

93.1　Extra Comfort, Extra Pleasure, No Extra Fare!
　　Advertising slogan for the train operated by the Western
　　Pacific; Denver & Rio Grande; and Chicago, Burlington &
　　Quincy Railroads, early 1950s. see also 30.1

94. CAMBRIDGE

94.1　The coming of the railway to Cambridge would be
highly displeasing both to God and myself.
　　George Neville Grenville, the Master of Magdelene,
　　c. 1844, quoted by Peter Burman in Railway Architecture,
　　(ed. by Binney and Pearce), 1979.

94.2　Again, Cambridge Station, which no-one in his
senses could characterise as being "beautiful", has weird
fascinations, happily peculiar to itself.
　　Victor Whitechurch, Notable Railway Stations, *within*
　　The Railway Magazine, vol. VII, *1900.*

94.3　One of those rare towns in which a man may lose
himself happily for weeks.
　　London & North Eastern Railway, The Holiday Hand-
　　book, *1940.*

94.4　A rich experience, an unforgettable town, sacred to
Knowledge, Beauty and Youth.
　　Ibid.

95. CAMBRIDGE UNIVERSITY

95.1　... the Great Eastern had prepared a Herculean
scheme for the complete transfiguration of its station
and network at Cambridge, where four rival companies
embrace a tangle of mutual inconvenience; but this godsend was stupidly declined, chiefly owing to opposition
fabricated by members of the University, who by tradition take the blind side in railway matters.
　　E Foxwell and T C Farrer, Express Trains English and
　　Foreign, *1889.*

96. CANADIAN NATIONAL RAILWAY

96.1　Courtesy and Service.
　　Advertising slogan, 1920s and 1930s.

97. CANADIAN PACIFIC RAILWAY

97.1　I will leave the Pacific Railway as a heritage to my
adopted country.
　　Alexander MacKenzie, quoted by Pierre Berton in The
　　National Dream, *1970.*

97.2　Canada has been unusually generous to the promoters of the Canadian Pacific Railway.
　　Sir Edward Watkin, Canada and the States, *1887.*

97.3　World's Greatest Travel System.
　　Advertising slogan, 1930s.

98. CANALS

98.1　I will be neutral on condition that you will faithfully
promise me one thing—it is, that you take care that your
railway, when established, shall ruin these infernal
canals.
　　Charles Waterton, naturalist, to the promoters of the
　　North Midland Railway, c. 1839. Quoted by C Hamilton
　　Ellis in British Railway History, vol. I, *1954.*

99. CAPABILITY

99.1　I have thirty-one miles of railway to make, and the
directors think that that is enough for any man at a time.
　　George Stephenson, excusing himself from becoming
　　engineer to the Leicester & Swannington Railway by refer-
　　ence to his contract with the directors of the Liverpool &
　　Manchester Railway, 1828.

99.2　Everything on wheels—from a wheelbarrow to a
first-class saloon.
　　Description of the Bristol Wagon & Carriage Company,
　　quoted in The Railway Magazine, vol. XI, *1902.*

100. CARDIFF STATION

100.1　... I was given to understand that it claimed most of
the architectural features of a moderately glorified fowl-
house.
　　E L Ahrons, of the original Taff Vale Railway station in
　　Cardiff, The Railway Magazine, *October 1922.*

101. CAREERS

101.1　Railroads are the career for a young man; there is
nothing in politics. Don't be a damned fool.
　　Cornelius Vanderbilt, to Chauncey Depew just before
　　the start of his employment as attorney for the New York
　　and Harlem Railroad. He later rose to be President of the
　　New York Central, but also became a US Senator for New
　　York. Quoted in Chauncey Mitchell Depew The Orator,
　　Welland Hayes Yeager, 1934.

101.2　Don't know anything about the LMS, boy, know
the LNER. Gentlemen at the top. Go there.
　　Career advice given by Col Pullein-Thompson to
　　Richard Hardy, who had been thinking of joining the LMS,
　　1940. Steam in the Blood, *1971.*

102. CARLISLE

102.1　... a junction which connects all parts of Scotland
with all parts of England.
　　Edwin A Pratt, British Railways in the Great War, *1921.*

102.2　Carlisle! The very name seems to hold the sound of
those long hammers with which men hit carriage wheels
in the still hours of the morning.
　　Attributed to H V Morton.

103. CARRIAGES

103.1 It is a very long carriage, supported on four low iron wheels, carries sixteen persons, exclusive of the driver, is drawn by one horse, and rolls along over an iron rail-road, at the rate of five miles an hour, and with the noise of twenty sledge hammers in full play. The passage is only four miles, but it is quite sufficient to make one reel from the car at the journey's end, in a state of dizziness and confusion of the senses that it is well if he recovers from in a week.

Richard Ayton, of a journey on the Swansea & Mumbles Railway, in A Voyage round Great Britain undertaken in the Year 1813, *1814. As quoted by Charles E Lee in* The First Passenger Railway, *1942.*

103.2 The only remaining point of consideration is that of conveying passengers with speed and convenience from place to place which may be done in long carriages resting on eight wheels and containing the means of providing the passengers with breakfast, dinner, etc., whilst the carriages are moving.

William Chapman, in evidence on London Northern Railway, 1825, quoted in The Diaries of Edward Pease, *1908.*

103.3 It rather resembled an Eastern pavilion than anything our northern idea considers a carriage.

Anon., correspondent of the Atheneum, *of the carriage prepared for the Duke of Wellington at the opening of the Liverpool & Manchester Railway, 1830.*

103.4 In all other positions of life there is egress where there is ingress. Man is universally the master of his own body, except when he chooses to go from Paddington to Bridgwater: there only the Habeas Corpus is refused.

Sydney Smith, of the practice of locking carriage doors by the Great Western Railway, letter to the Morning Chronicle, *1842.*

103.5 The seats with the back to the engine are the warmest and least exposed, but at some stations (York, for instance) they reverse the engine, and what goes in head comes out tail. Invalids should note this and change accordingly.

R S Surtees, Hints to Railway Travellers and Country Visitors to London, *in the* New Monthly Magazine, *1851.*

103.6 On those occasions the carriages are a species of horizontal shower-bath, from whose searching power there is no escape.

F S Williams, of open third class carriages, Our Iron Roads, *1852.*

103.7 Large, roomy, prebendal stall-fitted-up like vehicles, usurped the place of little stuffy, straw-bedded stages, into which people packed on the mutual accommodation principle, you letting me put my arm here, I letting you put your leg there.

R S Surtees, Plain or Ringlets? *1860.*

103.8 The one grand fault—there are other smaller faults—but the one grand fault is that they admit but one class.

Anthony Trollope, of passenger cars in America. North America, vol. 1, *1862.*

103.9 Two reasons for this are given. The first is that the finances of the companies will not admit of a divided accommodation; and the second is that the republican nature of the people will not brook a superior or aristocratic classification of travelling.

Ibid.

103.10 If a first-class railway carriage should be held as offensive, so should a first-class house, or a first-class horse, or a first-class dinner. But first-class houses, first-class horses, and first-class dinners are very rife in America.

Ibid.

103.11 The folly of hauling a house, with an audience-room capable of seating 60 people, through the country at the rate of 40 to 50 miles per hour, and calling it a vehicle for travelling, does not seem to have entered the mind of railway managers.

John B Jervis, stating the case for the small carriage, Railway Property, *1866.*

103.12 Perhaps the first thing that strikes an American engineer in railroad traveling in England is the inconvenience of the English passenger car. It should, however, be considered that this style of car suits the exclusive and retiring character of the English; and also that very long journeys, such as we are accustomed to make, are impossible in such a small island; moreover, the English do not travel nearly as much as we do, consequently the confinement in a small, locked-up compartment is not so much felt as it would be with us in our long journeys.

Edward Bates Dorsey, English and American Railroads Compared, *1887.*

103.13 The improvement has been vast, except on some of the south-country lines, where third-class curiosities are apparently kept going until such time as the Channel Tunnel allows of their being lost on the Continent.

W J Gordon, essay Building a Railway Carriage, *within* Foundry Forge and Factory, *1890. Formerly appeared in* Leisure Hour, *no date given.*

103.14 These carriages hold at a pinch four slim adults-a-side, and are innocent alike of racks, cushions, or communication cords. As, however, the pace never exceeds five miles per hour, nervous passengers need not be deterred from journeying on the line on this account, for it is quite within the bounds of safety to alight while the train is going at full speed.

Mary C Fair, of the carriages on the Ravenglass & Eskdale Railway before it was converted from narrow-gauge to miniature railway, Wide World, *19 December 1903.*

103.15 Doubtless this new development is but a tentative measure called forth by economic considerations, and we therefore refrain from expressing an opinion on the glaring ugliness of the vehicles running in the new livery, because we feel sure that the directors of the Great Western Railway will recognise that economy can be bought too dearly. . . .

Comment in The Railway Magazine, *in* Pertinent Paragraphs *about the then new all-brown livery of coaches on the GWR, vol. XVI, 1905.*

103.16 Happily the third class carriages on the London & North-Western are pretty comfortable.

Arnold Bennett, The Grim Smile of the Five Towns, *1907.*

103.17 The backs of the seats were somewhat too vertical to be comfortable, the door handles required a pipe-wrench to turn them, and finally, but by no means least, they were the "hardest-riding" coaches in the country. So much so, that I frequently suspected them of having octagonal wheels.

E L Ahrons, The Railway Magazine, *June 1915, of six-wheel coaches of the Great Northern Railway.*

103.18 Until the Pullman sleepers were introduced into Britain, the sight of a car resting on eight wheels was unprecedented, as no one thought of doubting the entire security from danger of a carriage with only four points of support. Indeed, the conservative Briton saw no more real necessity for a railway carriage having eight wheels, than for a horse to have more than four legs.

Joseph Husband, The Story of the Pullman Car, *1917.*

103.19 . . . they seemed to give the impression of moving castellated walls.

E L Ahrons, The Railway Magazine, *June 1917, of coaches of the South Eastern Railway.*

103.20 Most of the carriages were undersized and four-wheeled, nevertheless the wheel bases appeared to be many and various; but there were a number of six-wheelers to be seen, probably when all the possible dimensions of four-wheeled coaches had been exhausted. The styles of windows included Gothic, Norman and Early English, in great variety, intermingled with Elizabethan and Tudor samples.

Ibid.

103.21 The decoration of passenger equipment is rather a question of taste, and, therefore, of social efficiency;

though the use of paint as a preservative is a matter of economic importance.

Henry S Haines, Efficient Railway Operation, *1919.*

103.22 Though I believe Mr Gresley is on the high road to success, it does remain a fact that locomotive engineers have not yet wholly solved the problem of really heating those carriages which are supposed to be heated.

William Whitelaw, Chairman LNER, in jest to his CME, Annual Dinner, Institution of Locomotive Engineers, 1 March 1935.

103.23 The historic German car and the monument to French triumph will be brought to Berlin. The rails on which it had stood in 1918 will be destroyed so that no trace of Germany's defeat in 1918 will remain.

Adolf Hitler, proclamation in June 1940 concerning the Wagons-Lits Car 2419 in which the German surrender had been taken in 1918, and the French surrender in 1940. The proclamation ignored the fact that carriage was French, not German, having been built in the Wagons-Lits works at St Denis in 1918. Quoted in Orient Express, *by E H Cookridge, 1978.*

103.24 Pullman and Wagner, and half a dozen other men, started separately to build sleeping cars, then parlor cars and dining cars, gauding them all with all the art forms beloved of Victorians, including plush damask and paneled wood, mirrors on every hand and silver-plated cuspidors, thus achieving a true rolling horror on wheels that caught the fancy of the mass of new *eleganti* who were just emerging from the log-cabin era and were likely to confuse elaborateness with beauty whether in a house design by M. Mansard or a railroad coach by G. Pullman.

Stewart H Holbrook, The Story of American Railroads, *1947.*

103.25 . . . a picnic saloon was a railway carriage arranged like a giant tramcar, of such refined discomfort that no sober person could be expected to sleep in it.

C Hamilton Ellis, The Midland Railway, *1953.*

103.26 The American-type passenger car was widely heralded as an achievement of democracy, wherein all classes sat together in republican equality as a matter of choice.

Lucius Beebe, Mr Pullman's Elegant Palace Car, *1961.*

103.27 Horsehair upholstery tortured youthful legs and batswing gaslights, still in wide use, only served to make darkness visible.

E S Cox, of Lancashire & Yorkshire Railway electric trains, Locomotive Panorama, *1965.*

103.28 So many cruel armistices have been signed in railway carriages that it takes very little to give them an aura of fatality and gloom.

Peter Ustinov, God and the State Railways, *within* The Frontiers of the Sea, *1966.*

103.29 Gentlemen, we of the railroad industry have been immobile long enough. The only visible improvement in most of our passenger cars in the last 25 years is the small slot in the washrooms for disposing of razor blades, and many of our cars don't even have those.

Otto Kuhler, recalling his words to the New York Railroad Club, 1934, My Iron Journey, *1967.*

103.30 No mitred napkins with gilt menu decked,
No leather-mounted wine-lists, all erect;
No latticed luggage-racks, like gold, above,
And where steam heating was, a common stove,
Whose iron flue, jutting through the roof,
Has made the ceiling less than waterproof,
Which now with spots and cracks is stained and seamed,
Where, virginal, Lincrusta's whiteness gleamed.

Henry Maxwell, poem, Splendour in Decline, *of an old first class diner used as a mess hut, in* A Railway Rubaiyat, *1968.*

103.31 A rolling catalog of the physical discomforts that have turned U.S. passenger railroads into the nation's most unpopular form of mass travel.

Comment about the miscellaneous run-down passenger cars provided by the Pennsylvania Railroad for the funeral train of Sen. Robert Kennedy 8 June 1968, Newsweek Magazine, *June 1968. Quoted by Richard Saunders Jr,* Merging Lines, *2001.*

103.32 If it hasn't been before, let be said now: you can take the boy out of day coaches, but you can't take day coaches out of the boy. I know. I spent my youth in the dome-light, green-plush-walkover-seat, wooden armrest, 12-wheel, friction-bearing world exemplified here by Louisville & Nashville No. 1883. It was a cindery, 1½-cent-a-mile, *Azalean*/Corbin local/*Flamingo*/No. 104, 50 mph world that declined in the depression, knew its finest hour in World War II, then died.

David P Morgan, caption to photograph, Trains Magazine, *February 1970.*

103.33 The new Mk II carriages were comfy and warm, double-glazed and brightly upholstered, yet as sterile as a DHSS waiting room.

Nicholas Whittaker, Platform Souls, *1995. (DHSS is Department of Health and Social Security, Great Britain.)*

103.34 It was a tradition bordering on a religion that East Coast main line trains started their journeys clean.

Stanley Hall, Steam World, *December 2001.*

CARS (in passenger trains)
See Carriages

104. CARS (automobiles) AND TRAINS

104.1 It is perhaps knowledge of a sort to be able to state the number of cylinders in the new Buick or the new Ford, assuming, of course, that they are not already jet propelled. It may even be knowledge to know the short cut through Providence which, on the way to Cape Cod, avoids the traffic around the railroad station. But there is the rub, for the short cut also avoids the station. A pleasant childhood consists no more in circling a station at a distance than in circling a candy cane. Both should be sampled as often as possible in order that an agreeable child becomes an honourable man. A train is basic, while an automobile is turned in each year; a train is a friend, while a car is only a temporary possession.

Anon., The Lost Generation, *in the* New York Times, *reprinted in* Trains Magazine, *October 1949.*

104.2 The Bentley screamed down towards them like an express train.

Ian Fleming, Casino Royale, *1953.*

104.3 A car, however splendid, was a means of locomotion (he called the Continental 'The Locomotive'. . . 'I'll pick you up in my locomotive') and it must at all times be ready to locomote—no garage doors to break one's nails on, no pampering with mechanics except for the quick monthly service. The locomotive slept out of doors in front of his flat and was required to start immediately, in all weathers, and, after that, stay on the road.

Ian Fleming, Thunderball, *1961.*

104.4 JUDGE DOOM: They are calling it a freeway.
EDDIE VALIANT: Freeway? What the hell's a freeway?
DOOM: Eight lanes of shimmering cement running from here to Pasadena. Smooth. Safe. Fast. Traffic jams will be a thing of the past.
VALIANT: Nobody's going to drive this lousy freeway when they can take the Red Car for a nickel.
DOOM: Oh, they'll drive. They'll have to. You see, I bought the Red Car so I could dismantle it.

Gary K Wolf, exchange referring to the purchase of interurban companies by a collection of oil, automobile and rubber tire interests, in Who Framed Roger Rabbit, *1988.*

105. CARSON & COLORADO RAILROAD

105.1 Gentlemen, either we built this line 300 miles too long, or 300 years too soon.

Darius O Mills, on inspection trip, 12 July 1883.

105.2 It's the first railroad that begins nowhere, ends nowhere, and stops all night to think it over.

Anon. Quoted in Southern Pacific, *Neill C Wilson and Frank J Taylor, 1952.*

106. CASEY JR

See also Casey Jones

106.1 Make it like your locomotive, Ward, only cartoon it up a bit.

> *Walt Disney, instruction to Ward Kimball in designing Casey Jr, to appear in the animated film* Dumbo. *Kimball's locomotive was an 1881 Baldwin 2-6-0 he had acquired from the Nevada Central. Casey Jr was a 2-4-0.* Trains Magazine, *January 1942.*

107. CATCH POINTS

107.1 To prevent great Mischiefs by the Running of Waggons I intend at the Bottom of the Run to have a Branch laid and a switch Rail to be Shutt by an Old Man to sit in a Cabbin and with an Iron Rod thro' a Conduit fixt to a small Leaver (when he Sees a Waggon got loose) he may Shut the Switch Rail and turn the Waggon into the Branch which Riseing till it Runs against a Battery of Earth and Turf will take no harm and will fal back again gently and I'll have another Switch Rail which shall Shutt of itself and turn the Waggon (in its coming back) into another Branch where it will rest till brot into the Main Way below the first mentioned Branch which means if another Waggon or more shou^d follow they will Escape breaking one another.

> *Carlisle Spedding, of his concerns about the operation of a gradient of 1 in 11 on the Whingill waggonway at White-haven, 1755, as quoted by Michael Lewis in* Early Wooden Railways, *1970. It contains what is almost certainly the earliest proposal for what became known as the signal box.*

108. CATS

108.1 We found, too, that the status of the cats varied from place to place. At some stations their appointment could hardly be said to be official, though their presence was acknowledged. At others, they were regarded as definite members of the staff. There were also at one time several joint cats at the joint stations, but later they were assigned to one Company, or to the other under the Closer Working Arrangements.

> *Anon., stated to be by the LNER Cat Inspector for Scotland, of the cats on railway property used to control vermin in grain stores at stations and goods yards,* LNER Magazine, *May 1944.*

109. CENTENNIAL DIESEL

109.1 . . . it is truly the 800-pound gorilla of American railroading.

> *David Lustig, of the Union Pacific DDA40X twin-engine 6,600 hp diesel electric locomotive of 1969;* Trains Magazine, *September 2002.*

110. CENTRALIZED TRAFFIC CONTROL (CTC)

110.1 Here comes a non-stop meet!

> *Anon., dispatcher at Fostoria, Oh., on the occasion of the first meet of trains under CTC, and thus without train orders, 25 July 1927; quoted by Fred W Frailey in* Trains Magazine, *January 2000.*

111. CHANGING TRAINS

111.1 I wrote Books 2,3,4,5,6,7 and 10 in Paris between February and June of 1919. The Introduction was written between Paris and Egypt on my way out to Cairo by Handley-Page in July and August 1919. Afterwards in England I wrote Book 1: and then lost all but the Introduction and drafts of Books 9 and 10 at Reading Station, while changing trains.

> *T E Lawrence,* Some Notes on the Writing of The Seven Pillars of Wisdom by T E Shaw, *(as a Preface to* The Seven Pillars of Wisdom*) 1926. Perhaps the spirits of the Great Western Railway were taking revenge for the destruction Lawrence wrought on railways in the Middle East during the Great War.*

111.2 At Chengchow the Peking-Hankow 'express' was due to make connection with a train, similarly miscalled, on the Lunghai Railway. We had been warned that it was a point of honour on the former line to miss this connection by a matter of a very few minutes, and found this tradition faithfully observed.

> *Peter Fleming,* News from Tartary, *1936.*

111.3 . . . I spent at St Blazey Junction the forty odd minutes of repentance ever thoughtfully provided by our railway company for those who, living in Troy, are foolish enough to travel . . .

> *Sir Arthur Quiller-Couch,* Pipes in Arcady, *1944.*

112. CHANNEL TUNNEL

112.1 Sir Edward Watkin is one of those men who are wicked enough to desire that a tunnel should be constructed under the Channel to France, and I am compelled to confess publicly before you that I am one of those wicked enough to agree with him.

> *William Gladstone, 16 August 1887, quoted in* Our Railways, *J Pendleton, 1896.*

112.2 William Gladstone: If I live long enough, will you take me through without any change of carriage from Hawarden to Folkestone?
Sir Edward Watkin: Yes.

> *Exchange quoted in* Our Railways, *J Pendleton, 1896.*

112.3 The least intelligent thing which has been done in our generation was the refusal to build the Channel tunnel.

Sir Arthur Conan Doyle, quoted in The Railway Magazine, *vol. XXXI, 1912.*

113. CHARING CROSS

113.1　The terminus of Charing Cross
Is haunted, when it rains,
By Nymphs, who there a shelter seek,
And wait for mythic trains.
Arthur M Binstead, Pitcher in Paradise, *1903.*

113.2　... the merit of its elevation has been destroyed by the rebuilding of the two top storeys since the war, and British Railways have recently befouled it with a fascia hardly surpassed in vulgarity by anything else of the kind in London.
Jack Simmons, of the hotel above Charing Cross Station, St Pancras Station, *1968.*

114. CHARITY

114.1　The great unremarked American charity is that of arguing for the railroads. It involves a considerable expenditure of time, knowledge, patience and skill in exchange for which the donor derives merely the solace of self-satisfaction. No income tax deduction or box of cookies or button or red feather: just self-satisfaction.
David P Morgan, Trains Magazine, *October 1960.*

115. CHATTAHOOCHEE INDUSTRIAL RAILROAD

115.1　Better by a Dam Site.
Slogan, derived from the presence of a dam on the Chattahoochee River near the terminus of the line. Recorded in Railroad History, No 153, *1985.*

116. CHEMIN DE FER

116.1　The English know how to play at railway trains.
Ian Fleming, Thunderball, *1961. The reference, by Emilio Largo, is to the game of Chemin de Fer, which James Bond enjoyed.*

117. CHICAGO

117.1　If you journey in a Pullman from Mesa to Omaha without a waistcoat, and with a silk handkerchief knotted over the collar of your flannel shirt instead of a tie, wearing, besides, tall, high-heeled boots, a soft, gray hat with a splendid brim, a few people will notice you, but not the majority. New Mexico and Colorado are used to these things. As Iowa, with its immense rolling grain, encompasses you, people will stare a little more, for you're getting near the East, where cow-punchers are not understood. But in those days the line of cleavage came sharp-drawn at Chicago. West of there was still tolerably west, but east of there was east indeed, and the Atlantic Ocean was the next important stopping-place. In Lin's new train, good gloves, patent-leathers, and silence prevailed throughout the sleeping-car, which was for Boston without change.
Owen Wister, Lin McLean, *1897.*

117.2　It is a place that you're at home in right away: perhaps it is the train-smoke that does the thing to you. You smell the smoke the minute you get off the train, and it does something fierce and wonderful to you: it's better than a shot of gin, it's better than the breath of air off the prairies which for the most part you can't smell anyway, you just breath it in and you know that here comes everybody: Wops, Swedes, Jews, Dutchmen, Micks, and all of the rest of them, and you can take it, and you are in America, Chicago, U.S.A.
Thomas Wolfe, Oh Chicago, *within* Prologue to America, *1978.*

117.3　We swept along the many-railed track, and the straws and scraps of paper danced in our eddy as we passed. We overtook local trains, and they receded slowly in the great perspective, huge freight trains met us or were overtaken, long trains of doomed cattle passed northward, solitary engines went by—every engine, tolling a melancholy bell, contributed to a clanging that approached or receded but never ceased—open trucks crowded with workmen went cityward. By the side of the track, and over the level crossings, walked great swarms of common-looking people. So it goes on, mile after mile—Chicago.
H G Wells, travelling on the rear platform of the observation car of the Pennsylvania Limited; The Future in America, *1906.*

117.4　I am convinced that it is its many railroad stations that make Chicago so exciting a city, for railroad stations are alive, throbbing, dynamic, charged with the urgency of life. Airports, even the largest, are dull indeed compared to a steamcar depot; and bus stations are no better. For one thing, they don't smell right; there is no excitement in the fumes of gasoline, even high-test gasoline; and when and if the railroads go wholly over to Diesel, then railroad stations will lose much in the coal gases that no longer will remind the traveler of far places beyond the city limits. For another thing, Diesels don't sound right, for the snorting of motors contains nothing of the magic of a locomotive with its nervous breathing; nor is the bray of an air horn any substitute for the melody of steam through a reed or the music of hammer on brass. And finally, at least to my mind, the sleekest plane ever made and the shiniest bus on the road are rather dull objects compared to a locomotive drowsing at the head of a long string of passenger cars.
Stewart H Holbrook, The Story of American Railroading, *1947.*

118. CHICAGO & NORTH WESTERN RAILWAY

118.1 America's only southpaw railroad.
Anon. Advertising slogan, early 1920s, referring to its practice of running trains on the left track rather than the more common American practice of running on the right. The slogan was dropped after the C&NW had been told that in this respect it was not unique. One wonders how many members of the public understood the slogan.

118.2 The creation of Northwest Industries and its involvement with underwear, boots, chemicals and wire was unfortunate for the railroad.
Eugene M Lewis, of the creation of a holding company and resultant diversification for Chicago & North Western Railway. Email exchange *on RLHS newsgroup, 19 February 2005.*

119. CHICAGO BURLINGTON & QUINCY RAILROAD

119.1 How smoothly the trains run beyond the Missouri.
Willa Cather, poem, Going Home, Burlington Route.

119.2 Everywhere West.
Advertising slogan, 1930s.

120. CHICAGO MILWAUKEE ST PAUL & PACIFIC RAILROAD

120.1 The Route of the Hiawathas.
Advertising slogan, referring to its crack express trains. Late 1930s.

121. CHICAGO, ROCK ISLAND & PACIFIC RAILROAD

121.1 Well, the Rock Island Line she's a mighty good road
The Rock Island Line is a road to ride
Yeah, the Rock Island Line she's a mighty good road
And if you want to ride it
Got to ride it like you find it
Get your ticket at the station on the Rock Island Line.
Song, Rock Island Line, *attributed to Huddie Ledbetter, 1934.*

122. THE CHIEF

122.1 The flying boudoir of the Western continent.
Attributed to Lucius Beebe, of the Santa Fe's Chief *train between Chicago and Los Angeles;* Trains Magazine, *January 1992.*

123. CHILDHOOD

123.1 Childhood should be happy, as all the authorities agree. Childhood should move contentedly from given

point to given point—from Christmas to Christmas, for example, and from train to train.
Anon. The Lost Generation, *in the* New York Times, *reprinted in* Trains Magazine, *October 1949.*

124. CHILTERN HILLS (England)

124.1 A String of Pearls.
Great Central Railway Poster, c. 1912.

125. CHRISTMAS

125.1 'Twas the night before Christmas and through the roundhouse,
not a creature was stirring, not even a mouse.

The tools were all hung by the workbench with care,
in hopes that St. Clicktyclack soon would be there.

The engines were snuggled all safe in their beds,
while visions of sand towers danced in their heads.

And I in my hick'ries, and Ma in her cap,
(the one marked "Conductor"), were taking a nap.

When out in the yard there arose such a clatter,
I sprang from my bed to see what was the matter.

The moon on the roofs of the box cars and snow,
gave a luster of midday to objects below.

When what to my wondering eyes should appear,
but a tank engine pulled by 8 tiny traindeer.

With a coal-covered driver, so lively and quick,
I knew in a moment it must be St. Click.
Ben L Dibble, poem (part), A visit from St Click, *1989. www2pb.ip-soft.net/railinfo/poems/poetry.html, accessed 2004.*

126. THE CHURCH AND RAILWAYS

126.1 Both had their heyday in the mid-19th Century; both own a great deal of Gothic-style architecture, which is expensive to maintain; both are assailed by critics, and both are firmly convinced that they are the best means of getting man to his ultimate destination.
Rev Wilbert Vere Awdry. Often quoted, more or less extensively, and probably said by him on more than one occasion, this is the most comprehensive version found; Trains Magazine, *reporting his death, June 1997.*

127. CINCINNATI UNION TERMINAL

127.1 … of the railroads, by the railroads, for the railroads—pluralistic, private, posh, a monument to a vanished time of Pullman and Vanderbilt, the Van Swearingens and Willard, Tri-Motors, and Model A's, 1½-cent-a-mile fares and 10 per cent reductions on round trips.
David P Morgan, Trains Magazine, *February 1972.*

127.2 Lights wink across the track diagram over its 187-lever interlocking machine. Under the great rotunda tickets are being dated and sleeping car diagrams being inked in. Down the train concourse crowds dutifully assemble at the ramps and stairs leading to the platforms below. And below, at track level, smoke curls from diner galleys, blankets are tucked into lowers and uppers, and enginemen and trainmen compare Hamiltons.

　Ibid.

128. CIRCLE LINE (London)

128.1 In the old days they provided a sort of health resort for people who suffered from asthma, for which the sulphurous and other fumes were supposed to be beneficial, and there were several regular asthmatical customers who daily took one or two turns round the circle to enjoy the—to them—invigorating atmosphere.

　E L Ahrons, of steam-hauled underground trains, The Railway Magazine, *December 1924.*

129. CITY & SOUTH LONDON RAILWAY

129.1 A useful and promising novelty.

　Comment, Herepath's Railway Journal, *8 November 1890.*

129.2 The whole railway had a smell of wet feet or a changing room after games . . .

　John Betjeman, Coffee, Port and Cigars on the Inner Circle, *within* The Times, *24 May 1963.*

130. THE CITY OF NEW ORLEANS

130.1 This train's got the disappearin' railroad blues.

　Steve Goodman, song, City of New Orleans, *1970.*

131. CLAPHAM JUNCTION

131.1 . . . Clapham Junction, that flimsy collection of cast iron, glass, wood and brick set among so many shining rails.

　John Betjeman, Punch, *25 August 1954.*

131.2 . . . the station is really no more than a series of disconnected architectural incidents loosely linked either by bridge or subway and without any polarizing force by way of a terminal building.

　David Atwell, Railway Architecture *(ed. by Binney and Pearce), 1979.*

131.3 After discovering my means of escape to Brighton, just over an hour away—one hour to freedom—I never hated Clapham again in quite the same way.

　Lisa St Aubin de Terán, Off the Rails, *1989.*

132. CLASS DISTINCTION

See also First Class, Second Class, Third Class

132.1 The first class of train is the most fashionable, but the second or third are the most amusing. I travelled one day from Liverpool to Manchester in the lumber train. Many of the carriages were occupied by the swinish multitude, and others by a multitude of swine.

　Lord Brougham, at celebratory dinner for the opening of the Liverpool & Manchester Railway, 1830. Quoted in Railway Adventures and Anecdotes, *1888.*

132.2 Your Directors had been desirous to hold out every encouragement and facility to Third Class Passengers that might not interfere prejudicially with the amount of First or Second Class traffic, but the extent to which advantage of the accommodation thus afforded was taken by classes of persons for whom it was not intended, has obliged your Directors, in opening a further portion of the line to Totnes, to reduce the number of Third Class Trains, and so to regulate the time at which they run as to adapt them, as far as circumstances would admit, more exclusively to the wants of the labouring classes.

　Report *of South Devon Railway Directors, August 1847.*

132.3 Having on one occasion gone down by first-class, with an Oxford man who had just taken his M.A., an ensign of infantry in his first uniform, a clerk in Somerset House, and a Manchester man who had been visiting a whig Lord,—and returned third-class, with a tinker, a sailor just returned from Africa, a bird-catcher with his load, and a gentleman in velveteens, rather greasy, who seemed, probably on a private mission, to have visited the misdemeanour wards of all the prisons in England and Scotland; we preferred the return trip, that is to say, vulgar and amusing to dull and genteel.

　Samuel Sidney, Rides on Railways, *1851.*

132.4 By universal admission, there are, roughly speaking, three classes in all societies, and the existing arrangement of railway carriages appears to correspond very closely with the ordinary habits of life.

　Editorial comment on the decision of the Midland Railway to abolish second class accommodation and to upgrade the quality of third class, The Times, *12 October 1874.*

132.5 Gradually my mother consented to go in her own carriage, on a truck, by rail as far as Birmingham; farther she could not endure it. Later still, nearly the whole journey was effected by rail, but in our own chariot. At last we came to use the ordinary railway carriages, but then, for a long time, we used to have post-horses to meet us at some station near London: my mother would not be known to enter London in a railway carriage—'it was so excessively improper' (the sitting opposite strangers in the same carriage); so we entered the metropolis 'by land,' as it was called in those early days of railway travelling.

　Augustus Hare, of his mother Anne Paul Hare; The Years with Mother, *1896.*

132.6 At suburban railway stations—you may see them as
  you pass—
There are signboards on the platform saying 'Wait here
  second class':
And to me the whirr and thunder and the cluck of
  running gear
Seems to be forever saying, saying 'Second class wait
  here'—

Yes, the second class were waiting in the days of serf and
  prince.
And the second class are waiting—they've been waiting
  ever since,
There are gardens in the background, and the line is bare
  and drear,
Yet they wait beneath a signboard, sneering, 'Second
  class wait here'.
  *Henry Lawson, carriage painter in Sydney, Australia;
  poem, Second class wait here, first two verses, 1899.*

132.7 'I say,' began the money-lender, pursing his lips,
'that there is not one rule of right living which these
te-rains do not cause us to break. We sit, for example,
side by side with all castes and peoples.'
  Rudyard Kipling, Kim, 1901.

132.8 First-class passengers, keep your seats; second-class
passengers, get out and walk; third-class passengers, get
out and push.
  *Apocryphal instructions upon encountering heavy gra-
  dients on lesser railways, recorded by Thomas W Jackson,
  On a Slow Train Through Arkansaw, 1903.*

132.9 'What class are you going? I go second.'
'No second for me,' said Nicholas; 'you never know what
you may catch.'
  *John Galsworthy, of the days when two classes of travel
  were available on the London Underground. Exchange in
  The Man of Property, 1906.*

132.10 We have for some time been satisfied that for
long-distance traffic, and also on many subsidiary lines,
where the services are largely of an omnibus character,
the maintenance of three classes of accommodation,
which always means six, and sometimes nine, when we
come to consider smoking, non-smoking, and ladies'
compartments, is not justifiable. It means necessarily
longer trains, bigger engines, and longer platforms and
sidings, with all the attendant expense of dealing with
this sub-division of classes of passengers.
  *Viscount Churchill, Chairman, Great Western Railway,
  Meeting, February 1909.*

132.11 The gentry preferred to remain in their own carriages, loaded on an open car, rather than be associated
with the commonalty, until a compromise was effected
by assigning them, as first class passengers, to special
compartments more luxuriously appointed.
  Henry S Haines, Efficient Railway Operation, 1919.

132.12 There are two classes of passenger-accommoda-
tion—Goods and Cattle; between these the chief difference in comfort is that Goods is open, whereas Cattle has
a roof. In the latter class we were fortunate enough to
secure a corner seat.
  *Peter Fleming, of travel on the railway between Pinsiang
  and Changsha in central China, One's Company, 1934.*

132.13 The further a train is going, the higher the proportion of first-class carriages.
  *Paul Jennings, The Times Saturday Review, 5 February
  1972.*

132.14 Together with the fine designs of the royal saloons
and the first-class carriages, the comfort and loveliness of
the third has been swept away, and a new uniform has
been introduced. There are still first-class carriages and
first-class tickets, on every form of transport but the bus.
Despite all this, this chosen uniform is drab. Progress has
begun to shunt backwards. Even the jet-set have to elbow
their way to their jets through, nine times out of ten,
awesomely boring airports. There is no Grand Central
station of the air. What a pity that when the Paris mob
cried out for Liberty, Equality and Fraternity, it did not
add Quality to the list.
  Lisa St Aubin de Terán, Off the Rails, 1989.

132.15 Out of the long winter, through which the mass
desire to travel has never been so strong, nor the means
to do so as available, the international ostriches have
begun to pull their heads out of the grubby sand. A few
sparks of elegance can be seen again along certain lines,
though these are still as rare as fireflies on a summer's
night. The Orient Express glides backwards and forwards to Venice once more as a luxury venture. Yet still
the roll-call of ostriches is called, and the ministers of
transport raise their heads, and none of them has been
called Pullman or even Gresley. So grace and style are still
denied to the masses.
  Ibid.

132.16 Two classes of accommodation are provided but
most rolling stock is in such poor condition that classification is almost irrelevant.
  *Introduction to railways in Angola, Thomas Cook's
  Overseas Timetable, November-December 1997.*

133. CLEANERS (Locomotive)

133.1 There is no dirtier situation than that of cleaner.
  *Thomas Wilkinson Speight, Up and Down the Line,
  Household Words, 27 June 1857.*

134. CLERKS

134.1 He has gone off to seek his fortune in the wild,
adventurous knight-errant capacity of clerk on the Leeds
and Manchester railroad.
  Charlotte Brontë, Jane Eyre, 1847.

134.2 There's a class that's submerged in the railroad
   game,
      of which little is said or known,
Who should merit a niche in that hall of fame,
      which the railroads may claim for their own.
They are men and women, and each of them share
      in a multiple scheme of work,
That is wrought by all with an infinite care,
      in the role of the railroad clerk.
   Sidney Warren Mase, poem, February 1926, quoted in
Trains Magazine, *September 1981.*

135. CLOCKS

135.1 It has been found by experience that the passengers
are very likely to become nervous when at the windows if
there is not a large clock in sight.
   John A Droege, of the design and layout of ticket offices,
Passenger Trains and Terminals, *1916.*

COACHES
See Carriages

136. COAL

136.1 The strength of Britain lies in her iron and coal
beds; and the locomotive is destined, above all other
agencies, to bring it forth.
   George Stephenson, quoted by Samuel Smiles, Life of
George Stephenson, *1881.*

136.2 Whenever you put passenger-trains on a line, all the
other trains must be run at high speeds to keep out of
their way. But coal trains run at high speeds pull the road
to pieces, besides causing large expenditure in locomotive power; and I doubt very much whether they will pay
after all; but a succession of long coal trains, if run at
from ten to fourteen miles an hour, would pay very well.
   Ibid.

136.3 What! Coal by railway? They will be asking us to
carry dung next!
   *Capt W Bruyeres, Superintendent of the London & Bir-
mingham Railway, quoted in* Our Home Railways, *W J
Gordon, 1910. According to G P Neele, in* Railway Reminiscences, *George Stephenson commented "Tell Bruyeres that
when we carry him by railway we do carry dung."*

136.4 Coal, at least of the ordinary bituminous kinds,
ought not to be used as a staple fuel at all. It is the mere
raw material of fuel—the ore from which coke is
extracted, and contains some valuable compounds
which, on the one hand, ought never to be thrown into
the furnace, and, on the other, ought not to be wasted by
the prevailing method of coking.
   Daniel K Clark, quoted by Angus Sinclair, Development of the Locomotive Engine, *1907.*

136.5 Alec: It looks like a bit of grit.
Laura: It was when the express went through.
   Noel Coward, screenplay for film, Brief Encounter;
*exchange between Dr. Alec Harvey (Trevor Howard) and
Laura Jesson (Celia Johnson) in the refreshment room; the
occasion (and cause) of their first meeting, 1946.*

137. THE COAST DAYLIGHT

137.1 The most beautiful train in the World.
   Advertisement, Southern Pacific Railroad.

138. COATS OF ARMS AND LOGOS

138.1 It must be the goat that does it.
   *E L Ahrons, on the attractiveness of the Taff Vale Rail-
way coat of arms,* The Railway Magazine, *October 1922.*

138.2 It was promptly, and deservedly, dubbed "the ferret
and dartboard" by the railway fraternity.
   *George Dow, of the first coat of arms used on British
Railways locomotives, c. 1950,* Railway Heraldry, *1973.*

138.3 An artist's impression of a singularly severe
derailment.
   *Edmund Crispin, of the last corporate symbol of British
Railways.*

138.4 All railroads, really all industry, had wet noodle
envy.
   *Richard Saunders Jr, of the popularity of the Canadian
National Railways logo, composed of sinuous conjoined let-
ters CN (once called " a jagged squirt of toothpaste", anon);*
Merging Lines, *2001.*

139. COCK O' THE NORTH

139.1 I do not mind telling you—and my Chairman—one
thing, namely that the first "Cock o' the North" is the last
"Cock o' the North", and that further engines of the same
type will be slightly modified in the light of what I have
learned in France.
   *Sir Nigel Gresley, Annual Dinner, Institution of Loco-
motive Engineers, 1 March 1935. His Chairman, William
Whitelaw of the LNER, was present.*

139.2 Aye, she had a bonny appetite.
   *James Cunningham, one-time fireman of LNER class
P2 No. 2001, in conversation with John Crawley.*

140. COFFEE

140.1 Coffee is an institution aboard all cabooses, and a
pot of it is nearly always simmering on top of the stove.
As we sit there with a mug of this thick black liquid, we
appreciate the tradition that caboose coffee must be
strong enough to float a track spike.
   Eric A Grubb, Trains Magazine, *November 1947.*

141. C B COLLETT

141.1 . . . a narrow, suspicious man, who contributed little but an evolutionary development of Churchward's designs.

> *George W Hilton*, Railroad History, No 153, *1985.*

141.2 Had Collett's men mixed with other engineers freely, they might have questioned the theology of Swindon's locomotive department.

> *Ibid.*

142. COMMISSIONS, RAILWAY

142.1 In 1844 another commission was appointed with more specific powers. Their special duty was to make preliminary reports to Parliament on applications for railroad charters. They tried to do their work well, but were beset by difficulties on all sides. Railroad projectors hated them; Parliament itself was jealous of them. After a luckless existence of about a year, this board was abolished. It really died of too much work and too little pay. In 1846 Parliament tried the experiment of a railroad commission of another kind. It offered first-rate salaries, and secured well-known men; then it avoided all causes of offence by not giving them any powers. This lasted five years. Its fate was the reverse of that of its predecessors; it died of too much pay and too little work. Thus ended the first series of experiments in railway commissions.

> *Professor Arthur Hadley*, Railroad Transportation, *1906.*

143. COMMUNICATION CORD

143.1 There is, indeed, something almost ludicrously characteristic in the manner with which those interested in the railway management of Great Britain strain at their gnats while they swallow their camels. They have grappled with the great questions of city travel with a superb financial and engineering sagacity, which has left all other communities hopelessly distanced; but while carrying their passengers under and over the ebb and flow of the Thames and among the chimney pots of densest London to leave them on the very steps of the Royal Exchange, they have never been able to devise any satisfactory means for putting the traveller, in case of a disaster to the carriage in which he happens to be, in communication with the engine driver of his train. An English substitute for the American bell-cord has for more than thirty years set the ingenuity of Great Britain at defiance.

> *Charles Francis Adams*, Notes on Railroad Accidents, *1879.*

143.2 The awkward, clumsy and inefficient cord communication will soon be a relic of the past, and when compared with the many ingenious pneumatic and electric signals that have been devised, it is surprising to think that such a useless contrivance has been allowed to fill requirements for so long. Many will recall the numerous occasions on which the cord has been pulled, and after several yards have been hauled into the compartment there has been no response and resulting action; of this the accounts in the daily papers alone give sufficient proof.

> *Editorial comment*, The Locomotive Magazine, *6 June 1903.*

143.3 Commonly an electric button is placed high on the side of the carriage as an alarm signal, and it is unlawful to push it unless one is in serious need of assistance from the guard. But these bells also rang in the dining car, and were supposed to open negotiations for tea or whatever. A new function has been projected on an ancient custom. No genius has yet appeared to separate these two meanings. Each bell rings an alarm and a bid for tea or whatever. It is a perfect theory then, that, if one rings for tea, the guard comes to interrupt the murder, and that if one is being murdered, the attendant appears with tea.

> *Stephen Crane, of LNWR carriages*, The Scotch Express, *within* Men, Women and Boats, *1921.*

143.4 It is a curious thing that, in spite of the railway companies' sporting willingness to let their patrons have a tug at the extremely moderate price of five pounds a go, very few people have either pulled a communication-cord or seen one pulled.

> *P G Wodehouse*, Meet Mr Mulliner, *1927.*

143.5 If £5 you can afford,
Try your strength upon this cord.
If £5 you cannot pay,
Wait until another day.

> *Anon., found in the compartment of an LNER carriage, 1935, recorded in the* LNER Magazine, *July 1935.*

COMMUTERS

See Season Tickets and Season Ticket Holders, Suburban Services

144. COMPETITION

144.1 Competition in railways must lead to compromise, but I think the public would rather be in the hands of companies than of the government.

> *George Hudson, evidence before parliamentary committee of enquiry, 18 March 1844.*

144.2 Now, for my part, I would rather give my confidence to a Gracchus when speaking on the subject of sedition, than give my confidence to a railway director when speaking to the public on the effects of competition.

> *William Gladstone, Parliamentary debate*, Hansard vol. 76, *8 July 1844.*

144.3 Where combination is possible, competition is impossible.

Attributed to Robert Stephenson, c. 1854, Edward Cleveland-Stevens, English Railways and Their Development and Relation to the State, *1915.*

144.4 If you put your hand into our breadbasket, we will put our hands into your coal scuttle.

Anon., 1875. Stated to be a director of a company in competition with the Midland Railway, referring to the extensive coal traffic on the latter. Quoted by Frederick Williams, The Midland Railway, *1878.*

144.5 The tribe of Hinckleys, Cranes, Graves, Villards, and other cranks and thieves led by Hopkins and Gould, have built and will build Roads wherever fools with money will follow, and where three roads stimulated by contracts, are thus built to do the work of one, it, in the very near future, leads right up to a necessity for the nearest solvent Road to buy the other useless ones.

John M Forbes, Chicago, Burlington & Northern, letter to Charles Perkins, 25 August 1885, of the expansion of railroads in the West; quoted by Julius Grodinsky in Trans-Continental Railway Strategy 1869–1893, *1962.*

144.6 We are feeling our way, and until we are certain that the C&O are doing us any harm in our territory, we prefer to let them alone. Fortunately for us, the C&O people are not very active or aggressive.

Frederick J Kimball, President, Norfolk & Western Railroad, letter to Everett Gray, quoted by Thomas C Cochrane in Railroad Leaders, *1845–1890, 1953.*

144.7 There never was a more fallacious idea than that low rates could best be acquired by competition.

Robert Harris, President, Northern Pacific, letter to P B Johnson, 22 October 1887; quoted by Thomas C Cochrane in Railroad Leaders, *1845–1890, 1953.*

144.8 But Scotch railways deprived of the opportunity of fighting would scarcely know themselves again.

W M Acworth, The Railways of Scotland, *1890.*

144.9 "All Gaul," said Mr. Merrill—he was speaking to a literary man—"all Gaul is divided into five railroads. I am one, the Grand Gulf and Northern, the impecunious one. That is the reason I'm so nice to everybody, Mr Wetherell. The other day a conductor on my road had a shock of paralysis when a man paid his fare. Then there's Balch, president of the 'Down East' road, as we call it. Balch and I are out of this fight,—we don't care whether Isaac D. Worthington gets his franchise or not, or I wouldn't be telling you this. The two railroads which don't want him to get it, because the Truro would eventually become a competitor with them, are the Central and the Northwestern. Alexander Duncan is president of the Central."

Winston Churchill, Coniston, *1906.*

144.10 If unrestricted capitalization has increased the load which the railroads have had to bear, unrestricted competition has impaired their ability to support any load at all.

Stuart Daggett, Railroad Reorganization, *1908.*

144.11 Heretofore, we have sought to compel competition between railroads, yet now, while not abandoning the theory of forced competition, the tendency of our legislation is absolutely to destroy whatever competition there exists in rates.

Ivy Lee, of conflicting U.S. Government policies; address, the London School of Economics, 7 February 1910.

144.12 It is true, if there were advantages in competition, that those advantages will have to be foregone, but there was not much real competition before the War. Such competition as there was was the wrong kind of competition. It was the competition that tried to pinch somebody else's traffic. That is not the competition which makes traffic or gives prosperity. That is the competition which costs money, and the public pays in the end.

Sir Eric Geddes, on the prospect of the loss of competition between railways contemplated by the Railways Bill. This was a deliberate policy built into the bill. The reference to "pinching" of traffic was to the granting of running powers (trackage rights) to railway companies. Geddes, at that time Minister of Transport and therefore responsible for introducing the bill into Parliament, had been in the employ of the North Eastern Railway, a company which had enjoyed a monopoly almost throughout its territory, and had experienced little competition as a result. His comments upon "not much real competition" before the War may not have been much appreciated by other railwaymen. Parliamentary debate on the Railways Bill, 26 May 1921.

144.13 In theory, a monopoly has no competition.

Anon. Headline of advertisement, within advocacy advertising campaign by British Rail against government restrictions and policies, and underlining the more favourable treatment afforded to other, competing transport modes which showed that British Railways was not, in fact, a monopoly; 1980s.

144.14. The sixty-year-old Interstate Commerce Commission, whose only justifiable mission in life was to preserve competition in transportation, had fallen into the contradictory habit of protecting some of the competitors, at the expense of competition itself.

Charles O Morgret, Brosnan, vol 1, *1996.*

145. COMPUTERS

145.1 As the steam engine was to the first revolution, so the computer is to the second.

Geoffrey Crowther, Inaugural Ceremony of the Open University, 1969.

CONDUCTORS
See Guards and Conductors

CONNECTIONS
See Changing Trains

146. CONRAIL
(Consolidated Rail Corporation)

146.1 Not a step toward nationalization, with all its evils, but a step away from it.
> *Claude S Brinegar, Department of Transportation Secretary, describing the legislation that created Conrail. Quoted in* Trains Magazine, *December 1974.*

146.2 . . . an underdog living on borrowed time.
> *Anon., quoted by Nicholas Faith,* The Economist, *24 August 1985.*

CONSTRUCTION WORKERS
See Navvies, Railways—Construction

147. CONSULTANTS

147.1 They charge you a lot of money to tell you something you already know in a language that you do not understand.
> *Anon., current among senior railway officers in the 1950s and 1960s, related by George Dow to his son, the compiler.*

147.2 It was a case of the blind advising the deaf.
> *Christian Wolmar, of the many consultants advising government on privatization, early 1990s.* Broken Rails, *2001.*

148. CONTRACTORS

148.1 Still, it takes more than a couple of contractors, however enthusiastic, to construct a railway.
> *C P Gasquoine,* The Story of the Cambrian, *1922.*

149. CORONATION

149.1 There have been heavier or faster British trains than the Coronation, there have been trains travelling longer distances non-stop, but never has there been a train travelling so fast, so far, so frequently, so reliably.
> *A J Mullay,* Streamlined Steam, *1994.*

150. CORONATION SCOT

150.1 A symphony in red white and blue.
> *LMS description of the cocktail bar on the Coronation Scot train of 1937, quoted by A J Mullay,* Streamlined Steam, *1994.*

151. CORPORATE LIVERIES

151.1 We have simple structures along our route that serve the public's needs. By having them shaped alike and painted the same color, they become heralds. Since the public recognizes them as ours, we do not need to trouble ourselves by painting the railroad name on these depots.
> *Burlington & Missouri River Railroad official, early twentieth century, quoted by John Stilgoe in* Metropolitan Corridor, *1983.*

152. COSTS

152.1 In a railroad, some may be consumed because the railroad is in the business of carrying passengers, because a ticket was sold to a passenger, because that passenger rode on a train which moved between two particular cities, because the passenger rode in a specific type of car, because he tripped and fell when getting off the train, and because the temperature was 20° below zero when he took his journey.
> *Dwight R Ladd, of the complexities of cost recording for passenger service,* Cost Data for the Management of Railroad Passenger Service, *1957.*

152.2 The railroads, of course, have neither in-process nor finished inventory since production and consumption are simultaneous.
> *Ibid.*

152.3 Inventory valuation is not part of the determination of net income of a railroad. Thus the major impetus for the development of routine product costing in manufacturing industries was not felt in the railroad industry.
> *Ibid.*

152.4 It costs us $8 a mile to operate a train, but when we take trains off we save only $4 a mile.
> *Anon. president of large U.S. railroad, quoted by Dwight R Ladd, in* Cost Data for the Management of Railroad Passenger Service, *1957.*

153. COUPLINGS

153.1 So long as brakes cost more than brakemen, we may expect the present sacrificial method of car coupling to be continued.
> *Rev Dr Lyman Abbott, of link and pin couplers, quoted by Peter Lyon,* To Hell in a Day Coach, *1967.*

153.2 In truth, any of the American central couplers can be undone and linked-up again quicker and more easily than the venerable screw-link-coupling with side chains which Noah is believed to have designed during his time of enforced leisure on board the Ark.
> *W J Scott,* The New Competitor, *within* The Railway Magazine, vol. VII, *1900.*

153.3 Our system is imperfect because our automatic couplers are really only semiautomatic (and more semi than auto).

David P Morgan, Trains Magazine, *January 1970.*

154. COVENTRY

154.1 I waited for the train at Coventry;
I hung with grooms and porters on the bridge,
To watch the three tall spires; and there I shape
The city's ancient legend into this:-

Alfred Lord Tennyson, introductory verse of Godiva, *1842.*

155. COWS ON THE LINE

155.1 Member of Parliament: "Suppose, now, one of these engines to be going along a railroad at the rate of nine or ten miles an hour, and that a cow were to stray upon the line and get in the way of the engine; would not that be a very awkward circumstance?"

George Stephenson: "Yes, very awkward—for the coo!"

Exchange between a member of Parliament and George Stephenson. This oft-quoted story has yet to be found in any form which gives the name of the MP involved. It is feared that record of his identity perished in a fire at the House of Commons in 1834. It appears not to have been the practice for such proceedings to have been recorded in The Times *at that period. House of Commons Committee hearings into the Liverpool & Manchester Bill, April 1825.*

155.2 I can stand a hog but them cows are the devil to pay.

Anon., fireman of Baltimore-Washington train, after the locomotive had struck a cow lying on the line; comment to Alex Mackay, The Western World; *or, travels in the United States in 1846–47, 1850.*

155.3 Counsel: If a beast got on to the line as a train came along, what would happen to the beast?

Martin Ryan: It would exercise its running powers.

Martin Ryan, cattle dealer, in supporting evidence for an Irish railway bill, May 1892, as related in Fifty Years of Railway Life, *Joseph Tatlow, 1920.*

155.4 And when a cow is killed, it always happens to be a very valuable one. I believe that there is no better way of enhancing the value of a cow than to cross it with a CNR locomotive.

Sir Henry Thornton, quoted in The Tragedy of Sir Henry Thornton, *D'Arcy Marsh, 1935.*

155.5 In the Blue Ridge Mountains of Virginia,
On the Trail of the Lonesome Pine,
Lived a sweet brown cow with eyes so fine;
But you can't expect a cow to read a railroad sign!

Attributed to Cowper by Louis F Meyer Jr in an article in Trains Magazine, *November 1944.*

155.6 I got 22 cows what I chase every morning and every night over your railroad tracks here in Northport. Up until 2 weeks ago everything is fine, no trains is coming in the morning at 8 A.M. when we drive our cows over the crossing.

Then last Thursday comes a little pip squeak of a train with maybe 6 empty box cars going like a bat out of hell he comes at just 8 A.M. This I think is maybe a special so I hold my cows from crossing. Now day before yesterday comes the same dam train with those 6 empty box cars and I just get my 22 cows over the tracks when he comes barreling through.

What I want to know is who is this guy the railroad presidents son, so they give him his own little train to play with or some stupid conductor what forgets to take these box cars along on the regular run.

I would appreciate it very much if you would tell this hot shot engineer to kindly take another cup coffee in the morning so he should get here later than 8 A.M. and may be make hamburger out of my holsteins. Either that or he should stop at the Northport crossing and look both ways to see if any thing is coming what looks like cows. You can tell the people from the cows because the cows got a smarter look.

You got a pretty nice little railroad and I don't want to make you no trouble so you tell these guys they should send the little train through at maybe noon huh?

John Kraske, farmer, New London, Wisconsin, to the Traffic Manager, Green Bay & Western Railroad, quoted without a date in Railroad History, *No 115, October 1966.*

155.7 Their meeting it was sudden,
Their parting it was sad;
She gave her young life meekly—
'Twas the only life she had.

She's sleeping 'neath the willows,
And she's resting peaceful now,
For that's what always happens
When freight-train meets a cow.

Clifford Levin, poem In Memoriam, *within* Great Poems from Railroad Magazine, *1968; published earlier in that magazine at unstated date.*

156. THOMAS CRAMPTON

156.1 Dilettanti railway students of today, in this country at least, are regrettably liable to dismiss the memory of T R Crampton as that of a man who, a long time ago, designed a type of locomotive somewhat on the lines of a stern-wheel river steamer.

C Hamilton Ellis, The Locomotive Magazine, *15 March 1940.*

157. CREWE (England)

157.1 Crewe is the nursery, Wolverton the hospital for locomotives.
John Capper, writing of the London & North Western Railway, Household Words, *31 December 1853.*

157.2 There was a young lady in blue,
Who wanted to catch the 2.2;
Said the porter "Don't hurry,
Nor scurry, nor flurry,
It's a minute or 2 2 2.2!"
Anon., poem At Crewe, *within* The Railway Magazine, vol. IV, *1899.*

157.3 How oft since childhood have I wished I knew
What lay behind the mystery of Crewe;
It was a word on every mystic mouth
Of those who spoke of journeys North and South
And, long before I knew what "travel" meant
Crewe was a word of magical content.
"We change at Crewe" said grown-ups, yet they came
Back from adventure looking just the same!
Poem Crewe, Punch.

157.4 And when, at length, to travelling years I grew
Time but intensified the thrills of Crewe.
Crewe stood for haste, for working in the night,
For noise, for terror and for delight!
For loud-mouthed porters breaking up a dream,
For shrieking whistles and for hissing steam.
Crewe stood for trains in splendour, pomp and glow
For trains and ought else? Oh how I longed to know.
Ibid., second verse.

157.5 Crewe is a fantasy of travellers' brains
Born of the night, of Bradshaw, and—of trains
Ibid.

157.6 Once on the vast platforms of Crewe, the guilty couple would be safe from curiosity, lost in England, like needles in a haystack.
Arnold Bennett, These Twain, *1916.*

157.7 The new CME made several visits and was found to be a man with a great grasp of affairs, combined, more-over, with a courteous manner, which came as a surprise to some of the tougher elements who did not quite know how to deal with this characteristic.
T Lovatt Williams, of merger of the LNWR and LYR, 1922, when George Hughes of the LYR took responsibility for Crewe Works. The Railway Magazine, vol. 95, *1949.*

157.8 Crewe was ruled from the general office of the Locomotive Department.
W H Chaloner, The Social and Economic Development of Crewe, *1950.*

157.9 The place which *is* Crewe is not Crewe, and the place which is not Crewe *is* Crewe.
Anon., quoted by Brian Reed in Crewe Locomotive Works and Its Men, *1982.*

158. CRICKET FIELDS

158.1 You could lie on the bank above the cricket field and, between overs, watch the trains go by. There were the unceasing crossings and recrossings of green locals: the expresses to the South Coast with destinations recognisable from the composition of their buffet cars: the little steam line curving away between rising slate roofs and pious Victorian spires to the Surrey hills: four times during a day's play the gleaming coffee and cream of the "Belle": and, just before they drew stumps, the home-bound Newhaven boat express.
Bryan Morgan, The End of the Line, *1955.*

158.2 Over extra cover's head, then, as we lazed through those last strawberry summers, the Pullman expresses strummed plushily past: beyond long-leg boundary a mossy gravel way marked what there had been a century before.
Ibid., reflecting upon the disused formation of the ancient Surrey Iron Railway.

159. CRIME

159.1 Is it a crime to be a railroad?
David P Morgan, title of article about railroads' difficult and one-sided relationship with government; Trains Magazine, *August 1961.*

160. CRIMEA

160.1 The earliest instance of a purely military railway being constructed to serve the purpose of a campaign occurred in the Crimean War; and, although the line then made would to-day be regarded as little more than an especially inefficient apology for a railway, it was looked upon at the time as a remarkable innovation in warfare.
Edwin A Pratt, The Rise of Rail Power in War and Conquest, *1915.*

160.2 Most railway and army historians know that there was a railway of sorts at work in the Crimea during the war of 1854–6, and so far they have been content to leave it at that.
Michael Robbins, Journal of Transport History, *1954.*

161. CROMER

161.1 Gem of the Norfolk Coast.
London Midland & Scottish Railway poster, c. 1925.

162. CROSS COUNTRY

162.1 The service is a masterpiece of co-ordination, a vivid illustration of the public usefulness which the enterprise or railway companies can achieve when they renounce the stupid selfishness with which they at first parcelled out the land between them.

> *Anon., staff member of* The Times, *of the cross-country services offered by the Midland Railway, in* The Complete Railway Traveller, *part III,* The Times, *28 November 1905.*

162.2 No doubt all this is to the material benefit of the business community, and there may be occasions when I myself shall be glad not to spend the best part of a day in covering 150 miles. Still I recall the many hours of quiet happiness which, thanks to bad connections, have come to me when I have been left stranded and isolated upon the up platform of Dreamington-on-the-Marsh or the down platform of Sleepytown-in-the-Wold.

> *Gilbert Thomas,* The Manchester Guardian, *late 1920s, also in* Double Headed, *1963.*

162.3 It is true that, unless road competition succeeds some day in closing our railway branch lines, there will still be cross-country journeys in one sense. There will always be slow trains connecting villages with the nearest market-town. But the point is that you and I, the long-distance travellers, will no longer have to use them; and great will be our loss.

> *Ibid.*

163. CURVATURE

163.1 ... it can never be amiss to bear in mind that there is no case on record where a railway has been brought to bankruptcy by the expenses resulting from sharp curvature, nor is there any likelihood that there ever will be such a case, while the instances are many where companies have been bankrupted by their expenditures to obtain easy curvature.

> *Arthur M Wellington,* The Economic Theory of Railway Location, *1903.*

163.2 25 Miles to Next Curve.

> *Sign on Florida East Coast Railroad, beside cut-off line between Ft Pierce and Lake Okeechobee, illustrated in* Trains Magazine, *May 1947.*

164. CUTTINGS

164.1 You can't imagine how strange it seemed to be journeying on thus, without any visible cause of progress other than the magical machine, with its flying white breath and rhythmical, unvarying pace, between these rocky walls, which are already clothed with moss and ferns and grasses; and when I reflected that these great masses of stone had been cut asunder to allow our passage thus far below the surface of the earth, I felt as if no fairy tale was ever half so wonderful as what I saw. Bridges were thrown across from side to side across the top of these cliffs, and the people looking down upon us from them seemed like pigmies standing in the sky.

> *Fanny Kemble, of a journey through Olive Mount cutting, on the footplate with George Stephenson, letter, 26 August 1830.*

165. D1 AND E1 LOCOMOTIVES

165.1 Cor, guv, they were wonderful jobs. We got 3d a day extra for firing them on Boat Trains when we were knocking about as spare men but we never got mogadored with them.

Percy Tutt, describing ex-South Eastern Railway 4-4-0s, to Richard Hardy, 1954. To be "mogadored" was to be caught short of steam; apparently the word has cockney/gypsy origins.

166. DARLINGTON LOCOMOTIVE WORKS

166.1 Here are ten acres of intensive mechanical movement, and the tumult such activity breeds.

Dell Leigh, On The Line, 1928.

167. DEARBORN STATION (Chicago)

167.1 Dearborn is one of those stations that railroads do not illustrate in their gorgeous travel bulletins.

Linn H Westcott, Trains Magazine, March 1946.

168. DEATH

See also Suicide

168.1 Who is in charge of the clattering train?
The axles creak and the couplings strain,
And the pace is hot, and the points are near,
And Sleep has deadened the driver's ear;
And the signals flash through the night in vain,
For Death is in charge of the clattering train.

Anon., but understood to have been published in Punch, and also used by W S Churchill.

169. DECELERATION

169.1 But ten years ago, before Derby had taught us the use of the word "decelerate", we did not expect improvements of the best Midland express to advance at the rate of six seconds per annum.

Sir William Acworth, The Railways of England, 1899.

170. DEDICATION

(of books and other things)
See also Books on Railways, History—Railway, Writing about Railways

170.1 To the Grand Army of the West, which under command of General Sherman, in one hundred days of con-

tinuous battle, followed us over the line of the Georgia State Railroad from Chattanooga to Atlanta, and conquered where we only dared. This fragment of history is most respectfully and fraternally inscribed by the writer and his comrades of the Andrews Raid.

William Pittenger, Daring and Suffering, 1887.

170.2 To all those who with head, heart and hand toiled in the construction of this monument to the public service. . . .

Anon. Dedication at Grand Central Terminal, New York.

170.3 I venture to dedicate this work to the memory of A J Cassat and E H Harriman. Not only did these two men possess genius, a quality which Poe insists is far more abundant than is supposed; they possessed also the constructive ability, the faculty of analysis, the patience, the concentrativeness, that power of holding the attention steadily to the one purpose, self dependence and the contempt for all opinion that is opinion and no more, and in special that energy and industry without which Poe insists the works of genius cannot be brought forth. As though this were not enough, they had, too, the leaven of a noble enthusiasm, which acted as a ferment, stimulating to their utmost effort all who enjoyed their confidence.

L F Loree, Railroad Freight Transportation, 1922.

170.4 To the Memory of E. H. Harriman "whose services to the science of railroading will hardly be reckoned, by those who knew what his work was, as less than those rendered by George Stephenson himself."

George Kenman, quoting William Acworth in the Economic Journal, March 1916, in E. H. Harriman, 1922.

170.5 To Dorothy, who was courted with the occasional assistance of the North Shore Line, and who has traveled a good many interurban miles since then with remarkable forbearance.

William Middleton, The Interurban Era, 1961.

170.6 Because together we were pupped to the game; because together we experienced the martyrdom and the exaltation that are both so much a part of Andean railroading; it is my privilege to dedicate this work to my good friend and colleague for so many years, Charles Crofton Atkins.

Brian Fawcett, Railways of the Andes, 1963.

170.7 For Billy and Niko, who learned early that there was something very special about a GG-1.

William D Middleton, When the Steam Railroads Electrified, *1974.*

170.8 To the memory of Hugo Dyson who loved trains and never missed a connection.
Roger Green, Trains, *1982.*

170.9 For all the people I have met on trains.
Lisa St Aubin de Terán, Off the Rails, *1989.*

171. DEITIES

171.1 The Seven Euston Sleepers (one for each of the six British Railways districts and one for London) are the gods of the quiet country stations, where no sound is heard but the wind in the telephone wires and occasional signal bells; of the silent cuttings through woods; of the motionless rows of carriages at depots; of the pregnant silence when the train stops in a tunnel. They are silence at the heart of the solemn dance of public transport.
Paul Jennings, The Jenguin Pennings, *1963. See also Euston Station.*

171.2 The gods of high iron are a sardonic lot, although they are not without humor.
David P Morgan, Trains Magazine, *January 1972.*

171.3 God of concrete, God of steel,
God of piston and of wheel,
God of pylon, God of steam,
God of girder and of beam,
God of atom, God of mine,
All the world of power is thine.

Lord of cable, Lord of rail,
Lord of motorway and mail,
Lord of rocket, Lord of flight,
Lord of soaring satellite,
Lord of lightning's livid line,
All the world of speed is thine.
Richard G Jones, hymn, first two verses, 1983.

171.4 I am not angry. I am at peace. Lapses in service that at a normal restaurant would be considered appalling instead reinforce my faith in the underlying spirituality of the world around us. Just as the ancient native peoples knew that volcanoes erupted when the gods were feeling frisky, I know that I am only witnessing the ceaseless tug of war between the gods TuThuSu, MoWeSa, and WeFriMo. And WeFriMo is winning.
Charlie Hunter, after being subjected to poor dining car service on Amtrak's Texas Eagle; *article* The Gods of Amtrak, *in* Trains Magazine, *October 1995.*

171.5 TuThuSu, we determined, was a sober-minded, humorless, patriarchal sort, the god of operating efficiency on Amtrak. TuThuSu is the Zeus of Amtrak, powerful but not omniscient. TuThuSu has his work cut out for him—he's got the Sysyphean task of attempting to make aging equipment perform reliably on tracks of often uncooperative railroads. TuThuSu never smiles. His mightiest accomplishment is no accomplishment; an on-time performance with no untoward occurrence.
Ibid.

172. DELAWARE LACKAWANNA & WESTERN RAILROAD

172.1 The Lackawanna Railroad where does it go?
It goes from Jersey City to Buffalo.
Some of the trains stop at Maysville but they are few
Most of them go right through
Except the 8.22
Going west but the 10.12 bound for Jersey City
That is the train we like the best
As it takes you to Jersey City
Where you can take the ferry or tube for New York City.
Stephen Gale, poem, The Lackawanna Railroad, *within* The Poetry of Railways, *ed. Kenneth Hopkins, 1966.*

173. DELAY

See also Punctuality

173.1 I apologise for the delay to your train this morning. This is caused by wild sheep on the line ahead.
GNER Senior Conductor heard by Richard Hall, Stoke Bank, Spring 1997.

174. DELTIC LOCOMOTIVES

174.1 Once on a day-trip to York I did see
one of the last Deltics—I forget which—
but it roared like an oven and its wheels
squealed like a sack of kittens as it bit
its solid way south with a slow parcels.
As I tell the new lot, that was a real
sleepless-nights-of-wanting locomotive.
Jonathan Davidson, poem, The Train Spotter.

174.2 Burlington Zephyr
and Kenworths that pass in the night
Deltics and Steamers
and F88s in full flight
old train, take me back again
again to the one that I love.
Chris Rea, song Deltics, *1979.*

175. DENT

175.1 VISITOR: "Why did they build Dent station so far from the village?"
NATIVE: "'Appen they wanted it near t' railway lines."
Quoted in Settle-Carlisle Railway, *by W R Mitchell and David Joy, 1966.*

176. DENVER & RIO GRANDE RAILROAD

176.1 The history of the narrow gauge Rio Grande is extremely interesting; built as it was, through a territory without competition, a complete system in itself, with a large business at high rates, the road made money regardless of poor management.

Editorial comment, Railroad Gazette *22, 1890.*

177. DENVER & RIO GRANDE WESTERN RAILROAD

177.1 No man can consider himself qualified on the subject of American motive power without some knowledge of Denver & Rio Grande Western locomotives—both past and present.

David P Morgan, Trains Magazine, *January 1950.*

178. DENVER & SALT LAKE RAILROAD

178.1 The Denver & Salt Lake Railroad is not an ordinary railroad. A railroad is an ordinary railroad until it reaches the Rocky Mountains. When it reaches the Rocky Mountains it isn't an ordinary railroad in any sense of the word.

Jesse Fleming, quoted by Robert G Athearn in Rebel of the Rockies, *1962.*

179. DEPARTURE
See also Arrival, Stations, Whistles

179.1 A train of cars was just ready for the start; the locomotive was fretting and fuming, like a steed impatient for a headlong rush; and the bell rang out its hasty peal, so well expressing the brief summons which life vouchsafes to us, in its hurried career.

Nathaniel Hawthorne, The House of the Seven Gables, *1851.*

179.2 To attend the departure of this train, there arrive not only the republican omnibi and cabs, from the damp night crawler to the rattling Hansom, but carriages, with coronets and mitres emblazoned, guarded by the tallest and most obsequious of footmen, and driven by the fattest and most lordly of coachmen; also the neatest of Broughams, adorned internally with pale pink and blue butterfly bonnets; dashing dogcarts, with neat grooms behind, mustached guardsmen driving; and stately cabriolets prance in, under the guidance of fresh primrose-coloured gloves.

Samuel Sidney, of the departure of "the aristocratic Express" from Euston; Rides on Railways, *1851.*

179.3 Quick as a flash the valve was thrown open and the steam giant unchained!—but for an instant which seemed terribly long the locomotive seemed to stand still; Knight had thrown the full power on too suddenly, and the wheels slipped on the track, whirling with swift

revolutions and the hiss of escaping steam, before the inertia of the ponderous machine could be overcome. But this was an instant only; none of the soldiers had time to raise their muskets, give an alarm, or indeed to recover from their stupor before the wheels "bit," and the train shot away as if fired from a cannon!

William Pittenger, of the moment that the Andrews raiders stole a train, headed by the locomotive General, *at Big Shanty, Georgia, Saturday, 12 April 1862.* Daring and Suffering, *1887.*

179.4 From this point it was determined to have a race home. The start was even. Away went horse and engine, the snort of the one and the puff of the other keeping time and tune.

John H B Latrobe, of the race between Peter Cooper's Tom Thumb *and a horse, 1830; lecture delivered before the Maryland Institute, 23 March 1868, quoted by Richard Reinhardt in* Workin' on the Railroad, *1970.*

179.5 Sure, the next train has gone ten minutes ago.

Cartoon, Punch, *1871. This joke was used by J O C Orton, scriptwriter of the film* Oh! Mr Porter, *1937.*

179.6 This silent, casual departure of trains is a perpetually recurring surprise to me. Would it be contrary to republican principles to ring a bell for the warning of passengers?

Anon. English writer, of American practice, late 19th century, quoted by Dee Brown in Hear that Lonesome Whistle Blow, *1977.*

179.7 The busy lanterns wagged among the switches, the steady lights of the saloons shone along the town's wooden facade. From the bluffs that wall Green River the sweet, clean sage-brush wind blew down in currents freshly through the coal-smoke. A wrench passed through the train from locomotive to caboose, each fettered car in turn strained into motion and slowly rolled over the bridge and into silence from the steam and the bells of the railroad yard. Through the open windows of the caboose great dull-red cinders rattled in, and the whistles of distant Union Pacific locomotives sounded over the open plains ominous and long, like ships at sea.

Owen Wister, Lin McLean, *1897.*

179.8 But spinning futilely at first, the drivers of the engine at last caught the rails. The engine moved, advanced, travelled past the depot and the freight train, and gathering speed, rolled out on to the track beyond. Smoke, black and boiling, shot skyward from the stack; not a joint that did not shudder with the mighty strain of the steam; but the great iron brute—one of Baldwin's newest and best—came to call, obedient and docile as soon as ever the great pulsing heart of it felt a master hand upon its levers. It gathered speed, bracing its steel muscles, its thews of iron, and roared out upon the open track, filling the air with the rasp of its tempest-breath, blotting the sunshine with the belch of its hot, thick smoke.

Frank Norris, The Octopus, *1901.*

179.9 The next thing particularly noticeable is that everybody seems in a hurry to catch a train. This is a state of things which is not favourable to poetry or romance. Had Romeo or Juliet been in a constant state of anxiety about trains, or had their minds agitated by the question of return-tickets, Shakespeare could not have given us those lovely balcony scenes which are so full of poetry and pathos.

Oscar Wilde, Impressions of America, *1906.*

179.10 The only way of catching a train I have ever discovered is to miss the train before.

G K Chesterton, Tremendous Trifles, *1909.*

179.11 Far out, in the bright sunlight beyond the station, the engine can be seen pulling out, ejecting a stiff spire of smoke and horizontal billows of steam.

Christopher Morley, Broad Street Station, *within* Travels in Philadelphia, *1920.*

179.12 The monster roared suddenly and loudly, and sprang forward impetuously. A wrong-headed or maddened draft-horse will plunge in its collar sometimes when going up a hill. But this load of burdened carriages followed imperturbably at the gait of turtles. They were not to be stirred from their way of a dignified exit by the impatient engine.

Stephen Crane, The Scotch Express, *within* Men, Women and Boats, *1921.*

179.13 Those who went further than Chicago would gather in the old dim Union Station at six o'clock of a December evening, with a few Chicago friends, already caught up in their own holiday gaieties, to bid them a hasty good-bye. I remember the fur coats of the girls returning from Miss This-or-That's and the chatter of frozen breath and the hands waving overhead as we caught sight of old acquaintances, and the matchings of invitations: 'Are you going to the Ordways'? the Herseys'? the Schultzes'?' and the long green tickets clasped tight in our gloved hands. And last the murky yellow cars of the Chicago, Milwaukee and St Paul railroad looking cheerful as Christmas itself on the tracks beside the gate.

F Scott Fitzgerald, The Great Gatsby, *1926.*

179.14 Gently she steals out, along a corridor of that dusky underground forest where colored lights gleam like tropic birds.

Christopher Morley, of departure from Grand Central Terminal in New York, A Ride in the Cab of the Twentieth Century Limited, *1928, quoted by H Roger Grant,* We Took the Train, *1990.*

179.15 One does not rush to catch a Pennsylvania train—one proceeds to it in orderly but expeditious manner.

Charles Moore, of Pennsylvania Station, New York, The Life and Times of Charles Follen McKim, *(the architect of the station), 1929.*

179.16 Guard sounds a warning whistle, points to the clock
With brandished flag, and on his folded flock
Claps the last door: the monster grunts: "Enough!"
Tightening his load of links with pant and puff.

Siegfried Sassoon, poem, Morning Express.

179.17 In a slither of steam your train moves on.

Dell Leigh, On The Line, *1928.*

179.18 'Do you wish to travel, sir?' he asked gently, and when Donald had said that he was desirous of going as far as Aylesbury, the guard touched his hat and said in a most respectful manner, 'If you wish it, sir.' He reminded Donald of the immortal butler, Jeeves. Donald fancied, but he was not quite sure, that he heard the guard whisper to the engine-driver. 'I think we might make a start now, Gerald,' and he rather thinks the engine-driver replied in the same undertone. 'Just as you wish, Horace.'

A G Macdonnell, of a departure from Marylebone, England, their England, *1933.*

179.19 Conversations on the platform, before the departure of a train, are apt to be somewhat repetitive in character.

Agatha Christie, Murder on the Orient Express, *1934.*

179.20 And elsewhere there were the casual voices of the train men—conductors, porters, baggage masters, station men—greeting each other with friendly words, without surprise, speaking of weather, work, plans for the future, saying farewell in the same way. Then the bell tolled, the whistle blew, the slow panting of the engine came back to them, the train was again in motion; the station, and the station lights, a glimpse of streets, the thrilling, haunting, white-glazed incandescence of a cotton mill at night, the hard last lights of town, slid past the windows of the train. The train was in full speed now, and they were rushing on across the dark and lonely earth again.

Thomas Wolfe, Of Time and the River, *1935.*

179.21 The engine panted for a moment with a hoarse, metallic resonance, in the baggage-car someone threw mail-bags and thick bundles of evening papers off onto the platform, there was the swinging signal of a brakeman's lantern, the tolling of the engine bell, thick, hose-like jets of steam blew out of her, the terrific flanges spun for a moment, the short, squat funnel belched explosive thunders of hot smoke, the train rolled past with a slow, protesting creak of ties, a hard-pressed rumble of the heavy coaches, and was on her way again.

Ibid.

179.22 The ground began to thunder as the boys lay on it, near the tracks. The night freight was starting to move out. They craned back across their flat shoulders and saw the yellow beam of light come quivering along the way and race ahead of the engine in the shining steel tracks. The engine fumed and ground its way. Far, far back along the line the couplings jerked and rang and the squealing train was yanked along into increasing speed. The smoke tasted in the mouths of Tom and Danny where they lay as it went by. The fire-door flung a tiny hell at them and a mesh of sparks. The earth trembled and shook them. They could not hear anything but the train. It rollicked past above their very heads. They clenched their teeth and fists and squinted to resist the thundering sustained shock so near. But they didn't dare move off until all was quiet and they could free themselves safely from their fatal and secret point.

Paul Horgan, Main Line West, *1936.*

179.23 A change in the timbre of the tracks told them the end of the train was coming up. They opened their eyes and scrounged around and sure enough, there was the caboose, lighted up like a country church. Far gone was the engine, and fading was the clamor. The evening was about to hush again.

Ibid.

179.24 Then I find out from the indicator what platform the 11.53 departs from. It is Platform 3. Then I ask all the porters I meet what Platform the 11.53 departs from. They love it. Then I ask one or two guards, postmen, dining car attendants and naval warrant officers. When the weight of the evidence seems to tend towards Platform 3 I generously admit that the indicator may after all be right, and I feel free to amuse myself.

A P Herbert, Slow Train, *reproduced from* Punch *in the Great Western Railway Magazine, August 1938.*

179.25 When the train starts, and the passengers are settled
To fruit, periodicals and business letters
(And those who saw them off have left the platform)
Their faces relax from grief into relief,
To the sleepy rhythm of a hundred hours.
Fare forward, travellers! not escaping from the past
Into different lives, or into any future;
You are not the same people who left that station
Or who will arrive at any terminus,
While the narrowing rails slide together behind you; . . .

T S Eliot, poem, The Dry Salvages, *1943.*

179.26 The technique of loading the crowds is simple. A train of empty cars is backed into a platform and some of

the sailors get aboard. But most of them stay on the platform, bidding fond farewell to their girl-friends. Finally, at a word from the superintendent, trainmen call "All aboard, last train to Great Lakes, all aboard!" There is a jingle of the platform bell and the motorman edges the train slowly forward a few feet. The sailors on the platform quickly disengage, climb on board, and the train pulls out. Another train of empties is shunted to the platform and the process starts all over again.

A C Kalmbach, Trains Magazine, *January 1944.*

179.27 The guard blew his whistle and waved his flag—how weighted with ritual have the railways in their brief century become!—and the train crawled from the little station. The guard walked alongside through the snow-flakes, wistful for that jump-and-swing at an accelerating van that is the very core of the mystery of guarding trains.

Michael Innes, Appleby's End, *1945.*

179.28 In the 1940's the most romantic call in America still was " 'Booo-ard!" It was sung every day by a thousand conductors, echoed by ten thousand trainmen and Pullman porters, re-echoed by half a million passengers. It was almost the oldest call in the country and to most people it still meant adventure and hope and new horizons.

James Marshall, Santa Fe, *1945.*

179.29 It was six o'clock; possibly two or three seconds later. At the north end of the quay, visible to the conductor but not the engineman, the green light flashed, signal that the Concourse gate was closed. Instantly the conductor called the "All aboard!" The cry was taken up by the trainmen ranged along the entire length of the quay; and, just back of the observation car, a yard man dropped quickly to the tracks, and began rapidly "cutting her loose." Swiftly—one, two, three, four—the train valves were closed. The valves were closed upon the station reservoirs, and the air pipes were disconnected. The valves were closed upon the station boiler, and the steam pipes were disconnected. The switch was drawn, and the electric cable was unhooked. The *Century* henceforth drew light, heat and air-pressure from her locomotive.

It was 6:01. Slowly, smoothly, dead on the second, the *Century* began to roll. The yard man stood up and watched her move away. A last remaining cable, a thin flexible wire, was drawn taught, then fell abruptly to the track as the plug-in connection broke loose. Telephone service aboard the *Century* was disconnected.

Twenty-two seconds later the *Century* had cleared her dock and was out in the terminal fan. Like a liner on the bay, she lighted up the whole place about her, cheerily; and the gloom was all the blacker when she passed on.

David Marshall, of the evening departure of the New York Central's Twentieth Century Limited *from Grand Central Terminal. The local use of the word "quay" instead*

of platform is stated to originate in the French word "quai."
Grand Central, 1946.

179.30 From my narrow bench I saw Mr Thorpe finally
dispose of his watch: I saw his green flag raised and about
to be dropped; I saw him put his whistle to his mouth:
and I saw at the very moment that he did this, Powell
Spooner, the booking clerk, come galloping on the platform and heard his wild cry of "Hold hard! Hold hard!
There's a passenger coming up the hill."
    *Simon Dewes (as a small boy, sitting on the driver's seat
on the footplate), A Suffolk Childhood, 1959.*

179.31 'There,' he said, 'that's a bit of luck for you. Now
we'll be able'—and he patted the engine as he said so—
'now we'll be able to show you a bit of speed, making
up time.'
    Ibid.

179.32 After one of those squealing, juddering stomach-
dropping false starts with which trains so tactlessly articulate human emotion, we pulled ourselves out of the
great shed of Paddington and steamed west.
    Stephen Fry, Moab is My Washpot, 1998.

179.33 The 'right away' was quietly given and a start was
made that was so gentle, so perfect, so silent, so apt as to
be profoundly moving because one understood the care
that was needed to achieve such perfection.
    *R H N Hardy, of the departure from Waterloo of the
funeral train of Sir Winston Churchill, 30 January 1965.*
Steam World, *January 2003.*

179.34 Passengers are reminded that the big red slidey
things on the side of the train are doors. Let's try it again,
shall we? Please stand clear of the doors.
    *Announcement by London Underground driver
reported in* The Times, *26 July 2003.*

179.35 Summary Departures.
    *Anon., legend above television screens, seen at Waterloo
station March 2004.*

DEPOTS
See Stations

180. DERAILMENT
See also Accidents, Safety

180.1 I have adopted this word from the French: it
expresses an effect which is so often necessary to mention, but for which we have not yet had any term in our
railway nomenclature. By *déraillement* is meant the
escape of the wheels of the engine or carriage from the
rails; and the verb *to derail* or *to be derailed* may be used
in a corresponding sense.
    Dr Dionysius Lardner, footnote in Railway Economy,
1850.

181. DESIGN

181.1 In the history of mechanical art two modes of progress may be distinguished—the empirical and the scientific. Not the practical and the theoretic, for that distinction is fallacious; all real progress in mechanical art,
whether theoretic or not, must be practical. The true distinction is this: that the empirical mode of progress is
purely and simply practical; the scientific mode of progress is at once practical and theoretic.
    *William J Rankine, A Manual of the Steam Engine and
other Prime Movers, 1885. Quoted by J Parker Lamb in*
Perfecting the American Steam Locomotive, *2003.*

181.2 I only wish it was a recognised part of the cur-
riculum of every draughtsman engaged on locomotive
design to spend about six months as a running shed fit-
ter. He would at least learn by bitter experience how
some things should not be done.
    *R E L Maunsell, speech at Institution of Locomotive
Engineers, 29 January 1916.*

181.3 There is hardly a section in the new Silver Jubilee
train of the London and North Eastern railway which
shows in design or in decoration any evidence of quiet,
good, modern taste. It is a purely ostentatious travesty of
the modern spirit, and it is quite common to find in the
much trumpeted Royal Scot, with its shining new carriages, a prehistoric sleeping coach whose proper place is
in a museum of antiquities.
    Editorial comment, Design for Today, *February 1936.*

181.4 No frills, no childish excrescences, just good design.
    *Samuel Vauclain, of the first Atlantic Coast Line class
R-1 4-8-4, 1938, quoted in* The A, *Ed King, 1989.*

181.5 We have no sacred traditional standards, nor pre-
conceived ideas or preferences, with respect to the kinds
of motive power used, but have striven to the limits of
our collective abilities to provide units best suited to
meet the changing necessities of transportation by rail in
which numerous and varied problems are involved.
    *Paul Kiefer, of the motive power policy of the New York
Central System,* A Practical Evaluation of Railroad
Motive Power, *1947.*

181.6 Those who have taken pride in watching the growth
of designs on the drawing board and in later seeing them
take form and life in the shops feel a keen regret in the
passing of the steam locomotive. They concede the
streamlined cleanliness of the diesel-electric locomotive,
but they miss the throb and visual signs of life found in
steam. For them there is no creative inspiration in a piece
of wire. Even the whistle has lost its thrill.
    Alfred W Bruce, The Steam Locomotive in America,
1952.

181.7 Things which never fail may be over-designed.
Oliver Bulleid, quoted by H A V Bulleid, Bulleid of the
Southern, *1977.*

181.8 Most lines were designed in the counting-house,
not in the drafting room, and what would produce the
most for the smallest investment was considered the best
bargain.
*John H White Jr, of early American railroads built with
strap-reinforced wood rails. Chapter 6 of* Material Culture
of the Wooden Age, *1981.*

181.9 The requirements of efficient workshop performance were responsible to a large extent for the deliberate avoidance of anything which could be construed as
startlingly original or untried in the British Railways
standard rolling stock designs. It has been said that these
designs represented yesterday's answers to today's
questions.
John Johnson and Robert A Long, British Railways
Engineering, 1948–1980, *1981.*

181.10 Technology is 99% of the train—the way it grabs
the track, the on-time departure, the safe operation, the
smooth roadbed. What I do is 1%, the look of it. But
that's everything the passenger sees.
*Cesar Vergara, of his styling work on modern U.S. loco-
motives,* Trains Magazine, *October 2002.*

182. DETERMINATION

182.1 This is our last locomotive.
*Matthias Baldwin, after completing his first locomotive,
1832. His company went on to build many thousands of
locomotives. Quoted by J Parker Lamb in* Perfecting the
American Steam Locomotive, *2003.*

182.2 I never flinch from what I have once begun.
I K Brunel, quoted in The Great Western Magazine
and Temperance Union Record, *January 1889.*

182.3 I can't do it, I can't do it, I can't do it, puffed
Gordon.
I will do it, I will do it, I will do it, puffed Edward.
Rev W Awdry, The Three Railway Engines, *1945.*

182.4 He has written to tell me of this week's insuperable
problem and how he overcame it.
*George Dow, of Frank Bollins, modelmaker, who was
making a model locomotive for him. Conversation with his
son, the compiler, 1960s.*

182.5 When you've bought tickets for the pantomime you
want to see it through to the end.
*Sir Bob Reid, when asked if he would resign as Chair-
man of British Railways as a result of an unexpected Minis-
terial decision on the route of the Channel Tunnel Rail
Link, c. 1991.*

183 CHARLES DICKENS

183.1 Well, when treating of railways our beloved Dickens' pen is very little better than Turner's brush. The
word-picture and the water-colour are much on a par,
that is to say misty.
Alfred Rosling Bennett, The Locomotive Magazine,
April 1913.

184. DIESEL POWER

184.1 When it comes to the designs of Diesel power, however, most rail enthusiasts, and especially the
photographic-minded among them, are baffled and saddened. There is nothing significant of action in the passing of the Denver Zephyr except an *envoi* in the form of a
cloud of dust and the corpse of an occasional chicken
immolated on the altar of speed, and nothing so cheers
the traditionalist as the spectacle of the Super Chief being
assisted up the hillsides by one of the Santa Fe's North-
erns, all guts and exhibitionism and smoke . . .
Lucius Beebe, Introduction to High Iron, *1938.*

184.2 If ever there was a case of evolution, as opposed to
the term invention, the development of the Diesel locomotive is it.
Lucius Beebe, Trains in Transition, *1941.*

184.3 The Norfolk & Western is not allergic to Diesel
power plants.
Clarence Pond, Railway Age, *24 August 1946.*

184.4 When you can equal the Big Boy . . . come around
and we'll talk business.
*Anon. Union Pacific executive to representative of
EMD, recorded by Greg McDonnell in* Trains Magazine,
September 1997.

184.5 If you'll put our four-unit freight locomotive alongside your big boy on that hill out in Utah, we'll push your
big boy *so far back into Lionel's window with the rest of the
toys* that no one will ever talk about it again.
*Richard M Dilworth, Chief Engineer of EMD, in
response to UP, quoted in* The Dilworth Story, *1954.*

184.6 I had two dreams. The first was to make a locomotive so ugly in appearance that no railroad would want it
on the main line or anywhere near headquarters, but
would keep it as far as possible in the back country,
where it could do really useful work. My second dream
was to make it so simple in construction and so devoid of
Christmas-tree ornaments and other whimsey that the
price would be materially below our standard main-line
freight locomotives.
*Richard M Dilworth of EMD, of dream for the general
purpose ("GP") diesel road-switcher. Quoted by Franklin
M Reck in* The Dilworth Story, *1954.*

184.7 On the road to Santa Fe
Where the flying Diesels play,
Rootin, tootin on their whistles
As they claim the right of way,

Hear them screaming for the crossing,
Dashing past the old steam pots.
Stand aside, you puffing billies,
We can't wait while you get hot!

Watch them ride into the depot,
Rolling down with conscious pride.
We're Dick Dilworths's pups, they holler,
He's the man who made us stride!
Anon., but possibly Fred Shea, Clyde Engineering; verse inscribed on fly-leaf of book of Kipling's poems presented to Richard Dilworth. Quoted in The Dilworth Story, *1954.*

184.8 People may think they will quickly reconcile themselves to the loss of the visual beauty of steam, that they may even come in time to love the loathsome diesel—a contraption that doesn't even know which way it's pointing and has to have a driving cab at each end.
H F Ellis, Full Diesel Ahead, *about the B R Modernisation Plan, within* Punch, *2 February 1955.*

184.9 The diesel that did it.
David P Morgan, title of article about GM unit 103, Trains Magazine, *February 1960.*

184.10 In retrospect, the FT is a locomotive that exhausts the dictionary of superlatives. It must be ranked as perhaps the most influential piece of motive power since Stephenson's Rocket, for in one stroke it broke steam's historic monopoly of freight traffic and thereby forecast total dieselization, here and abroad.
David P Morgan, Trains Magazine, *March 1962.*

184.11 A hammering rumble
That shakes the hills—
Crescendoing roar the still night fills.

A frenzied white light—
A chorus of horns
A crossing on the outskirts warns

Red signal discs—
A clanging bell—
Brute steel monster unleashed from hell.
Don Buchholz, poem, Extra 728 east, *first three verses;* Trains Magazine, *September 1962.*

184.12 . . . not in the public interest . . .
Attributed to Robert Young without explanation; Trains Magazine, *November 1965.*

185. DILTON MARSH HALT

185.1 There isn't a porter. The platform is made of sleepers
The guard of the last up-train puts out the light
And high over lorries and cattle the Halt unwinking
Waits through the Wiltshire night.
John Betjeman, poem, Dilton Marsh Halt.

186. DISCIPLINE

186.1 We find no difficulty in that respect; we fine a man twice, the third time we discharge him.
Edward Pease, of the company's authority over their servants. This predated the formal organization of baseball in 1846, which has given us the current expression "three strikes and you are out." Evidence to Select Committee on Railways, 1839.

186.2 The element of discipline is especially strong. The railway employee is continually conscious of some higher authority to which he must "report" and must render faithful obedience. It may be that he is sometimes too much warped by this consciousness. But in some branches of railway activity a quasi-military authority with corresponding submission seems indispensable, no doubt. Great human interests demand it, even though it may occasionally be felt to have the appearance of an infringement of human liberty. The peculiar combination of knowledge and discipline necessary to railway service gives it a character of its own, one which has its advantages and from which men in certain other walks of life may perhaps learn something of value.
B C Burt, Railway Station Service, *1911.*

186.3 It can hardly be doubted that the influence exerted by the railways towards the formation of character of a rather concrete and substantial type is something insignificant neither in strength nor in extent. Certainly there are many positions in the railway service, to fill which competently, men of such a type of character are required.
Ibid.

186.4 In an imperfect world perfection is not instantly available. Railroad safety, for instance, cannot be secured by mechanical devices alone. It is primarily a resultant of care and discipline.
Ivy Lee, address before the Traffic Club of Pittsburgh, 8 December 1913, reproduced in Human Nature and Railroads, *1915.*

187. DISILLUSION

187.1 Once I built a railroad, I made it run,
Made it race against time.
Once I built a railroad, now it's done—
Brother, can you spare a dime?
E Y Harburg, second verse, song, Brother Can you Spare a Dime? *notably recorded by Bing Crosby, 1932.*

188. DISNEYLAND

188.1 I just want it to look like nothing else in the world. And I want it surrounded by a train.
Walt Disney c. 1953, quoted by Michael Broggie in Walt Disney's Railroad Story, *1997.*

189. DISPATCHERS

189.1 A train dispatcher is a strange combination of elements, most of which are foreign to other people. His prime asset, probably, is a sense of humour, the ability to laugh at himself as well as others. Without this, his mental equipment may crack up under the strain. He must be a lawyer of sorts, a psychologist, an expert on operating rules. He must be capable of making swift but accurate mental calculations, grasping situations quickly, and acting instantly. He must be able to look at least a few hours into the future, and he must know his railroad thoroughly. Last, but not least, he must be a maker of holeproof alibis.
Peter Josserand, Trains Magazine, *February 1948.*

189.2 Unseen by the public, his existence even ignored, is the Train Dispatcher, whose responsibility is far greater than an engineer's, and whose mind requires for the efficient performance of his duties the comprehensive perception of a general's, the instantaneous and correct reaction of a racing driver's, and the foresight and planning power of a chess champion's. For these talents he is awarded a salary that no air pilot would deem worth while picking up. But then, remember, a dispatcher is quite out of the public's sight!
Brian Fawcett, Railways of the Andes, *1963.*

189.3 All of a sudden dispatchers were getting orders to run trains to West Jockstrap.
Anon. Penn Central official, of the confusion resulting from the merger of the New York Central and the Pennsylvania Railroads, quoted in The Wreck of the Penn Central, *Joseph Daughen and Peter Bintzen, 1971.*

190. DISS

190.1 Dear Mary
        Yes it will be bliss
To go with you by train to Diss,
Your walking shoes upon your feet
We'll meet, my sweet, at Liverpool Street.
That levellers we may be reckoned
Perhaps we'd better travel second;
Or lest reporters on us burst,
Perhaps we'd better travel first.
John Betjeman, poem A Mind's Journey to Diss, *addressed to Mary Wilson.*

191. DIVIDENDS

191.1 It is the straphangers who pay the dividends.
Charles Yerkes, quoted in A History of London Transport, vol. II, *1974.*

192. DOCTOR'S ORDERS

192.1 My doctor has forbidden me to pull.
The Fat Director (in later books the Fat Controller), who was also forbidden to push, in both cases failed trains. Rev Wilbert Awdry, The Three Railway Engines, *1945.*

193. DOME CARS
See also Observation Cars

193.1 After all, the purpose of the new equipment is to attract and keep the trade. If we have to build a penthouse into a structural roof, well, that is only another engineering headache, but if it keeps the passenger happy, it is well worth while and aspirin doesn't cost very much.
Col E J W Ragsdale, 13 February 1946, of the incorporation of vista domes into stainless steel cars. Quoted by Bruce A MacGregor and Ted Benson in Portrait of a Silver Lady, *1977.*

193.2 One of the genuine thrills of railroading, now or at any other time, is holding down the front seat in the front dome of the *California Zephyr* out of Chicago as the hoghead takes that sweeping curve at 16th and Halstead in high gear, winding up his three units for a low flight through the suburbs to Aurora.
David P Morgan, Trains Magazine, *October 1955.*

193.3 A dome permits you to breathe down the engineer's neck, examine the train from stem to stern, see *all* the railroad from signal bridges to switch shanties, sense the roll of the trucks, and stay clean and cool in the process.
Ibid.

194. DOORS

194.1 . . . wherever we have gone we have never found but one undeviating and unalterable rule among the door-opening and door-closing portion of the railway community. Is it an instinct or an abstract idea with them that the door of a compartment cannot be closed unless the closing be accompanied with a loud and violent bang, which pleases nobody and sets nervous persons into a state of glowing trepidation? Or is there a masonry among the officials of this grade, from participation in which the higher classed portion of the railway fraternity is excluded, the secret of the craft being that all railway doors must be banged as they close them? Dear colleagues of the class of carriage door shutters, country station masters, inspectors, ticket collectors, guards of the first class, guards of the second, and temporary guards, foremen porters, ordinary porters and good-looking porters, porters of the strong back, and porters of brachial muscle much developed, let us entreat and implore you to retire from the brotherhood and learn to close carriage doors gently, quietly, as becomes your gentle natures.

　　Sir Cusack Roney, Rambles on Railways, *1868.*

195. DRACULA

195.1　Pardon me, Boy
　　Is this the Transylvania Station?
　　　Anon., spoof version of Chattanooga Choo-choo, *telling of Count Dracula's attempt to get home from New York at Christmas. Fragmentally remembered by David Elliot, 2003.*

196. DREAMS

196.1　Well, you get some repose in the form of a doze,
　　　with hot eye-balls and head ever-aching,
　　But your slumbering teems with such horrible dreams
　　　that you'd very much better be waking;
　　For you dream you are crossing the Channel, and tossing
　　　about in a steamer from Harwich
　　Which is something between a large bathing machine
　　　and a very small second class carriage And you're
　　　giving a treat (penny ice and cold meat) to a party of
　　　friends and relations—
　　They're a ravenous horde—and they all came on board at
　　　Sloane Square and South Kensington Stations.
　　W S Gilbert, Lord Chancellor's song, Iolanthe, *1882.*

196.2　Underneath the arches, I dream my dreams away.
　　　Reg Connelly, Bud Flanagan and Joseph McCarthy, song, Underneath the Arches, *referring to undesirable places under urban railway viaducts; 1933.*

196.3　Have you ever sat by the railroad track
　　And watched the empties coming back?
　　Lumbering along with a groan and a whine,
　　Smoke strung out in a long gray line,
　　Belched from the panting engine's stack—
　　Just empties coming back!

　　I have—and to me the empties seem
　　Like dreams I sometimes dream
　　Of a girl—or money—or maybe fame;
　　My dreams have returned the same,
　　Swinging along the homeward track—
　　Just empties coming back.
　　　C B Clark, quoted within a short story in Railroad Magazine, *March 1936.*

197. DANIEL DREW

197.1　He had been at once a good friend of the road and the worst enemy it had as yet known.
　　　Charles Francis Adams, Some Chapters of Erie, *1871.*

197.2　Shrewd, unscrupulous, and very illiterate,—a strange combination of superstition and faithlessness, of daring and timidity,—often good-natured and sometimes generous,—he ever regarded his fiduciary position of director in a railroad as a means of manipulating its stock for his own advantage.
　　　William Z Ripley, Railway Problems, *1913.*

DRIVERS (and Engineers)
See Enginemen

198. DUGALD DRUMMOND

198.1　Men have come and gone, in locomotive departments, as in other organisations; big men, little men, deep men, shallow men, swart men, sallow men, fathers and taskmasters, aesthetes and Philistines. But Dugald Drummond was a Personage.
　　　C Hamilton Ellis, The North British Railway, *1955.*

199. EAST ANGLIA

199.1 Who would want to travel to York from Kings Cross when the same money could buy a much longer ride through a fascinatingly alien East Anglia?
John Gibbons, The Railway Magazine, vol. 90, *1945.*

200. EAST BROAD TOP RAILROAD

200.1 On misty October mornings, the East Broad Top Railroad, nestled deep in the central Pennsylvania valley formed by Aughwick Creek, is the Brigadoon of steam railroading.
Karl Zimmerman, Trains Magazine, *April 1993.*

201. EAST COAST MAIN LINE

201.1 They stood on the footbridge over Boathouse station and looked at the metals gleaming coldly.
"You should see the Flying Scotsman come through at half past six!" said Leonard, whose father was a signalman. "Lad, but she doesn't half buzz!" and the little party looked up the lines one way, to London, and the other way, to Scotland, and they felt the touch of these two magical places.
D H Lawrence, (with an imperfect knowledge of the timing of the Flying Scotsman express), Sons and Lovers, *1913.*

202. EASTERN COUNTIES RAILWAY

202.1 Even a journey on the Eastern Counties must have an end at last.
Attributed to William Makepeace Thackeray by George Ottley in Introduction to A Bibliography of British Railway History, *1965.*

202.2 . . . it was certainly no honour to be chairman of such a company, for it was ever in trouble; and there could be no pride in being its head for its profits were ever in inverse proportion to its promises.
John Francis, of George Hudson, A History of the English Railway, *1851.*

202.3 On Wednesday last a respectably-dressed young man was seen to go to the Shoreditch terminus of the Eastern Counties Railway and deliberately take a ticket for Cambridge. No motive has been assigned to this rash act!
Anon., in Punch; quoted J Pendleton, Our Railways, *1896.*

202.4 It was on the old Eastern Counties railway that the tale was originally told of how a ticket collector who was expostulating that a strapping lad of sixteen could scarcely be entitled to travel at half rate was met by the crushing reply that he was under fourteen when the train started.
John Aye, Railway Humour, *1931.*

203. EATING ON TRAINS

203.1 Carry your own provisions; thus you can dine when you are hungry instead of when the railway directors think you ought to be.
R S Surtees, of the days before dining cars, when trains stopped at selected stations for passengers to eat in refreshment rooms; Hints to Railway Travellers and Country Visitors to London, *in the* New Monthly Magazine, *1851.*

203.2 Luncheon or dinner on board a Pullman palace-car will surely banish Boredom from railway journeys.
Comment, The Daily Telegraph, *following the introduction of dining cars on the Great Northern Railway, between Leeds and London, November 1879.*

203.3 If you have made your trans-Missouri journeys only since the new era of dining cars, there is a quantity of things you have come too late for, and will never know. Three times a day in the brave days of old you sprang from your scarce-halted car at the summons of a gong. You discerned by instinct the right direction, and, passing steadily through doorways, had taken, before you knew it, one of some sixty chairs in a room of tables and catsup bottles. Behind the chairs, standing attention, a platoon of Amazons, thick-wristed, pink-and-blue, began immediately a swift chant. It hymned the total bill-of-fare at a blow. In this inexpressible ceremony the name of every dish went hurtling into the next, telescoped into shapelessness. Moreover, if you stopped your Amazon in the middle, it dislocated her, and she merely went back and took a fresh start. The chant was always the same, but you never learned it. As soon as it began, your mind snapped shut like the upper berth in a Pullman. You must have uttered appropriate words—even a parrot will—for next you were eating things—pie, ham, hot cakes—as fast as you could. Twenty minutes of swallowing, and all aboard for Ogden, with your pile-driven stomach dumb with amazement.
Owen Wister, Lin McLean, *1897.*

203.4 ... it is not given to everyone to balance a mutton-chop and potatoes gracefully on his knees the while he pours himself out a glass of claret with his hands.
Sir William Acworth, of eating in compartments, The Railways of England, *1900.*

203.5 No man, within twenty-four hours after eating a meal aboard a Pennsylvania Railroad dining car, could conceivably write anything worth reading.
Henry L Mencken, The Divine Afflatus, *1921.*

203.6 There is no pleasanter courtesy than to be invited into the diner for afternoon tea and to have the steward suggest and provide chess, checkers, or dominoes for games.
Missouri Kansas Texas Railroad, entry in The Official Guide, *found by Christopher Morley and related in* Human Being, *1932.*

203.7 Discretion in mixing cocktails and serving crusted port—regard for our passengers' eupeptic welfare—these are the qualities that distinguish the LNER waiter.
London & North Eastern Railway poster, 1933.

203.8 The dining car was certainly unchanged. On each table there still stood two opulent black bottles of some unthinkable wine, false pledges of conviviality. They were never opened, and rarely dusted. They may contain ink, they may contain the elixir of life. I do not know. I doubt if anyone does.
Peter Fleming, on the Trans-Siberian Railway, One's Company, *1934.*

203.9 Lavish but faded paper frills still clustered round the pots of paper flowers, from whose sad petals the dust of two continents perpetually threatened the specific gravity of the soup.
Ibid.

203.10 The cruet, as before, was of more interest to the geologist than to the gourmet. Coal dust from the Donetz basin, tiny flakes of granite from the Urals, sand whipped by the wind all the way from the Gobi Desert—what a fascinating story that salt-cellar could have told under the microscope!
Ibid.

203.11 Sign-language they interpreted with more eagerness than apprehension: as when my desire for a hard-boiled egg—no easy request, when you come to think of it, to make in pantomime—was fulfilled, three-quarters of an hour after it had been expressed, by the appearance of a whole roast fowl.
Ibid.

203.12 There is no meal that I enjoy more than my weekly breakfast on the train, and the line to Kemble, the last stop before Paddington, is punctuated in my mind not by stations but by stages I have reached in my break-fast. Clear of Gloucester comes the porridge stage, at Stonehouse I reach the fresh fish, and moving out of Stroud, I enter the long run through bacon, kidney and potatoes. Kemble, running to time, sees the end of the last slices of toast and the last cup of coffee. I shall always associate with kidney and bacon the long, five-mile rise of the Golden Valley, after Stroud.
Colin Howard, Cotswold Days, *1937.*

203.13 A man can put in a lot of time in a dining-car if he is experienced. He can order item by item as he eats, and then eat very slowly, with full pauses now and then to read two or three consecutive pages in some interesting book, and with other pauses for the passing landscape.
Rollo Walter Brown, I Travel by Train, *1939, quoted by H Roger Grant,* We Took the Train, *1990.*

203.14 I refuse to be made sentimental in the middle of a Great Western lunch.
Noel Coward, screenplay of In Which We Serve*; spoken by Alix Kinross (Celia Johnson), wife of destroyer captain played by Noel Coward, 1942.*

203.15 Those were such breakfasts as made history.
Canon Roger Lloyd, of the 7.20 am Manchester Central to St Pancras, in the 1930s; The Fascination of Railways, *1957.*

203.16 Why is it that however old one gets it is *always* exciting to have breakfast in a train?
Ibid.

203.17 Meals in L.N.E.R. trains are different from all others. They are conducted with a decent and seemly ritual. Before you begin, the attendant comes down the car with a tray of glasses full of sherry, so that all kinds of people have one who did not really intend to do so when they sat down. Thus the sale of sherry on LNER dining cars must be at least twice what it is on any others.
Ibid.

203.18 The L.N.E.R. caterers have minds which are not tied and bound to rabbit, fish and sausage; and, until quite recently, this was more than could be claimed for any other line. When it arrives it is well cooked and served, and its appearance causes no one to look dubiously at it, and wonder how much of it can be left at the side of the plate without hurting the attendant's feelings.
Ibid.

203.19 In a time when the best food in America was being served on its dining cars, the carriers expected to lose money on meals and counted it money well spent.
Lucius Beebe, The Trains We Rode, *1965.*

203.20 The Missouri Pacific was not, truth to tell, an abode of transcendental gastronomy. To be sure, its diners in no way reflected the dyspepsia that was one of the several ills to afflict Jay Gould, architect of its original destinies; quite the opposite; Mopac fare was notoriously for heroes at the table and its plenitude legendary, reflecting the outsize tastes of its Texas clientele and the bounty of the Ozarks through which it ran.

Ibid.

203.21 There is no way that you can lose or pick up freight customers so fast as by the quality of the meal or the cup of coffee that you provide the traveler.

Harry A de Butts, president, Southern Railway (U.S.), quoted by Professor John Stover, American Railroads, *1968.*

203.22 A Pullman breakfast
probably makes the most
perfect start to the day.
Sliding down the west coast
on buttery kippers, tea,
marmalade and toast.

Malcolm Taylor, poem, Breakfast Call.

203.23 Odd it is, this affair of dining on the move—an institution assumed by travelers, beloved of publicists, deplored by accountants.

David P Morgan, Trains Magazine, *September 1975.*

203.24 I doubt if we'd have had an Amtrak if the railroads had knocked off all their dining cars before 1971, such is the public affection for sit-down meals en route.

Ibid.

203.25 . . . and then you got a cook car. See, there's no restaurants any place around, so you got a cook car; pots and pans and coal or wood stove and a long table down the middle to eat at. Only thing they don't hire is a cook. That's 'cos they're cheap.

Bruce "Utah" Phillips, monologue Mooseturd Pie, *about his days as a gandy dancer south of Las Vegas, Nevada.*

203.26 While crossing Kansas, one pays much attention to food. Meals become an important break in the scenery.

Don Phillips, Trains Magazine, *October 1994.*

203.27 Let me tell you one thing. The most beautiful thing you could ever see is a dining room on a train. You got the white tablecloths, you got the silver that's shiny as gold, see, and it had a good atmosphere to it.

James T Steele, Pullman attendant, within Those Pullman Blues, *David D Perata, 1996.*

204. ECONOMY and ECONOMICS

204.1 I assure you, in completing the undertaking, I will act with that economy which would influence me if the whole of the work was my own.

George Stephenson, letter to Edward Pease, 2 August 1821.

204.2 Railroads have brought temptations in the way of many who can neither afford to go to London nor resist it. People hear of Time being Money, which it undoubtedly is to active business men; but every idler adopts the idea, and because he gets to London much more quickly and easily than he used to do, flatters himself with the notion that he is therefore economising.

R S Surtees, Town & Country Papers, *1860.*

204.3 I might put £500 into my pocket by getting you to buy my patent cast-iron rails. But I know them. Take my advice, and don't lay down a single cast-iron rail.

George Stephenson, letter to Stockton & Darlington Railway directors, no date given; quoted by Jeaffreson in The Life of Robert Stephenson, *1864.*

204.4 Quantity is the essential element of Railway success.

Attributed to Mark Huish by Terry Gourvish.

204.5 It costs $2.50 to stop a train.

Warning notice in the signal cabins of the Missouri Kansas Texas Railroad, 1920s. Quoted by Robert Bell, in paper The Railways in 1927, *delivered to the LNER Debating & Lecture Society, 13 March 1928.*

204.6 The merits of any front end arrangement must be measured by two standards; the discharge of the required amount of steam at the lowest possible back pressure, and the drawing through the front end of an adequate volume of combustion gases. The purpose of varying the front end arrangement is to secure the desired high ratio of steam flow to exhaust pressure. These are conflicting conditions, but the closer they can be met, the greater will be the operating economy of the locomotive.

Ralph P Johnson, The Steam Locomotive, *1945.*

204.7 A dangerous myth created by railway economics writings of the last forty years has been that a 'cost-based' charging system would have been preferable to charging what the traffic would bear. Both theory and experience disprove this assertion, however closely it may appear to match text-book notions of optimal pricing rules for public utilities. If these critics, at leisure and with hindsight, could not get the economics of the problem right, it is hardly fair to expect that the railway managers of the Thirties would have attained technically elegant solutions to their pricing problems. In fact, the railwaymen's intuition had more validity than their critics' misplaced use of the theory.

Stewart Joy, The Train that Ran Away, *1973.*

204.8 All of this contradicted the most basic truth in railroad engineering. And that is the seeming paradox that the most expensive railway is the cheapest, in the long run.

John H White, Railroad History, *No 141, 1979.*

204.9 The best way to the poor house—as some railroads discovered too late—is poor track maintenance.

John Kenefick, meeting of the Lexington Group in Transportation History, Milwaukee, 1992.

204.10 I'm delighted I'm going to be replaced by a board of eight people. . . . it's part of the Government's war on waste and over-manning.

Tom Winsor, Railway Regulator, at a meeting of an all-party group of Members of Parliament; on the impending expiry of his contract. Recalled by Nigel Harris to the compiler, 19 January 2004.

205. EFFICIENCY

205.1 In the operations of a railroad, efficiency must never be sacrificed for the sake of economy . . .

James O Fagan, Confessions of a Railroad Signalman, 1908.

205.2 It shall be my endeavor, therefore, to demonstrate to you that the popular clamor about inefficiency in the management of our railways is unjustified; and by doing so try to show you why you should bring the light of reason to bear on all questions pertaining to railways that may be presented to you in future, resisting temptation to tinker with or attempt to repair a piece of machinery until it is thoroughly understood, heeding the wise injunction that I remember, as a child, to have seen prominently displayed on French mechanical toys: "Quoi qu'elle soit trés solidement montée il ne faut pas brutaliser la machine."

Julius Kruttschnitt, Union Pacific. The French may be translated "Although it is solidly built, do not handle this machine roughly." Address to the Graduate School of Business Administration, Harvard University, 26 April 1911, quoted in The Railway Library, 1911, *1912.*

205.3 Arithmetic we'll briefly try:
What a railroad must do,
When costs and wages both are high,
To guard its revenue.

"Raise rates!" the class in chorus yell,
"Nay, not at all," cries he;
Just use my formula, and swell
Two dollars into three.

Expense plus waste two millions net;
The second strike away;
Fifty per cent, the more you get,
A million dollars a day!

"The proof? It needs no learned lore,
'Tis clear from laying bricks,
A man who used to lay but four
Today is laying six.

Anon., poem The Pedagogue, *verses six to nine. The reference to a million dollars is to the campaign waged by a lawyer, Louis Brandeis, to demonstrate unconscionable waste by the nation's railroads. Quoted by Warren S Stone, from a Boston newspaper, Eleventh Annual Meeting of Grand Chief Brotherhood of Locomotive Engineers, quoted in* The Railway Library, 1911, *1912.*

206. ELECTRICAL TRAINING

206.1 I think it is a mistake for any lad to do electrical work pure and simple, he should start by obtaining a thorough grounding in Mechanical Engineering and specialise afterwards if he thinks proper—really electrical matters to my mind consist of about 75% Mechanical Engineering, 20% common sense and 5% purely electrical work. . . .

George Hughes, CME of Lancashire & Yorkshire Railway, letter to Nigel Gresley, GNR, April 1914.

207. ELECTRIC TRAINS

207.1 The "H", the "O"s, the "V"s, the "T"s,
The "Esses" and the "I"s,
All hustle on to Bridge or Cross;
The youth, the maid, the boy, the boss,
En masse to business ties.

Percy Morris, a Southern Railway substation worker at Lewisham, of the headcodes carried by Southern Railway electric trains, indicating their routes; poem of seven verses, Southern Railway Magazine, *February 1927.*

207.2 What a gloomy sight is an electric train, even one of the posher sort—engineless, graceless, worm-like things, without distinction or romance!

Rev Ralph Gardner, vicar of St Mary's South Benfleet, in his parish magazine. Quoted in The Railway Magazine, *vol. LXXVIII, 1936.*

207.3 . . . a paragon of rather tedious efficiency. . . .

C Hamilton Ellis, Italian Railways 1839–1939, in The Railway Magazine, *vol. LXXXV, 1939.*

208. ELECTRICITY

208.1 Steam in kid gloves.

Frederick Westing, Trains Magazine, April 1954.

208.2 Steam by wire.

Editorial comment, Trains Magazine, 2002.

208.3 An attractive way to generate electricity in Switzerland.

Roger Waller, of a steam locomotive propelling an electric locomotive which was regenerating power into the wire as a means of providing a load. Talk to Institution of Mechanical Engineers, 15 March 2004.

209. ELECTRIFICATION

209.1 Soviet Russia equals Socialism plus Electrification.
  Maxim attributed to Vladimir Lenin in Railroads of
the USSR, *Association of American Railroads, 1960.*

209.2 I have the credit of being the inventor of the locomotive and it is true that I have done something to
improve the action of steam for that purpose. But I tell
you, young man, that the day will come when electricity
will be the great motive power of the world.
  *Attributed to George Stephenson, booklet, British Elec-
trical Development Association, 1930s, quoted by John
Johnson and Robert A Long,* British Railways Engineering,
1948–1980, 1981.

209.3 Electrification is carried out not to save money, but
to make more.
  Attributed to Sir John Aspinall, by John Glover in
Southern Electric, *2001.*

209.4 By looking with optimism upon all phases of electrical working and predicating the most efficient installa-
tion, and looking with pessimism upon steam operation,
and taking for contrast a line poorly equipped and badly
worked, a very favorable case may be drawn up for the
use of electricity.
  L F Loree, Railroad Freight Transportation, *1922.*

209.5 . . . no grinding, no jerking, no puffing, no pulling,
no straining, no disturbed slumbers—just a keen sense
of moving swiftly, of being propelled by power vastly in
excess of requirements. You ride with ease—you are at
ease—it is the very last word in transportation.
  Thomas Edison, quoted by William D Middleton,
When the Steam Railroads Electrified, *1974.*

209.6 A switchman sat in a Southern sub.,
  And viewed with practised eye
The metered toll, for the public needs
The "CC" load on his traction feeds,
  As the "Tizzies" scuttled by.

His gaze was fixed on the panel tops
Where the well-placed flashguards showed;
The hour has struck when breakers break,
When brushes spark and panels shake,
  Through the urge of an overload.

The needles crept to their peak time line,
  The voltage dropped and soared,
While the Kents advance to wield their pen,
Full twenty packed in a coach for ten,
  On a "P" train all aboard.
  *Percy Morris, a Southern Railway substation worker at
Lewisham, poem of seven verses,* Southern Railway Maga-
zine, *February 1927.*

209.7 The Sparks Effect.
  *Anon., of the benefits of electrification to trade and pas-
sengers, c. 1962 when the West Coast Main Line was being
electrified.*

210. ELEVATED RAILWAYS
See also Underground

210.1 The track that found and lost itself a thousand
times in the flare and tremor of the innumerable lights;
the moony sheen of the electrics mixing with the reddish
points and blots of gas far and near; the architectural
shapes of houses and churches and towers, rescued by
the obscurity from all that was ignoble in them, and the
coming and going of the trains marking the stations with
vivider or fainter plumes of flame-shot steam—formed
an incomparable perspective. They often talked afterward of the superb spectacle, which in a city full of
painters nightly works its unrecorded miracles; and they
were just to the Arachne roof spun in iron over the cross
street on which they ran to the depot; but for the present
they were mostly inarticulate before it.
  William Dean Howells, A Hazard of New Fortunes,
1889.

210.2 The El in Chicago is *L*,
Proofreaders at Trains mark it well!
In New York, I dare say,
They might well spell it that way,
But here in Chicago its *L*.
  Alan Lind, Trains Magazine, *May 1963.*

210.3 We've spread the word and marked it well,
That in Chicago El is *L*.
We ask you, too, if the shoe fits,
To note that sometimes *its* is *it's*.
  Response to the above verse by Trains Magazine, *ibid.*

211. C HAMILTON ELLIS

211.1 England's Beebe, and more . . .
  David P Morgan, Trains Magazine, *January 1959.*

212. ELOPEMENT
See also Marriage

212.1 O, rather listen to the boiler singing;
List to the Railway bell, so loudly ringing;
Quit, quit with me this antiquated scene,
And fly on railroad wings to Gretna Green.
  Basil Montague. Spoken by Stoker to Delia; Railroad
Eclogues, *1846.*

212.2 In the case of Dumfries and Galloway, the new stations being jointly planned include Gretna, so noble
Lords will soon be able to elope for no more than the
cost of a day return ticket.

Earl of Lindsay, of recent improvements on railways in Scotland, House of Lords debate on the Railways Bill, Hansard, 15 June 1993.

213. EMPLOYMENT LAW

213.1 Moreover, it hardly need be said that railroad labor legislation, like much other legislation, may represent in part the thought and effort of disinterested idealists and may be in part the result of negative and affirmative pressure of organized interests.

Professor Emory Johnson, Government Regulation of Transportation, 1938.

214. ENCOURAGEMENT

214.1 Our Railway. Come to the meeting to-night!! Citizens: The enemy is in the field. Arouse ye! Arouse ye! And do battle to the enemies of your sacred rights. Having pledged yourselves at two public meetings to support of the Launceston and South Devon Railway, now, in the hour of trial, stand forth as a Samson to crush tyranny and despotism. Now or never!! But be a united band, and act on the Cornish motto—"a long pull, a strong pull, and a pull together"—and the victory will be yours!!!

Handbill for a meeting held on 23 April 1862, quoted by Michael Robbins in Railways and Politics in East Cornwall, within The Railway Magazine, vol. 104, 1958.

214.2 Come along, my beauty, come along.

Driver Jack Stuckey, to Southern Railway No. 850 Lord Nelson, quoted by Cecil J Allen in The Railway Magazine, July 1927.

214.3 Come on old girl, I thought, we can do better than this!

Joseph Duddington, driver of Mallard on world speed record run, recalling his thoughts at 116 mph, in subsequent BBC interview, 1938.

214.4 Steady on, old girl!

LNER driver of locomotive derailed at Hatfield, July 1946, recounting what happened, at the subsequent inquiry. Reported by Canon Roger Lloyd, The Railway Magazine, vol. 92, 1946.

215. ENGINEERS and ENGINEERING

(for locomotive engineers, see *Enginemen*)

215.1 No engineer could succeed without having men about him as highly gifted as himself.

Robert Stephenson, quoted in The Story of the Cambrian, C P Gasquoine, 1922.

215.2 Robert, my faith in engineers is wonderfully shaken. I hope that when you return to us your accuracy will redeem their character.

Michael Longridge, letter 27 February 1826 to Robert Stephenson while the latter was in South America.

215.3 I can engineer matter very well, but my great difficulty is in engineering men.

Attributed to George Stephenson by Robert Stephenson; quoted in Robert Stephenson—The Eminent Engineer, ed Michael Bailey, 2003.

215.4 You are now appointed to the locomotive department at a salary of £500 which I hope will be satisfactory to you, and if you don't turn out one of the cleverest fellows that ever ruled a company I have been talking and protesting too much in your favour.

Joseph Locke, letter of appointment to W B Buddicom, Grand Junction Railway, 3 January 1840. Quoted by N W Webster in Joseph Locke, 1970.

215.5 Just as the Italians are musicians, and the Germans metaphysicians—by birth—so Englishmen are engineers, and there is nothing more natural than that Englishmen should apply their faculties to the science of locomotion.

Michael Reynolds, The Model Locomotive Engineer, 1879.

215.6 Thank God you have been so obstinate.

Edwin Clark, letter to Robert Stephenson, of his insistence upon contingencies in case a tube was dropped during lifting into place, Britannia Bridge, Menai Straits, 1849. Quoted by Samuel Smiles, Life of George Stephenson, 1881.

215.7 Who are the greatest men of the present age? Not your warriors, not your statesmen. They are your engineers.

Attributed to John Bright, House of Commons speech.

215.8 This parliamentary business is very good for the English engineers; probably many of them make as much, or more, engineering in the Houses of Parliament as they do in the field.

Edward Bates Dorsey, English and American Railroads Compared, 1887.

215.9 It is evident that the English engineer always had in view that there was behind him plenty of money, and abundance of time to spend it.

Ibid.

215.10 Engineers have nothing to do with mere assertions. Leave them to the lawyers, whose business it is to befog a jury and make black appear to be white.

Walton Evans, quoted in discussion papers in English and American Railroads Compared, 1887.

215.11 The key to the evolution of the American railway is the contempt for authority displayed by our engineers, and the untrammelled way in which they invented and applied whatever they thought would answer the best purpose, regardless of precedent. When we began to build our railways, in 1831, we followed English patterns for a short time. Our engineers saw that unless vital changes were made our money would not hold out and our railway system would be very short. Necessity truly became the mother of invention.

Thomas Curtis Clarke, essay in Scribner's Magazine, *1888, reprinted in* The American Railway, *1889.*

215.12 It would be well if engineering were less generally thought of, and even defined, as the art of constructing. In a certain important sense it is rather the art of not constructing; or, to define it rudely but not inaptly, it is the art of doing that well with one dollar, which any bungler can do with two after a fashion.

A M Wellington, The Economic Theory of the Location of Railways, *1903.*

215.13 Stroudley aired his science in these engines, and had his way to a bolt.

David Joy, concerning William Stroudley's involvement in the design of machinery for steamers for the London Brighton & South Coast Railway, quoted in Extracts from the Diaries of David Joy, *within* The Railway Magazine, vol. XXIII, *1908.*

215.14 A Roman gentleman—now deceased—who effused Latin poetry, largely, it would appear, for the benefit of the public schoolboy, made the sapient remark, "Quot homines tot sententiae," which, being broadly interpreted, means "So many locomotive engineers, so many opinions."

E L Ahrons, The Railway Magazine, *October 1918.*

215.15 Other nations used to smile at the thoroughness of British railway engineers and to declare that they built their permanent way and rolling stock as though for all eternity.

Comment, The Times Railway Supplement, *15 August 1921.*

215.16 A hundred engineers were now engaged in improving and perfecting the power which had been introduced through the labours of one or two.

Robert Young, Timothy Hackworth and the Locomotive, *1923.*

215.17 . . . I can assure you, gentlemen, that we have a great deal to learn from the French.

Sir Nigel Gresley, Annual Dinner, Institution of Locomotive Engineers, 28 February 1936.

215.18 I suppose that at a very early age most healthy boys want to be an engine driver. It is at a later and less healthy age that they want to design them.

Leslie Burgin, Annual Dinner, Institution of Locomotive Engineers, 25 March 1938.

215.19 Yet the railway network of Switzerland today witnesses to the fact that in the mind of the railway engineer the word "impossible" does not exist . . .

Cecil J Allen, Switzerland's Amazing Railways, *1959.*

215.20 If you cannot think of anything original yourself you ought to be intelligent enough to copy a good thing when you see it.

Sir Nigel Gresley, attributed by Oliver Bulleid, quoted in The London & North Eastern Railway, *Cecil J Allen, 1966.*

215.21 The key to the evolution of the American railway is the contempt for authority displayed by our engineers.

John Hoyt Williams, A Great and Shining Road: The Epic Story of the Transcontinental Railroad, *1988.*

215.22 I am an Engineer; anything is possible.

Bob Clarke, Railtrack Zonal Director, dinner table conversation, 1993.

215.23 Railway engineering is long dead . . .

David Newland, on the studies in various forms of engineering in British universities; letter to The Times, *5 September 2003.*

215.24 Engineering is, of course, an art, not a science.

Roger Ford, Modern Railways, *January 2004.*

215.25 And since engineers are driven both professionally and commercially to explore the art of the possible, nothing new can be expected to work perfectly out of the box.

Ibid.

216. ENGINEMEN

216.1 It is very true that the locomotive engine, or any other kind of engine, may be shaken to pieces; but such accidents are in a great measure under the control of enginemen, which are, by and by, not the most manageable class of beings. They perhaps want improvement as much as the engines.

Robert Stephenson, letter to Michael Longridge, 1 January 1828.

216.2 It is impossible that a man that indulges in reading should make a good engine-driver; it requires a species of machine, an intelligent man, an honest man, a sober man, a steady man; but I would much rather not have a thinking man. I never dare drive an engine, although I always go upon the engine; because if I go upon a bit of line without anything to attract my attention I begin thinking of something else.

I K Brunel, Evidence to Select Committee on Railways, para. 762, Parliamentary Papers, *22 March 1841.*

216.3 . . . in fact, a gentleman cannot be an engine driver, or any man who can earn a livelihood in any quiet, comfortable way; he must know something of machinery.
Ibid., para. 763.

216.4 He is to keep himself clean, and take care that the fireman is also.
Extract from Byelaws, Rules, Orders and Regulations, *Great Western Railway, 1842.*

216.5 O the engineer's joys! to go with a locomotive!
To hear the hiss of steam, the merry shriek, the steam-whistle, the laughing locomotive!
to push with resistless way and speed off in the distance.
Walt Whitman, poem, Poem of Joys, *lines 10–13, in* Leaves of Grass, *1860.*

216.6 There are people who think that a driver is a man who pulls a handle to turn on steam, and then stands and looks at the result till it is time to turn it off again, and that a fireman has only to shovel on coal with no more intelligence than is displayed by the domestic footman. They would know better if they had ever been on the footplate of an engine that was booked to run for two or three hours without a stop.
W M Acworth, The Railways of England, *1889.*

216.7 Two leading pioneers of railway progress in Europe took diametrically opposite views of the intellectual qualities best calculated to make a good engineer. George Stephenson preferred intelligent men, well educated and read up in mechanical and physical science; Brunel recommended illiterate men for taking charge of engines, on the novel hypothesis that, having nothing else in their heads, there would be abundant room for the acquirement of knowledge respecting their work. In every test of skill, the intelligent proved victors.
Angus Sinclair, Locomotive Engine Running and Management, *1902.*

216.8 No demand for illiterate or ignorant engineers has ever arisen in America.
Ibid.

216.9 Miss Snow draws near
The cab to cheer
The level-headed Engineer.
Whose watchful sight
Makes safe her flight
Upon the Road of Anthracite.
Earnest Elmo Calkins. One of the many jingles he composed about Phoebe Snow, who always travelled in spotless white linen in Delaware Lackawanna & Western Railroad advertisements, 1900–1917.

216.10 With tireless hands he feeds the coal in the thundering monster's maw,

And hour by hour he trusts his soul to the God whom he never saw,
And hour by hour his life depends on the care of the other man,
Who, scanning the track where it slopes or bends, keeps vigil as best he can.
S E Kiser, first verse, The Railroad Fireman, *reprinted from* Chicago Record-Herald *within* Santa Fe Employes' Magazine, *February 1908.*

216.11 The hiss of steam is the sweetest song that ever he hears or knows,
And in every throb as they rush along the worth of his toiling shows;
With tireless hands he feeds the coal in the thundering monster's maw,
And hour by hour he trusts his soul to the God whom he never saw.
Ibid., last (fourth) verse.

216.12 It should be a well-known fact that, all over the world, the engine-driver is the finest type of man that is grown. He is the pick of the earth. He is altogether more worthy than the soldier, and better than the men who work on the sea in ships. He is not paid much; nor do his glories weight his brow; but for outright performance, carried as constantly, coolly, and without elation, by a temperate, honest, clear-minded man, he is the further point. And as the lone human at his station in a cab, guarding money, lives, and the honor of the road, is a beautiful sight. The whole thing is aesthetic.
Stephen Crane, The Scotch Express, *within* Men, Women and Boats, *1921.*

216.13 The fireman presents the same charm, but in a less degree, in that he is bound to appear as an apprentice to the finished manhood of the driver. In his eyes, turned always in question and confidence toward his superior, one finds this quality; but his aspirations are so direct that one sees the same type in evolution.
Ibid.

216.14 The safe and punctual operation of our railroads has always depended to a great extent, and will always depend, upon the discriminating care and judgement of the engineer.
Ibid.

216.15 There is a mute but complete understanding between these two; bred of a side-by-side association in ten square feet of working space wherein a man's mouth moves inaudibly. They each have their jobs; mentally intensive in the one case, physically stressful in the other—and they get on with them. They are both poised masters of the immediate situation: deaf to the pandemonium that is now beating upon their ears: placid

within the swaying monster that now stampedes bellowing through the night at their behest.
Dell Leigh, On The Line, *1928.*

216.16 I'd rather be a commando, or drive a railway train,
I'd rather be a commando, Lord! drive a railway train,
Than sort dem Fuel Forms into streets again.
Philip Larkin, song, Fuel Form Blues, *third verse, of tedious work as a civil servant, 1942, within a letter to Kingsley Amis, published in* The Times, *10 May 2003.*

216.17 During their boyhood years
All nice little boys want to grow up to be locomotive engineers,
But as they approach maturity,
The ambitions of all except the very nicest lose their purity,
And only in the breasts of the creme de la creme de la creme
Still burns the hard, gem-like flame,
And that's the reason why locomotive engineers number in their ranks no cads or poltroons or snobs or hoi pollois,
Because their ranks are recruited only from the nicest of all the nice little boys.
Ogden Nash, poem Ding Dong, Toot Toot, All Aboard.

216.18 This is Willie Bain, I warrant!
Cecil J Allen, of the driver who kept class A1 pacific St Johnstoun *in immaculate condition in the early 1950s.* Trains Illustrated, *September 1953.*

216.19 On the engine, you're the boss. What you say goes.
Walter J Tuohy, President, C&O, to an engineer in the latter's cab. Recalled by David P Morgan in Trains Magazine, *November 1960.*

216.20 ... there are probably no set of craftsmen in this country so universally admired as the men who drive our railway engines, and hold all our lives in their safe hands.
Canon Roger Lloyd, The Fascination of Railways, *1957.*

216.21 Wait 'til you get to be a driver and you'll find that you go home more tired than you did as a fireman.
Anon., railwayman, The Ballad of John Axon, *broadcast 2 July 1958.*

216.22 He may be bright and keen, performing his rhythm of 'little and often' plus keeping his footplate swept and watered in an inspired ballet dance which is a joy to behold, or he may be in love or out of sorts or couldn't care less, unable or unwilling to achieve good combustion, or through laziness filling up his box and causing blowing off at the safety valves for miles.
E S Cox, of firemen, Locomotive Panorama, *vol. 2, 1966.*

216.23 Don't think that what you are about to see is the only way to drive a locomotive.

Norfolk Southern engineer (driver), comment to the compiler, immediately prior to company Chief Executive Officer Robert Claytor driving preserved class J locomotive 611 on a special train, May 1984.

216.24 Here is his watch, which was given to him on the day that he joined the company. It strikes me as strange that in this country we give him a watch on the day that he retires.
Prof. Roderick Smith, of an illustration of a Shinkansen driver in Japan, and of practice in the United Kingdom. Lecture: Two and a half Millennia of Railways, *York, 5 June 2001.*

217. ENGINE SHEDS

217.1 In our black stable by the sea,
Five and twenty stalls you see—
Five and twenty strong are we:
The lanterns tossed the shadows round,
Live coals were scattered on the ground,
The swarthy ostlers echoing stept,
But silent all night long we slept.
Robert Louis Stevenson, poem, The Iron Steed.

217.2 Corraled in the sooty roundhouse the steel-ribbed Titan stands,
Blind, inanimate, biding the touch of his master's hands;
The sun draws down to the skyline, the sudden switchlights gleam,
And the Titan rolls from the inner gloom in a swelter of singing steam.
Henry Herbert Knibbs, poem, Right of Way *within* Songs of the Trail, *1920.*

217.3 Who guess that some sequestered single line
Could fashion moments touched by the divine,
That engine-sheds at evening could unfold
A Rembrandt mystery of dark and gold,

That steam, massed white against a changing sky,
Was of the breath of lyric poetry:
Now that this miracle of steel and fire
Is one, for us, with Nineveh and Tyre?

Something that was more England than its earth,
And which of English talent had its birth,
Is dead. And in its death there also dies
Another of this England's majesties.
Henry Maxwell, poem, Ave Atque Vale, *verses 26–28, in* A Railway Rubaiyyat, *1968.*

218. ENTHUSIASM and ENTHUSIASTS

218.1 There is one more type of railway traveller that must be mentioned here, if only for the guidance of the young and simple. He is usually an elderly man, neatly dressed but a little tobacco-stained, always seated in a corner, and he opens the conversation by pulling out a gold hunter and remarking that the train is at least three minutes behind time. Then with the slightest encouragement, he will begin to talk, and his talk will be all of trains.

J B Priestley, Papers from Lilliput, *1922.*

218.2 As some men discuss their acquaintances, or others speak of violins or roses, so he talks of trains, their history, their quality, their destiny. All his days and nights seem to have been passed in railway carriages, all his reading seems to have been in timetables.

Ibid.

218.3 Beware of the elderly man who sits in the corner of the carriage and says that the train is two minutes behind time, for he is the Ancient Mariner of railway travellers, and will hold you with his glittering eye.

Ibid.

218.4 One of the qualifications of what is popularly termed the "railwayac,"—the man who, though not in the railway service, is keenly interested in the running and working of trains,—is that he should be able to recite, on demand, an accurate catalogue of engine names.

C P Gasquoine, The Story of the Cambrian, *1922.*

218.5 The railway does exercise its fascination upon all ages, but it is a lamentable delusion that those who succumb to it suffer a stunting of mental growth that causes them to go through life regarding it with the uncreative admiration of a small boy with his box of trains.

Comment in Pertinent Paragraphs, *within* The Railway Magazine, *vol. LXXIX, 1936.*

218.6 He likes the Central Vermont all right, but he is somewhat disappointed in the Rutland because the brakeman wants to talk baseball instead of block signals.

Doug Welch, short story, Mrs Union Station, *within* Saturday Evening Post, *1937.*

218.7 The collecting of locomotive numbers is a perfectly legitimate hobby for railway enthusiasts and pursued intelligently may serve useful and instructive purposes.

Editorial comment, The Railway Magazine, *vol. 90, 1945.*

218.8 If you can sit for hours on draughty stations,
With Job-like patience till the trains go through;
If you can put behind you all temptations,
To cross the line and get a "better view";
If you can wait and not get tired by waiting,
Or, being shouted at, not answer back,
Or being hated, not give way to hating,
And slyly try to "trespass on the track";
If you can brave the righteous wrath of porters,
And still observe the rulings of the game;

If you can take a hand with those "defaulters",
And keep your schoolboy honour just the same;
If you can fill each unforgiving minute
With sixty seconds' worth of "Spotting" done,
Yours is the world and everything that's in it,
And, what is more, you'll be a "Spotter", son.

Margaret Brannagan, former secretary to Ian Allan. With apologies to Kipling, within Trains Illustrated *No 4, 1947.*

218.9 And thus, as is but right and proper, we end this catalogue of the variety of railway lovers with the small boys, who are the basis and the recruiting ground of the whole army. For on what does the whole vast and varied membership of the craft rest? It rests, of course, on the little boys whom you see any day loitering about on the far end of station platforms in every part of the British Isles, each one with his grubby notebook and blunt pencil, and his list of all the engines on the railway system, collecting their numbers and names in the vain hope that one day he will have collected them all. They identify unerringly every engine when it is still far away. They know the names, and can recognise the more famous drivers. They are generally noisy, and sometimes they can be rather a nuisance, but the officials wisely interfere with them very little. The officials know that the mysterious but valuable spirit of the loyalty of the people to the company is kept in existence by just such boys as these, and their numbers grow steadily each year, and not all the aeroplanes in the world can diminish it.

Canon Roger Lloyd, The Art and Mystery of the Railway, *within* The Railway Magazine, *vol. 94, 1948.*

218.10 We are not "cranks" although we take an interest in big ends!

R S McNaught, Railway Enthusiasts, *within* The Railway Magazine, *vol. 97, 1951.*

218.11 The curious but intense pleasure that is given to many people by the watching and the study of trains, their engines, and the detail of their organization is both an art and a mystery. It is an art because the pleasure to be had is in exact proportion to the informed enthusiasm one puts into it. It is a mystery because, try as one will, it is impossible to explain to others exactly in what the pleasure consists.

Canon Roger Lloyd, The Fascination of Railways, *1951.*

218.12 HUBERT: It is my lost youth that you have given back to me. I gazed at the smoky station and the hurrying people with new eyes. I listened to the escaping steam, the porters shouting, the whistles blowing, with new ears, I felt I wanted to shout and sing. It was only out of consideration for you that I refrained from bribing the engineer to allow me to drive the locomotive.
CHARLOTTE: Please do not speak of trains. I loathe them.
HUBERT: Forgive me. I had forgotten. How tactless of me.

CHARLOTTE: Railways are my husband's first love, last love and only love. I realised quite soon after our wedding that in marrying me he was committing bigamy. His heart and his allegiance belong to the Illinois Central. My married years have been flattened and deafened by the railways of the world. I too have, at last, escaped, only to find that my deliverer wishes to drive an engine. It is more than I can bear.

HUBERT: What can I say to retrieve my hideous blunder? What excuse can I give? The signals were against me and I crashed over the points.

CHARLOTTE: Hubert!

Noel Coward, exchange in act I scene II, Quadrille, *1952.*

218.13 He loved trains and he looked forward to the rest of the journey.

Ian Fleming, of James Bond, Live and Let Die, *1954.*

218.14 Did he have moments of longing for the peaceful simple things of life? Of course not. He liked Paris and Berlin and New York and trains and aeroplanes and expensive food, and, yes certainly, expensive women.

Ian Fleming, of James Bond, Moonraker, *1955.*

218.15 In the last analysis, there are probably two types of railway enthusiast, the man who is fascinated by mechanical devices, and the one who is intrigued by systems. The first is likely to be interested, mainly, or exclusively, in locomotives, though his addictions may spread beyond the railway field to clocks or tinkerable motorbikes. Your system-fancier, on the other hand, finds satisfaction in the railway as an organic, working whole—its growth, its operation, and the inter-relation and classification of all its parts. In extreme cases, such folk may not even be interested in the steam locomotive at all.

R K Kirkland, Railway World, vol. XVII, *1956.*

218.16 I first came across the word ferro-equinologist on a boy's record card under the heading "Hobbies". It is a lovely word and could only have been coined by a railway enthusiast.

Dr Sidney Best, Headmaster of Doncaster Grammar School, Introduction to Catalogue of Doncaster Grammar School Railway Society Museum, *1957.*

218.17 Pity for a soul so drugged with the conventional that it does not respond to the sigh of air brake cylinders or the wallop of a crossing taken at 85. Pity for the eyes that see not the beauty of a smoke plume, ears that hear noise instead of stack music, hands that grip Pullman handrails without touching magic.

David P Morgan, Trains Magazine, *May 1957.*

218.18 'Well, he was at Norwich, or somewhere, getting up steam, or whatever one does, before he took his engine out for a trip, when some of the railwaymen who were on strike molested him. He was polishing the brass on the nameplate, I believe, when they got into the cab and took the brake off. The engine started to move, and as Bertie tried to clamber on to the footplate he fell under the wheels.'

'How ghastly!'

'Yes, run over by his own engine!'

'Was he badly hurt?'

'He had to have both legs amputated. Never walked again, except on crutches.'

'What a horrible end to one's hobby!'

John Hadfield, conversation between Jasper Pye and Miss Tidy about the misfortune of Lord Flamborough during the General Strike, Love on a Branch Line, *1959.*

218.19 Unfortunately railway enthusiasts are mostly people interested in engines and speeds of trains and the history of railway development. There is very little written about railway buildings. The names of the architects of some of our grandest stations such as Bristol Temple Meads are forgotten.

John Betjeman, The Daily Telegraph, *quoted in* Railway World, *May 1960.*

218.20 A fan might be described as anyone who enjoys the railroad without or beyond the considerations of employment or ownership.

David P Morgan, Trains Magazine, *March 1962.*

218.21 The fan is proving himself an indefatigable historian, doing a first-rate job of preserving the mechanical detail if not yet the overall economic sweep of the past.

Ibid.

218.22 . . . there is a breed of railroad *aficionados* to whom the heady smells of hot valve oil and the sound of engine bells are the sum of all happiness and who care not to what terminals the trains may come so long as they do it enveloped in satisfactory clouds of the best soot and are behind the most radiantly obsolete motive power.

Lucius Beebe, The Central Pacific and the Southern Pacific Railroads, *1963.*

218.23 Well, we've preserved nearly every movable object (steam, that is) on British Railways, but perhaps the greatest unpreserved loss has been the *gricer*, or full time railway enthusiast.

Attributed to N B A. The earliest use of the word gricer, *as recognised by the Oxford English Dictionary;* Railway World, *December 1969. The word* gricing *appeared in the* Scarborough Mercury *on 26 September 1968.*

218.24 I mean, a grease-drenched crosshead, coupling between piston rod and main rod, bathed in steam, laid against engine truck and driving wheel: that's it brother, A-to-Z.

David P Morgan, Introduction to Portrait of the Rails, *Don Ball Jr, c. 1972.*

218.25 There was time en route to study and chat with the British fan, more intense if anything than the U S species. He approaches his love laden not only with the customary camera and tape recorder, but also with books—books of maps, history, schedules. He is nothing if not knowledgeable.

David P Morgan, Trains Magazine, *July 1975.*

218.26 The difference between men and boys is the price of their toys.

Doris Rowland, wife of Ross Rowland, owner and operator of an ex C&O 4-8-4. Quoted in Trains Magazine, *October 1976.*

218.27 We just didn't know that it couldn't be done. So we went ahead and did it.

Bob Anderson, who led the restoration of an 1878 Rogers locomotive delivered new to New Zealand, dumped in a river in 1927, and recovered in 1974. It was steamed again in 1981. Quoted in Trains Magazine, *September 1983.*

218.28 Fortunate is the man who can fix his enthusiasm to the diesel locomotive.

George W Hilton, Trains Magazine, *September 1984.*

218.29 The serious railfan has two favorite railroads—the B&O and another one.

Anon., quoted by David P Morgan in Trains Magazine, *December 1985.*

218.30 There is no known cure for an obsession with trains. It may strike early in life as was evident in Lisa St Aubin de Terán, infant progidy of the District Line. Prep school on the railway embankment of East Dulwich Station and the lure of Brighton undoubtedly placed temptation in her way, but it was a childhood journey through Russia on the Occident Express that began a lifetime's addiction.

Anon., note on flyleaf of Off the Rails, *by Lisa St Aubin de Terán, 1989.*

218.31 Their knowledge of railways is minimal.

Neil Cossons, of enthusiasts, after-dinner speech, 9 October 1993.

218.32 It is difficult for me to understand why it is that if one collects 17th century Chinese figurines, she or he is cultured; if one picks up a guitar and sings folk songs, she or he is well-rounded; but to collect Alco PA builder plates makes you a yob.

Douglas F Kydd, Trains Magazine, *August 1994.*

218.33 We appreciate our fans, but like football coaches, we don't consult them on our strategies.

Catherine Wheeler, BNSF, Wall Street Journal, *18 September 1995.*

218.34 The trainspotter has become everyone's favourite wally. With blacks, gays and big-boobed women all off

the right-on comic's agenda, here's a man you can titter at in safety, political integrity unblemished. The Identikit is hideous: a gormless loner with dandruff and halitosis, a sad case obsessed by numbers, timetables and signalling procedures. He has no interest in girls, and girls have even less interest in him.

Nicholas Whittaker, Platform Souls, *1996.*

218.35 Unlike the golf club or the Chamber of Trade, no-one cares about your income, your profession, or who you know. What you know about trains is what counts. It's one of the few areas where you'll see professors comparing notes with latch-key kids.

Ibid.

218.36 If it seems odd to find solace in an old steam engine after a month at sea, on the way to God knew where and in the middle of a world war, all I can say is, well, you weren't there and didn't see it, and I had my passion to tend.

Eric Lomax, of his discovery of a locomotive on a plinth in Cape Town during his journey to the Far East, 1941. The Railway Man, *1996.*

218.37 They wanted to know why I had drawn the railway on my map. I tried to convince them that I was a railway enthusiast, that I had made the map so that I would have a souvenir of Siam and the railway, and know where the stations were. They could not imagine that this was partly true: I had not lost my instinct to record and list and trace. I spoke to them about trains, loaded them up with information about British standard gauge and how interesting it was to see a metre-gauge railway in operation, and the problems of exporting locomotives designed for one system to countries with different systems. The interpreter struggled to find the right terms, about gauges, and boiler sizes and engine weights.

Ibid., shortly before he is tortured by Japanese soldiers for a second time.

218.38 He kept saying to me: 'You are railway mania?', meaning, I think, 'maniac' or 'fanatic', his voice expressing genuine, angry puzzlement, and then he would try to explain this incredible excuse to his colleague, who looked darker and more brutal by the minute.

Ibid.

218.39 I don't want to sound like a trainspotter, but I am a believer in rail and given a chance I'd always go by rail.

David Harding, newly appointed Finance Director of Railtrack, quoted in Rail *magazine, 16 May 2001.*

218.40 If you see a railroader hold up a Katy trainsheet from June 3, 1946, as if it were the first draft of the New Testament, you're looking at a railfan.

Mark W Hemphill, Editorial comment, Trains Magazine, *April 2004.*

219. ENTHUSIAST SPECIALS

219.1 Excursion programs face many obstacles, from paralyzing indifference to absolute hostility. Therefore nothing is accomplished without a sense of trust between railroad management and railroad enthusiasts. Nothing is done without maturity and sound judgment on the part of the enthusiast. Enthusiasts must be practical and careful not to push on the railroads something that just doesn't make sense. For it's vital that society and groups interested in railroading keep a learning process going with the railroads. This is what's good for the railroads, for the country, for our heritage.

James Bistline, quoted in Locomotive & Railway Preservation, *September/October 1987.*

219.2 Remember the smell of donuts and hot dogs wafting through the commissary car? Remember how hot it was trackside when you stepped down for that runby at Lynchburg? And the cool of the shade when your train dove into the tunnel above New Albany?

Remember the reunions with old friends from the Heart of Dixie and the Piedmont Chapter and the Roanoke Chapter? And the hospitality of Bill Purdie and Jim Bistline, or Carl Jensen and the Claytors? And the camaraderie that prevailed as everyone on *Lookout Mountain's* platform fell silent to hear 4501's whistle echo down the French Broad River valley, or to instinctively grab the handrail as 610 dug into the grade on Christiansburg Mountain? And remember how beautiful the 611 looked as she sailed over the big CNO&TP viaduct on the "Rathole" at South Fork, Kentucky?

Summer just isn't the same.

Kevin P Keefe, of Norfolk Southern steam excursions, Trains Magazine, *May 1997.*

220. ENVIRONMENT

220.1 Railways are safer, more fuel-efficient, less noisome than road or air. But they cannot live on environmental godliness alone.

Subtitle, part 2 of series, Nicholas Faith, The Economist, *24 August 1985.*

221. EPITAPHS AND EULOGIES

221.1 This Tablet
a tribute of personal respect and affection has been placed here *to mark the spot* where on the 15th of Septr 1830 the day of the opening of this rail road the Right Honble. William Huskisson M.P. singled out by the decree of an inscrutable providence from the midst of the distinguished multitude that surrounded him. In the full pride of his talents and perfection of his usefulness met with the accident that occasioned his death: which deprived England of an illustrious statesman and Liverpool of its most honored Representative which changed a moment of the noblest exultation and triumph that science and genius had ever achieved into one of desolation and mourning: and striking terror into the hearts of assembled thousands. Brought home to every bosom for the forgotten truth that

"In the midst of life we are in Death"

Memorial to William Huskisson, Newton-le-Willows, Lancashire. The original is in the National Railway Museum in York, and a replica is at the site in Lancashire.

221.2 My Engine now is cold and still,
No water does my boiler fill:
My coke affords its flame no more,
My days of usefulnefs are o'er.
My wheels deny their noted speed,
No more my guiding hand they heed.
My whistle, too, has lost its tone.
Its shrill and thrilling sounds are gone.
My valves are now thrown open wide.
My flanges all refuse to guide
My clacks, also, tho' once so strong,
Refuse their aid in the busy throng.
No more I feel each urging breath,
My steam is now condens'd in death.
Life's railway's o'er, each station's past.
In death I'm stopp'd, and rest at last.
Farewell dear friends, and cease to weep!
In Christ I'm SAFE, in Him I sleep.

Attributed to Thomas Codling, on the headstone of Oswald Gardner, St Mary the Virgin parish church, Whickham, Co. Durham, killed at Stocksfield station at the age of 27 by a broken connecting rod, 15 August 1840. He was employed by the Newcastle & Carlisle Railway, and the headstone records that it was erected at the expense of workmen on that railway. It also records that the epitaph "was composed by an unknowing friend to commemorate his worthiness and left at the Blaydon Station." See appendix A for a discussion on this poem.

221.3 The Line to heaven by Christ was made
With heavenly truth the Rails are laid,
From Earth to Heaven the Line extends,
To Life Eternal where it ends.

Repentance is the Station then
Where Passengers are taken in,
No Fee for them is there to pay,
For Jesus is himself the way.
God's Word is the first Engineer
It points the way to Heaven so clear,
Through tunnels dark and dreary here
It does the way to Glory steer.

God's Love the Fire, his Truth the Steam,
Which drives the Engine and the Train,
All you who would to Glory ride,
Must come to Christ, in him abide

In First, and Second, and Third Class,
Repentance, Faith and Holiness,
You must the way to Glory gain
Or you with Christ will not remain.

Come then poor Sinners, now's the time
At any Station on the Line,
If you'll repent and turn from sin
The Train will stop and take you in.

*Anon., poem on tombstone in the south porch of Ely
Cathedral, entitled* The Spiritual Railway, *in memory of
the lives of William Pickering (30 years old) and Richard
Edger (24 years old), both died 24 December 1845, in a railway accident. Later developed and expanded into a song of
the same name and found in various forms.*

221.4 The click of steel battering steel and the wind
a-whining through the dewberry briers—and Huey with
his hand on the throttle and the scream of the whistle—
and us down behind the willows waiting to wave at
Huey—all wanted to be like him. But now—the silent
hands—the steady quiet eye and we stand and shed our
tears unashamed of tears for our engineer that once
pulled the train where no track now is—nothing but the
wind and the dents in the earth and cinders ground
down and old bridges—but we remember—we'll always
remember—and Huey—our engineer—we wonder on
what silent train and to what silent land our engineer
has gone.

*Jesse Stuart, of his friend Huey, an engineer on the East-
ern Kentucky Railroad, which became the East Kentucky
Southern in 1929 and was abandoned in 1933;* Huey, the
Engineer, *within* Headlights and Markers, *ed. by Frank P
Donovan and Robert Selph Henry, 1946.*

221.5 In a calling and occupation which has given few
names to the American lexicon, that of Casey Jones will
be forever brightest and best known as synonymous with
the daring and romance of the high iron in its golden
age. He lived and died in the fullest flower of railroad
expansion and in the age of steam. It is safe to say that the
wonder and glory of that saga is now beginning to be
obscured, diminished and cheapened by the devisings of
modernity. There will be no songs written about, nor
monuments raised to the memory of, any railroad man
living or dying by the diesel locomotive.

*Lucius Beebe, at a ceremony to dedicate a memorial on
the grave of Casey Jones, 1947, as recorded in* Trains Magazine, *October 1947.*

221.6 Farewell, my friends of the thundering clan,
On the mountain heights, farewell.
The honking geese in the wintry sky,
In their cries my sadness tell.

Once more I saw you hulky and strong,
Enveloped in hissing steam:

Explosive eruptions of masterminds,
Creations of mankind's dream.

Like the pines on the ridge beyond the track
the pillars of clouds arose.
Immortal now, immersed in the skies,
The final picture they pose.

Othmar Tobisch, to steam locomotives, poem, Trains
Magazine, *April 1956.*

222. ERIE RAILROAD

222.1 Go East by Erie.

*Slogan, particularly on a huge billboard on an island at
Niagara, which was locally regarded as an eyesore and was
eventually demolished by a local resident, 1887. Related by
G P Neele in* Railway Reminiscences, *1904.*

222.2 There will be icicles in hell when Erie common pays
a dividend.

Attributed to Jay Gould, by Albro Martin, Railroads
Triumphant, *1992.*

222.3 Father asked me what road I was getting a ticket on.
He said the Lake Shore was the best. I made some vague
answer to that. I didn't like to tell him, after he had been
so generous to me, that I had bought a cut-rate ticket to
Chicago and back, for eleven dollars, on an Erie Special
Excursion. The Erie was so awful in those days that it was
a joke. It didn't go nearly as far as Chicago, of course, but
it had arranged for trackage rights over a number of
other one-horse railroads for its Special Excursions.

*Clarence Day, of his effort to get to the World's Fair in
Chicago in 1893 at the lowest possible cost,* Life with Father,
1935.

222.4 It took that train three days and two nights, if I
remember correctly, to get to Chicago. We stopped at
every small station. We waited for hours on sidings. Most
of the time I had very little idea of where we were. The
Excursion wandered around here and there, in various
parts of this country and Canada, trying to pick up extra
passengers. Of course, the train had no sleeping cars or
diner—only day coaches.

Ibid.

222.5 The Scarlet Woman of Wall Street.

*Unattributed, but dating from a battle for control, 1867,
quoted in* Unfinished Business, *Maury Klein, 1994.*

222.6 You don't want the Erie. Try the Pennsylvania or the
Central.

*Anon., ticket agent in Chicago to Linn H Westcott, who
was determined to travel on the Erie to New York—and did.*
Trains Magazine, *March 1946.*

222.7 The Erie is very courteous to its passengers, excep-
tionally so for an Eastern road. This was evident immediately as we boarded the car. For every passenger there

was a smile 4 feet 8 ½ inches wide on the face of the train-man who helped us up the steps and directed us politely as to just what car to take.

Linn H Westcott, ibid.

223. ESCALATORS

223.1 I've always wondered how difficult it was.

Sir Ralph Wedgwood, upon being found trying to walk up a descending escalator at Liverpool Street Station, Cecil J Allen, The London & North Eastern Railway, *1966.*

224. ESTIMATES

224.1 It will be close work to keep them there, but you know a good deal can be done by pinching, long hours, short grub time, some cussin, and a fair amount of per-suasion and praying.

James C Clarke, of efforts to keep within estimates agreed by his board; letter to William Ackerman, quoted by Thomas C Cochrane in Railroad Leaders, 1845–1890, *1953.*

224.2 Self-confessed bullshit.

Sir Alistair Morton, on the sharply increased Railtrack estimates for the East Coast Main Line upgrade work, The Times, *15 February 2001.*

225. ETON COLLEGE

225.1 Notwithstanding anything in this Act contained it shall not be lawful for any person whomsoever to form, make, or lay down any branch railway or tramroad or other road or way whatever passing or approaching within 3 miles of the said College of Eton and connecting with the said railway hereby authorised to be made with-out the consent of the Provost and Fellows for the time being of the said College of Eton, to be signified by same in writing under the corporate seal.

Section 100, Great Western Railway Act, 1835.

226. EUSTON HOTEL

226.1 But look first at the dimensions of the Euston—why it's a town in itself!

R S Surtees, a Jorrocks novel, date not known.

227. EUSTON STATION

227.1 As Melrose should be seen by the fair moonlight, so Euston, to be viewed to advantage, should be visited by the gray light of a summer or spring morning, about a quarter to six o'clock, three-quarters of an hour before the starting of the parliamentary train. . . .

Samuel Sidney, Rides on Railways, *1851.*

227.2 At the hour mentioned, the Railway passenger-yard is vacant, silent, and as spotlessly clean as a Dutchman's kitchen; nothing is to be seen but a tall soldier-like policeman in green, on watch under the wooden shed, and a few sparrows industriously yet vainly trying to get breakfast from between the closely-packed paving-stones. How different from the fat debauched-looking sparrows who throve upon the dirt and waste of the old coach yards!

Ibid.

227.3 We cannot bestow unqualified praise upon the sta-tion arrangements at Euston. Comfort has been sacri-ficed to magnificence.

Ibid.

227.4 Those who would see a station work at full power should look in at Euston on the eve of St. Grouse, or an evening or so before, between, say, seven and eight.

W J Gordon, Everyday Life on the Railroad, *1898.*

227.5 What London is in the world of cities Euston is amongst railway stations.

James Scott, Railway Romance and other Essays, *1913.*

227.6 As for me, I can never understand why people should seek Egypt in search of the Sphinx and the pyra-mids, when they can visit Euston station and survey the wonders of the stone arch.

Aubrey Beardsley, quoted in Round the London Ter-mini, *Basil Mercer, within* The Railway Magazine, *vol. XLVII, 1920.*

227.7 The frieze, where of old would prance an exuberant processional of gods is, in this case, bare of decoration, but upon the epistyle is written in simple, stern letters the word "EUSTON." The legend reared high by the gloomy Pelagic columns stares down the wide avenue. In short, this entrance to a railway station does not in any way resemble the entrance to a railway station. It is more the front of some venerable bank.

Stephen Crane, The Scotch Express, *within* Men, Women and Boats, *1921.*

227.8 Anon, I descry the cavernous open jaws of Euston. The monster swallows me, and soon I am being digested into Scotland.

Max Beerbohm, Ichabod, *1923.*

227.9 Oh, the Euston Station! What botheration, And tribulation at the Christmas time . . .

Anon., opening lines of poem quoted in Old Euston, *1938.*

227.10 The termini of other railways are straightforward and rational; big glass halls with news theatres and flower shops. Not so Euston, which is like a Midland ware-house, littered with new bicycles half wrapped up in cardboard. Indeed, I have a suspicion that Euston is also a market, with regular auction sales of corrugated iron, and hens.

Paul Jennings, The Observer, *30 October 1949.*

227.11 Departing passenger trains go from old, disused parts of the market. Arrivals are met with an air of vexation. Here's another of these passenger trains, they say, and they make it wait for 20 minutes in that tunnel just outside while they crossly remove the crates of bicycles from some obscure platform to make room for it.
   Ibid.

227.12 Our own modernization and rebuilding, in contrast with that of the Continent, lags far behind, but I suppose that being the people we are there would be an outcry if there was a suggestion of rebuilding Euston station on lines appropriate to the twentieth century.
   Peter Allen, On the Old Lines, *1957.*

227.13 . . . I was answered immediately by a sepulchral, curiously dignified voice which simply said: "Euston Sleepers!". I couldn't have been more awed if it had said: "Delphic Oracle, at your service" or "Vestal Virgins, good morning". So *that's* what they've got upstairs in that central hall at Euston, with its pillars and statues, the tutelary deities of British Railways. How extraordinary that we should not have guessed; for we have all been faintly troubled by the feeling that in some way Euston is more than a station . . .
   Paul Jennings, The Jenguin Pennings, *1963.*

227.14 When we look at the early prints of Euston, showing it in all the glory of cream stone against a Canaletto sky, it is easy to imagine the people coming through the Propylaeum with their gifts to the old railway gods. The houses that now huddle round Euston were a later development, as innkeepers and traders moved in to cater for the pilgrims.
   Ibid.

227.15 Holidays which involve Euston have a habit of starting on a depressing note.
   David St John Thomas, Double Headed, *1963.*

227.16 From the point of view of the people, however, it was less than entirely satisfactory, as was seen when someone pointed out rather sharply that the designers had entirely forgotten to provide any seating whatever in the huge concourse, so that passengers waiting for trains or other people must sit on their suitcases, lie on the floor, stand up, or go away.
   Bernard Levin, of the new Euston, Now We Are Seventy, *within* Pendulum Years, *1970.*

227.17 It was expected that, when this oversight was pointed out the responsible authorities would, with red faces, hastily concoct an excuse for their forgetfulness and remedy it at once. No such thing; to the general stupefaction, the authorities declared that the lack of seating was entirely deliberate, and would not be remedied at all, and, when asked why, gave as their reason that 'We like a tidy station, and passengers make it untidy.'
   Ibid.

227.18 The symbol of the new age is the new Euston, an all-purpose combination of airport lounge and open-plan public lavatory.
   J Richards and J M MacKenzie, The Railway Station, *1986.*

227.19 Its grandeur was quite disproportionate to the accommodation provided for passengers behind it . . .
   Jack Simmons, of the Doric portico, The Victorian Railway, *1991.*

227.20 Mostly underground and unphotographable.
   George H Drury, of the modern replacement station, Trains Magazine, *December 1997.*

228. EXETER ST DAVID'S

228.1 Strictly speaking, one can neither use the terms "East" and "West" nor "Up" and "Down", seeing that the Great Western Railway proceeds here approximately north and south, while down Great Western trains travel the same way as up London and South Western trains, and *vice versa.* As, however, Great Western traffic proceeds eastwards and westwards, these may be adhered to, occasional necessary reference to north and south being bracketed thus: "East (north)".
   J F Gairns, footnote in Notable Railway Stations and their Traffic, *within* The Railway Magazine, vol. XLVI, *1920.*

228.2 St. David's Station at Exeter was then almost as great a terror to far western travellers as was Reading to those of the nearer west.
   A G Bradley, Exmoor Memories, *1926.*

228.3 If you were leaving the main line at either for other routes, there you stayed for unconscionable periods of time. Every detail of their features was burnt into your brain for life. But, hated as they were, I seem to see them now, despite structural transformation, as all aglow with youth's fond memories. For a railway station is, after all, significant of half life's pleasures, of its memories and anticipations, and some of its sorrows.
   Ibid.

228.4 St David's Station—fifteen after seven,
A splendid structure; really, 'tis a pity
They didn't build it closer to the city,
For one can almost fancy on the spot
He's in the Midlands, where of course he's not.
   Frederick Thomas, poem, St David's Station to Starcross. *Within* The Poetry of Railways, *ed. Kenneth Hopkins, 1966.*

229. EXPERIENCE

229.1 When I am at Derby, I am Chairman of the Midland Railway Company; when I am at Glasgow, I am Chairman of the Glasgow and South Western Railway

Company; when I am at Bradford, I am a brewer; and when I am in London I am a barrister, like you, sir.

Sir Matthew Thompson, Bart., under cross examination before a committee of the House of Commons, c. 1885. Related in The History of the Midland Railway, *Clement Stretton, 1901.*

229.2 It's great to ask old railroaders how they used to do things, but it's not a good idea to ask two of them.

Aarne Frobom, conversation with the compiler, on the perils of being given conflicting advice from an earlier generation of railroad engineers, 20 April 2002.

229.3 When I started in this job I didn't know a fishplate from a fish supper. I do now.

Tom Winsor, Railway Regulator, at a meeting of an all-party group of Members of Parliament; reflecting shortly before the expiry of his contract. Recalled by Nigel Harris to the compiler. 19 January 2004.

230. EXPERIMENT

230.1 The United States is trying an experiment which never has been successfully worked out yet in the history of the world. It is trying to build, develop, and operate its railroads by private capital under rates and regulations fixed, not by the owners of the capital, but by the public.

Hon. Charles H Prouty, ICC, address before the National Association of Railway Commissioners, reproduced in The Railway Library, 1913, *1914.*

231. FACTORY LOCATION

231.1 Put your plant where it will grow.
London Midland & Scottish Railway poster, c. 1928.

231.2 Factories on LNER lines are on the right lines.
London & North Eastern Railway poster, c. 1930.

232. FAITH

232.1 A religion that can't stand a railroad isn't worth its salt.
Brigham Young, 1860s, of fears that the coming railroad in Utah would adversely affect the Mormon faith. Quoted by Oscar Lewis in The Big Four, *1938.*

232.2 Inspiration shall be looked for more in engine cabs than in pulpits,—the vestibule trains shall say deeper things than sermons say. In the rhythm of the anthem of them, singing along the rails, we shall find again the worship we have lost in church.
Gerald Stanley Lee, Atlantic Monthly, *1900. Quoted by John R Stilgoe in* Metropolitan Corridor, *1983.*

232.3 *Vicar*: They can't possibly close it.
Town Clerk: What about the Old Canterbury and Whitstable line? They closed that.
Vicar: Perhaps there were not sufficient men of faith in Canterbury.
T E B Clarke; exchange in film The Titfield Thunderbolt, *1952.*

233. FARE DODGERS

See also Tickets, Ticket Collectors and Collecting

233.1 Every passenger who refuses to pay his fare or produce and deliver up his ticket upon the request of the conductor of the train and the train servants of the company, be expelled from and put out of the train, with his baggage, at any usual stopping place: Provided that the conductor shall first stop the train and use no unnecessary force.
Provision of a Canadian Railway Act, quoted in Trains Magazine, *February 1954.*

233.2 And fare-dodgers, well, they're the lowest of the low, and should be strung up, put in the stocks and exposed to public disdain on the concourse at Euston. There ought to be a specific place, a railway Tyburn between the Sock Shop concession and the Casey Jones stand-up buffet.
Nicholas Whittaker, Platform Souls, *1995.*

234. FARES and CHARGES

234.1 The reduction of fares produces a maximum of income.
Robert Stephenson, evidence to Select Committee on Railway Promotion, *1846.*

234.2 If this goes on, I should think that the great majority of my West Riding constituents, washed and unwashed, will visit London in the course of next week.
Edmund Denison, MP, Chairman GNR, of the low fares in the price war with the LNWR, January 1856. Quoted in C H Grinling, The History of the Great Northern Railway, *1903.*

234.3 The fares of railways—the fixed prices at which these great monopolies carry passengers—are as accidental, as much the result of inconclusive experiments, as any prices can be.
Walter Bagehot, Economic Studies, *1879.*

234.4 That it shall be unlawful for any common carrier subject to the provisions of this act to charge or receive any greater compensation in the aggregate for the transportation of passengers or of like kind of property, under substantially similar circumstances and conditions, for a shorter than for a longer distance over the same line, in the same direction, the shorter being included in the longer distance.
An Act to Regulate Commerce, 1887, section IV (part), by which the Interstate Commerce Commission sought to prevent rebates of any kind for many years, and known as the Long and Short Haul clause; reproduced in the First Annual Report of the ICC, *1887.*

234.5 In England, at that time, the solemn importance that was attached by railway operators to their work could be gauged by the interminable extracts from Acts of Parliament about tolls, printed in big type, which adorned the walls of country stations, but which even those who had missed one train and were waiting for another never had the courage to read.
D B Hanna, Trains of Recollection, *1924.*

234.6 Railroad transportation after the war between the States and until the Interstate Commerce Act became a law was sold pretty much as are automobiles today. If you can persuade the salesman that your used car has a greater trade-in value than shown in the "blue book" you have gained an advantage over the person whose car is possibly better but who does not have the ability to reach such a good bargain.

Clyde H Freed, The Story of Railroad Passenger Fares, *1942.*

234.7 The railway managers, as railway managers do, explained that the peculiar nature of their business made it inevitable that they acted as they did; but the grievance remained.

Michael Robbins, of fares on the Isle of Wight Railways, which seemed to change every day. The Isle of Wight Railways, *1953.*

234.8 *Passenger at Oxford*: Day return Banbury, please.
Booking clerk: Eight shillings and threepence.
Passenger: Well, look, I booked one from Banbury to Oxford only ten days ago, and they charged me seven shillings and ninepence.
Booking clerk: But it's up hill.

Anon., quoted by Gerard Fiennes in Fiennes on Rails, *1986.*

235. FARMING

235.1 A railway ride presents a rapid panorama of agriculture . . .
R S Surtees, Ringlets or Plain? *1860.*

236. THE FAST FLYING VIRGINIAN

236.1 Along came the F.F.V., the fastest on the line,
A-running o'er the C & O Road, twenty minutes behind the time;
A-running into Sewell Yard, was quartered on the line,
A-waiting for strict orders and in the cab to ride.
Anon. poem, The wreck on the C & O.

237. FATHERS

237.1 My Dad, though, was the highest authority and the most respected. He, too, loved engines as a boy—strange engines painted crimson and blue and yellow and "black with red lining," engines with inside cylinders and no bells or headlights—English engines. He'd even been an apprentice at the Crewe Works of the London & North Western before entering the ministry. He bought *Railroad Man's Magazine* each month and helped us spike home American Flyer empires all over the attic and told us of other locomotives—Central Hudsons, Pennsylvania K4's, and Seaboard 4-6-2's.
David P Morgan, Trains Magazine, *February 1982.*

238. SIR SAM FAY

238.1 He made an empty sack stand upright.
Attributed to a Cheltenham Editor, quoted by George Dow in Great Central, *vol. III, 1965.*

239. FEAR

239.1 And it is one of the reasons why the train-robbing profession is not so pleasant a one as either of its collateral branches—politics or cornering the market.
O Henry, (William Sydney Porter), Holding up a Train, *1916, within* Short Lines, *ed. Rob Johnson, 1996.*

240. FENCES

240.1 The railroads in this country are constructed under the impression that the people who move about amongst them are in possession of their senses. Why shouldn't a man have to exercise his faculties in saving himself from being run down by a locomotive as well as from being run over by a wagon or a coach? The engine makes enough noise, surely.
Anon., American to James Burnley, upon his comment that American railroads are generally unfenced; Two Sides of the Atlantic, *1880.*

241. FERRY SERVICES

241.1 Two rivers have to be crossed, and each time the whole train is banged aboard a big steamer. The steamer rises and falls with the river, which the railroad don't do; and the train is either banged up hill or down hill. In coming off the steamer at one of these crossings yesterday, we were banged up such a height that the rope broke, and our carriage rushed back with a run down hill into the boat again. I whisked out in a moment, and two or three others after me, but nobody else seemed to care about it.
Charles Dickens, August 1867. Quoted by Alvin F Harlow in Steelways of New England, *1946.*

241.2 It is a pity that Dante, when he wrote his *Inferno*, had no knowledge of the torture of the transfer between St Louis and East St Louis in those times. Had he known of it he would have let the condemned be taken across the dark waters by that method, instead of having them rowed over by Charon in a comparatively peaceful way.
Anon., of the ferry service prior to the opening of the Eads Bridge at St Louis, serving among others the Illinois Central Railroad. Quoted in Main Line of Mid America, *Carlton Corliss, 1950.*

242. FESTINIOG RAILWAY

242.1 Its success as a steam passenger carrying line was due to the initiative and mechanical ability of the Spooner family, and to the trumpet-blowing of Robert Fairlie, which last was world-wide.
C Hamilton Ellis, Nineteenth Century Railway Carriages, *1949.*

242.2 Of course, the one we would like to have got our hands on was the Festiniog, but it was too far gone.

Earl of Northesk, soon after the rescue of the Talyllyn Railway, related by Alan Pegler.

243. FEUDAL SYSTEM

243.1 I rejoice to see it, and think that feudality is gone for ever. It is so great a blessing to think that any one evil is really extinct.

Dr. Thomas Arnold, of the coming of railway travel. Quoted in Our Railways, *J Pendleton, 1896.*

244. FIRING

244.1 Give her plenty of diamonds, if you please.

Frank H Spearman, The Nerve of Foley, *1901; spoken by an engineer from the Philadelphia & Reading, accustomed to anthracite, or "black diamonds". Within* Short Lines, *ed. Rob Johnson, 1996.*

244.2 Reynolds says that when heat, water and fuel are being thrown away in steam through the safety valves it is positive proof that either the locomotive is too small for its work, or too great for its man.

L F Loree, referring presumably to Michael Reynolds, author of The Model Locomotive Engineer, *but not found in this work.* Railroad Freight Transportation, *1922.*

244.3 Hello, Billy Boy, are you coming with me? Off we go then, and all you need do is to put as much coal as you can through that hole and I'll sort it out for you.

Driver Sam Gingell, to sixteen year old cleaner Peter Wensley, with no main line experience, sent in emergency to fire for Gingell, April 1956.

244.4 You had to shovel coal right off the level of the footplate, and it were real hard work because you had to raunge down with your back, to get hold of your coal, to get it into t' firebox. But then they vou t' married man's tender out. Oh, and weren't that a dream. You hadn't got t' bend as far; you could just shovel coal and into t' firebox it went. And by gad owt married men took another lease on life.

Anon. railwayman; "raunge" is almost certainly "range", meaning to stretch or reach, while "vou" is probably "found". The Ballad of John Axon, *broadcast 2 July 1958.*

244.5 Quelle panache, quelle panache!

Jacques Deseigne, Chauffeur de Route, SNCF, of the LNER style of firing used by Richard Hardy, c. 1965.

245. FIRST CLASS

See also Class Distinction, Second Class, Third Class

245.1 Obedient to authority, and at its expense, I was travelling first-class, and authority saw to it that I did so

by presenting me with the ticket and by paying the extra shilling to reserve a seat into the bargain. Normally if I travel first I have a stupid but quite unbiddable conscience about the extravagance about it, with the result that I commit the extravagance and then cannot enjoy the journey, thus making the worst of both worlds.

Canon Roger Lloyd, The Fascination of Railways, *1957.*

246. FISH TRAFFIC

246.1 The North Western's chief customer is the lordly salmon. The mainstay of the Great Eastern is the humble bloater.

W M Acworth, The Railways of England, *1889.*

247. FLATTERY

247.1 Flattery is like the Rome Express, it takes my breath away and wins my heart.

Lisa St Aubin de Terán, Off the Rails, *1989.*

248. FLATULENCE

248.1 Farting in public is one of the few acts that is still taboo in Western Protestant countries, outside boys' boarding prep schools, changing rooms in the cheaper rugby clubs and the Northern Line on the London Underground.

Philip Howard, The Times, *30 September 2000.*

249. FLORIDA EAST COAST RAILWAY

249.1 The main line will go out to sea, as far as Key West.

Henry M Flagler, owner of the railroad, quoted in Railroad Avenue *by Freeman Hubbard, 1945.*

250. FLYING

250.1 If God had intended us to fly he would never have given us the railways.

Michael Flanders, At the Drop of Another Hat, *1964.*

251. FLYING SCOTSMAN and THE FLYING SCOTSMAN

(*The London & North Eastern Railway generally used the definite article for the train, but not for the locomotive.*)

251.1 The war is not yet over, and further changes may be made to this train, but it cannot be robbed of its greatness, and nothing is more certain than that when peace returns the "Flying Scotsman" will once more come into its own, with paintwork shining and amenities restored, and will continue, as the years and the miles roll by, to be the pride of the L.N.E.R. and a mirror of all that is best in railway travel.

Of the train, The Locomotive Magazine, *15 July 1942.*

251.2 Except for her shrill whistle, she scarcely spoke at
all except when an American was invited to take her regulator (throttle) on the footplate (deck) and he, in turn,
deliberately notched her down (with a vertical screw
reverse). . . . Then, and only then, would the lady from
Doncaster bark. Oh, she talked it up a bit ascending
Georgia and Alabama hills, but even then with a reserved
BBC voice, capable but not excitable.

Of the locomotive, David P Morgan, Trains Magazine,
February 1970.

251.3 Pegler had left his heart and his first love in San
Francisco.

*Ted Benson, of Alan Pegler's financial difficulties in the
ownership of the locomotive, reported in* Trains Magazine,
November 1972.

251.4 . . . she can certainly be likened to a beautiful
blonde—marvellous to look at but incredibly expensive
to keep in the manner to which she's accustomed.

Of the locomotive, Alan Pegler, The Railway Magazine,
March 2001.

252. FOOTBRIDGES

252.1 It seems to be of a Railtrack standard design, by an
engineer who the night before ate a bowl of spaghetti
Bolognese and who had a nightmare afterwards.

*Allan Patmore, of the new footbridge at Brough station,
conversation with the compiler, 19 January 2002.*

253. FOOTWARMERS

253.1 At first I loved thee—thou wast warm,—
The porter called thee " 'ot,", nay, "bilin'."
I tipped him as thy welcome form
He carried, with a grateful smile, in.

Alas! thou art a faithless friend,
Thy warmth was but dissimulation;
Thy tepid glow is at an end,
And I am nowhere near my station!

I shiver, cold in feet and hands,
It is a legal form of slaughter,
They don't warm (!) trains in other lands
With half a pint of tepid water.

I spurn thy coldness with a kick,
And pile on rugs as my protectors,
I'd send—to warm them—to Old Nick,
Thy parsimonious directors!

To a Railway Footwarmer, Punch, *reprinted in* Mr
Punch's Railway Book, *c. 1900.*

254. FOREIGNERS

254.1 It is interesting to note that those responsible for
the working of the Metropolitan District Railway have

now realised that foreign designs, foreign workmanship and foreign material are not suitable for use on
our railways. They have consequently given an order to
an English firm for four hundred axles to replace the
defective German ones.

*Editorial comment upon the findings of a report on two
accidents on 12 March 1906,* The Railway Magazine, vol.
XIX, *1906.*

255. FORTH BRIDGE

255.1 . . . especially when one remembers what is likely to
be the cost of the maintenance of the bridge, with its
acres of exposed ironwork ready to absorb tons upon
tons of paint, and to employ the labour of whole gangs of
painters all the year round from January to December.

*W M Acworth. Probably the earliest recognition of the
feature of the bridge that passed into the language as a simile for a never-ending task.* Railways of Scotland, *1890.*

255.2 On physical valuation principles our Parliament
makes the Forth Bridge fifteen miles long for local, and
twenty-five miles long for long-distance traffic.

William Acworth, The Economic Journal, *September
1915, reproduced in* The Railway Library, *1915, ed. Slason
Thompson, 1916.*

255.3 Its beauty is unconscious, as in so many Victorian
structures. Not one of its designers, builders or sponsors
remarked on any ornamental quality, yet there it stands in
all its giant grace, today part of the land-and-seascape on
which it was imposed, and scenically as full of moods as a
mountain. It should be seen at sunrise; it should be seen
in the evening; it should be seen in a storm; it should be
seen when a white sea mist drifts up the firth, hiding all
but the tops of the towers; it should be seen at night, when
the fireman of a crossing engine opens his firebox door
and floods the girders momentarily with an orange glare
up to the topmost booms. To sailors in two great wars, it
has been one of the best-loved landmarks in Great Britain.

C Hamilton Ellis, The North British Railway, *1955.*

255.4 Oh, beautiful Railway Bridge of the silv'ry Forth!
Taking the trains from South to North
From Edinburgh to the county of Fife
Alas, a journey that risks your life
Or at least makes your heart jump into your mouth
When going South to North, or North to South.

Ivan Shakespeare, The Forth Bridge Disaster, *after
William McGonagall.*

255.5 'Twas when Railtrack acquired the bridge I hear
They decided routine maintenance was too dear
And now it's no longer continually painted I fear
And soon became covered in dead birds and grime
Which will be there for a very long time.

Ibid., second verse.

256. SIR HENRY FOWLER

256.1 'E's great on boilers, is Sir 'Enry.
*A mechanic at Kentish Town shed, date unknown.
Recalled by Charles S Lake,* The Railway Magazine,
vol. 88, *1942.*

FREIGHT CARS

See Wagons

257. FREIGHT SERVICE AND TRAINS

257.1 The freight agents of American railways are a peculiar class, trained and experienced in a long course of
ingenuity and subtlety. They have by long practice
acquired extraordinary perfection in those arts. They
and their soliciting agents have long since arrived at the
conclusion that tonnage irrespective of profit is their
legitimate object.
*Sir Henry Tyler, Presidential address, Grand Trunk
Railway, 1877.*

257.2 Nowhere in the world is there a larger or more
important business that that of handling freight transportation. Nowhere in freight transportation better than
here can the rule of the lawn tennis player be followed,
"accuracy first, then put on your speed," though this will
be found to apply to almost everything in freight movement. Before you start anything it is well to make sure
where you will finish.
L F Loree, Railroad Freight Transportation, *1922.*

257.3 The freight traffic of England is said to be decidedly
retail in character, while that of America is *wholesale.*
John Droege, Freight Trains and Terminals, *1925.*

257.4 The LNER can solve your freight transport prob-
lems as effectively as the magic carpet solved those of
Prince Hussein.
London & North Eastern Railway poster, 1930.

257.5 He brought to it a memory of the loaded box-cars
slatting past at fifty miles an hour, of swift breaks like
openings in a wall when coal cars came between, and the
sudden feeling of release and freedom when the last
caboose whipped past.
Thomas Wolfe, Of Time and the River, *1935. Also published as part of a short story,* The Train and The City,
changed only to put it in the first person, 1935.

257.6 He remembered the dull rusty red, like dried blood,
of the freight cars, the lettering on them, and their huge
gaping emptiness and joy as they curved in among raw
piney land upon a rusty track, waiting for great destinies
in the old red light of evening upon the lonely, savage,
and indifferent earth; and he remembered the cindery
look of the road-beds, and the raw and barren spaces in
the land that ended nowhere; the red clay of railway cuts,

and the small hard lights of semaphores—green, red, and
yellow—as in the heart of the enormous dark they shone,
for great trains smashing at the rails, their small and pas-
sionate assurances.
Ibid.

257.7 As for the freight cars, they were companion to
these things, and they belonged to all the rest of it, as
well. Even their crude raw color—the color of dried
ox-blood, grimed and darkened to the variation of their
age—seemed to have been derived from some essential
pigment of America, somehow to express the whole
weather of her life.
Thomas Wolfe, No More Rivers, *1983.*

257.8 "A goods train! a goods train! a goods train!" he
grumbled. "The shame of it, the shame of it, the shame
of it."
*Rev W Awdry; spoken by Gordon, a passenger engine in
freight service;* The Three Railway Engines, *c. 1947.*

257.9 At three thirty-seven I opened the gate
For Danny O'Toole and his manifest freight
The hogger is Irish by habit and birth
And hasn't a chance to inherit the earth.
His engine was doggedly beating the main
Like a Kansas tornado lambasting the plain.
The reefers went swooshing and swaying in sections;
Hell would tear loose if they missed their connections.
The engine screamed frantically half a mile east
And shortly thereafter the tempo increased.
My heart was a-pounding and jerking my vest
As all that vast tonnage poured out of the west.
My tonsils were hampered as though by a noose
When whipple-de-click went the demon caboose.
The moon was full bright and the prairie was clear
As an op softly swore at a damned engineer.
Charles D Dulin, poem, Redball, *in* Railroad Maga-
zine, *1948.*

257.10 Even the *clank-clonk-clink* of goods trains is to go.
The present system for stopping trucks is to let each one
hit the one in front a healthy wallop, so that the guard of
a long goods train can count on eighty or ninety separate
and distinct warnings that he is going to get a hell of a jolt
any moment now. The Commission's plan to fit continuous brakes to freight cars means that the guard won't
get any warning at all; he won't even get a jolt.
H F Ellis, Full Diesel Ahead, *about the BR Modernisation Plan, within* Punch, *2 February 1955.*

257.11 Save time and money by missing bedlam and
confusion.
Statement in The Official Guide *within the entry of the
Bonhomie and Hattiesburg Southern Railroad, reported in*
Trains Magazine, *August 1957.*

257.12 In the long spring evening's twilight, when the sun
    is setting low,
And the smoke from all the engines flushes up, a rosy
    glow,
Then I come up to the bridge-head, watch the lights and
    net-work rails,
Think of when I rode the freighters—engines spouting
    steam like whales,
D.L.W., *Jersey Central*, old *Rock Island*, *Pere Marquette*,
Reading coal cars down from Scranton, piled with
    anthracite like jet.
    Phoebe Hoffman, poem, The Freight Yards.

N.&W., the *Great Northern*, *Lehigh Valley*, *B.&O.*,
Like a giant earth-worm twisting, slowly 'round the
    curve they flow.
Caravans of freight move westward, bearing eastern
    goods away—
To come back with hogs and cattle, bales of sweet
    Kentucky hay.
Brakemen walk along the roof-tops, lingering for a
    moment's chat:
There an engineer, while smoking, long and eloquently
    spat.
    Ibid.

257.13 Freight trains seem beautiful to me: the square,
No-nonsense sturdiness of soft barn-red,
Insignia flaunted with a jaunty air,
As though a frosting had been put on bread!
How many times at railroad crossings, I
Have waited in my car and read aloud
The painted emblems as boxcars squeaked by.
Way of the Zephyrs! Like a waltz, it sings
Across the mind. . . Route of the Southern Belle,
And Pere Marquette, The Rebel Route—each brings
The feeling that it has a tale to tell!
Collectors range from bottle caps to art,
And I have freight trains running through my heart!
    Lee Avery, The Saturday Evening Post, *14 September*
1957.

257.14 Freight trains, according to railroad propaganda,
make money. Passenger trains do not. And, of course,
freight trains do not mind being delayed an hour here
and there, nor do they talk back, nor form militant
groups, nor demand that they all be transported along
the tracks in the same direction at the same time during
two very brief periods each weekday morning and night.
    Jerome Beatty Jr, Show Me the Way to go Home, *1959.*

257.15 Look at those high cars roll. There's nothing so
beautiful in the world as a money making train going
places fast on a spring evening!
    F E Williamson, President, New York Central, in con-
versation with Lucius Beebe on the observation platform of
The Twentieth Century Limited. Recounted by Lucius
Beebe, 20th Century, *1962.*

257.16 A carload of coal northbound in the fall, and a
carload of ashes southbound in the spring.
    Roger Arcaro, of the sparse freight traffic on the New
York, Westchester and Boston Railroad. Westchester's Forgotten Railway, *1962.*

257.17 We carried oil to New York at eight cents less than
nothing.
    A J Cassatt, Pennsylvania Railroad, of rate wars insti-
gated by John D Rockefeller, quoted by Peter Lyon, To Hell
in a Day Coach, *1967.*

257.18 all freight, the sudden trains that uncouple my
    passage home
like flash floods, stranding me in these winter afternoon
    rains
counting carloads of lumber, flatcars of heavy
    equipment, sealed
boxcars heading out of the city, cities themselves, miles
    long
and full of industry, but with only a grim mayor at the
    throttle
and a handful of sleepy maintenance workers for
    inhabitants:
where have all the passengers, the rightful citizens, gone?
    Alvin Greenberg, poem Freight Train, Freight Train,
1989.

257.19 Freight railroading is a stealth industry. Forget the
old songs about trains, and forget the days when boys
went down to the tracks to admire engineers. The only
time people think about freight trains today is when one
blocks a crossing or derails.
    Don Phillips, Trains Magazine, *September 1994.*

Gg

258. GAMBLING

258.1 The Pullman Company calls the attention of its patrons to the fact that "Card Sharks" and "Con Men" have started their winter campaign on railroad trains.

Passengers can protect themselves by refusing to play with strangers.

Sign, Pullman Company, date not known.

GANDY DANCERS

See Permanent Way Men

259. CHRISTOPHER GARNETT

259.1 Platform Soul.

Anon., of the practice of Christopher Garnett, GNER, of Management by Walking About. Caption to photograph, The Daily Telegraph, 25 July 2002.

260. GAUGE

See also Broad Gauge

260.1 It will be a great loss if these railways, like the common road, should require to be altered that they may communicate with each other.

Robert Stevenson, civil engineer, of the need for a common gauge, letter to Lord Melville, 29 January 1825.

260.2 Make them of the same width: though they may be a long way apart now, depend upon it they will be joined together some day.

George Stephenson, when consulted about the gauge of the Leicester & Swannington and Canterbury & Whitstable Railways. Quoted by Samuel Smiles in Life of George Stephenson, *1881.*

260.3 It has been asserted that 4 feet 8 inches, the width adopted on the Liverpool and Manchester Railway, is exactly the proper width for all railways, and that to adopt any other dimension is to deviate from a positive rule which experience has proved correct; but such an assertion can be maintained by no reasoning.

I K Brunel, letter to directors of the Great Western Railway, 15 August 1838.

260.4 The objections to a material increase in the gauge beyond that in general use are—

1st That the necessary increased weight of the carriages consequent upon increased axles, which must also be increased in diameter to preserve equal strength, will only increase the cost, but the proportion of weight of cars to goods carried, which upon a road like this between St Petersburg and Moscow intended as it is for the transportation of large quantities of goods should be avoided as much as possible.

2nd A material increase of gauge involves an increased cost in the formation of an increased width of road bed—the precise amount of which I am not at present prepared to say—but it is evident that in a line of such extent it must be very considerable.

George Whistler, letter to Count Kleinmichel, 9 September 1842, quoted in Bulletin 117, *Railroad & Locomotive Historical Society, October 1967.*

260.5 These are the principal disadvantages of an increased gauge as immediately applicable to the St Petersburg and Moscow Railroad, and I think it must be admitted that they are not met by any equivalent advantages.—I would therefore recommend—that five feet be adopted as the width or "gauge" of the track.—

I have adopted the additional 3½ inches, because it is not necessary in this case to adhere to the precise dimension of 4 feet 8½ inches for the purpose of connecting with other roads of that gauge, and because it is too small an increase to affect materially the important question of cost—and will be available for convenience in the construction of Engines.

Ibid.

260.6 It shall not be lawful (except as hereinafter excepted) to construct any railway for the conveyance of passengers on any gauge other than four feet eight inches and one half an inch in Great Britain, and five feet three inches in Ireland: Provided, always, that nothing hereinbefore contained shall be deemed to forbid the maintenance and repair of any railway constructed before the passing of this Act on any gauge other than those hereinbefore specified, or to forbid the laying of new rails on the same gauge on which such railway is constructed within the limits of deviation authorized by the several Acts under the authority of which such railways are severally constructed.

Clause 1, Gauge of Railways Act, (9 & 10 Vict. c. 57), 1846.

260.7 Bifocal.

Anon. lawyer referring to mixed gauge on the East Tennessee & Western North Carolina Railroad, noted by Archie Robertson in Slow Train to Yesterday, *1945.*

260.8 . . . the gauges from 4 feet 9⅜ inches to 4 feet 8 inches (both inclusive) may be counted as standard, as rolling stock used upon either is interchanged without objection.

Comment in Tenth Census of the United States, 1880, quoted by George Rogers Taylor and Irene D Neu, in The American Railroad Network, 1861–1890. According to The Statistics of Railways in the United States, 1889, published by the Interstate Commerce Commission, in 1880 21 track gauges were in use, of which 7 fell in the dimensions noted above, accounting for about 78 percent of the track miles in the country at the time.

260.9 You are not a country, but only five separate islands.

Alleged statement by Japanese government, taunting Australia, on account of the several gauges of its railways, immediately prior to the Second World War. Recounted by Harold A Edmondson, Trains Magazine, April 1971.

260.10 . . . even to this day, industrial railways are notorious for the variety and inconsequence of their gauges. It is often tempting to suppose that rails were laid to fit a waggon built at random, rather than that waggons were built to fit a deliberately chosen gauge.

Michael Lewis, Early Wooden Railways, 1970.

260.11 There is no real mystery, however, about the basic reason why the Tyneside waggonway gauges fall into the range from 4 to 5 ft. Charles E Lee has pointed out that a figure of this sort is the optimum axle-length for a one-horse vehicle; if the axle is much shorter, the vehicle is small and the horse does not pull an economic load; if it is much longer, the cart becomes too heavy for one horse. Times change, but the horse does not. This seems to be the simple reason why Greek and Roman and even pre-classical vehicles, as evidenced by the ruts they left behind, were of much the same gauge as the Tyneside waggons of the seventeenth and eighteenth centuries and British Railways today.

Ibid.

260.12 For this reason there is a mite of truth in the oft-repeated explanation of our standard gauge of 4 ft 8½ in., that George Stephenson copied it from Roman chariot ruts found on Hadrian's Wall: both were following the same natural law. But to assert that there is any direct connection between the two is nonsense. Stephenson did not build Killingworth waggonway; indeed, when he first went there at the age of twenty-three the 4-ft. 8-in. gauge was already at least forty years old. When he built his first locomotives for Killingworth, they were of necessity of 4-ft. 8-in. gauge, and when, with reputation established, he went on to build new railways elsewhere, he not unnaturally gave them the gauge he was accustomed to. The Hetton and Springwell colliery railways of 1822 and 1826, the Stockton and Darlington of 1825 and the Liverpool and Manchester of 1830, all were built to 4-ft. 8-in. gauge. The extra ½-in. which makes up our present standard gauge seems to have crept in during the 1830s, through the desire for greater side-play for the flanges. If Stephenson had found his mission in life at Heaton, our standard gauge would probably be 4 ft. 3 in.; if at Wylam, about 5 ft. It was largely a question of chance.

Ibid.

260.13 Of all the big losers in railroad engineering, however, none promised to do more than the narrow gauge.

John H White, Railroad History No 141, 1979.

260.14 As American railroading began its long slide into decline after 1907, Baldwin responded to falling sales by broadening its product line, building on its strength in custom work. The company's 1908 catalog listed 379 different sizes and types of engines, while the 1915 edition offered 492 varieties. If none of these five-hundred-odd types suited a customer, Baldwin would build to its drawings or create a new design on demand, guaranteed to perform on any of the world's fifty-seven different railway track gauges.

John K Brown, The Baldwin Locomotive Works 1831–1915, 1995.

261. GAUGE, BROAD AND STANDARD

261.1 Mr Robert Stephenson will tell you there is no difficulty in making an engine as powerful on a 4 feet 8½ inch gauge as can be used safely upon any other gauge. Now his opinion and that of Mr George Stephenson I would rather take than any other man's.

Capt. J M Laws, General Manager of Manchester & Leeds Railway, evidence to the Gauge Commissioners, 1845.

261.2 The meaning of these phrases depends, in great measure, upon whether you happen to have shares in the Great Western or the North Western Line. If you are a proprietor of the former, the narrow gauge is a paltry humbug;—if of the latter, the broad gauge is an extravagant quackery.

Angus B Reach, in the days before true narrow gauge, when what was eventually known as standard gauge was called narrow; The Comic Bradshaw, or Bubbles from the Boiler, 1848.

261.3 A broad gauge tail on a narrow gauge dog.

Erl H Ellis, of the Farmington branch of the Denver & Rio Grande, built as a standard gauge line, isolated from other standard gauge tracks and eventually converted to three foot gauge. Denver Westerners, 1954.

262. GERMAN RAILWAYS

262.1 . . . there is a grain of truth in the English sarcasm that on a German railway "it almost seems as if beer-drinking at the stations were the principal business, and travelling a mere accessory."

Charles Francis Adams, Notes on Railway Accidents, *1879.*

263. WILLIAM GLADSTONE

263.1 An impressive guest list was headed by Gladstone, then Prime Minister, who was accompanied by a veritable bag of Gladstones, ten all told, ranging from Herbert J. (M.P.) to grandson Master William.

George Dow, of the ceremony for the cutting of the first sod of the Bidston-Hawarden line of the Manchester Sheffield and Lincolnshire Railway, 21 October 1892, Great Central, vol. II, *1962.*

264. GOVERNMENT

See also Parliament, Politicians

264.1 I do not conceive, myself, that it is of advantage to the public, in the management of the railway system, that any power should be given to the Board of Trade, or any central body, to issue regulations for the management of the concern.

I K Brunel, Evidence before Select Committee on Railways, 22 March 1841. Parliamentary Papers, Transport, *para. 563.*

264.2 The continental system is a paternal system in which the Government overlooks and controls all the acts of the companies. The American system is one of complete freedom. Neither system is exactly suited to our requirements or our character.

Report of Royal Commission on Railways, 1867, quoted in Edward Cleveland-Stevens, English Railways and their Development and their Relation to the State, *1915.*

264.3 No government is capable of properly executing or administering commercial enterprises.

Comment, Poor's Manual of Railroads, 1868–69, *1868.*

264.4 France and Germany are essentially executive in their Government, while England and America are legislative. The executive may design, construct, or operate a railroad; the legislative never can.

Charles Francis Adams, Railroads; their Origins and Problems, *1878.*

264.5 We will pass the anti-railroad legislation and hear the railroad men speak afterwards. The condemned are always allowed to speak from the scaffold.

Anon. Iowa Senator, quoted in Iowa State Register, *27 February 1888.*

264.6 It is an object lesson that tells the whole story between the enterprise and intelligence with which the American railway system is managed, and the ignorance with which the Government system of Europe is managed.

Chauncey Depew, of the cheapness of fresh fruit which has travelled 3,000 miles in America compared with the high cost of fresh fruit which has travelled far shorter distances in Europe. Speech upon return from Europe, 10 September 1890; Chauncey Mitchell Depew, Orator, *Welland Hayes Yeager, 1934.*

264.7 Regulation is as essential to railways as to the public. In depriving the roads of the power to hurt each other they will be divested of the power to injure the public . . .

Augustus Schoonmaker, Commissioner, ICC, in 1891; quoted thus by George R Blanchard in The Year Book of Railway Literature, *1897.*

264.8 Fellow citizens, we have come here to day to remove the seat of the government of Kansas from the Santa Fe offices back to the statehouse where it belongs. You have beaten the Santa Fe railroad and you must organize the legislature tomorrow, and I wouldn't let the technicalities of the law stand in the way.

Lorenzo Lwelling, newly elected Governor of Kansas, recorded in Topeka Daily Capital, *10 January 1893, quoted by Keith L Bryant Jr, in* History of the Atchison Topeka and Santa Fe Railway, *1974.*

264.9 In Continental Europe the regulation of railway rates by public authority has reduced the railway manager largely to a man who sits in his office and orders his subordinates to run trains back and forth. In America the absence of restriction upon the railway manager—beyond that imposed by common law—has allowed the American railway manager to become the most powerful single factor in our national life for the discovery and the development of the resources of our country, and the promotion of trade and industry.

Hugo R Meyer, Government Regulation of Railway Rates, *1905.*

264.10 The Government has recently undertaken to do something quite different from that which it has ever undertaken to do before. It has undertaken to exercise a controlling influence upon the administration of railway properties through the agency of their accounts. The aim of the supervision of accounting is to exercise influence upon the administration and management of railway property.

Professor Henry C Adams, statistician, Interstate Commerce Commission, to his staff, c. 1 July 1907, quoted by L F Loree, Railroad Freight Transportation, *1922.*

264.11 . . . the inconvenience of dealing with economic problems by legal machinery.

Attributed to W M Acworth, of the regulation of railways by government. Ivy Lee, address, the London School of Economics, 7 February 1910.

264.12 We must realize, as I think we all do (after a series of very hard knocks), that the railroads are not strictly private property, but subject to regulation by the public through its regularly constituted authorities—that the Government may reduce our earnings and increase our expenses has been sufficiently proved.

Edward P Ripley, speech at the annual dinner of The Railway Business Association, 1909; The Railroad Library 1909, *1910.*

264.13 The people must first realise that regulation must not be strangulation.

James J Hill, address at Annual Dinner of The Railway Business Association, 10 December 1912. Quoted in The Railway Library, 1912, *1913.*

264.14 The future of the great corporations will lie more and more in the Government offices and less in Parliament. It is hopeless to expect to influence Parliamentary majorities, who will act under prejudice when they do not act upon orders. But it may be possible to persuade administrators . . .

Anon., possibly A B Cane; internal memorandum of the Railway Association, c. 1913. Quoted by Geoffrey Alderman in The Railway Interest, *1973.*

264.15 It must be remembered that railroads are the private property of their owners; that while from the public character of the work in which they are engaged the public has the power to prescribe rules for securing faithful and efficient service and equality between shippers and communities, yet in no proper sense is the public a general manager.

U.S. Supreme Court; quoted in The Railway Library, 1913, *1914.*

264.16 The attitude of Governments to railways may be described as positive or negative. The positive attitude is that of the chief continental states; it consists in aid to railway construction, definite assumption of responsibility for finance, of rights of interference and of dictation as to management; in its logical sequence it extends to State-ownership and working. The negative attitude is English; no assistance is afforded to companies; they are given charters which lay stress rather on what they may not do than what they may do; interference takes the form of legislating against certain possible evils, not of planning general schemes for harmonious progress.

Edward Cleveland-Stevens, English Railways and Their Development and Relation to the State, *1915.*

264.17 Fifty stripes save one.

Anon. railroad president, referring to the (oppressive) regulation of American railroads by forty-eight separate states and the Federal Government. Noted by William Acworth in his review of a book in The Economic Journal, *September 1915, reproduced in* The Railway Library, 1915, *1916.*

264.18 Locomotor Ataxia.

Frank Trumbull, part title of address on U.S. Government regulation of railroads. He cited the definition of this as "a disease of the spinal chord characterized by peculiar disturbances of gait, and difficulty in co-ordinating voluntary movement." He added: "Surely Webster must have had railroads in mind when he wrote that!" Cedar Point Ohio, 12 July 1916.

264.19 Court plasters, not blood remedies.

Frank Trumbull, of the nature of U.S. Government regulation hitherto; address to The National Hay Association, Railway Regulation and Locomotor Ataxia, Cedar Point Ohio, 12 July 1916.

264.20 Railway regulation has been more conspicuous for quantity than for quality, and "legislation" and "regulation" are not synonymous terms.

Ibid.

264.21 Immediately upon the reassembling of Congress I shall recommend that these definite guarantees be given: First, of course, that the railway properties will be maintained during the period of Federal Control in as good repair as when taken over by the Government; and second, that the roads shall receive a net operating income equal in each case to the average net income of the three years preceding June 30, 1917; and I am entirely confident that the Congress will be disposed in this case, as in others, to see that justice is done and full security assured to the owners and creditors of the great systems which the Government must now use under its own direction or else suffer serious embarrassment.

President Woodrow Wilson, statement upon making Proclamation of Federal Control of Railroads during the First World War, 26 December 1917.

264.22 It is highly desirable if private operation of railways is to be resumed that the retaliatory character of present railroad laws be eliminated, but such a desire will be difficult of attainment unless railroad authorities definitely abandon all lines of conduct which tend to keep alive the retaliatory spirit of the public.

T W Van Metre, Failures and Possibilities in Railroad Regulation, *within* War Adjustments in Railroad Regulation, *published as* The Annals of the American Academy of Political and Social Science, vol. LXXVI, *March 1918.*

264.23 We do not have too much regulation or too little regulation; what we suffer from is an unwholesome combination of good legislation, bad legislation, and no legislation.

Ibid.

264.24 The President of the United States has aptly said that there should be more business in government and less government in business and of all lines of business there is none in which the hand of government is felt more than in railroad management and operation.

William H Bremner, President, Minneapolis & St Louis Railroad, Annual Report, 1921.

264.25 The Railroad Administration has erred more frequently in overstatement than in under-achievement.

I L Scharfman, of the United States Railroad Administration under William McAdoo, The American Railroad Problem, 1921.

264.26 With the passing of the glacial age of railway control a decided and most welcome advance has been made in the facilities provided by the companies for the travelling and trading public. The fact that it is only a month since they came again into their own makes the improvement all the more remarkable.

Editorial Comment, of the end of government control, imposed in 1914, The Times, 21 September 1921.

264.27 It must not, however, be forgotten that the railways are the creation of this House. They have been granted privileges, rights, powers by this House, and the House is entitled to see that these rights, privileges and powers are not used perversely.

Sir Eric Geddes, Transport Minister, Commons debate on the Transport Bill, Hansard, 26 May 1921.

264.28 The future of the Government ownership of railways in my opinion, depends entirely on their operation being carried on free from any political or other influence. The operation should be carried on the same as a private corporation, keeping in view at all times economy and the interests of the shareholders who, of course, are the people of Canada.

Dr John D Reid, Minister for Railways and Canals, Canada, 1917–21 in letter to D B Hanna, c. 1918, quoted in D B Hanna, Trains of Recollection, 1924.

264.29 The essential requisite of government regulation of railroad services is to bring about maximum coöperation both of the government with the carriers and of the carriers with each other. The regulation of the railroad service by the Government must always be partly corrective in purpose, but its primary aim must be constructive.

Professor Emory Johnson, Government Regulation of Transportation, 1938.

264.30 The nation's transportation system is suffering from a disease called subsidy.

Wayne A Johnston, President, Illinois Central Railroad. Quoted in Trains Magazine, March 1952.

264.31 The best carvers, it has been remarked, pay some attention to bone structure when they dismember a chicken.

Editorial Comment on the British Government's approach to what became the 1953 Transport Act, abolishing the Railway Executive, The Times, 6 November 1952.

264.32 I am convinced that less Federal regulation and subsidization is in the long run a prime requisite of a healthy intercity transportation network.

John F Kennedy, message on transportation reported in Trains Magazine, August 1962.

264.33 There is a certain irony in the fact that we must seek aid from the very hand that has had at least a significant part in forcing us to our present posture.

Anon. Penn Central vice president, at Congressional hearing into the collapse of Penn Central, 1970. Quoted in Trains Magazine, October 1970.

264.34 Given the economic atmosphere and the trends of the present, it matters little that the Norfolk & Western, or the Southern, or any other railroad, enjoys financial strength and relative prosperity. In the same way, the well-being of the other modes becomes academic if Government extends its controls over the railroads. All transportation is woven into a common fabric, and a weakness anywhere degrades the usefulness of the whole.

John P Fishwick, President, Norfolk & Western, address, 18 November 1974, reported in Trains Magazine, March 1975.

264.35 The Hepburn Act of 1906, but more especially the Mann-Elkins Act of 1910, effectively froze railroad rates at a time when all other prices were rising as much as 10 per cent a year. The result, as any economist will understand, was to *arbitrarily*, and *irrationally*, siphon profits out of the railroad system into the rest of the economy. We bled the railroads white, and we have been paying for it ever since.

Albro Martin, address to the Railroad Public Relations Association, 29 June 1975, quoted in Trains Magazine, September 1975.

264.36 But however you define price and service, it didn't matter under regulation, because regulations allowed railroads very little latitude in coming up with ways to meet their customers' needs. For example, the rates had to be the same for all customers, and they had to be published in unwieldy tariff documents that made sure all railroad shippers paid the same rate for the same service. Entering into a private contract with an individual shipper was against the law.

Harry J Bruce, Chairman & Chief Executive, Illinois Central Gulf, address at University of Pennsylvania, 16 April 1987, quoted in Trains Magazine, June 1987.

264.37 Nor could rates be changed easily. Any change, upward or downward, had to be submitted in a written request to the ICC; the request for a revision itself had to be published, and competing carriers and shippers were given the opportunity to object to it at a public hearing. Thirty days elapsed before new rates became effective.
These restrictions made it virtually impossible for

railroads to compete for certain types of traffic, such as produce. An owner-operator trucker with a refrigerated trailer could drive up to a grower's property and literally make a word-of-mouth deal to haul a load of strawberries or oranges to a market 2000 miles away. But a railroad capable of beating the trucker's price had little chance of getting the business; the produce would be shipped, delivered, sold and consumed before the ICC finished processing the rail carrier's rate application.
> *Ibid.*

264.38 The pudding of American railroad policy has been stirred, steamed, sauced, eaten, and found inedible.
> *Albro Martin,* Railroads Triumphant, *1992.*

264.39 At one time, near the end of the old century, it was widely assumed that every mile of important rail route in the nation would have to be double-tracked. The rapid development of automatic block signals and, later, centralized traffic control that permits "meets" on single track without either train coming to a halt, made such an undertaking unnecessary, which is just as well, for repressive government regulation made it impossible.
> *Ibid.*

264.40 Britain has been well served by railways for a century and a half. In 1992 central Government went mad and stayed mad. The Tories privatised trains in a way that destroyed their managerial coherence. Labour renationalised them and is determined to maul them to death by "health-and-safety".
> *Simon Jenkins,* The Times, *11 July 2003.*

264.41 Sometimes I do wonder when the government will stop reviewing the railway industry and let it get on with the job.
> *Tom Winsor, following the announcement of a review of the industry, 21 January 2004, reported in* Rail, *4–17 February 2004.*

264.42 . . . the Cassandras will parade their ignorance, their vendettas and their prejudices.
> *Ibid.; his expectations of the review.*

264.43 He has chosen to replace a nut with a Bolt.
> *Roger Ford, of Alistair Darling's choice of Chris Bolt to replace retiring Tom Winsor, Rail Regulator; Railway Forum Innovation Awards, 29 March 2004.*

264.44 . . . the attempt to make a plastic carrier bag out of a sow's ear continues.
> *Roger Ford, of government's continuing involvement in railway organization; editorial comment,* Modern Railways, *July 2004.*

265. GRACE

265.1 For food, for friendship, for railways, we give thanks.

> *John Hope, Archbishop of York, twenty-first anniversary dinner of the Friends of the National Railway Museum, 16 May 1998.*

GRADE CROSSINGS
See Level Crossings

266. GRADIENTS

266.1 Some short time before this conveyance a new method was invented for carrying Coals to the River in large machines called Waggons made to run on Frames of Timber fixt in the Ground for that purpose and since called a Waggon Way which frames must of necessity lye very near, if not altogether upon a level from the Colliery to the River and therefore wherever there are any hills or vales between the colliery and the river and the same cannot be avoided, it is necessary in order to the laying of such waggon ways, then to make cutts through the Hill or level the same, and raise or fill up the Vales so that such Waggon Way may lye upon a level as near as possible.
> *Anon., counsel's opinion, 1763, upon a conveyance dated 1672, quoted in* Edgar Allen News, *1925 and repeated by Michael Lewis in* Early Wooden Railways, *1970.*

266.2 The resistance produced by steeper gradients can be compensated by slackening the speed, so that the power shall be relieved from as much atmospheric resistance as is equal to the increased resistance produced by the gravity of the plane which is ascended. And, on the other hand, in descending the plane the speed may be increased until the resistance of the atmosphere is increased to the same amount as that by which the train is relieved of resistance by the declivity down which it moves.
> *Dr Dionysius Lardner, of the alleged benefits of the undulating railway principle, in which steep gradients are permitted.* The Steam Engine, *1840, quoted by C F Dendy Marshall in* A History of the Southern Railway, *1936.*

266.3 If there is one thing the Utah Railway isn't, it's level.
> *Donald Sims,* Trains Magazine, *January 1956.*

266.4 Shorter + Steeper = Faster + Cheaper.
> *David P Morgan, title of article,* Trains Magazine, *June 1974.*

266.5 From an engineer's perspective, uphill is easy, and even boring. It's mostly a matter of putting the throttle in Run 8, pouring a cup of coffee, and watching the signals slowly appear for the next hour. Coming downhill—now that's the test.
> *Ed King,* Trains Magazine, *April 2004.*

267. GRAND CENTRAL TERMINAL

267.1 They had another moment of rich silence when they paused in the gallery that leads from the elevated station to the waiting rooms in the Central Depot and looked down upon the great night trains lying on the tracks dim under the rain of gas-lights that starred without dispersing the vast darkness of the place. What forces, what fates, slept in those bulks which would soon be hurling themselves north and east and west through the night! Now they waited there like fabled monsters of Arab story ready for the magician's touch, tractable, reckless, will-less—organized lifelessness full of strange semblance of life.

> *William Dean Howells, of the predecessor station to the present Grand Central,* A Hazard of New Fortunes, *1889.*

267.2 Crossroads of a million lives! Gigantic stage on which are played a thousand dramas daily!

> *Grand Central Station radio show, New York, which was clearly based upon Grand Central* Terminal *but which used sounds of steam locomotives in presentations. Quoted in* Locomotive & Railway Preservation, *September/October 1993.*

268. THE *GREAT BEAR*

268.1 Why did that young man want to build a Pacific? We could have sold him ours!

> *G J Churchward of the Great Western Railway, of Nigel Gresley and the failure of Churchward's Pacific to live up to expectations. Quoted by F A S Brown in* Nigel Gresley, Locomotive Engineer, *1961.*

269. GREAT CENTRAL RAILWAY

269.1 There's a flutter at St. Pancras, there's a bustle at King's Cross,
They are posting bills and tearing timesheets down,
There's uneasiness at Euston in the office of the "boss"—
For the Central is a-coming up to Town.

> *A B S, first verse of* The New Comer, *within* The Railway Magazine, *vol. IV, 1899.*

269.2 . . . a line for the conveyance not only of passengers, but of coal, manure, fish, and other abominations.

> *Petition by the artistic community of St Johns Wood, London, against the construction of the London Extension. Quoted by George Dow in* Great Central, *vol. II, 1962.*

269.3 . . . a belated, and almost entirely superfluous, product of the original era of fighting construction.

> *Sir John Clapham, of the London Extension, quoted by George Dow in* Great Central, *vol. II, 1962.*

269.4 The Great Central will be in receivership before the year is out. I know. I am their banker.

> *Col Robert Williams, who in the event was wrong, to Sam Fay, 1902.*

269.5 Rapid Travel in Luxury.

> *Advertising slogan, 1904.*

269.6 It goes to places that do not need a railway, that never use a railway, that probably do not yet know that they have a railway. It goes to way-side halts where the only passengers are milk-churns. It visits lonely platforms where the only tickets are bought by geese and ducks. It stops in the middle of buttercup meadows to pick up eggs and flowers. It glides past the great pile of willow branches that are maturing to make England's cricket bats. It is a dreamer among railways, a poet, kindly, absurd and lovely.

> *A G Macdonnell,* England, Their England, *1933.*

269.7 The genesis of the Great Central Railway can be traced to the need for better communication between Manchester and Sheffield than that provided by the canal system which, hampered by the physical barrier of the Pennines, circuitously linked the two places via Ashton-under-Lyne, Huddersfield, Wakefield, Barnsley and Rotherham and involved goods in a journey time of eight days.

> *George Dow, opening words of chapter 1, vol. I, Great Central, 1959.*

269.8 In truth was Watkin's legacy paid for on the h.p.!

> *George Dow, of the use of a financing company ("hire purchase") to fund the acquisition of locomotives and rolling stock needed by the Great Central for the London Extension.* Great Central, *vol. II, 1962.*

269.9 But the MS&L—the Railway Flirt—was no more. The Great Central—the Becky Sharpe, poor and fated to live on its wits—was now controlled by men who knew how to keep their word.

> *Ibid.*

269.10 What a line that was for romance! On one of those warm still evenings which March seldom fails to bring to the Pennines, the western sky would still have its mantle of smoky pink spread above the hills which form the backbone of England, whilst in the valley, in the eastward dusk, a steady procession of Great Central coal trains would be blasting their way, one after another, up to Woodhead tunnel, the deep-throated bellow of the 2-8-0s booming in the crannies and crevices of Wharncliffe Crags, whilst occasional shafts of light from an open firebox door would pick out the white trunks of the forest of white birches alongside the line. As one train passed out of earshot round the curve to Thurgoland another would be developing a hearty crescendo in the woods by Oughty Bridge; and so it went on, hour by hour.

> *C B Harley,* Railway World, *September 1964.*

269.11 Years later, in the dark nights of 1940 and 1941, when we stood on the edge of defeat, when all we valued was thrown in the melting-pot and all our hopes seemed to be dashed, Home Guards on their nightly duties would listen to the same old sound from the same old engines, the G. C. Consolidations pounding up the bank as of yore, as if all was well, as if the nightmare had vanished, the one sane reassuring note in a mad and hostile world. These were the engines which helped us to win the first world war and, with a bit of luck, would see us through to win the second. Good old Great Central! Peace to its ashes!

Ibid.

270. GREAT EASTERN RAILWAY

270.1 The Poor Man's Line.

Anon., quoted by C Hamilton Ellis in Nineteenth Century Railway Carriages, *1949.*

270.2 The Great Eastern is especially the workmen's London railway—the one above all others which appears to welcome him as a desirable customer, whose requirements it accordingly makes the subject of special study and provision to an extent and in a variety of ways that no other London line seems to do.

London County Council report 1892, quoted in Passenger Class Distinctions, *Charles E Lee, 1946.*

271. GREAT NORTHERN RAILWAY
(England)

271.1 The Great Northern Railway ends in a ploughed field four miles north of Doncaster.

Edmund Denison, quoted by W J Gordon, Our Home Railways, *1910.*

271.2 . . . you could not work on a Great Northern engine unless you had flat feet.

Richard Hardy, summarising the view of Herbert Harrison, Steam in the Blood, *1971.*

272. GREAT NORTHERN RAILWAY (U.S.)

272.1 Most men who have really lived have had, in some shape, their great adventure. This railway is mine. I feel that a labor and a service so called into being, touching at so many points the lives of so many millions with its ability to serve the country, and its firmly established credit and reputation, will be the best evidence of its permanent value and that it no longer depends upon the life or labor of any single individual.

James J Hill, closing words of valedictory address to stockholders, 1 July 1912. Quoted in The Railway Library, *1912, 1913.*

272.2 The Short—Clean—Cinderless—Scenic Route.

Advertising slogan, 1930s.

273. GREAT WESTERN RAILWAY

273.1 Why not make it longer, and have a steamboat go from Bristol to New York and call it the Great Western?

Attributed to I K Brunel, of the Great Western Railway, at a meeting with Great Western Railway directors, 1853, by L T C Rolt in Isambard Kingdom Brunel, *1957. Quoted by Adrian Vaughan in* Isambard Kingdom Brunel: Engineering Knight-Errant, *1991, as "Why not make it longer? Build a steamship to go to New York and call it the Great Western."*

273.2 The broad gauge is not the only peculiarity of the Great Western Railway Company. The whole of its management is peculiar. In no railway is there so little of the mercantile element, and so much of the political element, as in the Great Western Railway.

"£.s.d.," author of The Broad Gauge. The Bane of the Great Western Railway, *1846.*

273.3 I do not know of any Company so difficult to come to an agreement with as the Great Western Railway, or one which keeps so honourably to an agreement when once made.

Captain Mark Huish, evidence before Commons Committee on South Wales Railway Bill, 1861.

273.4 The Great Way Round.

Anon., origin obscure, referring to the original route from London to Bristol via Chippenham and Bath, quoted in E Foxwell and T C Farrer, Express Trains English and Foreign, *1889, some years before the more direct route via Badminton was proposed.*

273.5 The Great Western is a very solid line, and makes it progress in a solid style: doing some great things and many small, but all alike with the immovability of Jove.

E Foxwell and T C Farrer, Express Trains English and Foreign, *1889.*

273.6 Unprogressiveness seemed the distinguishing mark of Paddington at that time!

George P Neele, of the GWR's failure to agree to excursion arrangements for the Exhibition of 1862; Railway Reminiscences, *1904.*

273.7 God's Wonderful Railway.

Anon., stated by some to have been conceived in sarcasm but subsequently adopted by admirers of the railway as a plaudit. Date of origin not known.

273.8 There was for many years a deep-rooted idea in Paddington minds that it was utterly impossible for a standard gauge train to run as fast as a broad gauge train, and the authorities could not for a long time be brought to reduce the 95 or 97 min. allowance for the Paddington-Swindon runs.

E L Ahrons, The Railway Magazine, *March 1916.*

273.9 There may have been another reason lurking in the Paddington mind. All trains except the "Dutchman" and "Zulu" conveyed third class passengers, and this type of being was not *persona grata* with the Great Western Rail-

way. Consequently to convey him from Swindon to London at first class speed was a thing not to be thought of under any consideration.
Ibid.

273.10 The Great Western is the line. So smooth. So polite. West Country politeness. So fast.
Arnold Bennett, letter 13 June 1924.

273.11 The old Great Western Railway shakes
The old Great Western Railway spins
The old Great Western Railway makes
Me very sorry for my sins.
John Betjeman, poem, Distant View of a Provincial Town.

273.12 In the amalgamation of great railway systems some years ago the Great Western alone kept its identity. It is still the Great Western; no muddling outsiders have come to introduce damnable improvements, economies, and modernizations. It is conspicuously more the Great Western than ever.
Arnold Bennett, The Railway Gazette, 15 October 1926.

273.13 . . . so that it took the Great Western two years to make up its mind: that being so, I am not surprised that it took the remaining Companies four.
Felix Pole, letter 1933, commenting upon the Registered Transit scheme for encouraging traders to send consignments by railway. Reproduced in Felix Pole: His Book, 1954.

273.14 Until the turn of the twentieth century the Great Western looked forward. Thereafter it looked Churchward.
Anon. A reference to the lack of progress in locomotive design after the retirement of G J Churchward, Chief Mechanical Engineer 1902–21.

273.15 . . . the Great Western company had its eye on the Bristol to Birmingham route, but its overtures were ponderous, and its financial inducements discouraging.
C Hamilton Ellis, The Midland Railway, 1954.

273.16 It was better at producing an epic than a sonnet.
C Hamilton Ellis, British Railway History, vol. II, 1959.

273.17 Now, remember this; wherever you may go, or whatever you may do, always stick up for the Great Western.
George Churchward, farewell words to Henry Holcroft upon his leaving Swindon works for the South Eastern & Chatham Railway, April 1914. Railway Adventure, 1962.

273.18 Left hand, right hand, each other never knew,
But to the dear old Broad Gauge, we'll ever still be true.
Anon., described as "Opening bars of GWR March" *by Derek Barrie in letter to George Dow, April 1978.*

273.19 Trains tended to drowse amid gentle hissings at countrified junctions—six of them in some forty miles

along the main line between Carmarthen and Fishguard —while patient passengers savoured that vintage aroma of fish, milk, damp Welsh coal and warm engine-oil which some mystique made peculiarly Great Western, like the monogrammed 'GWR' gold-braided cap which on state occasions following nationalisation appeared from nowhere on the heads of senior inspectors, in silent defiance of the latest Marylebonian phillipic about the 'standard uniform'.
Derek Barrie, A Regional History of the Railways of Great Britain, vol. 12, 1980.

273.20 Many people have not noticed how appropriate it was that the Great Western Railway museum was established in a Non-Conformist chapel.
George Dow, comment to his son, the compiler, c. 1980.

273.21 Paddington speaks only to Swindon, and Swindon speaks only to God.
Anon., of an attitude described by Michael Bonavia, The Four Great Railways, 1980.

273.22 At Paddington no man can denigrate the company any more than he can speak sharply about Brunel.
Gerard Fiennes, Fiennes on Rails, 1986.

273.23 The Great Western Railway of England shared with the Pennsylvania Railroad—the Standard Railroad of the World—an absolute inability to do anything in the fashion of any other railway if human ingenuity could rise to an alternative.
George W Hilton, book review, Railroad History, No. 153, *Autumn 1985.*

273.24 Broad Gauge Bastards.
Collective name for ex–Great Western Railway staff, coined in early days of Nationalisation by staff from other regions who tried to get their cooperation. Noted by George Dow in a letter to Railroad History, No. 154, *Spring 1986.*

273.25 It would have been correct for the ancients to believe that Swindon was the centre of the universe for it was the home of the Great Western Railway around which, of course, all railways revolved.
Tony Atkins, letter to The Times, *8 August 1998.*

274. SIR NIGEL GRESLEY
(man and locomotive)

274.1 I present to you a designing man—the man of the articulated bogie and the funnelless railway engine.
Professor J L Stocks, University of Manchester, upon conferment of honorary degree of Doctor of Science upon Nigel Gresley, 20 May 1936.

274.2 But in him science is paired with imagination; he has also something of the artist. When the engine-driver has gone the way of the highwayman, and the railway has become a romantic memory, the museums of the world

will compete for his masterpieces, and the silver ghost of his Jubilee train will vie with Dick Turpin for pride of place in the storied memories of the road to York.
    Ibid.

274.3 And so the hands upon the clock moved on,
    Counting the anxious minutes one by one,
    And now and then a momentary stir
    Would sway the watchers as a harbinger,

Till, in what seemed the lethargy of Time,
    There sudden sounded an authentic 'chime',
    And, on the instant, every lens and eye
    Was focused on that tense proximity.

And still we felt it could not really be
    That in another moment we should see,
    Surge from behind the angle of that wall
    That which seemed no longer terrestrial.

Then instantly, it shone there in our view,
    A kingly Herald, clad in black and blue,
    With silver blazonings and crimson feet,
    And all a shimmer in the summer heat.

Strange as some splendid visitant from Space,
    It moved towards us with a royal grace,
    Accepting homage but demanding none,
    Servant and Lord alike of everyone.
        Henry Maxwell, poem, Mammoth Preserved, *verses eleven to fifteen, of the locomotive when first preserved in garter blue, seen at Newcastle Central, in* A Railway Rubaiyyat, *1968.*

274.4 Many people think that the initials G W R stand for Great Western Railway. They don't. They stand for Gresley Was Right.
        Head of National Railway Museum in conversation recalled by Julian Riddick, 1993.

275. GRIMSBY

275.1 It's a fine dock, fifty acre if it's an inch. But it's not big enough for Sir Edward yonder, smiling at t' Princess. If truth were known, he'd like to make a dock of the North Sea, and put a goods warehouse on the Dogger.
        Grimsby fishing skipper, 22 July 1879, referring to Sir Edward Watkin of the MS&LR, which developed the port (talking to the Princess of Wales), quoted by J Pendleton, Our Railways, *1896.*

276. GROUPING OF RAILWAYS
See also Mergers

276.1 We believe the safest and surest way out of the present impasse is to restore the railways to their *status quo ante bellum* and afford them facilities, denied them in the past, for forming amalgamations. The railways of this country are quite capable of working out their own sal-

vation and their operations should not be retarded by the dead hand of officialdom.
        Editorial Comment, Modern Transport, *25 September 1920.*

276.2 The great risk of too big amalgamations is that general headquarters may get out of touch with local communities and the possibilities of development. There is a tendency toward Civil Service standards—correctness, not improvement.
        Comment, The Railway Gazette, *1 April 1921.*

276.3 Sir J D Rees: Is it a fact that there has been some reconsideration of this lateral grouping in favour of the longitudinal principle?
Sir E Geddes: It is a little difficult to say what is lateral and what is longitudinal.
        Exchange in Parliamentary Questions before the debate on the Railways Bill, 1921, Hansard, 11 April 1921.

276.4 As I listened to the Minister of Transport speaking on behalf of the proposals of the Bill, I could not get away from the impression of a grown man, knowing a great deal about railways, going back to the nursery and playing trains.
        G Balfour, M P, Commons debate on the Railways Bill, 1921, Hansard, 30 May 1921.

276.5 The assumption of the Bill is that men who have risen to a great position in the control of industry are weak and are of no ability, or even possibly dishonest; but this bureaucratic Ministry will provide the brain, will provide the initiative, will provide the honesty and, forsooth, by magic will guide and control and inspire great enterprises which cannot be done by the tested, proved ability of men who have built up those great industries.
        Ibid., of the creation of the Ministry of Transport.

276.6 Hooray! Never even blew me cap off!
        Dick German, cartoon caption, in a Cardiff newspaper, 1921, commenting on the fact that the Great Western Railway had managed to ensure that it kept its corporate identity in the Railways Bill 1921. The cartoon showed a porter rejoicing. Reproduced by Felix Pole in His Book, *1954.*

276.7 A sort of bastard nationalization.
        Sir Frederick Banbury, of the grouping proposals in the Railways Bill; House of Commons debate, recorded by Howard C Kidd, A New Era for British Railways, *1929.*

277. GUARDS and CONDUCTORS

277.1 He is expected to keep up a friendly and harmonious intercourse with the enginemen; but should occasion require it, no false delicacy is to intimidate him from reporting any want of cooperation on the part of that or any other individual to the master of transportation.
        First book of Rules, Baltimore & Ohio Railroad, 1827.

277.2 An American conductor is a nondescript being, half clerk, half guard, with a dash of the gentleman. He is generally well dressed, sometimes wears a beard, and when off duty passes for a respectable personage at any of the hotels and may be seen lounging about in the best company with a fashionable wife. No one would be surprised to find that he is a colonel in the militia. At all events, the conductor would need to be a person of some integrity, for the check upon his transactions is infinitesimally small.

William Chambers, Things as They Are in America, *1854, quoted by Richard Reinhardt in* Workin' on the Railroad, *1970.*

277.3 All he says is "Ticket!" and he utters the word in a dry, callous tone, as if it would cost something to be cheerful.

Ibid.

277.4 . . . it is obvious that the conductor should be a man combining fidelity with courteous manners, general intelligence, and sound practical sense; careful to avoid danger, and skilful to extricate his train in case of difficulty; he must be unknown at saloons of dissipation—an indulgence which would inevitably impair his character for fidelity of trust, and should be carefully avoided.

John B Jervis, Railway Property, *1866.*

277.5 Larry Donovan was a passenger conductor, one of those train-crew aristocrats who are always afraid that some one may ask them to put up a car window, and who, if requested to perform such a menial service, silently point to the button that calls the porter. Larry wore this air of official aloofness even on the street, where there were no car-windows to compromise his dignity.

Willa Cather, My Antonia, *1918.*

277.6 With the wisdom of experience he realises that he exists for two objects, one being to take care of the luggage, and the other to father the oft-times childish travelling public. He stands as regards the latter in a somewhat similar position to what is known at Army courts-martial as "Prisoners' Friend." Moreover, unlike the porter, he expects no fee for his advice and labour, and when such does come his way he is correspondingly grateful.

John Aye, Railway Humour, *1931.*

277.7 Let me sit on the right hand side,
A-hold of the throttle and Johnson bar,
And make our rough wild hogger ride

At the other end in the old way car.
Just watch him try to stay in the hack
When I start the train with the air and the slack,
And hear him holler when his head I drove
Right in behind the crummy stove,
And as the train goes into the sag,
Knock him out with the slack in that drag.
And let him lay there on the floor,
Afraid to stand up, for fear he'd get more.
I'd head right in on the longest track,
And cut her off a long way back;
When the hogger walks in, to him I'd say:
"Well, how was the ride you got today?"

B H Terry, poem Put the Hogger in the Crummy, *in* Railroad Magazine, *December 1941.*

277.8 I'm not allowed to run the train;
The whistle I can't blow.
I'm not even allowed to say
how far the train can go.
I'm not allowed to shovel coal,
No'r e'en ring the bell.
But let the damn thing jump the track,
 and see who catches hell!

Anon., poem, The Poor Conductor, *found on the internet at www2pb.ip-soft.net/railinfo/poems/poetry.html, without attribution.*

277.9 The Guard is a man, who sits in a van,
The van at the back of the train,
The driver, up front, says the Guard is a ****,
And the guard says the driver's the same!

Anon., poem, remembered by Nigel Harris, and thought to be many years old.

278. GUAYAQUIL & QUITO RAILWAY

278.1 Oh Collectors, oh Inspectors, hear the travelling
 public roar,
And when we are gone forever, they'll forget us
 nevermore!
Oh the happy days we passed there,
On numbers one, two, three and four,
Chuckin' drunks and checkin' chickens
On the trains in Ecuador.

Anon., song to be sung to the tune of Clementine, *recounted by John L McIntyre, who joined the G&Q from the Caledonian Railway in 1910, in* Railways of the Andes, *Brian Fawcett, 1963.*

Hh

279. TIMOTHY HACKWORTH

279.1 It appears that no single individual in this country had, up till the year 1830, done so much for the improvement of the locomotive, and for its establishment as a permanent railway motor, as Mr Timothy Hackworth.
Daniel Clark, Railway Machinery, 1860.

279.2 Timothy Hackworth is original, is of himself improving the locomotive in essentials as no other man is doing, and is incomparably in advance of George Stephenson in everything which may be truly said to lay claim to distinction. He has and is stamping a character upon the structure of the locomotive of the very highest importance, judging from the practical results following his efforts.
J G Pangborn, The World's Railway, 1894. This book was written in the present historic tense; in spite of appearances this was not a contemporary comment and should instead be regarded as an appreciation several years after Hackworth's death.

280. PETER HANDFORD

280.1 The Man the Engines Talk To.
John Gale, title of article in The Observer, about Britain's leading railway sound recording expert.

281. RICHARD HARDY

281.1 Richard Hardy, that determined stoker of Chapelon Pacifics between Calais and Paris and in his spare time Divisional Manager at Kings Cross, has once every six months a meeting of his staff representatives for a "State of the Union" message from him. He does this very well indeed.
Gerard Fiennes, I Tried to Run a Railway, 1967.

282. EDWARD HARRIMAN

282.1 Harriman was intense and combative—the forerunner of an age when speed and efficiency would replace grace and charm.
Maury Klein, The Life and Legend of E H Harriman, 2000.

283. FRED HARVEY

283.1 Fred Harvey attained a position as leader in the art of serving the railway traveler inner man akin to that Webb C Ball established for himself as the pioneer and pace maker in the work of creating, maintaining and inspecting railroad watches.
John A Droege, Passenger Trains and Terminals, 1916.

283.2 Fred Harvey dining service is a distinctive feature of this distinctive railway.
Legend on posters, Atchison Topeka & Santa Fe Railway, 1929.

284. HARWICH

284.1 Harwich for the Continent.
London & North Eastern Railway poster, c. 1930. This slogan is known to have been followed, by music hall comics, with the statement Southend for the incontinent.

285. F W HAWKESWORTH

285.1 Hawkesworth was a true son of Swindon, a master of uncommunication.
E S Cox, Locomotive Panorama, 1965.

286. HAYLING

286.1 Here is the train so very late
That we poor travellers have long to wait;
So we write on the wall, with weary finger,—
" 'Hayling' in future should be called 'Haylinger'."
L J M de Michele, poem On the Wall at Hayling Station, After Waiting an Hour for a Train, within The Poetry of Railways, ed. Kenneth Hopkins, 1966.

287. HEADQUARTERS

287.1 Close by those reeds, of Mimram's oozing stream,
L.N.E.R. has well preserved its cream.
There, stands a structure of majestic frame,
Which from the neighb'ring Hampden takes its fame.
Here Railway Rulers oft the fate decree
Of Specials, and the likes of you and me.
Here thou Great Newton! whom three Realms respect,
Doth sometimes counsel take—sometimes reject.
Hither the typists and the clerks resort,
To plan awhile the torments of transport.
"Buckeye", poem Hampden Court, first of three verses. Buckeye was a member of staff at the wartime headquarters of the London & North Eastern Railway at The Hoo, Whitwell, owned by the Hampden family. The reference to Newton is Charles Newton, Chief General Manager of the LNER, under whom three Divisional Managers served.

Published "A long way after Alexander Pope" in the Souvenir Programme *for the Farewell Party for the Hoo, 2 November 1945.*

287.2 At lunchtime on that same day a somewhat bewildered group of people from the different departments assembled in the senior officers' mess, where they were joined for a drink by Sir William Slim, as he then was, whose wartime experiences in the Burma Jungle had not prepared him for discovering his route to the Members' Mess in this complex and rambling building.

E S Cox, of New Year's Day, 1948, the first day of British Railways' existence at their new Headquarters. Sir William Slim was Deputy Chairman. Locomotive Panorama vol. 2, *1966.*

287.3 . . . a refreshingly vulgar, purse-proud building.

Sir John Betjeman, of 222 Marylebone Road, quoted in Modern Railways, *January 1987.*

288. HEALTH

288.1 I am refreshed and expanded when the freight train rattles past me . . .

Henry David Thoreau, Walden, *1854.*

288.2 To the jaded and the toilworn, therefore, we would say, close your books, leave the desk, fly the study, hasten to the nearest railway station, and take a return ticket for twelve or fifteen miles. On arriving at your destination, scud down the green lanes and across the fields, setting out at a brisk pace and maintaining it until you return to the departure station in a couple of hours' time. The journey homeward will appropriately cap the achievement; which, although we do not vaunt as a panacea for all the ills of life, we nevertheless declare, from experiences to be one of the very best repairers of health and restorers of spirits.

Anon., The Railway Traveller's Handy Book, *1862.*

288.3 Weak, tall, loosely-knit persons, and those suffering under various affections, more especially of the head, heart, and lungs, are very unsuited for habitual railway travelling.

Dr Walter Lewis, medical officer of the London Post Office, quoted in Good Things for Railway Readers, *1863.*

288.4 I can assure you that the draughts during the winter months are enough to kill a bronze rhinoceros.

Passenger writing to the Metropolitan Railway, to complain of the facilities at Notting Hill Gate station, January 1880. Quoted in London's Metropolitan Railway, *Alan A Jackson, 1986.*

288.5 A journey from King's Cross to Baker Street is a form of mild torture, which no person would undergo if he could conveniently help it. Passengers have been consoled by the assurance that self-asphyxiation by sul-

phurous fumes is not an injurious thing even for the asthmatic; but this is a point on which coughing sufferers cannot be expected to agree with railway directors.

Editorial comment, The Times, *7 October 1884.*

288.6 The fast express trains which the Midland Railway have put on to "compete" with the London & North Western service to Manchester may be strongly recommended to those who have the misfortune to have a sluggish liver, and want it agitated.

Comment, Railway Engineer, *vol. 23, 1902.*

288.7 No liver can get in at Putney and remain sluggish beyond Walham Green.

Owen Seaman, of the District Line on the London Underground, Punch, *1906.*

288.8 . . . the deceased died from cold and exposure from travelling in a second-class carriage on the Great Western Railway.

Coroner's verdict recorded in History of the Great Western Railway, *MacDermot, 1927.*

288.9 Hike for Health.

Southern Railway poster, c. 1930.

288.10 I would feel better if I didn't have to spend most of my life on trains.

Collis P Huntington, who at the time was spending four nights a week traveling between his various business interests. Quoted by Oscar Lewis, The Big Four, *1938.*

288.11 Health is the reason for the travel of many. To escape the rigor of a northern winter and to enjoy the health rays from the semi-tropical sun is worth much to many. Any normal person would rather pay for a railroad ticket than a doctors fee.

Clyde H Freed, The Story of Railroad Passenger Fares, *1942.*

288.12 Another big problem of the daily train rider is known as Commuter's Elbow. This has nothing to do with drinking or exercise. The problem arises in railroad coaches where two people, usually strangers, have no armrest between them. After a while one becomes increasingly aware that the other is moving an arm and bumping against his elbow. This is distracting because it can mean (1) the other fellow is occupying more than his fair share of the double seat, or (2) someone is annoying someone and it may lead to words.

Jerome Beatty Jr, Show Me the Way to go Home, *1959.*

288.13 I never felt the need for a work out at the health club after a day at the Spring Grove Avenue freight house.

John H White, Railroad History, *No 141, 1979.*

288.14 Arrival of the Fittest.

Great North Eastern Railway poster, Peterborough station, June 2000.

288.15 Those concerned about the development of deep-vein thrombosis on delayed train and Tube journeys can immediately decrease the risk by giving up their seat to the person standing nearest to them.

Liz Murray, letter to The Times, *23 January 2001.*

289. HEAVEN

289.1 Last night I lay a-sleeping
there came a dream to me—
I stood within a Steam Shed,
A marvellous Shed to see.
The walls were clean and spotless,
And smoke troughs white as snow,
And not a spot of grease was seen
Upon the pits below.
O Loco Men, O Loco Men,
Shout loud for well you may—
'Twas the Blessed Steam Shed of Paradise
We all shall see some day.

John Bowen, The Crewe Steam Shed's Song, *first verse, sung to the tune of* The Holy City, *probably 1896.*

289.2 A man stood at the pearly gates,
His face was worn and old
And meekly asked the man of fate,
Admission to the 'fold'.
"What deed can you account for
To gain admission here?"
"Why I worked at Eveleigh Loco
Until my dying year."
The gate swung open sharp,
As St. Peter touched the bell,
"Come in," he said "and take a Harp,
You've had enough of 'Hell'. . .

Anon. poem, published in Eveleigh News, *the news sheet of the Eveleigh Locomotive Shop Committee, New South Wales, 1954.*

289.3 Of Jimmies in a blast pipe
No engineman did dream,
For every engine in that Shed
Was guaranteed to steam.
There were no Washers-out there
No boilers e'er did prime,
And with a load of thirty-six
Each engine did keep time.
O Enginemen, O Enginemen:
'Twas the Blessed Steam Shed of Paradise
You'll go there when you die.

Ibid., verse three.

289.4 God a great railway to heaven has planned.
He staked out the line with His dear, loving hand;
Away back in Eden the grant was first given,
On Calvary's cross the last spike was driven.
The road was surveyed with a special design,

To make it a practical Holiness line;
The grade was thrown up with the greatest of care,
Directly through Canaan, a country most fair.

Anon., The Beulah Railway, *first verse, quoted by Robert Hedin,* The Great Machines, *1996.*

289.5 No accident has this railway yet known,
The Dispatcher is He who sits on the throne.
Trains only move at Jehovah's command,
He holds the throttle with Omnipotent hand.
The Holy Spirit is the headlight so clear,
Revealing the track to the wise engineer.
The angels are brakesmen, so kind and urbane,
Adding much to the comfort of all on the train.

Ibid., verse seven.

290. HELL

290.1 A Reward for Sabbath Breaking. People taken safely and swiftly to Hell, next Lord's Day, by the Carlisle Railway for 7s. 6d.

Sabbatarian handbill, August 1841, under the name of William C Burns, reproduced in The Railway Magazine, *February 1923. The railway referred to was the Newcastle & Carlisle Railway.*

290.2 There is a railway downward laid,
Which God the Father never made,
But it was laid when Adam fell—
What numbers it conveys to Hell.

Six thousand years are nearly gone,
Since first this railway was begun,
The road is wide, and smooth, and gay,
And there are stations on the way.

Song, The Down Line, *anon., a parody on* The Spiritual Railway, *quoted in* Long Steel Rail, *Norm Cohen, 1990.*

290.3 I had my first experience of Hades today, and if the real thing is to be like that I shall never again do anything wrong. I got into the Underground railway at Baker Street.

R D Blumenfeld, R D B's Diary, *1887–1914, 1930. R D Blumenfeld's son, Sir John Elliot, was Chairman of London Transport many years later.*

291. JOHN HENRY

291.1 Now John Henry told his little woman,
Goin' to cook my supper soon,
I got ninety miles of track to line right tonight,
Goin' to line 'em by the light of the Moon,
Oh Lord!,
Goin' to line 'em by the light of the Moon.

Song, John Henry, *as sung by Lonnie Donegan. It bears little resemblance to the many versions recorded in* Long Steel Rail, *Norm Cohen, 1990.*

292. GEORGE HILTON

292.1 It could only have been done by a person who was at once a competent scholar, a serious railfan, a historian with a broad and a narrow sense of perspective, and a writer able to communicate simultaneously with the professional and amateur alike. And Professor Hilton is just that.
J William Vigrass, in review of The Cable Car in America, *by Hilton, RLHS Bulletin 126, April 1972.*

293. HISTORY—RAILWAY
See also Dedication, Writing about Railways

293.1 Well then—Had pilgrimage been made by rails
In Chaucer's days, his Canterbury Tales
Had ne'er been written—in bluff Harry's age,
Had people gone by broad or narrow gauge,
The gorgeous story of the field of gold,
Where France and England met, had ne'er been told;
The Monarch might well have whirled to Paris one day,
(Perhaps, economizing time, on Sunday,)
Ate his ragout and hurried back on Monday.
Basil Montague, with astonishing prescience of what speedy travel would bring; Railroad Eclogues, *1846.*

293.2 In few fields of investigation are the sources of information so widely scattered as that of transportation.
Dr Howard Douglas Dozier, A History of the Atlantic Coast Line Railroad, *1920.*

293.3 For its true history was always the history of transportation, in which the names of railroad presidents are more significant than those of Presidents of the United States. Those names emerged—Gould, Vanderbilt, Hill, Huntington and Harriman.
Philip Guedalla, of the history of the United States. The Hundred Years, *1936.*

293.4 The history of a railway is like that of the life of a philosopher—rich in important results but poor in interesting narrative.
Drake's Road Book, quoted in Old Euston, *published by* Country Life, *1938.*

293.5 Any big book on railroading published over the joint byline of Lucius Beebe and Charles Clegg is bound to be rich as chocolate fudge cake—rich in illustration, rich in text, and, understandably, rich in price.
David P Morgan, opening paragraph of review of Hear the Train Blow, *in* Trains Magazine, *December 1952.*

293.6 It is a startling thing about the written history of English railways that so much of it is not quite right.
Michael Robbins, Author's Note in The Isle of Wight Railways, *1953.*

293.7 Its story has been engrossed for perpetuity in a book by M C Poor of 500 quarto pages, embracing well

over half-a million words of letterpress together with half-tones, maps, graphs, statistical tables, and an index and other critical apparatus that would do credit to the mythology of the Holy Grail or the Arthurian Legend.
Lucius Beebe, of Denver South Park & Pacific, *first published 1949.* Narrow Gauge in the Rockies, *1958.*

293.8 Summed up: a lot of meat, but it should have been left in the oven longer.
David P Morgan, of a book reviewed in Trains Magazine, *January 1959.*

293.9 The railway is the beginning of all history in Kenya.
Attributed to Sir Edward Grigg, Governor of Kenya 1925–30, by Trains Magazine, *May 1961.*

293.10 If you find any skeletons in our closet, drag 'em out.
G B Aydelott, President of the Denver & Rio Grande Western, to Robert G Athearn, author of Rebel of the Rockies, *1962.*

293.11 No branch of history can exist in isolation: transport history least of all.
T C Barker, A History of London Transport, vol. 1, *1963.*

293.12 This book is not an inspired court portrait of a great corporate enterprise written with the approval of management for the satisfaction of its executives, stockholders or the descendants of the pioneer railroad moguls who laid the rails and built the fires under the crown sheets alike of its finances and its primeval teapot locomotives.
Still less is it available to classification as a "social document," the accepted description of a latter-day appraisal, usually in terms of envy and malice, of a generation of builders and doers and, like as not great rascals to boot, confected by experts at defamation of their betters attired in the black bombazine of professional scholarship.
Lucius Beebe, opening paragraphs of Foreword, The Central Pacific and the Southern Pacific Railroads, *1963.*

293.13 The prevailing interpretation of the influence of railroads on American economic growth during the nineteenth century is still dominated by hypotheses spawned during that era. This long rule has been facilitated by three features of historiography: first, the acceptance of certain propositions propounded during the Gilded Age as obvious truths that did not require critical examination; second, a tendency to analyze issues connected with the construction and operation of railroads within the conceptual framework established during the post-Civil War debates on railroad policy; and third, a failure to exploit the quantitative techniques and theoretical tools of modern economics.
Robert William Fogel, Railroads and American Economic Growth, *1964.*

293.14 While many distinguished treatises have been written on railroad history and related topics in recent decades, it is no slight to the importance of these works to point out that their central focus has not usually been analysis of the effects of railroads on economic growth.
Ibid.

293.15 We steam diehards don't know how much we don't know about our pet subject!
Robert Le Massena, Trains Magazine, *March 1966.*

293.16 The last train to Curtainsville.
Lucius Beebe, of his book The Trains We Rode, *expecting it to be his last. It was; he died before the second volume was published. Referred to in* Trains Magazine, *April 1966.*

293.17 The nature of man is that he never fully appreciates what is about him until it is either declining or gone. Not that one can believe railroads are in this state, but the years of transition have been sudden and far-reaching in their consequences. Who among us cannot find regrets in lost opportunities of the past, in not having ridden some railroad now abandoned, or passenger service that is no more; or does not become angry with himself for not photographing some little known steam locomotive that for a little extra effort was waiting to be pictured for posterity. But crying over lost opportunities is a poor pastime, though this reminds us to appreciate the present. What we see today is tomorrow's history.
James Plomer, The Golden Age of Railroad Books, *within* Bulletin 116, Railroad & Locomotive Historical Society, *April 1967.*

293.18 If times have been changing rapidly, we seem to be on the edge of bigger and faster changes. I believe rails and flanges will always be with us. They are as fundamental as the wheel itself, but who can guess future forms and styles.
Ibid.

293.19 Not in the same class as MacDermot on the GWR or Tomlinson on the NER, and not to be named in the same breath as Dow on the GCR, Grinling nevertheless is a minor railway classic.
Henry Stanton (pseudonym of Paul Drew), book reviewer, of Grinling's History of the Great Northern Railway, *new edition,* Modern Railways, *May 1967.*

293.20 Railroad history is a fragile commodity.
Robert L Frey, Railroad History, *No. 135, 1976.*

293.21 Proud past, bright future.
Slogan of Indiana Harbor Belt Railroad.

293.22 Railway history, I profoundly believe, involves a great deal more than "nuts and bolts." It can, and does, embrace the marketing of rail service, the competitive situation, the status and impact of regulation, (not to mention the political climate), the formulation of financial policy (increasingly a responsibility of railway law departments), the organization of medical services, the long-standing work of the agricultural and industrial development agencies—and countless other activities. Over all these undertakings preside the decision-making executives who formulate and integrate the company's strategy.
Richard C Overton, Trains Magazine, *February 1982.*

293.23 . . . who may fairly be described as the Edward Pease of railway bibliography . . .
George Ottley, of Professor Jack Simmons, Introduction, second edition, A Bibliography of British Railway History, *1983.*

293.24 The clamorous confusion of parochial loyalties that enliven but muddy railway history.
Charles Wilson, First with the News, *1985.*

293.25 *Best Friend of Charleston, Wabash Cannonball,*
Are legend now; but emblems, gold and sable,
Recall them: beaver, mountain goat, the tall
Sequoia or snowy Shasta, robed in fable,
War bonnets of the Salt Lake Road, or torrid
Many-rayed sunset, blazoning their desire
To wrestle the jeweled tiara from Time's forehead,
Set records: *Express* or *Limited, Special, Flyer.*

In Promontory, Utah, '69,
The golden spike uniting East and West
Was sledged in creosote: high iron's sign,
Past roundhouse, shoo-fly shunting, toward their vast
Envisioning. And the spike held, tight as true
Fingers that lock in love and won't undo.
John Frederick Nims, poem, Freight, *1989.*

293.26 Anyone reviewing railroading's past is likely to take the historian's lofty hindsight and see more or less logical successions of events marching toward inevitable conclusions. They'll note "secular trends," as the higher-grade historians call them, and clear causes and effects. As an amateur historian, I do these things too, and I'm constantly humbled by how unpredictable and perverse history is when you're slogging through it day by day. I'm also amazed and encouraged to see that it really operates in cycles—sort of.
Herbert H Harwood Jr, Trains Magazine, *January 1991.*

293.27 This book is an effort to resolve what I have come to think of as the Grand Paradox in the history of American material civilization. How was it that an innovation as clearly revolutionary as the steam railway in its potential to do so many things better and to do so many other things that could not be done at all without railroads came to be "despised and rejected" by a politically contentious people who made it the chief scapegoat of their discontents? Why did the making of basic business decisions become a matter for political rather than economic resolution on the railroads and few other places in American society? Why did a great people, watching the cost of transportation rise ever higher in real economic terms, fail to see that less economic modes and methods of hauling goods and people were flourishing under laws repressive of the railroads' efforts to adjust to new conditions? Why did we fail to heed the warnings of two generations of railroad leaders and, finally, another generation of transportation experts who were beginning to adjust to the new realities, until disaster was on the doorstep of Congress? What did it take to revive the faithful old servant, how complete has been his convalescence, and how well prepared is he for a future that depends upon his vigorous rebirth?

> *Albro Martin,* Railroads Triumphant, *1992.*

293.28 This book contains no general index. I make no apology for this; it would have added considerably to both cost and time.

> *T R Pearce, Foreword to* The Locomotives of the Stockton & Darlington Railway, *1994.*

293.29 One of the most neglected topics of locomotive history is the tender.

> *John White Jr,* American Locomotives—An Engineering History, *1997.*

293.30 John H White, Jr, has the habit of writing definitive works.

> *Comment,* Journal of American History, *quoted on dust-jacket of the second edition of John White's* American Locomotives—An Engineering History, *1997.*

293.31 It is customary for authors on this topic to retire to their studies with a stack of 15 or 20 previous books and emerge with one more.

> *Walter P Gray III, reviewing a new book on the construction of the CP/UP transcontinental railroad,* Trains Magazine, *December 2000.*

293.32 Since the Santa Fe was such a perfect railroad, it will be absent from most of the stories that follow. There has to be a problem for there to be a story, and the Santa Fe had few problems until the 1980s. That is quite a tribute to business success.

> *Richard Saunders, Jr,* Merging Lines, *2001.*

293.33 For example, he and I argued about how far the spread of fish and chips could be put down to the railway. There was fish and chips in Oldham in the 1860s; where did the fish come from, and how?

> *Michael Robbins, of discussions early in his friendship with Prof Jack Simmons,* The Impact of the Railway on Society in Britain, *(ed. Evans & Gough), 2003.*

293.34 ... 19th century railways are less history and more the first part of a sustained evolution.

> *Gordon Biddle,* Britain's Historic Railway Buildings, *2003.*

293.35 Too much railroad research has fooled with the railroad's artifacts—its locomotives and its stations—and too little with what it did for a living and what happened to that living.

> *Richard Saunders Jr,* Main Lines, *2003.*

294. HOBOES

294.1   Behind
My father's cannery works I used to see
Rail-squatters ranged in nomad raillery,
The ancient men—wifeless or runaway
Hobo-trekkers that forever search
An empire wilderness of freight and rails.
Each seemed a child, like me, on a loose perch,
Holding to childhood like some termless play.
John, Jake or Charley, hopping the slow freight
—Memphis to Tallahassee—riding the rods,
Blind fists of nothing, humpty-dumpty clods.

> *Hart Crane,* The River, *April 1930.*

294.2 Who is this fellow J B King,
Who writes his name on everything?
'J B King' on every wall,
On flat cars and boxcars tall,
Whether he does it for money or fun,
He sure is a scribbling son of a gun.

> *Anon., of the identity of the person whose name was to be found chalked on freight cars all over America: a railroad equivalent of Kilroy, quoted (with another version) in* A Treasury of Railroad Folklore, *B A Botkin and Alvin Harlow, 1953.*

294.3 Third boxcar, midnight train,
destination Bangor, Maine
Old worn out suit and shoes,
I don't pay no union dues
I smoke old stogies I have found,
short but not too big around
I'm a man of means by no means,
king of the road.

I know every engineer on every train
All of the children and all of their names
And every handout in every town,
And every lock that ain't locked when no-one's around.

> *Roger Miller, song,* King of the Road.

294.4 I was there when they opened the boxcar,
And found him stone dead on the floor,
Though thumbin' and bummin' was all of our trade,
No one had seen him before.
He wore the face of a stranger,
A lost and unseen in a crowd.
He looked so small as we carried him down,
Wrapped in a newspaper shroud.
The wind blows cold in Wyoming,

The stars shine clear and bright,
If you don't wake up tomorrow at all
I guess it's "Old buddy, goodnight."
Bruce "Utah" Phillips, song, Old Buddy Goodnight,
c. 1970.

295. HOMESICK THOUGHTS

295.1 I'm a-walking down the track,
I got tears in my eyes,
Tryin' to read a letter from my home.—
If that train runs right,
I'll be home tomorrow night,
'Cause I'm nine hundred miles from my home,—
An' I hate to hear that lonesome whistle blow.
Anon., song, 900 miles, *date unknown.*

296. HORNSEA

296.1 Lakeland by the Sea.
North Eastern Railway Poster, 1911.

297. HORSE RACING

297.1 The Jockey Club feels that a railway to Newmarket
will not only be a great convenience to parties anxious
to participate in the truly British sport of racing, but
will enable Members of Parliament to superintend a race
and run back to London in time for the same night's
debate.
*John Villers Shelley, of the Jockey Club, at the luncheon
following the cutting of the first sod of the Newmarket and
Chesterford Railway, 30 September 1846.*

298. HORSES

298.1 Behind the coal-waggons, on the last carriage of all,
a low truck, stood an old cart-horse quietly eating hay
out of a basket. The sagacious animal, thus left to himself, on a bare platform without side-rail or guard of any
description, displayed a consciousness of the danger of
jumping out, by the mode in which he cautiously rested
on his haunches, prepared by his attitude against the
sudden possible contingency of a halt.
Sir George Head, A Home Tour through the Manufacturing Districts and other parts of England, Scotland and
Ireland, *1835.*

298.2 . . . and for shunting carriages and waggons, and
other work at stations, no less than 619 horses, all of good
breeds, well-fed, intelligent, and *well-tutored*.
*Sir Cusack Roney, of the equipment of the London &
North Western Railway,* Rambles on Railways, *1868.*

298.3 How he conducted himself on the return journey I
never heard, though the G.W.R. made no claims for
damage to their property, so far as I know. It was his only
railway trip in the twenty-five years of his existence, and

no doubt remained the event of his sheltered and easy life
to its end.
A G Bradley, of his pony, Exmoor Memories, *1926.*

298.4 In late Victorian London the horse took revenge on
the locomotive.
Jack Simmons, referring to horse-drawn trams, The
Railway in Town and Country, 1830–1914, *1986.*

299. HORWICH WORKS

299.1 The stores there was located in the vaults beneath
the Erecting Shop and was reputed to have been formed
out of the materials discovered in cupboards and under
the benches during a drive undertaken by the
management.
T Lovatt Williams, The Railway Magazine, vol. 95, *1949.*

300. HOTBOXES

300.1 "A hotbox," began the Compound, picking and
choosing his words as thought they were coal, "a hotbox
is the penalty exacted from inexperience by haste.
Ahem!"
"Hotbox!" said the Jersey Suburban. "It's the price
you pay for going on the tear. It's years since I've had one.
It's a disease that don't attack shorthaulers, as a rule."
"We never have hotboxes on the Pennsylvania," said
the Consolidation. "They get 'em in New York—same as
nervous prostration."
"Ah, go home on a ferryboat," said the Mogul. "You
think because you use worse grades than our road'd
allow, you're a kind of Allegheny angel. . ."
*Rudyard Kipling, conversation between locomotives in
the roundhouse,* .007, *1898.*

301. HOTELS

301.1 One of the greatest comforts and luxuries in a good
railroad is a first-class hotel station where a good, well-
prepared meal can be enjoyed.
Correspondent, Cincinnati Enquirer, *1857.*

301.2 Twenty-two miles an hour, a splendid road, magnificent scenery—in fact, everything tip-top, save the
hotels. If you except the McKissock House at Sedalia,
which is really a first-class hotel—the hotels on the
Pacific railroad are literally whitewashed Mugby Junctions, built after the prevailing American style of hotel
architecture, which has been denominated by a surly
bachelor friend of ours, as "long barn style."
Henry Morton Stanley, 22 July 1867, republished in My
Early Travels and Adventures in America and Asia, vol. I,
1895.

301.3 And there's the Greenbrier, whose hushed opulence
makes the Chesapeake & Ohio coal drags seen from its
windows seem like animation in another world rather

than an operation directly responsible for the thick carpeting and faultless service.

David P Morgan, Trains Magazine, *June 1956.*

301.4 When in doubt, Bond always chose station hotels. They were adequate, there was plenty of room to park the car and it was better than even chances that the Buffet de la Gare would be excellent. And at the station one could hear the heartbeat of the town. The night-sounds of the trains were full of its tragedy and romance.

Ian Fleming, Goldfinger, *1959.*

302. HOWRAH STATION (Calcutta)

302.1 So Howrah Station looked like a fortified version of a vast circumlocution office, an impression that buying a ticket there only confirms.

Paul Theroux, The Great Railway Bazaar, *1975.*

303. GEORGE HUDSON

303.1 When Railways and railway shares were dark as night,
Men said that Hudson ruled, and all was right.

Sir Thomas Leggard, opening of the Filey branch of York & North Midland Railway, October 1846.

303.2 I find a burning disgust arising in my mind—a sort of morbid canker of the most frightful description—against Mister Hudson. His position seems to me to be such a monstrous one, and so illustrative of the breeches pocket side of the English character, that I can't bear it.

Charles Dickens, Letters, vol. IV.

303.3 Let him be thought of as one who, more sinned against than sinning, has been a scapegoat for the sins of the many; and let it, too, be considered that he has done the state some service, and may yet do it much more.

John Francis, History of the English Railway, *1851.*

303.4 There was a time when not to know him was to argue one's self unknown; now he is only a tradition.

Obituary, The Times, *16 December 1871.*

303.5 It is a great mistake to look back upon him as a speculator. He was a man of great discernment, possessing a great deal of courage and rich enterprise.

William Gladstone, quoted in The North Eastern Railway, *William Tomlinson, 1914.*

303.6 . . . a man who united largeness of view with wonderful speculative courage—the kind of man who leads the world.

Ibid.

303.7 George Hudson is perhaps the only business man of a hundred years ago whom we can easily imagine at home in the modern world. Other giants of the early stages of our Industrial Revolution pale into insignificance beside him in respect of personal energy and capacity for simultaneously directing many distinct and complex enterprises. The slow-moving world of 1845 was fascinated above all by the speed with which he worked.

Richard Lambert, The Railway King, *1934.*

303.8 Hudson was a man of vision, courage and enterprise, who went on and got things done while others knitted their brows and worried about the risks. His schemes provided employment for thousands of men and brought huge benefits to north-east England as well as facilitating greater freedom and movement for working people.

Brian Bailey, George Hudson, *1995.*

304. MARK HUISH

304.1 In fairness to him it can truthfully be said that he wore himself out in the service of the LNWR. It is equally true to say that the railway world was a better place without him.

George Dow, Great Central, vol. 1, *1959.*

305. HULL

305.1 Britain's third and cheapest port.

London & North Eastern Railway poster, c. 1930.

306. HUNGER

See also Eating on Trains

306.1 Are you sure that rumble is the next train approaching?

Advertisement for McDonald's, a fast food chain, seen at Paddington (Circle Line) station, Summer 1999.

307. HUNTING

307.1 Gentlemen, I detest railroads; nothing is more distasteful to me than to hear the echo of our hills reverberating with the noise of hissing railroad engines, running through the heart of our hunting country, and destroying that noble sport to which I have been accustomed from my childhood.

Anon. Cheltenham MP, c. 1840, quoted in Railway Adventures and Anecdotes, *1888.*

307.2 Railroads were denounced as detrimental to hunting; but they do far more harm by drawing the superfluous cash into London than by any impediments they present to the chase.

R S Surtees, Town & Country Papers, *1929.*

308. ILFRACOMBE

308.1 The point about Ilfracombe, surely, was that it had not been designed for cars. It was a classic railway resort, with tall hotels and sloping streets.
Paul Theroux, The Kingdom by the Sea, *1983.*

309. ILLINOIS CENTRAL RAILROAD

309.1 Main Line of Mid-America.
Advertising slogan.

309.2 A north-south railroad in an east-west world.
Comment, Trains Magazine, *December 1996.*

310. IMMATURITY

310.1 I arrived at King's Cross behind a B17. I remember giving the loco a cursory glance as I went to the barrier; I had other things on my mind. That's immaturity for you.
Barry Fleming, Musical Notes, *within the* Gresley Observer No 119, *Autumn 1999.*

311. INDUCEMENTS

311.1 However desirous one may be to deliver an impartial judgment, complimentary tickets, choice luncheons, and a variety and abundance of excellent wines, are apt to tinge a descriptive account with somewhat too much of the *couleur de rose.*
Editorial comment, The Railway Times, *1838, quoted in* Railway World vol. XXI, *1960.*

312. INFORMATION BOOTHS

312.1 Cried a Voice from the depths of a Booth
"My job is to guide and to soothe.
So whether I'm routing
To Barking or Tooting,
I'm smiling, I'm smart, and I'm smooth."
Anon., concerning the Information Booths established by railways in London during the Second World War to give advice to travellers at the time of air raids. LNER Magazine, *January 1941.*

313. INJECTORS

313.1 I remember committing what we regarded as a heinous sin, in trying to find out the secret of the injector. On a Sunday forenoon in a Scotch town, where it was considered next to a crime to miss going to church, three of us stole quietly into an engine-house and took an injector apart, when we knew the foreman, like a pious man, was listening to the sermon. All our plans had been laid the previous day, and the necessary tools laid conveniently at hand. The feeling of disappointment we experienced at finding nothing to explain the working of the thing is still vivid in my memory.
Angus Sinclair, Locomotive Engine Running and Management, *1886.*

314. INNOVATION

314.1 The small wheels, with their frame, work on the road the same as an independent wagon; and being geared short, they go round a curve with as much ease as a common wagon, and being leaders, they bring round the working wheels, and the large frame on which the whole machinery of the engine rests, with as much ease as practicable.
John B Jervis, letter to American Railroad Journal, *vol. ii, 1833, of his claim to have invented the leading bogie for steam locomotives. Quoted by Jervis in* Railroad Property, *1866.*

314.2 The danger of too much ingenuity.
Attributed to George Stephenson in D K Clark's Railway Machinery, *1855, and repeated by J G H Warren in* A Century of Locomotive Building, *1923.*

314.3 We are all practical men on this road, and don't believe in thy gimcracks.
Anon., within Journal of the Franklin Institute, *Jan 1868. Quoted by John H White in* American Locomotives, *1997.*

314.4 Pride alone should prevent them from allowing any outsider from surpassing them in inventions to subserve the interests of this Company, their employer.
Board Minute, Pennsylvania Railroad, concerning the attitude of its mechanics and engineers, 1886 or 1887. Quoted by Steven W Usselman in Regulating Railroad Innovation, *2002.*

314.5 Of course, every person familiar with machinery knows that many worthless or even ridiculous things are made for a time successful when some influential personage is interested, as in the case of the Webb three cylinder compound locomotives; but the test of time passes only things that are fit for purpose.
Angus Sinclair, Development of the Locomotive Engine, *1907.*

314.6 Old customs and systems die hard at the works and, whatever their own opinion of the matter may be, the officials are not considered by the workmen to be of a very progressive type. Many of the methods employed, both in manufacture and administration, are extremely old-fashioned and antiquated; an idea has to be old and hoary before it stands a chance of being adopted here.

Alfred Williams, of practice at Swindon Works, Life in a Railway Factory, *1915.*

314.7 Advances are always initiated as a result of comparison.

W E Woodard, locomotive designer, 1928, quoted in Trains Magazine, *March 1952.*

314.8 The scientific and technical advances that have made Britain famous in other fields have not in recent years borne full fruit in the railways.

Conclusion, Report of Select Committee on Nationalised Industries, *paragraph 419, July 1960.*

314.9 Southern gives a green light to innovations.

Advertising slogan, Southern Railway (U.S.). Also rendered as The railway system that gives. . . , *1960–c. 1982.*

314.10 . . . the march of progress suffers when engineering development outstrips legislative and commercial practice.

John Johnson and Robert A Long, British Railways Engineering, 1948–1980, *1981.*

314.11 The tgv was that rarest of phenomena, a major step forward which employed proven technology and a pioneering effort which made a profit from day one.

Nicholas Faith, The Economist, *24 August 1985.*

314.12 The future is not the past. We are re-inventing the railroad.

Liviu Alston of the World Bank, quoted by Nicholas Faith, The Economist, *24 August 1985.*

314.13 Technology is costly, but outdated technology is even costlier.

Robert Claytor, Railway Age, *July 1986.*

315. INSPECTION SALOONS

315.1 A trip in a saloon is a most wonderful opportunity for persuasion.

Sir William McAlpine, in conversation with the compiler, on board his saloon GE1, 1993.

316. INTER-CITY

316.1 Decades before the term "inter-city" became an advertising gimmick of British Railways it was to be found in essence on the Midland, which followed a policy of running medium weight frequent passenger trains between the towns and cities it served so well.

George Dow, Midland Style, *1975.*

317. INTERLOCKING
See also Signal Boxes

317.1 Interlocking is a simple system by which the signalman has to do something before he can do something else, and if he doesn't he can't.

W Mills, 4 ft 8½ and All That, *1964.*

318. INTERSTATE COMMERCE COMMISSION

318.1 The interstate commerce law was passed in 1887. It was crude in its provisions and was the result of compromises between radicals and conservatives; it sought both to foster competition and to abolish it, and in that respect remains still contradictory and impossible.

Edward P Ripley, speech at annual dinner of the Railway Business Association, 10 November 1909; The Railway Library, *1909, 1910.*

318.2 With the best intentions in the world the present Interstate Commerce Commission is so enmeshed in its own anti-railway traditions, so enamoured of the administrative control theories of its statistician, so covetous of unbridled, irresponsible authority to tear down where it has no constructive capacity, that anything like co-operation between the Commission and the railway management for the public good seems out of the question.

Slason Thompson; the statistician of the ICC at that time was Professor H C Adams. The Railway Library, *1909, 1910.*

318.3 How can the Interstate Commerce Commission build up its patient if "bleeding" is to be the only remedy permitted?

H U Mudge, President, Chicago Rock Island & Pacific; speech at Commerce Club of Topeka, 11 April 1911; The Railway Library, *1910, 1911.*

318.4 Gentlemen, the railroads of this country cannot endure half slave, half free.

Ibid., paraphrasing Abraham Lincoln and referring to the contradictions between private enterprise (the railroads) and state regulation (the ICC).

318.5 Dazzled and possibly alarmed over the magnitude of the vast business it was created "to conserve and protect," as well as regulate, the Commission seems to have set itself the task of levelling all railway rates down to the point where their entire income is divided among the employe, the tax collector, and the creditor, ignoring the claims of the public for improved, up to date facilities, and of the owners for reasonable remuneration.

Slason Thompson, The Railway Library, *1911, 1912.*

318.6 There is no parallel I know of in any other country to its greatest industry being placed, down to its minutest details, under the almost autocratic power of seven men owing defined accountability to no one, selected for relatively short terms and according to no particular standard of training or qualifications, and being practically free from control, restraint, or appeal.

Otto H Kahn, The World's Work, *February 1916. Reproduced in* The Railway Library, *1915, 1916.*

318.7 The ICC, conceived in the sweat and tears of the Granger Movement, finds today that there is a great public benefit in the ability of shippers to deal with one railroad rather than two. *Sic transit gloria mundi.*

ICC Commissioners William Tucker and Charles A Webb, dissenting from ICC findings in the merger case for Atlantic Coast Line and Seaboard Air Line, 1963. Quoted by Richard Saunders Jr, Merging Lines, *2001.*

319. INTERURBAN SERVICE

319.1 When it comes to cheap, irresponsible, and satisfactory recreation, the trolley is certainly the very best thing.
Statement in World's Work, *1903, quoted by William D Middleton in* Interurban Era, *1961.*

319.2 They say this regional transit was junked because it was unprogressive, and now we're trying to progress right back to it.
Paul Crume, Dallas Morning News, *21 June 1966.*

320. INVESTMENT IN RAILWAYS

See also Shareholders

320.1 A man who commences railway undertakings incurs a serious responsibility, and he ought to be very careful, and to weigh well and anxiously the details of his measure, before he projects such schemes and induces others to embark upon them. There are many poor people who embark their property in railways, and if they prove unsuccessful, the consequences can hardly be calculated.
George Hudson, evidence before Select Committee on Railways, 1844.

320.2 O Hobnail, to a king this case I owe,
A king, for Railway Specs have made him so.
"Buy shares, get scrip," he said, your pockets fill;
Be rich, and play the devil as you will."
Basil Montague; stated by Flunkey to Hobnail. The reference to "king" is almost certainly to George Hudson. Railroad Eclogues, *1846.*

320.3 In the poorer countries of Europe, the rage for railway construction might have had worse consequences than in England, were it not that in those countries such enterprises are in a great measure carried on by foreign capital. The railway operations of the various nations of the world may be looked upon as a sort of competition for the overflowing capital of the countries where profit is low and capital abundant, as in England and Holland. The English railway speculations are a struggle to keep our annual increase of capital at home; those of foreign countries are an effort to obtain it.
John Stuart Mill, Political Economy, *1848.*

320.4 It is hardly needful to point out how fully the remarks in the text have been verified by subsequent facts. The capital of the country, far from having been in any degree impaired by the large amount sunk in railway construction, was soon again overflowing.
Ibid., footnote.

320.5 Ten per cent talks louder than five.
Attributed to Cornelius Vanderbilt, when explaining why, at the age of sixty-nine, he was getting involved in railroads after a life in shipping. Quoted by David Marshall in Grand Central, *1946.*

320.6 If brilliantly printed programmes might avail anything, with gorgeous maps, and beautiful little pictures of trains running into tunnels beneath snowy mountains and coming out of them on to the margin of sunlit lakes, Mr Fisker had certainly done much. But Paul, when he saw well these pretty things, could not keep his mind from thinking whence had come the money to pay for them.
Anthony Trollope, of the promotion of the fictitious South Central Pacific and Mexican Railway, The Way We Live Now, *1875.*

320.7 Oor new clock lichted shows the 'oor,
The parish kirk has got a toor,
And better still, up to our door
Has come the Wigtown Railway.

Lord Galloway, gude, worthy man,
The enterprise at first did plan,
An' great and sma' pit to their han'
To help mak the Railway.

And while it stood at Sorbie Mill,
For want of cash to mount the hill—
But Johnston Stewart, with richt gudewill,
Has brocht us up the Railway.
O B, poem on the railway from Newton Stewart to Whithorn, first three verses; recorded by H V Morton, In Scotland Again, 1933.

320.8 The extension of railways had, up to the year 1844, been mainly effected by men of the commercial classes, and the shareholders in them principally belonged to the manufacturing districts,—the capitalists of the metropolis as yet holding aloof, and prophesying disaster to all concerned in railway projects. The Stock Exchange looked askance upon them, and it was with difficulty that respectable brokers could be found to do business in the shares. But when the lugubrious anticipations of the city men were found to be so entirely falsified by the results, —when, after the lapse of years it was ascertained that railway traffic rapidly increased and dividends steadily improved—a change came over the spirit of the London speculators. They then invested largely in railways, the shares in which became a leading branch of business on the Stock Exchange, and the prices of some rose to nearly double their original value.

Samuel Smiles, Life of George Stephenson, *1881.*

320.9 It was a railroad era; most of the issues that we sold—or tried to sell—were railroad securities. I became a walking encyclopedia of railroad statistics. When I declare that I could give off-hand the capitalization, earnings, and general characteristics of every well-known railroad in the United States I am not exaggerating in the least. The drawers of my desk were crammed full of railroad maps. I studied them day and night. After a while I had a mental picture of every important railroad system; I could take a map of the United States and mark with a pencil the main line and principal branches of any railroad one might name.

William McAdoo, of his time selling bonds, early 1890s, Crowded Years, *1931.*

320.10 The necessary development of railroad facilities is now endangered by the reluctance of investors to purchase new issues of railroad securities in the amounts required. This reluctance is likely to continue until the American public understands the essential community of interest between shipper and investor, and the folly of attempting to protect the one by taking away the rewards of good management from the other.

Howard Elliott, address to the Minneapolis Chapter of the American Institute of Banking, Minneapolis, 26 April 1913, published in The Truth about the Railroads, *1913.*

320.11 They are indispensable to our whole economic life, and railway securities are at the very heart of most investments, large and small, public and private, by individuals and by institutions.

Woodrow Wilson, writing in 1914; quoted by Frank Trumbull in address to The National Hay Association, Railway Regulation and Locomotor Ataxia, *Cedar Point Ohio, 12 July 1916.*

320.12 We must recognise that the man who is willing to invest his money at a moderate rate of interest in railroad securities is not exploiting the public but is a public benefactor.

Oscar W Underwood, Senator, Alabama, address in Chicago 4 February 1916, quoted by Frank Trumbull in address to The National Hay Association, Railway Regulation and Locomotor Ataxia, *Cedar Point Ohio, 12 July 1916.*

320.13 Railway speculation has become the sole object of the world—cupidity is aroused and roguery shields itself under its name, as a more safe and rapid way of gaining its ends. Abroad, as well as at home, has it proved the rallying point of all rascality—the honest man is carried away by the current and becomes absorbed in the vortex; the timid, the quiet, the moral are, after some hesitation, caught in the whirlpool and follow those whom they have watched with pity and derision.

Quoted from The Illustrated London News, *c. 1845, in* Fifty Years of Railway Life, *Joseph Tatlow, 1920.*

320.14 Probably there are few homes in England whose financial security has not tottered or collapsed at some time in the past century as a consequence of the financial panic and "hard times" in the United States, and it is fairly certain that England has poured more money across the sea into American railroads than ever went back in profits and dividends.

Roland E Collons, The Railway Magazine, *vol. LXXX, 1937.*

320.15 There was a period of railroad financing in the nineteenth century when no rail stock was considered respectable unless it contained the word "Pacific" even if its charter never provided for it to cross the Mississippi.

Lucius Beebe, Trains in Transition, *1941.*

320.16 The Southern stocks are better than those of any of the other railways, and if they were left alone they might have a better future than most of the other companies. The best stocks of all—if you do not mind my giving you advice—are those of the London Passenger Transport Board.

Lord Walkden, answering a point from Lord Monkswell, Lords debate on the 1947 Transport Bill, 1947, Hansard, *20 May 1947.*

320.17 The railroads—viewed from the standpoint of what they do rather than from that of who owns them—are not an industry separate from manufacturing, mining and agriculture. They are, rather, the principal part of the inter-plant transportation facilities of all other industry. The denial of means to the railroads to maintain the same rate of progress in plant improvement as that in the rest of industry served by the railroads will, in the long run, be just as costly to industry and the consuming public as if this dearth of capital investment were occurring in industry itself.

Editorial comment, Salisbury Post, *(N.C.), reprinted in* Trains Magazine, *November 1948.*

320.18 The record has been maintained, but as brokers say, this is no longer a stake for widows and orphans—a far cry from the days when good Philadelphia children were taught to pray for the Republican Party, the Girard Trust and the Pennsylvania Railroad.

Nathaniel Burt, The Perennial Philadelphians, *1963.*

320.19 The railroad industry was like the dodo bird with its head where its tail feathers ought to be, who can see very clearly where it has been, but only with difficulty where it is going.
  Attributed to Al Perlman, of the view taken by security analysts in 1964. Trains Magazine, *January 1966.*

320.20 We do not think BLE members would consider such investment of their funds a wise risk, particularly in view of the shaky financial condition of some railroads.
  Editorial Comment, Locomotive Engineer, *the journal of the Brotherhood of Locomotive Engineers, c. 1971, as recorded in* Trains Magazine, *August 1971.*

320.21 There are, broadly speaking, two ways of viewing a railway company as an investment. One is that the maximisation of profit and dividend is the major objective, and all decisions on the company management and development should be governed by this consideration. The other is the wider view that, although a loss-making company is not likely to remain operating in any sense for very long, the objectives of the concern embrace a constellation of economic, social and psychological factors with a wide influence on its operating *esprit* and its cultural environment. Such an approach by investors and shareholders requires a longer and broader view than its alternative.
  John Neville Greaves, Sir Edward Watkin, 1819–1901, *2005.*

320.22 Investors aren't public servants.
  Mark W Hemphill, title of article concerning the expectations of the public after the deregulation of U.S. railroads; Trains Magazine, *April 2005.*

321. ISLE OF MAN RAILWAY

321.1 . . . the diminutive and absurd train which by breathless plunges annihilates the sixteen miles between Douglas and Port Erin in sixty-five minutes.
  Arnold Bennett, Anna of the Five Towns, *1902.*

322. THE ISLE OF WIGHT

322.1 The Isle of Wight is at first disappointing. I wondered why it should be, and then I found the reason in the influence of the detestable little railway. There can be no doubt that a railway in the Isle of Wight is a gross impertinence, is in evident contravention to the natural style of the place.
  Henry James, English Hours, *1905.*

323. ISLE OF WIGHT (NEWPORT JUNCTION) RAILWAY

323.1 Some railways have certainly been more fortunate than others. Cases of contractors going bankrupt, though not common, are not unknown; nor are quarrels with neighbours, serious breakdowns of engines, collapses of bridges, disputes between directors and management, delays in getting certificates for opening to traffic, chronic penury, or penury, or receivership; but a line less than ten miles long must be accounted uncommonly unlucky to have incurred every one of these mischances.
  Michael Robbins, The Railway Magazine, *October 1959.*

324. IVOR THE ENGINE
See also Thomas the Tank Engine

324.1 Not very long ago, in the top left-hand corner of Wales, there was a railway. It wasn't a very long railway or a very important railway, but it was called The Merioneth and Llantisilly Rail Traction Company Limited, and it was all there was.
  And in a shed, in a siding at the end of the railway, lives the Locomotive of the Merioneth and Llantisilly Rail Traction Company Limited, which was a long name for a little engine so his friends just called him Ivor.
  Oliver Postgate and Peter Fimin, Ivor the Engine, *1959.*

325. J CLASS LOCOMOTIVES
(Norfolk & Western Railway)

325.1 In their sheer size and massive proportions the Nor-folk & Western's J-class 4-8-4s were super steam locomo-tives suggesting the swaggering might of a Krupp artil-lery piece.
Robert C Reed, The Streamline Era, *1975.*

326. JAMAICA

326.1 Without No. 54 . . . the bikinis aren't enough.
David P Morgan, cover caption referring to Jamaica Railway 4-8-0 No. 54 that had been retired. Trains Maga-zine, *May 1975.*

327. JAPANESE TRAINS

327.1 . . . the Japanese train relies on aircraft comforts: silence, leg room, a reading light—charging an extra ten dollars to sit two (instead of three) abreast, and discour-aging passengers from standing and gabbing at the exits.
Paul Theroux, of the high speed services, The Great Railway Bazaar, *1975.*

328. SAMUEL JOHNSON

328.1 The precision of a Breguet watch and the beauty and finish of an Adam House went into a Johnson engine on the Midland.
C Hamilton Ellis, The Locomotive Magazine, vol. XLV, *1939.*

328.2 No locomotives now have a finish by Johnsonian standards.
Ibid.

329. JOINT STATIONS

329.1 Two companies using the same tracks and buildings at the same time, it may be, are exposing themselves to conflict if not positively courting it. Conflict, greater or less, is inevitable if two companies at a joint station have different rules as to equipment, so that what passes inspection on one side is rejected on the other. The same is true if there are different rules as to sealing cars or other matters. And even if no conflict break out between the companies it goes on clandestinely, as it were, in the minds and experience of the joint employees. It necessarily costs effort of a self-contradictory, non-harmonious sort to deal at one and the same time with two distinct sets of accounts. Similarly with regard to still other matters. It may be remarked, in closing, that while transfer and junction-point stations are absolutely unavoidable, joint stations are justifiable only as mea-sures of distinct or considerable economy in operation; in themselves they are rather absurd.
B C Burt, who in a footnote excluded union stations from these views (presumably because of their special man-agement arrangements); Railway Station Service, *1911.*

330. CASEY JONES
See also Casey Jr

330.1 Come all you rounders, for I want you to hear
The story told of an engineer;
Casey Jones was the rounder's name,
A heavy right-wheeler of a mighty fame.
Wallace Saunders, song, The Ballad of Casey Jones.
First verse of the earliest published of many versions, 1908.

331. JUNCTIONS

331.1 It was a Junction-Station, where the wooden razors before mentioned shaved the air very often, and where the sharp electric-telegraph bell was in a very restless condition. All manner of cross-lines of rails came zig-zagging into it, like a Congress of iron vipers; and, a lit-tle way out of it, a pointsman in an elevated signal-box was constantly going through the motions of drawing immense quantities of beer at a public-house bar.
Charles Dickens. The "wooden razors" were the arms of semaphore signals; the Lazy Tour of Two Idle Apprentices, Household Words, *17 October 1857.*

331.2 Sidings were there, in which empty luggage-vans and cattle-boxes often butted against each other as if they couldn't agree; and warehouses were there, in which great quantities of goods seemed to have taken the veil (of the consistency of tarpaulin), and to have retired from the world without any hope of getting back to it. Refreshment-rooms were there; one, for the hungry and thirsty Iron Locomotives where their coke and water were ready, and of good quality, for they were dangerous to play tricks with; the other, for the hungry and thirsty human Locomotives, who might take what they could get, and whose chief consolation was provided in the form of three terrific urns or vases of white metal, con-taining nothing, each forming a breastwork for a defiant and apparently much-injured woman.
Ibid.

331.3 But there were so many Lines. Gazing down upon them from a bridge at the Junction, it was as if the concentrating Companies formed a great Industrial Exhibition of the works of extraordinary ground spiders that spun iron.

Charles Dickens, Mugby Junction, 1866.

331.4 And then so many of the Lines went such wonderful ways, so crossing and curving among one another, that the eye lost them. And then some of them appeared to start with the fixed intention of going five hundred miles, and all of a sudden gave it up at an insignificant barrier, or turned off into a workshop. And then others, like intoxicated men, went a little way very straight, and surprisingly slued round and came back again. And then others were so chock-full of trucks of coal, others were so blocked with trucks of casks, others were so gorged with trucks of ballast, others were so set apart for wheeled objects like immense iron cotton-reels; while others were so bright and clear, and others were so delivered over to rust and ashes and idle wheelbarrows out of work, with their legs in the air (looking much like their masters on strike) that there was no beginning, middle, or end to the bewilderment.

Ibid.

331.5 There was an Old Man at a Junction,
Whose feelings were wrung with compunction,
    When they said "The train's gone!"
    He exclaimed "How forlorn!"
But remained on the rails of the Junction.

Edward Lear, poem, Nonsense Rhymes.

331.6 Slow and woeful Junction Town,
Where devils laugh and angels frown
To see a passenger set down;
Where trains run only with a view
To help a restaurant or two;
Where rusty rails and barren boards
Are all the point of view affords.
But O, the barren board of all
Is that within that eating stall!
Yes, stall I said, and well-deserved
The name! where beastly feed is served.
And so I say without compunction,
My curses on this Railroad Junction.

Edmund Vance Cooke, poem concerning South Berwick, where the Eastern Railroad had a junction with the Boston & Maine, and refused to be cooperative about making connections. Date not known: quoted by Alvin F Harlow in Steelways of New England, 1946.

331.7 With saddened face and battered hat
And eye that told of blank despair,
On wooden bench the traveler sat,
Cursing the fate that brought him there.
"Nine hours," he cried, "we've lingered here,
With thoughts intent on distant homes
Waiting for that delusive train
That, always coming, never comes,
Till weary, worn, distressed, forlorn
And paralyzed in every function!
I hope in hell, their souls may dwell
Who first invented Essex Junction!

"I've traveled east, I've traveled west
Over mountain, valley, plain and river;
Midst whirlwind's wrath and tempest's blast
Through railroad's crash and steamboat's shiver
And faith and courage faltered not,
Until I reached this dismal spot,
Of man accursed, of God, forsaken!
Where strange, new forms of misery
Assail men's souls without compunction
And I hope in hell his soul may dwell
Who first invented Essex Junction!

"Here Boston waits for Ogdensburg
And Ogdensburg for Montreal,
And late New York tarrieth
And Saratoga hindereth all!
From far Atlantic's wave-swept bays
To Mississippi's turbid tide
All accidents, mishaps, delays
Are gathered here and multiplied!
Oh, fellow man, avoid this spot
As you would plague or Peter Funk shun!
And I hope in hell his soul may dwell
Who first invented Essex Junction!

"And long and late conductors tell
Of trains delayed or late or slow,
Till e'en the very engine's bell
Takes up the cry, 'No go! No go!'
Oh! let me from this hole depart
By any route, so't be a long one,"
He cried with madness in his heart,
And jumped aboard a train—the wrong one,
And as he vanished in the smoke
He shouted with redoubled unction,
"I hope in hell his soul may dwell
Who first invented Essex Junction!"

Edward John Phelps, poem, The Lay of the Lost Traveler, c. 1890. Written following his having changed trains at Essex Junction (near Burlington, Vt.) and having unwittingly boarded a train which took him back whence he had come. The significance of the mention of Peter Funk is not understood. Reproduced in Trains Magazine, August 1958.

332. K4S LOCOMOTIVES

332.1 Unmistakably there was power in the shapeless boiler with its bumpy Belpaire firebox, speed in the close-coupled drivers and heavily muscled cylinders and rods.

What disappointed me was their lack of grace. Their clumsy proportions bore no resemblance to the lean Pacifics I had seen in Sam Vauclain's catalogue.

Otto Kuhler, of his first sight of a Pennsylvania Railroad K4s at Manhattan Transfer station, May 1923; My Iron Journey, *1967.*

333. KING'S CROSS STATION

333.1 ... Kings Cross had always suggested Infinity. Its very situation—withdrawn a little behind the facile splendours of St Pancras—implied a comment on the materialism of life. Those two great arches, colourless, indifferent, shouldering between them an unlovely clock, were fit portals for some eternal adventure, whose issue might be prosperous but would certainly not be expressed in the language of prosperity.

E M Forster, Howards End, *1910.*

333.2 This circled cosmos whereof man is god
Has suns and stars of green and gold and red,
And cloudlands of great smoke, that range o'er range
Far floating, hide its iron heavens overhead.

God! shall we ever honour what we are,
And see one moment ere the age expire,
The vision of man shouting and erect,
Whirled by the shrieking steeds of flood and fire?

Or must Fate act the same grey farce again,
And wait, till one, amid Time's wrecks and scars,
Speaks to a ruin here, "What poet-race
Shot such cyclopean arches at the stars?"

G K Chesterton, Kings Cross station, *Collected Poems, 1927.*

333.3 Whoever considers King's Cross Station today? Yet it is one of the finest buildings in the world. Two enormous brick arches filled with glass, divided by a plain tower with no superfluous decoration. The offices, blocks and the crescent-shaped hotel form part of the same scheme, simple buildings with an appropriate veneer of classical decoration. And inside the station are those two great receding tunnels of glass, with their rhythmical pattern of iron girders and supports.

John Betjeman, Ghastly Good Taste, *1933.*

333.4 The atmosphere of Kings Cross was so completely unalike that at St Pancras that one almost expected to see placards up to the effect that there was no connection with the establishment next door.

C Hamilton Ellis, The Trains We Loved, *1947.*

333.5 For the best of all sights from the signal "box" is one of Gresley's quiet-spoken, three cylinder Pacifics getting the right of way, beclouding her front end with cylinder-cock steam (to the delight of the inevitable "engine-spotter" schoolboys), feeling her way through the crossovers, then shoving her smooth green barrel into Gas Works Tunnel, and sucking her trailing cars into the boiling smoke and steam at the portal.

That is the best of all sights.

David P Morgan, Trains Magazine, *February 1961.*

333.6 King's Cross station may indeed be described as one of the forerunners of modern architecture.

Architectural Correspondent, The Times, *3 September 1966.*

333.7 Dingy and grand, an ochre-tinged arcade,
Seeming less apt for railway than for mart,
With square clock tower above a colonnade,
It stood from other termini apart.

Henry Maxwell, poem, King's Cross, *verse three, in* A Railway Rubaiyyat, *1968.*

333.8 And there beyond its platforms, mouth by mouth
The smoking tunnels gaped and closed it in:
Symbolic termination to The South,
And of The North the sable paladin!

Sepulchral focus, where all lines converged
And trains departing vanished out of sight,
While those arriving startlingly emerged,
As from the grave, incongruously bright!

Ibid., verses six and seven.

333.9 Nevertheless, the interior of the station carries an atmosphere of mystery and expectancy about it; travelling from King's Cross is still an exciting experience, a quality curiously absent from St Pancras next door.

David Atwell, Railway Architecture, *(ed. by Binney and Pearce), 1979.*

334. KISSING

See also Love-Making

334.1 The Footplate Regulations don't allow for kissing the engine-driver.

John Hadfield, Love on a Branch Line, *1959.*

335. LADIES CARRIAGES

335.1 Ladies' cars are barbarisms. There is no more seclusion, nor safety against tobacco indecencies, where a lady journeys with married gentlemen or gallants, than where she may chance to have the company of bachelors or stray benedicts. Every passenger car should have a 'saloon,' as the two by three foot closet is commonly called, but an exclusive ladies' car is usually a nuisance.
Comment, Holly's Railroad Advocate, *20 September 1856, quoted by August Mencken in* The Railroad Passenger Car, *1957.*

336. LANCASHIRE

336.1 . . . the home and haunt of railway enterprise. . .
John Francis, History of the English Railway, *1851.*

337. LANCASHIRE & YORKSHIRE RAILWAY

337.1 In the middle of the seventies it was probably the most degenerate railway in the kingdom, to which even the South Eastern or the London, Chatham & Dover could have run only a bad second.
E L Ahrons, The Railway Magazine, *August 1917.*

337.2 . . . if there was any doubt as to which took over which, the average Lanky man could be relied upon to give an emphatic and characteristic answer!
George Dow, upon the amalgamation of the L&Y with the L&NW in 1922, Railway Heraldry, *1973.*

338. LANCASHIRE DERBYSHIRE & EAST COAST RAILWAY

338.1 . . . as mad a scheme as was ever presented to Parliament.
Sir Edward Watkin, evidence before Parliamentary Committee, April/May 1891, quoted by George Dow in Great Central, vol. II, *1962.*

338.2 The Dukeries Route.
Advertising slogan.

338.3 Neither Lancashire nor East Coast.
George Dow, chapter heading in Great Central, vol. III, *1965.*

339. LAND

339.1 The roads, railways, and canals were not constructed to give value to land: on the contrary, their natural effect was to lower its value, by rendering other and rival lands accessible . . .
John Stuart Mill, Political Economy, *1848.*

340. LANDSCAPE, RAILWAYS IN

340.1 Silence and stillness reign within its precincts, and harmonize with the grandeur of the spectacle; the rails converging in perspective form the track of the terrestrial zodiac,—lines terminating in points in the horizon, whence at prescribed periods earthly objects rise and perform their transit, while many a muscular arm toils in preparation for the phenomenon, which appears and passes away.
Sir George Head, A Home Tour through the Manufacturing Districts and other parts of England, Scotland and Ireland, *1835.*

340.2 . . . the wreaths of white smoke that float above the deep foliage of the Weald marking the progress of the trains across the old country of the Iguanadon and the Plesiosaurus.
Murray's Handbook for Surrey, *1865.*

340.3 There comes a crowd of burly navvies, with pickaxes and barrows, and while hardly a wrinkle is made in the mother's face, or a new curve of health in the blooming girl's, the hills are cut through, or the breaches between them spanned, we choose our level, and the white steam-pennon flies along it.
George Eliot, George Eliot's Life, Letters and Journals, *1887.*

340.4 East to west, along a ridge bounding the lower desert, ran the railroad, a line as harshly compromising as the cold mathematics of the engineers who had mapped it. To the north spread unfathomably a forest of scrub pine and piñon, rising, here and there, into loftier growth. It was if man, with his imperious interventions, had set those thin steel parallels as an irrefragable boundary to the mutual encroachments of forest and desert, tree and cactus.
Samuel Hopkins Adams, Success, *1921.*

340.5 . . . the railwayman who can ride in a bus up Ludgate Hill without resenting the railway bridge which hides St Paul's from him must have lost his sense of proportion.
Canon Roger Lloyd, The Fascination of Railways, *1957.*

340.6 The railway has taken its place in the landscape with all the other artificial elements that man has put there:

fields, hedges, farms, roads, canals. It has taken its place because it fits in—it rarely dominates in any view—and because, unlike the airfield which must obliterate existing features to create its shaven emptiness, the railway etches in fresh detail to the scene.

Michael Robbins, The Railway Age, *1962.*

340.7 Being unable to find anyone who can talk to the antelope and ask how they feel about love's ability to find its own way, we are reduced to pointing out to the Government that within 30 or 40 miles east and west of the proposed line we have tracks that have been in operation since 1890. And we see substantial evidence that the sex lives of the antelope continue fruitful indeed, thank you.

Louis W Menk, Chairman of Burlington Northern, commenting upon fears that a new coal line in Wyoming would disrupt the breeding of antelope. Quoted in Trains Magazine, *March 1975.*

340.8 It was man's best machine traversing earth's best feature—the train tracking in the narrow angle between vertical rock and horizontal water.

Paul Theroux, The Kingdom by the Sea, *1983, about Shakespeare's Cliff, Dover.*

340.9 There was nothing in the world more restful; the train seemed like the highest stage of civilization. Nothing was disturbed by it, or spoiled; it did not alter the landscape; it was the machine in the garden, but it was a gentle machine. It was fast and economical and as safe as a vehicle could possibly be.

Ibid.

341. DR DIONYSIUS LARDNER

341.1 His book tells exactly how to set a rail in its proper place, but not how to set an assistant superintendent in his office.

Thomas C Cochrane, of Lardner's book Railway Economy, *in* Railroad Leaders, 1845–1890, *1953.*

341.2 Lardner was a prolific writer on Scientific and technical subjects, on which he sometimes espressed himself without knowledge or discretion . . .

W O Skeat, George Stephenson, the Engineer and his Letters, *1973.*

342. LAW

342.1 The present state of the law in regard to railways is perfectly abominable.

Comment, Railway Chronicle, *20 December 1845.*

342.2 It is evidently impossible, for example, to obey the laws of two states when one requires cinder deflectors, screens, and cuspidors, and the other forbids their use; when one kind of headlight is demanded by one state, and forbidden by another.

Dr Howard Douglas Dozier, A History of the Atlantic Coast Line Railroad, *1920.*

342.3 The carriers can not serve two masters.

Rep. Alfred L Bulwinkle, North Carolina (11th District) in promoting a bill to resolve a conflict between the obligations of cooperation between railroads in the Transportation Act of 1940 and the prevention of such practices by the Sherman Anti-Trust Act of 1890. Address in the House of Representatives, reported by the Association of American Railroads in leaflet "The Carriers CANNOT Serve Two Masters," *May 1945.*

343. LAWYERS

343.1 . . . Provided always, that no practising Solicitor shall be appointed an Officer of the said Company.

Provision, section 143, London & Birmingham Railway Act, *1833.*

343.2 What can a mere sucking barrister know about the practical details of railway matters?

Comment, Bradshaw's Railway Gazette, *20 May 1848.*

343.3 Oh dear, Oh dear, now what shall I do,
For Law is so flat and my clients so few,
Sure the world is quite full of vexations and troubles
And nought now is heard of but new schemes and
    bubbles.

My Professional brother outrivals a few
With his feasible scheme of the BUXTON AND SCREWE,
I'll oppose him—that's flat—'tis sure to pay well
Like the Jew with his razors, I'll make one to sell.

A flaming prospectus shall forthwith come out,
'Twill be swallowed instanter, I haven't a doubt,
If great meetings be called, resolutions debated
By a batch of new squires who have just been created.

We'll talk most immensely of coal and of lime,
Of plans and of sections, of savings of time;
We'll puzzle the natives with much show of sense,
It signifies nothing if we get the pence.

You're a SHARPE; and you've also a sharper than he
But they're dust in the balance when pitted with me,
A superlative lawyer (I'll not boast at all),
Though ones SHARPE, and one's sharper, I'm sharpest
    of all.

Gervase Forrester, The Rival Railways v. Bubbles; *the reference to Sharpe is to Robert C Sharpe of Manchester, quoted in* The North Staffordshire Railway, *Manifold, 1952.*

343.4 I don't want you to tell me what I can't do; I pay you to tell what I *can* do.

Attributed to J P Morgan, to his chief counsel Elbert

Grey, concerning rival railroads. Quoted by Albro Martin,
Railroads Triumphant, *1992.*

343.5 So infinitesimal did I find the knowledge of Art,
west of the Rocky Mountains, that an art patron—one
who had in his day been a miner—actually sued the railroad company for damages because the plaster cast of
Venus of Milo, which he had imported from Paris, had
been delivered minus its arms. And, what is more surprising, he gained his case and the damages.
Oscar Wilde, Impressions of America, *1906.*

343.6 The important departments often spoken of as the
"Clerical" ones, i.e., the Secretary's and the Accountant's
were run on much the same lines in Utopia as in Great
Britain that no details concerning them seem needful:
much the same may be said of the Solicitor's office. But
solicitors of Utopian railways never were called into
council, nor did they offer advice outside of strictly legal
matters. They had no voice in the general policy of the
company.
Anon., Railway Management in Utopia, *within* Great
Western Railway Magazine, *January 1914.*

344. LEAVES ON THE TRACK

344.1 MR MALINS: Does the Minister know why or how
leaves on a railway line can apparently render the whole
line non-working? Can the Government suggest a remedy? Connex South Central could not get a train into
London from Surrey today, because of leaves on the line.
Is there an explanation?
MR SKINNER: New Labour, New Leaves.
*Exchange in the House of Commons during Oral Ques-
tions, 18 November 1997. The Minister being questioned
was Glenda Jackson; the interruption was from Dennis
Skinner, Labour MP for Bolsover.*

345. LEEK & MANIFOLD VALLEY RAILWAY

345.1 On the whole, there is little to regret in the advent of
the railway. From the utilitarian standpoint it has placed
a once very remote pastoral district within reach of a
market for its milk, and even the most conservative lover
of nature undefiled, is bound to admit that the making of
the line has done wonderfully little to desecrate the
scenery.
Charles Masefield, Staffordshire, *1910.*

346. LEGAL EXPENSES

346.1 The curse of railway enterprise in England.
G W J Potter, The Easingwold Railway, *within* The
Railway Magazine, *vol. XLI, 1917.*

347. LEICESTER & SWANNINGTON RAILWAY

347.1 . . . a line sixteen miles in length—a director to the
mile.
Frederick Williams, The Midland Railway, *1878.*

348. LEVEL CROSSINGS

348.1 Stop Look and Listen.
*Attributed to Thomas Gray, who promoted the idea of
such signs after the death of a nephew at a grade crossing on
the Southern Pacific, 1884;* Trains Magazine, *February
1947.*

348.2 Grade crossings on single track dangerous, on double track they should seldom be permitted; on four track
road, never. They must be abolished as fast as possible.
*Charles Peter Clark, President, New York, New Haven
& Hartford Railroad. Quoted in* RLHS Bulletin 110, *April
1964.*

348.3 With a quavering voice he whistled for his first
grade crossing (an event in the life of a locomotive), and
his nerves were in no way restored by the sight of a frantic horse and a white-faced man in a buggy less than a
yard from his right shoulder.
Rudyard Kipling, .007, within Scribner's Magazine,
August 1897.

348.4 The present, the positive, was mainly represented,
ever, by the level railway-crossing, gaining expression
from its localization of possible death and destruction,
where the great stilted, strident, yet almost comically
impersonal train, which, with its so often undesignated
and so often unservanted stations, and its general air of
"bossing" the neighbourhoods it warns, for climax of its
characteristic curtness, to "look out" for its rush, is
everywhere a large contribution to one's impression of a
kind of monotony of acquiescence.
Henry James, The American Scene, *1907.*

348.5 Ride in the engine of a train, hurtling across the
country at from fifty to seventy miles an hour, roaring
over bridges, grinding and crashing over crossings and
switches, and your idea of the intelligence of the human
race changes.
Rex Stuart, article People Act as if They Wanted to be
Killed, *within* American Magazine, *September 1921.*

348.6 Some people are always cussing the railroads and
saying we are careless and risking the lives of people. But
they certainly don't *act* as if they thought we were
careless.
*Jack Rigney, New York Central locomotive engineer,
quoted by Rex Stuart, ibid.*

348.7 Train Approaching
Whistle Squealing
Pause!
Avoid That
Run-down Feeling
Burma-Shave

Anon., typical sequential roadside advertisement for Burma-Shave toilet products, this one highlighting grade crossing dangers, 1930s. Quoted by John R Stilgoe in Metropolitan Corridor, *1983.*

348.8 Stop—Death—Stop

Alonzo Billups, who installed a gantry sign across the road on a Highway 7 grade crossing with the Illinois Central Railroad, Grenada, Miss., 1930s. The words were supplemented by flashing lights and an air raid siren. Illustrated in Trains Magazine, *May 2003.*

348.9 All the safety equipment installed in the last decades of the nineteenth century, and the massive publicity campaigns of the 1900–15 era aimed at convincing the public that semaphores, interlocking plants, and automatic train stop devices made collisions between trains and derailments of trains almost impossible, backfired at every crossing.

John R Stilgoe, of the failure of the public to learn the dangers of taking risks at level crossings. Metropolitan Corridor, *1983.*

348.10 The average time it takes a train to pass this crossing is 10 seconds, whether your car is on it or not.

Warning sign at a crossing on the Seminole Gulf Railroad, Nokomis, Florida. Noted in Road & Track Magazine, *July 1988.*

349. LIFE

349.1 If we do not get out sleepers, and forge rails, and devote days and nights to the work, but go to tinkering upon our *lives* to improve *them*, who will build railroads? And if railroads are not built, how shall we get to heaven in season? But if we stay at home and mind our business, who will want railroads? We do not ride on the railroad; it rides upon us. Did you ever think what those sleepers are that underlie the railroad? Each one is a man, an Irishman, or a Yankee man. The rails are laid on them, and they are covered with sand, and the cars run smoothly over them. They are sound sleepers, I assure you. And every few years a new lot is laid down and run over; so that, if some have the pleasure of riding on a rail, others have the misfortune to be ridden upon. And when they run over a man that is walking in his sleep, a supernumerary sleeper in the position, and wake him up, they suddenly stop the cars, and make a hue and cry about it, as if this were an exception. I am glad to know that it takes a gang of men for every five miles to keep the sleepers down and level in their beds as it is, for this is a sign that they may sometime get up again.

Henry David Thoreau, Walden, *1854.*

349.2 All through the ghostly stillness of the land the train made on forever its tremendous noise fused of a thousand sounds and haunted by the spell of time. And that sound evoked for him a million images: old songs, old faces and forgotten memories, and all strange, wordless, and unspoken things men know and live and feel, and never find a language for, the legend of dark time, the sad brevity of their days, the strange and bitter miracle of life itself. And through the thousand rhythms of this one design he heard again, as he had heard ten thousand times in childhood, the pounding wheel, the tolling bell, the whistle wail. Far, faint, and lonely as a dream, it came to him again through that huge spell of time and silence and the earth, evoking for him, as it had always done, its tongueless prophecy of life, its wild and secret cry of joy and pain, and its intolerable promises of the new lands, morning and a shining city.

Thomas Wolfe, Boom Town, *1934.*

349.3 And all these other ones—those unnumbered motes of life who now were being hurled into this tortured rock through tunnels roaring with the blind energy of the subway trains, who were roaring in across the bridges, rushing in by train, sliding in packed in a dense wall across the blunted snouts of ferries—who were pouring out of seven million sleeping cells in all the dense compacted warrens of the city life, to be rushed to seven million other *waking* cells of work—did *they* know?

Thomas Wolfe, of the fragility of life, No More Rivers, *1984.*

349.4 They were being hurled in from every spot upon the compass—the crack trains of the nation were crashing up from Georgia, flashing down out of New England, the crack trains thirty coaches long that had smashed their way all through the night across the continent—from Chicago, St. Louis, from Montreal, Atlanta, New Orleans, and Texas. These great projectiles of velocity that had bridged America with the pistoned stroke, the hot and furious breath of their terrific drive, were now pounding at the river's edge upon the very lintels of the city. They were filled with people getting up—at sixty miles an hour—filled with people getting dressed—at sixty miles an hour—filled with people getting shaved—at sixty miles an hour—walking down carpeted narrow aisles between the green baize curtains of the Pullman berths—at sixty miles an hour—and sitting down to eat substantial breakfasts in splendid windowed dining cars, while the great train smashes at the edges of that noble wink, the enchanted serpent of the Hudson River. The hot breath of the tremendous locomotive fairly pants at

the lintels of the terrific city—all at sixty miles an hour—and all for *what*? for *what*?
Ibid.

349.5 I'll go through Life first class or third, but never second.
Attributed to Noel Coward, BBC television, 13 December 2003.

349.6 In the wider issues of the world I am a democrat; in the microcosm of my motherhood, and when it comes to trains, I am a despot.
Lisa St Aubin de Terán, of her determination that her family will always travel by train; Off the Rails, *1989.*

349.7 When I saw the ad in the Sunday paper—BRAKE-MEN WANTED—I thought of it as a chance to clean up my act and get away. In a strategy of extreme imitation, I felt that by doing work this dangerous, I would have to make a decision to stay alive every day, to hang on to the sides of those freightcars for dear life. The railroad transformed the metaphor of my life. Nine thousand tons moving at sixty miles an hour into the fearful night.
Linda Niemann, who had used drink and drugs, and who took a job as a brakeman on the Southern Pacific. Boomer, *1990.*

350. LIGHTING UP

350.1 Last Saturday we lighted the fire in the Tram Waggon and work'd it without the wheels to try the engine; on Monday we put it on the tram road. It work'd very well, and ran up hill and down with great ease, and very manageable. We had plenty of steam and power.
Richard Trevithick, of the first trial of the first steam locomotive in the world; letter to Davies Giddy, 15 February 1804.

350.2 I took me pipe glass and let me pipe I thought to myself I would try to put fire to Jimmy ockam it blaaze away well the fire going rapidly lantern and candle was to no use so No 1 fire was put to her on line by the pour of the sun.
Robert Metcalfe, quoted in George & Robert Stephenson, L T C Rolt, *1960. This describes the use of oakum, lit by the sun's rays through a glass, (thus rendering unnecessary a candle for which someone had been sent) to light the fire of Stockton & Darlington Railway No 1,* Locomotion, *on the very first occasion that she was set on the rails of the railway.*

350.3 The stabler of the iron horse was up early this winter morning by the light of the stars amid the mountains, to fodder and harness his steed. Fire, too, was awakened thus early to put the vital heat in him and get him off.
Henry David Thoreau, Walden, *1854.*

351. LIMERICK JUNCTION

351.1 The principal architectural features of the place, other than the station buildings, consisted of an engine shed on the left front, a fairly large gasometer in the centre, which protruded violently into the view of the Galtee mountains, and a couple of haystacks *en echelon* on the distant right.
E L Ahrons, The Railway Magazine, *June 1924.*

351.2 The effect on the map was a sort of four-sided Irish triangle, the chief geometrical property of which is that the longest way round is the shortest way there.
Ibid.

352. LINE

352.1 A railway is a clear, linear thing and needs equally clear ownership and lines of responsibility.
Libby Purves, The Times, *14 May 2002.*

353. LITERATURE
See also Reading on the Train, Writing on Trains

353.1 One of the peculiarities of modern travel is the great demand there is for books, a book to prevent people seeing the country being quite as essential as a bun to prevent their being hungry.
R S Surtees, Plain or Ringlets? *1860.*

353.2 What a boon to literature railway carriages are!
Morris Bishop, opening line of poem Railroad-Coach Seating-Arrangement, I Cannot But Deplore You. *Date not known.*

353.3 "—perturbed Spirit!" I finished the sentence for her "Yes, that describes a railway-traveller exactly!"
Lewis Carroll, of Hamlet's words, Sylvie and Bruno, *1889.*

353.4 Shakespeare *must* have travelled by rail, if only in a dream: 'perturbed Spirit' is such a happy phrase.
Ibid.

353.5 If Steam has done nothing else, it has at least added a whole new species to English Literature!
Ibid.

353.6 "But the booklets—the little thrilling romance, where murder comes at page fifteen, and the Wedding at page forty—surely *they* are due to Steam?"
"And when we travel by electricity—if I may venture to develop your theory—we shall have leaflets instead of booklets, and the Murder and the Wedding will come on the same page."
Ibid.

354. LIVERPOOL STREET STATION
(London)

354.1 At one time the suburban platforms at Liverpool Street were a forest of tall hats during the morning and evening rush hours, but we did not call them rush hours, because one does not rush in a top hat.

Julian Kaye, LNER Magazine, *March 1933.*

354.2 Through crystal roofs the sunlight fell,
And pencilled beams to gloss renewed
On iron rafters balanced wee
On iron struts; though dimly hued,
With smoke o'erlaid, with dust endued,
The walls and beams like beryl shone;
And dappled light the platforms strewed
With yellow foliage of the dawn
That withered by the porch of the day's divan.

John Davidson, poem, Liverpool Street Station.

354.3 . . . the station which never sleeps.

Langley Aldrich, Railways, vol. 3, *1942.*

354.4 Liverpool Street was unknown to the genteel.

Gwen Raverat, commenting upon the popularity of the Great Northern Railway route from Cambridge into London. Period Piece, *1952.*

355. LLANGYNOG

355.1 Y ddel gerbydres welir—yn rhedeg
Ar hyd ein dyyffryn-dir,
Ac yn gynt ar ei hynt hir
Y fellten ni theithia filltir.

O ganol tre Llangynog—am naw
Cychwyn wneir yn dalog,
Fe'n ceir cyn tri'n fwy gwisgi na'r gog,
A hoenus yn Llundain enwog.

(*The little train so smoothly glides
Along our lovely valley,
And faster than the lightning flash
It travels on its journey.*

*We leave Llangynog town at nine
Without a darkening frown,
And fleeter than the cuckoo's flight
At three reach London town.*)

Anon., quoted in The Story of the Cambrian, *C P Gasquoine, 1922.*

356. LOADING GAUGE

356.1 Loading gauge—it won't fit.

John Elliot, in a conversation with an American Colonel about a Pullman Car made available for General Eisenhower in England, when the Colonel told Elliot to shove the Pullman Car up his backside. Related by David Elliot in Transport Digest, *Spring 1997.*

356.2 We have no trouble crossing the Low Countries; it is only when we get to France that we run into difficulty.

Anon. Deutsche Bundesbahn official describing the potential of high loading-gauge vehicles running from Germany to the Channel Tunnel. Conversation with Alan Stevens of Central Railways, c. 1996.

357. LOCAL HISTORY

357.1 The arrival of the railway is usually recorded in such works, but it is then apt to disappear. Once established, it has become part of the furniture of the place and is taken entirely for granted. In recent studies the closing of the railway line or the station, at some time during the past 30 years, may receive more attention than its opening or its operation in the Victorian age.

Jack Simmons, The Railway in Town and Country, *1830–1914, 1986.*

358. JOSEPH LOCKE

358.1 He may be termed the Commercial Engineer, one who made the money go as far as possible. He was not ambitious of expensive and thrilling works.

R B Dockray, diary, 21 September 1860; quoted by R M Robbins in Journal of Transport History, *November 1965.*

LOCOMOTIVES

Entries on Locomotives are divided into subheadings, thus: Beauty and Otherwise, In the Cab, Character, Chimney (Stack) Design, Cleanliness, Compound, Cylinders, Eternal Qualities, Invention, Manufacturers, Midland Railway, Musing, Names, Need, Operating, Power of Achievement, Principles, Success and Failure, Utility, and Vulnerability.

359. LOCOMOTIVES— Beauty and Otherwise

359.1 . . . I have been talking a great deal to my father about endeavouring to reduce the size and ugliness of our travelling-engines, by applying the engine either on the side of the boiler or beneath it entirely, somewhat similarly to Gurney's steam coach.

Robert Stephenson, letter to Michael Longridge, 1 January 1828.

359.2 We have a splendid engine of Stephenson's it would be a beautyfull ornament in the most elegant drawing room and we have another of Quaker-like simplicity carried even to shabbyness but very possibly as good an engine but the difference in the care bestowed by the engine man the favour in which it is held by others and even oneself not to mention the public—is striking—a *plain* young lady however amiable is apt to be neglected —now your engine is capable of being made very handsome—and *it ought to be so.*

I K Brunel, letter to Thomas Harrison, 6 March 1838.

359.3　In all that relates to thorough building, and in avoiding all tinsel in their engines, the English are our superiors.

John B Jervis, Railway Property, 1866.

359.4　It would be much better for us, if we substituted wrought iron for much of our cast iron material, and left off the brass ornaments, which are not in sound taste on a machine that is made to perform work, and not for show.

Ibid.

359.5　. . . bright engines will cost less in the long run than the plainer one.

Gustavus Weissenborn, American Locomotive Engineering, 1871, quoted by Jack White Jr in American Locomotives, 1997.

359.6　Improved Engine Green.

Description given to the rich gamboge used on Stroudley locomotives on the London, Brighton & South Coast Railway. Perhaps it was a pigment of his imagination.

359.7　I like all my head officers to personally decide matters relating to their departments. Now, as Locomotive Superintendent, you have to decide what colour the engines are painted. It's a matter I don't want to be troubled about, so, as long as it's black I do not mind what colour you choose for them.

W Cawkwell, General Manager, London & North Western Railway, in conversation with Francis Webb, to whom the remark has often been attributed. It has also been attributed to Henry Ford, in respect of his motor cars, but clearly this instance, in 1873, when the LNWR adopted black as the standard livery of its locomotives, predates anything said about cars by Henry Ford, who was ten years old at the time. Quoted by W L Steel, The History of the London & North Western Railway, (stated therein to be a legend and without guarantee of its truth), 1914. (Ford's statement, quoted in the Oxford Dictionary of Quotations, [1999], was "Any customer can have a car painted any colour that he wants so long as it is black." This is stated to have been said in 1909 but not published until after the Cawkwell statement, in 1922.)

359.8　Thee for my recitative,
Thee in the driving storm even as now, the snow, the winter-day declining,
Thee in thy panoply, thy measur'd dual throbbing and thy beat convulsive,
Thy black cylindric body, golden brass and silvery steel,
Thy ponderous side-bars, parallel and connecting rods, gyrating, shuttling at thy sides,
Thy metrical, now swelling pant and roar, now tapering in the distance,
Thy great protruding headlight fix'd in front,
Thy long, pale, floating vapor-pennants, tinged with delicate purple,
Thy dense and murky clouds out-belching from thy smoke-stack,

Thy knitted frame, thy springs and valves, the tremulous twinkle of thy wheels. . . .

Walt Whitman, poem, To a Locomotive in Winter, lines 1–10, 1876.

359.9　Fierce-throated beauty!
Roll through my chant with all thy lawless music, thy swinging lamps at night,
Thy madly-whistled laughter, echoing, rumbling like an earthquake, rousing all,
Law of thyself complete, thine own track firmly holding,
(No sweetness debonair of tearful harp or glib piano thine,)
Thy trills of shrieks by rocks and hills return'd,
Launch'd o'er the prairies wide, across the lakes,
To the free skies unpent and glad and strong.

Ibid., lines 18–25.

359.10　Naa, mon, I canna spile my grand engine with the likes o' that machinery outside o' her.

Patrick Stirling, concerning the proposal to apply Joy's valve gear to a "Single" locomotive of his design. Quoted in Extracts of the Diaries of David Joy, in The Railway Magazine, vol. XXIII, 1908.

359.11　A laddie runnin' wi' his breeks doun.

Patrick Stirling, of the appearance of a coupled engine running at high speed quoted by George Dow in British Steam Horses, 1950.

359.12　Where that shade of green was invented I do not know; possibly some member of the locomotive department had been making investigations on the somewhat small 800-ton steamers plying between Grimsby and the Continent, on which passengers were wont to exhibit various shades of it.

E L Ahrons, of the experimental bilious sea-green paint used on Manchester Sheffield & Lincolnshire locomotives in the mid 1880s, The Railway Magazine, vol. XXXVII, 1915.

359.13　. . . after a few moments of silent observation, he suddenly dived into the small refreshment bar at the back of the buffer stops, for as he afterwards naively explained, "A 'Black and White' was absolutely necessary to take away the taste of that green."

Ibid.

359.14　I have sometimes, when examining locomotive designs, have been driven to the conclusion that the artistic instincts of the engineer were altogether too highly developed. The proportions were perfect and the lines were pleasing to the eye. Wonderful ingenuity had been exercised in arranging out of sight and in most inaccessible positions any excrescence, no matter how important its function might be, that might interrupt the symmetry of the design.

R E L Maunsell, speech at Institution of Locomotive Engineers, 29 January 1916.

359.15 . . . the modern tendency towards ultra commercialism is rapidly proving fatal to the artistic appearance of the modern locomotive, not only on the Midland, but also on many other railways.

E L Ahrons, The Railway Magazine, *March 1920.*

359.16 A Committee of Royal Academicians might not have unanimously approved of the design of the tank, perhaps, but it is understood to have been improvised in a great hurry. It rather looks as if practical Mr I W Boulton had remorselessly confiscated the domestic cistern in his anxiety to fill the order.

A R Bennett, of a former L&NWR Sharp Stewart locomotive rebuilt by him; The Chronicles of Boulton's Siding, *1927.*

359.17 This is definitely the *machine de resistance,* a lovely piece of functional architecture.

Comment, Liverpool Post and Mercury, *c. September 1830, quoted in* LNER Magazine, *October 1930.*

359.18 They grew larger and larger, and that magnificent shape we admired so much as boys, the "fair round belly" of Shakespeare, gradually spread to the knees, and the locomotive began to become as Moynihan would have the human body, with all the pipes outside. We sometimes see at present a dreary looking thing driven by that mysterious and useless element, electricity, looking for all the world like a rather dull furniture van, and then we have been delighted in what you might call our ten thousandth moment to see the Land Zeppelin of Mr Gresley.

Lt Col E Kitson Clark, Annual Dinner, Institution of Locomotive Engineers, 27 February 1931. "Moynihan" was Lord Moynihan, an eminent surgeon of the day; the "Land Zeppelin" was the LNER high pressure compound 4-6-4 No 10000.

359.19 They would always regard the engine with awe, trying to realize that this up close was the same locomotive that from across the plain looked like a shining new toy. Seen nearly, the engine had only the poetry of immense utility about it, unlovely and crude in its details, with warts of steamy sweat on its iron hide, oozing rust from its throbbing plates, pissing water with violence in some places and in others, leaks with slovenly dripping; The plates crazily bolted and loosely jointed, that whined and sang when the engine rocked along the track.

Paul Horgan, Main Line West, *1936.*

359.20 . . . a colour scheme so rich as to be something of an acquired taste to strangers, like the cookery of Tudor England.

C Hamilton Ellis, of the early livery of Great Eastern Railway locomotives, The Trains We Loved, *1947.*

359.21 I lit another Camel and crossed the cab to take in the glory of what Al Kalmbach would call a "soul-disturbing sight." Headed west for Sherman was a smoke-deflectored 4-6-6-4 leading a 4-8-8-4 on long tonnage—eight high-pressure cylinders talking themselves hoarse through four stacks on the big hill. Hallelujah and damn the diesels!

David P Morgan, Trains Magazine, *January 1952.*

359.22 At Liverpool Street Station on a Sunday
I gloried in an engine, shiny black,
And, though for late July it was a dun day,
The ribboned red and white across the back
And down the sides—I never saw a crack
In all the painting—made a jolly showing.
I did not think it would be there on Monday
So, careless of its coming and going,
I quenched my thirsty soul with wonder overflowing.

Terence Greenidge, verse one of The Nationalisation of the Railways, *within* Girls and Stations, *1952.*

359.23 What will ever equal the action of rods and valve gear flashing in the sunlight?

A C Kalmbach, Trains Magazine, *May 1954.*

359.24 As a boy I had access to enough copies of *Railway Wonders of the World* to form the opinion that the European steam locomotive, seldom a handsome animal at best, declined steadily in esthetic appeal as one moved eastward from France. Or southward, for that matter.

David P Morgan, Trains Magazine, *March 1964.*

359.25 Bellissima locomotiva!

Rivarossi catalogue, advertising model of German class 52 2-10-0 locomotive, c. 1965.

359.26 The engine was a fine example of Austrian individuality. It appeared as if the builder had mixed up his blueprints and got the boiler and cab backwards on the frame, but that did not detract from the locomotive's appearance to a lover of railroads.

Otto Kuhler, almost certainly of the Golsdorf 2-6-4 passenger express tender locomotives built at Floridsdorf and elsewhere from 1911. My Iron Journey, *1967.*

359.27 The majestic balloon-stacks had shrunk away. Sand and steam domes had deteriorated to mere warts, because that was all that was needed. Ornamentation for ornamentation's sake disappeared. The engine's power needs led naturally to clean, functional lines, for in the new age of utility, a locomotive did not need to acquire the character of a conquering hero of the vast western lands.

Otto Kuhler, My Iron Journey, *1967.*

359.28 As locomotive builders continued to build bigger and more powerful boilers, engines grew taller, and other features were submerged into the whole. The once proud cabs of earlier years became an incidental shed at one end, and the important-looking bonnet chimney disappeared inside the smokebox. The domes were eventually truncated for clearance purposes, making the engine a great solid mass, fairly straight on top with merely a few innocuous protruberances to break the monotony of its profile.

Then came the plumbing.

Robert C Reed, The Streamline Era, *1975.*

359.29 Where's the key?

W A Stanier, of the crude toy-like appearance of the Bulleid class Q1 locomotive, 1942. Quoted by H A V Bulleid in Bulleid of the Southern, *1977.*

359.30 You loved them too: those locos motley gay
That once seemed permanent as their own way?—
The Midland "lake", the Caledonia blue;
The Brighton "Stroudleys" in their umber hue;
North Western "Jumbos", shimmeringly black,
That sped, shrill-whistled, on their "Premier" track;
And all a forest's tints of green—GC,
GN, GW, LTS, HB,
South Western, Highland, "Chatham": many more
Both on our own and on the Emerald shore?

Gilbert Thomas, poem, Nostalgia. *Published in* Punch Almanack, *7 November 1949, but not in complete form. The complete version appeared in* Selected Poems *by Gilbert Thomas, 1951. It also appeared in* Double Headed, *co-authored by Gilbert Thomas and his son David, 1963. In the latter book, the poem has many minor differences of punctuation, one difference of spelling (Caledonian in the third line above) and two additional lines at the opening of the fifth (last) stanza.*

359.31 The engine was one of the old locomotives of the 'Highland Light' class of around 1870 which Bond had heard called the handsomest steam locomotives ever built. Its polished brass handrails and the fluted sand-dome and heavy warning bell above the long, gleaming barrel of the boiler glittered under the hissing gaslights of the station. A wisp of steam came from the towering balloon smoke-stack of the old wood-burner.

Ian Fleming, Diamonds are Forever, *1956.*

359.32 The Englishman's reaction on seeing a French locomotive for the first time is usually anything but favourable. Its bewildering complexity and apparent ugliness suggest a blatant disregard for beauty altogether at variance with his own notions. On the other hand if he lingers long enough to grow familiar with these brutish monsters, the day will surely come when he begins to realise why his Gallic friends complain that the sturdy simplicity of the British steam engine strikes them as uninteresting and insipid!

P Winding, Railway World, *vol. XXI, 1960.*

359.33 Except among the unreconstructed fringe of churls and curmudgeons, everybody fell under the spell of steam locomotion.

Lucius Beebe, Mr Pullman's Elegant Palace Car, *1961.*

359.34 The Atlantics were amongst the loveliest engines Robinson ever designed. Their curvaceous bodies, opulent livery and excellent performance captured the fancy of railway enthusiast and layman alike.

George Dow, Great Central, *vol. III, 1965.*

359.35 The engine, gleaming in black and yellow varnish and polished brass, was a gem. It stood, panting quietly in the sunshine, a wisp of black smoke curling up from the tall stack behind the big brass headlight. The engine's name 'The Belle' was on a proud brass plate on the gleaming black barrel and its number, 'No. 1', on a similar plate below the headlight.

Ian Fleming, The Man with the Golden Gun, *1965.*

359.36 I fully understand Southern Pacific's operational reasons for turning a Mallet around so that it could operate cab-first. I can also appreciate their consideration in running the result in sparsely populated areas.

Ed King, of SP's cab-forward locomotives, Trains Magazine, *February 1985.*

359.37 This isn't a locomotive. It's the box the locomotive came in.

Christopher Knapton, characterising reactions to the then new Amtrak AMD-103 diesel-electric locomotive. Quoted in Trains Magazine, *November 1993.*

360. LOCOMOTIVES—In the Cab

360.1 My railway carriage tonight, which is a compound of the coal-cellar, the bakehouse oven, and the fiery dragon, is the conductor, the ruler, the guardian and the leader of all these—it is the engine.

John Hollingshead, Riding the Whirlwind, *Household Words, 12 December 1857.*

360.2 . . . offices of professional gentlemen rather than the posts of duty of hard-handed engine-drivers.

Comment, Engineering, *July 26 1867, quoted by John White Jr in* American Locomotives, *1997.*

360.3   To bear
The pelting brunt of the tempestuous night,
With half-shut eyes, and puckered cheeks, and teeth
Presented bare against the storm; . . .

Anon., quoted by Frederick Williams in The Midland Railway, *1878.*

360.4 In Great Britain the ingenious theory that superior appliances of greater personal comfort in some definable way lead to carelessness in employés was carried to such an extent that only within the last few years has any protection against wind, rain and sunshine been furnished on locomotives for the engine-drivers and stokers. The old stage-coach driver faced the elements, and why should not his successor on the locomotive do the same?

Charles Francis Adams, Notes on Railway Accidents, *1879.*

360.5 Our locomotive was under a full head of steam. The engineer stood with his hand on the lever, with the valve wide open. It was frightful to see how the powerful iron monster under us would leap forward under the revolutions of her great wheels. Brown would scream to me ever and anon., 'Give her more wood, Alf,' which command was promptly obeyed. She rocked and reeled like a drunken man, while we tumbled from side to side like grains of popcorn in a hot frying-pan. It was bewildering to look at the ground or objects on the roadside. A constant stream of fire ran from the great wheels, and to this day I shudder as I reflect on that, my first and last locomotive ride.

J A Wilson, the fireman to engineer Wilson W Brown during the Andrews Raid, April 1862; Adventures of Alf. Wilson, *1880.*

360.6 We sped past houses, stations, and fields, and out of sight, almost like a meteor, while the bystanders who scarcely caught a glimpse of us as we passed, looked on as if in both fear and amazement. It has always been a wonder to me that our locomotive and cars kept the track at all, or how they could possibly stay on the track. At times the iron horse seemed literally to fly over the course, the driving wheels of one side being lifted from the rails much of the distance over which we now sped with a velocity fearful to contemplate. We took little thought of this matter then. Death in a railroad smash-up would have been preferred by us to capture.

Ibid.

360.7 The engine-cabs of England, as of all Europe, are seldom made for the comfort of the man. One finds very often this apparent disregard for the man who does the work—this indifference to the man who occupies a position for the exercise of temperance, of courage, of honesty, has no equal at the altitude of prime ministers.

Stephen Crane, The Scotch Express, *within* Men, Women and Boats, *1921.*

360.8 This valkyric journey on the back of the vermilion engine, with the shouting of the wind, the deep mighty panting of the steed, the gray blur at the trackside, the flowing quicksilver of the other rails, the sudden clash as a switch intersects, all the din and fury of this ride, was of a splendor that caused one to look abroad at the quiet, green landscape and believe that it was of a phlegm quiet beyond patience. It should have been dark, rain-shot, and windy; thunder should have rolled across the sky.

Ibid. Stephen Crane lived in England in the last years of

his life before he died in 1900. He went from Euston to Glasgow on the London & North Western and Caledonian railways, and in the cab of the locomotive from Euston to Crewe, and then again from Carlisle to Glasgow. His reference to the "vermilion engine" can only mean that he travelled on the locomotive Greater Britain, which was painted scarlet from May 1897 to July 1898 to celebrate the Diamond Jubilee of Queen Victoria. His reference elsewhere in this essay to the coaches being white and bottle green is not understood.

360.9 I suppose the greatest moments in life are those when you don't believe it's yourself. It *can't* be you, in that holy of holies of small-boy imagination, the cab of an engine—and such an engine.

Christopher Morley, referring to NYC 5217; A Ride in the Cab of the Twentieth Century Limited, *1928.*

360.10 Alive, shouting, fluttering her little green flags, she divided the clear cool afternoon. Looking out into that stream of space I could have lapsed into a dream. I came closer than ever before to the actual texture of Time where our minds are made. This was not just air or earth that we flew upon. This was the seamless reality of Now. We were abreast of the *Instant.*

Ibid.

360.11 . . . something like riding on an enormously heavy solid-tyred bicycle.

Eric Gill, LNER Magazine, *after riding on the* Flying Scotsman, *in the cab of an LNER A1 Pacific, 1933.*

360.12 The other train came slowly on with that huge banging movement of the terrific locomotive, eating its way up past the windows, until the engine cab was level with Eugene and he could look across two or three scant feet of space and see the engineer. He was a young man cleanly jacketed in striped blue and wearing goggles. He had a ruddy colour and his strong, pleasant face, which bore on it the character of courage, dignity, and the immense and expert knowledge that these men have, was set in a good-natured and determined grin, as with one gloved hand held steady on the throttle, he leaned upon his sill, with every energy and perception in him fixed with a focal concentration on the rails. Behind him his fireman, balanced on the swaying floor, his face black and grinning, his eyes goggled like a demon, and lit by the savage flare of his terrific furnace, was shovelling coal with all his might. Meanwhile, the train came on, came on, eating its way past, foot by foot, until the engine cab had disappeared from sight, and the first coaches of the train drew by.

Thomas Wolfe, describing an overtaking train, Of Time and the River, *1935. See also 361.25.*

360.13 Received into the warm of the cab, these four blinked at the red throat of the firebox and felt the rich reprieve of some convalescence.

The enormous night held the little train. The enormous sky held the little storm. Standing in the gangway Elmar Dasher threw back his head and took the hour. Motion, roar, electricity blue on the metal, thunder of wheel and of cloud, and he in their midst. He felt in flight over farther spaces. They swept by a steel plant, saw the belch and glare, made out men and mounds of slag. One of the sheds stood drenched in the red beauty of the pouring, and naked figures flowered from wild color. In that flash were the tumult of farther industry, roar of innumerable trains, glitter of all the cities, their veins flowing with men and women in sin, joy, anguish. He was feeling all that there was to feel! He could have cried a challenge as the engine went challenging the lit blackness.

Zona Gale, A Far Cry, *1937.*

360.14 Caution should be used in opening the throttle, as it has been proved these locomotives will start whether the train does or not.

Advice from a Santa Fe engine foreman to someone inexperienced in the new diesel locomotives, 1940. Recalled by David P Morgan, Trains Magazine, *November 1970.*

360.15 Smoke filters in and tries to fog the lighted cab gauges, and you feel your breathing rough and strained. This is the opposite end of the railroad book from vista domes and tip-top taverns. It's railroading stripped of all glamor, yet robed in a glory unknown to parlor car passengers.

David P Morgan, of a passage through Tennessee Pass tunnel on a 4-8-4 behind a 2-8-8-2 helper; Trains Magazine, *September 1949.*

360.16 ... a magnificent experience.

Princess Elizabeth, after a ride in the cab of a Canadian locomotive, 1951, as noted in Trains Magazine, *December 1952.*

360.17 Each revolution of her ten-coupled 57-inch drivers shakes the seatbox and jars cocks and gauges. The heat from over three score square feet of inflamed coal crust seeps through the Butterfly firedoors until sweat dampens your socks and trickles down the small of your back. In the tunnel blackness her stack hammers hard against raw rock, and the thick, pungent, gassy smoke swirls back into the cab to slow your breathing and film your face; always the cinders are creeping past the red bandanna round your neck. The noise is rhythmic, constant, deafening: the pound of the harness of main and side rods, crossheads and valve motion; the monotone of the stoker screw; the asthmatic whine of the injector; the exhalation of brass valves; the rattle of deck plates.

Yet for this a man would leave the security of a shoe store, the smell of fresh-plowed earth, the regularity of factory hours.

David P Morgan, caption of photograph, Trains Magazine, *May 1953.*

360.18 When I say stop, you stop, or you get some more.

Douglas Goody, train robber, to driver Jack Mills, who had just been beaten on the head with an iron bar, and who was then instructed to drive his locomotive; 8 August 1963. Recounted in The Robbers' Tale, *Peta Fordham, 1965.*

360.19 Once you sling a boiler capable of several thousand horsepower high up over driving wheels as tall as a basketball player and exhaust the energy through a transmission of reciprocating steel rods, it's patently clear that the men aboard operate in a world of heat and metal all their own.

David P Morgan, The Mohawk that Refused to Abdicate, *1975.*

360.20 I simply cannot understand why steam engines have to have such clumping great levers to control them.

Prince Philip, quoted by Terence Cuneo in The Railway Painting of Terence Cuneo, *1985.*

360.21 The fireman cools off in the gap where other railways have cab doors.

Ron White, in a dig at the Great Western Railway for being different; caption to transparency of ex-Great Western Railway locomotive Earl Cairns, in Colour-Rail catalogue, *1996.*

360.22 Are you alright with one of these?

Anon. driver to Ray Towell, when asking him to mind his class 8F locomotive for a while.

360.23 I suppose it was a bit like being asked if one minded going on a date with Brigitte Bardot.

Alan Pegler, on an invitation to ride in the cab of A4 Sir Nigel Gresley on the day of its special working for the Stephenson Locomotive Society, 23 May 1959, recorded in The Railway Magazine, *March 2001.*

361. LOCOMOTIVES—Character

361.1 As soon as the engine has performed this task (which shall be equal to the travelling from Liverpool to Manchester) there shall be a fresh supply of water and fuel delivered to her, and as soon as she can be got ready to set out again, she shall go up to the starting post, and make ten trips more, which shall be equal to the journey from Manchester back again to Liverpool.

Extract from Regulations for the conduct of locomotive trials at Rainhill, October 1829.

361.2 We were introduced to the little engine which was to drag us along the rails. She (for they make these curious little firehorses all mares) consisted of a boiler, a stove, a small platform, a bench, and behind the bench a barrel containing enough water to prevent her being thirsty for fifteen miles,—the whole machine not bigger than a common fire engine.

Fanny Kemble, letter, after a journey on the Liverpool & Manchester Railway; 26 August 1830.

361.3 She goes upon two wheels, which are her feet, and are moved by bright steel legs called pistons; these are propelled by steam, and in proportion as more steam is applied to the upper extremities (the hip-joints, I suppose) of these pistons, the faster they move the wheels; and when it is desirable to diminish the speed, the steam, which unless suffered to escape would burst the boiler, evaporates through a safety-valve into the air. The reins, bit, and bridle of this wonderful beast is a small steel handle, which applies or withdraws the steam from its legs or pistons, so that a child might manage it.

Ibid.

361.4 The coals, which are its oats, were under the bench, and there was a small glass tube affixed to the boiler, with water in it, which indicates by its fullness or emptiness when the creature wants water, which is immediately conveyed to it from its reservoirs. There is a chimney to the stove, but as they burn coke there is none of the dreadful black smoke which accompanies the progress of a steam vessel. This snorting little animal, which I felt rather inclined to pat, was then harnessed to our carriage, and Mr Stephenson having taken me on the bench of the engine with him, we started at about ten miles an hour.

Ibid.

361.5 There were noble steeds in days of old,
They were fierce in battle, in danger bold;
They clanked in armor, and shone in gold,
And they bore their riders with lordly pride;
But the Iron Horse, there were none like him!
He whirls you along till your eye is dim,
Till your brain is crazed, and your senses swim,
With the dizzy landscape on either side!

He springs away with a sudden bound,
His hoof, unshodden, spurns the ground,
His nostril dashes its foam around,
Like the first faint clouds of a thunder shower:
And a stated moment he ever hath,
When he rushes forth on his iron path,
And wo to him who shall rouse his wrath,
By curbing him in, beyond the hour!

Daniel March, poem, The Iron Horse, *first two verses. Written as a college exercise at Yale, 1839.*

361.6 We believe that the steam engine, upon land, is to be one of the most valuable agents of the present age, because it is swifter than the greyhound, as powerful as a thousand horses; because it has no passions and no motives; because it is guided by its directors; because it runs and never tires; because it may be applied to so

many uses, and expanded to any strength.

Comment, Merchants' Magazine, *October 1840.*

361.7 And as to the ingein,—a nasty, wheezin', creakin', gaspin', puffin', bustin', monster, always out o' breath with a shiny green-and-gold back, like a unpleasant beetle in that 'ere gas magnifier,—as to the ingein as is alvays a pourin' out red-hot coals at night, and black smoke in the day, the sensiblest thing it does, in my opinion, is, ven there's somethin' in the vay, and it sets up that 'ere frightful scream vich seems to say, "Now here's two hundred and forty passengers in the wery greatest extremity o' danger, and here's their two hundred and forty screams in vun!"

Charles Dickens, Master Humphrey's Clock, *1841.*

361.8 I love to see one of these huge creatures, with sinews of brass and muscles of iron, strut forth from his smoky stable, saluting the long train of cars with a dozen sonorous puffs from his iron nostrils, fall gently back into his harness. There he stands, champing and foaming upon the iron track, his great heart a furnace of glowing coals; his lymphatic blood is boiling in his veins; the strength of a thousand horses is nerving his sinews—he pants to be gone.

Elihu Burritt, American Railroad Journal, *July 1844, quoted in* Railroads and the Character of America, *James A Ward, 1986.*

361.9 A locomotive engine must be put together as carefully as a watch.

Attributed to Robert Stephenson by Ralph Waldo Emerson, Journal RS, *page 234, 1849.*

361.10 I really think that there must be some natural affinity between Yankee 'keep moving' nature and a locomotive engine . . . Whatever the cause, it is certain that the 'humans' seem to treat the 'ingine', as they call it, more like a familiar friend than as the dangerous and desperate thing it really is.

Mrs Houstoun, Hesperos: or, Travels in the West, *1850.*

361.11 This locomotive was possessed of a certain inborn cussedness, which could hardly be the attribute of a mere machine—her spiritual nature was a sort of Mephistophelian cross with a Colorado mule—and as to her physical constitution and membership, a cotton-factory 'mule' was simple in comparison.

Alexander L Holley, of the locomotive Advance, *built in 1851 to demonstrate the use of the Corliss valve gear on locomotives. Quoted by Jack White Jr in* American Locomotives, *1997.*

361.12 All day the fire-steed flies over the country, stopping only that his master may rest, and I am awakened by his tramp and defiant snort at midnight, when in some remote glen in the woods he fronts the elements incased in ice and snow; and he will reach his stall only with the morning star, to start once more on his travels without rest or slumber. Or perchance at evening, I hear him in his stable blowing off the superfluous energy of the day, that he may calm his nerves and cool his liver and brain for a few hours of iron slumber. If the enterprise were as heroic and commanding as it is protracted and unwearied!

Henry David Thoreau, Walden, *1854.*

361.13 The Locomotive, naturally, was my chief object of study; and to the investigation of that admirable machine, its construction, its working, and its performance, my attention principally was directed. I may add that the peculiar union of beauties, mechanical and aesthetical, to be found in the locomotive:—the association of compact, concentrated, mechanical action, with freedom and elegance of form, and with the most graceful of movements,—made me conscious that there was not alone a theorem to investigate, or a problem to solve, but an object, besides, of enduring admiration. Whatever the absolute magnitude, whatever the form into which it was moulded by eccentricity of genius, quaintness of originality, depravity of taste, or lack of the mechanical faculty, the essence at least was there, embodied in the admired combination of the multitubular boiler, the blastpipe, and, in later times, the link-motion.

Daniel K Clark, Introductory Preface to Railway Machinery, *1855.*

361.14 If you talk of locomotives and would like to know the star,
Step up here on the footplate for a trip to Waratah.
Oh, I drive the finest engine—I can prove the statement true,
They've neither man or engine equals me and Twenty-Two.

There's the four-wheeled coupled Fairburns, Numbers One, and Two, and Three,
They're as fleet as Flying Dutchmen, but they're wake as any flea:
For speed and strength and steaming, and likewise for running true,
There's a happy combination in old Number Twenty-Two.

Anon. poem, first two verses, published under the name Javey *in* The Locomotive Journal, *Australia, 1880.*

361.15 There's the Thirties and the Forties, they are Beyer Peacock's make,
They're easy on the lever, they're handy on the brake,
With improvements and inventions, and with everything that's new;
But the bully engine of them all is Number Twenty-Two.
Ibid., verse five.

361.16 Whirling along its living freight, it came,

Hot, panting, fierce, yet docile to command—
The roaring monster, blazing through the land
Athwart the night, with crest of smoke and flame;
Like those weird bulls Medea learned to tame
By sorcery, yoked to plough the Colchian strand
In forced obedience under Jason's hand.
Yet modern skill outstripped this antique fame,
When o'er our plains and through the rocky bar
Of hills it pushed its ever-lengthening line
Of iron roads, with gain far more divine
Than when the daring Argonauts from far
Came for the golden fleece, which like a star
Hung clouded in the dragon-guarded shrine.

Christopher Cranch, poem, The Locomotive, *c. 1890.*

361.17 Then, faint and prolonged, across the levels of the ranch, he heard the engine whistling for Bonneville. Again and again, at rapid intervals in its flying course, it whistled for road crossings, for sharp curves, for trestles; ominous notes, hoarse, bellowing, ringing with the accents of menace and defiance; and abruptly Presley saw again, in his imagination, the galloping monster, the terror of steel and steam, with its single eye, cyclopean, red, shooting from horizon to horizon; but saw it now as the symbol of a vast power, huge, terrible, flinging the echo of its thunder over all the reaches of the valley, leaving blood and destruction in its path; the leviathan, with tentacles of steel clutching into the soil, the soulless Force, the iron-hearted Power, the monster, the Colossus, the Octopus.

Frank Norris, The Octopus, *1901.*

361.18 But the moving train no longer carried with it that impression of terror and destruction that had so thrilled Presley's imagination the night before. It passed slowly on its way with a mournful roll of wheels, like the passing of a cortege, like a file of artillery-caissons charioting dead bodies; the engine's smoke enveloping it in a mournful veil, leaving a sense of melancholy in its wake, moving past there, lugubrious, lamentable, infinitely sad, under the grey sky and under the grey mist of rain which continued to fall with a subdued, rustling sound, steady, persistent, a vast monotonous murmur that seemed to come from all quarters of the horizon at once.
Ibid.

361.19 France has given to the world a fair share of the freaks designed to send the ordinary forms of locomotives prematurely to the scrapheap, and, incidentally, to demonstrate what amateur designers could do in wandering away from the well trodden paths of engineering rectitude. Gallic sentiment leans kindly to things that look new.

Angus Sinclair, Development of the Locomotive Engine, *1907.*

361.20 It came into use a crude apparatus of wood and iron, without comliness, elimental in form, material and in operation; but the hands of master mechanics by degrees converted it into the shape that required merely additions in dimensions and weight to produce the locomotive of today.

Ibid.

361.21 It may safely be stated, without exaggeration, that in the whole history of British Railways there has never existed such an extraordinary collection of freak locomotives as those which were built for the Great Western and delivered during a period of about eighteen months from November 1837.

E L Ahrons, contribution to History of the Great Western Railway, *(E T MacDermot), 1927.*

361.22 I have a theory about railway engines being bad tempered. Well, when I say 'bad tempered', that's putting it mildly. They're actually livid, furious beasts, and they loathe humanity.

Reginald Gardiner, monologue on gramophone record Trains, *1934.*

361.23 I was vastly impressed, because it seemed to me to be four times as big and eight times as livid as any engine I'd ever seen before. To begin with it had eight of everything: cowcatchers, bells, and everything but the kitchen stove, hanging all over it. Added to which it had a very bad-tempered word written across the front. It just said NORD, which I know is horrid to start with.

Ibid., of the first French locomotive he saw, at Calais.

361.24 . . . little locomotives could curve on the brim of a sombrero.

Cy Warman, of the narrow gauge locomotives of the Denver South Park & Pacific, quoted by Lucius Beebe in Narrow Gauge in the Rockies, *1958.*

361.25 Then the locomotive drew in upon them, loomed enormously above them, and slowly swept by them with a terrific drive of eight-locked pistoned wheels, all higher than their heads, a savage furnace-flare of heat, a hard hose-thick hiss of steam, a moment's vision of a lean old head, an old gloved hand of cunning on the throttle, a glint of demon hawk-eyes fixed forever on the rails, a huge tangle of gauges, levers, valves, and throttles, and the goggled blackened face of the fireman, lit by an intermittent hell of flame, as he bent and swayed with rhythmic swing of laden shovel at his furnace doors.

Thomas Wolfe, Of Time and the River, *1935. See also 360.12.*

361.26 He has paused by night in the yards at Cleveland where the feline Hudsons of the New York Central panted in the hot darkness, mousing for green lights among the switchpoints, and the yard engines throbbed with a subdued tensity waiting for the Century, in three sections, to thunder through like the implications of destiny itself.

Lucius Beebe, of his own observations, as recorded in High Iron, *1938.*

361.27 Big Boy.

Anon. shop worker at American Locomotive Company 1941. Chalked sign on smokebox of a Union Pacific 4-8-8-4, which gave the class its nickname.

361.28 Twenty-two cents a pound.

Anon., description of Big Boy locomotive in Railroad Magazine, *December 1941.*

361.29 Behold a steed with thews of iron,
A heart and brain of fire;
His voice a thousand trumpets shames;
His sinews never tire.

E B Rittenhouse, poem, The Iron Horse, A Study of Railway Transportation, vol. 2, *Association of American Railroads, 1942.*

361.30 Of body dark, gigantic, vast,
His way no arm can bar;
Restless as the battle gods,
His flight is like a star.

Ibid., second verse.

361.31 His path, twin bands of virgin steel,
That stretch from East to West;
O'er beams the invaded forest gave,
Now fixed in nature's breast.

Ibid., third verse.

361.32 We do not use trailing trucks in England.

Major R Hart-Davies, giving a reason (not wholly correctly) why standard U.S. Army locomotive designs for use overseas in World War II should not have the 2-8-2 wheel arrangement. Hart-Davies had worked in the drawing office at Doncaster locomotive works, and knew that some British locomotives had trailing axles even if they were not carried in trucks as understood by American practice; to Col Howard G Hill, U.S. Army. Recalled by Hill in Trains Magazine, *December 1964.*

361.33 With lungs of fire and ribs of steel,
With shrieking valve and groaning wheel,
With startling scream and giant stroke,
Swift showers of sparks and clouds of smoke,
The iron horse the train is bringing,
So look out while the bell is ringing.

George W Bungay, poem, The Locomotive.

361.34 The steam locomotive is one of the few mechanical inventions of the last hundred years which has not been prostituted to purposes of lying, hatred and slaughter.

Canon Roger Lloyd, The Fascination of Railways. *1957.*

361.35 Scaling is the arterio-sclerosis of the steam locomotive.

Louis Armand, CME of SNCF and expert on water treatment, in conversation with George Carpenter, c. 1960.

361.36 The personification of a mountain of misbegotten machinery—in this case a diesel, if I may talk dirty—appears over the august initials of the editor himself in the October 1965 Trains, and I fear the impact of its example on less mature and more impressionable minds.

Lucius Beebe, of the practice of referring to locomotives as "she"; Trains Magazine, March 1966.

361.37 The breath which transformed this mere machine into a personality was steam. Steam enabled it to show—visibly and audibly—willingness, disapprobation or repose. The locomotive engine moved with joy and grace and a merry blast of whistle when the train was light and there was a schedule to be kept. It slipped its wheels or puffed in exasperated, labored tones when its load was heavy. It panted impatiently at station stops, waiting to be given the reins for the next stage of the journey. Its safety valves blew off exultantly as it topped a long gradient and the demands upon it suddenly ceased. When its work was done, it hissed to itself in contented repose, awaiting the next call to duty.

Omer Lavalee, at valedictory for Canadian National Railway locomotive 6218, upon its retirement in 1971.

361.38 You haven't seen a steam locomotive until you've seen a woodburning Garratt at night.

Allen Jorgensen, quoted by Charles P Lewis in Trains Magazine, *April 1972.*

361.39 It was a steam train, and for the first time since leaving home I wished I had brought a camera, to take its picture. It was a kind of demented samovar on wheels, with iron patches on its boiler and leaking pipes on its underside and dribbling valves and metal elbows that shot jets of vapour sideways. It was fuelled by oil, so it did not belch black smoke, but it had bronchial trouble, respirating in chokes and gasps on grades and wheezing oddly down the slopes when it seemed out of control.

Paul Theroux, The Old Patagonian Express, 1979.

361.40 They sent it down to see if we could make an engine out of it. We couldn't.

Anon. vice president of Norfolk & Western Railway, of the Pennsylvania Railroad duplex drive steam locomotive 5515, which visited the N&W in June 1948. Quoted in Trains Magazine, *January 1983.*

361.41 I am electric, feel my attractions, feel my magnetism, you will agree.
I am electric, I have the contacts, I am electric, the future is me.

I am electric, mind how you touch me, I can shock you, I can set you on fire,
I can reach up and pluck down the lightning, watch the conductor, see the live wire.

Richard Stilgoe, song, AC/DC, Starlight Express, 1993.

361.42 The more I learn about diesels, the more I respect steam.

Neil Jordison, steam locomotive engineer, conversation with compiler, 23 May 2001.

362. LOCOMOTIVE—Chimney (Stack) Design

362.1 If indeed chimneys are thought of as hats, and if a good hat makes a well-dressed engine, the Great Western bears away the palm every time, for their engines have chimneys which look as though they fitted, and they all have that resplendent copper rim by way of a little seemly flourish.

Canon Roger Lloyd, The Fascination of Railways, 1957.

362.2 . . . remember, as the hat is to a well-dressed man, so is the chimney to the locomotive.

Robert Weatherburn, Ajax Loquitur, 1899.

362.3 Topping the locomotives was a ridiculously effeminate flared stack.

Otto Kuhler, of his first sight of a pair of Pennsylvania Railroad K4s Pacifics, My Iron Journey, 1967.

363. LOCOMOTIVES—Cleanliness

363.1 Everything shone from the bufferheads to the seat of the driver's overalls!

Kenneth Perry, of Stanier Pacific Coronation prepared for hauling a Royal train, Railway World, vol. XVIII, 1957.

364. LOCOMOTIVES—Compound

364.1 She's economical (I call it mean) in her coal, but she takes it out in repairs.

Rudyard Kipling, .007, Scribner's Magazine, August 1897.

364.2 Of the thousands of engines built by the Webb at Crewe, many were called Compounds, because they had two pairs of driving wheels which could turn in opposite directions simultaneously. No other C.M.E. ever succeeded in making ambidextrous locomotives.

C Hamilton Ellis, Rapidly Round the Bend, 1959.

365. LOCOMOTIVES—Construction

365.1 Give us this day our daily rebuild.

Anon., "Daily morning prayer", Doncaster Works Locomotive Drawing Office; response to Edward Thompson's seeming desire to rebuild one of everything in the LNER stock; recalled by Malcolm Crawley to the compiler.

366. LOCOMOTIVES—Cylinders

366.1 ... an engine with outside cylinders judiciously constructed may be a better engine than the inside cylinder engine on the narrow gauge, but it is very easy to make a bad engine with outside cylinders.

William Fernihough, (using the contemporary expression narrow gauge *to distinguish from the broad gauge, before the term* standard gauge *came into general use); evidence to the Gauge Commissioners, August 1845.*

366.2 But these had outside cylinders, and the motion of the connecting rods and cranks produced a loping effect, rather like that of a fine games mistress leading her side on the hockey field—all long legs and suspenders—while the Midland singles swept along in quite a different way. They seemed to be so effortless!

C Hamilton Ellis, The Midland Railway, 1953.

367. LOCOMOTIVES—Eternal Qualities

367.1 I cannot express the amazed awe, the crushed humility, with which I sometimes watch a locomotive take its breath at a railway station, and think what work there is in its bars and wheels, and what manner of men they must be who dig brown iron-stone out of the ground, and forge it into THAT!

John Ruskin, The Cestus of Aglaia, 1860.

367.2 I have a sort of touch-the-brim-of-the-hat respect for the thing, and am never so busy that I cannot give it a civil look as it goes by.

Benjamin Taylor, The World on Wheels, 1874.

367.3 A machine easy to make, easy to run, easy to repair, never weary from its birth in mint condition to the days that saw it worn, dirty and old; wasteful as nature and as inefficient as man, very human in characteristic, far from ideally economic in action but, like our race, ever in a stage of development, master in emergencies, its possibilities of improvement inexhaustible.

Col Kitson Clark, quoted in George & Robert Stephenson, *L T C Rolt, 1960.*

367.4 Breathes there a man with soul so dead that he does not thrill at the sight of a modern steam locomotive at work? Somewhere in the breast of every normal homo sapiens there stretches a cord that vibrates only to the sight and sound of a fine steam locomotive.

Anon., New York Tribune, 1920s, quoted by H Roger Grant, We Took the Train, *1990.*

367.5 It is a regret that the fussy devisings of men, forever enlarging and perfecting yesterday's folly in favor of new and more fearsome complexities of inconvenience, did not cease its search for frustration when, for one moment in eternity, it chanced upon a machine at once useful and beautiful.

Lucius Beebe, Mixed Train Daily, 1947.

367.6 There is not the slightest reason to believe that we have reached the ultimate in design and operation of the steam locomotive.

Anon. member of Norfolk & Western Railway engineering staff, quoted in Trains Magazine, *January 1949.*

367.7 Will there ever be another superb achievement like this? Will the diesel-electric locomotive show its paces in this country as it has in the United States? And if it does, where will be the heralding plume of smoke and steam— the deep-throated beat of the exhaust—the characteristic warning yet friendly note of the whistle—the fascinating flash of piston rod, crosshead, valve gear and side rods, thrusting, swinging, rising and falling so rapidly and purposefully in unison?

Those were the days.

George Dow, reflecting on Mallard's achievement, British Steam Horses, *1950.*

367.8 There is something about the alliance of man and steam horse—of human skill with huge mechanical strength—that never fails to capture the imagination of those who admire an intricate piece of machinery, brilliantly designed, skilfully put together, and beautifully handled.

Arthur Peppercorn, Foreword to British Steam Horses, *1950.*

367.9 In the end doubtless I shall re-visit Germany, but I think they will have to produce something very interesting in locomotives before I do.

Dr P Ransome-Wallis, On Railways, 1951.

367.10 Typically American, even in the sixties, was the emphasis upon headlight and bell. The former represents one item of locomotalia which has not grown in stature with the years, for if the oil lamp that adorned the motive power of the Civil War was not so efficient as its electric descendant, it was incomparably more magnificent, as a glance at any engine picture will show. On the other hand, of all the features of steam locomotive equipment, the bell has changed probably the least. Mechanical aid may now do its ringing, and the modern decline in human pride may have reduced its polish, but bell it essentially remains, and its presence upon a locomotive has ever been a distinguishing mark of North America.

Robert C Black, Railroads of the Confederacy, 1952.

367.11 A steam engine does something which no other piece of mechanism succeeds in doing. It makes energy visibly and audibly impressive, and it seems to have an endless hold over human imagination as it does so.

Canon Roger Lloyd, The Fascination of Railways, 1957.

367.12 As for diesels, I respect what they will do, but a world without steam engines would appear almost as bad as a world without women, Beethoven, autumn foliage or peppermint-stick ice cream.
O Winston Link, Trains Magazine, *March 1957.*

367.13 We've adjusted your greasy wedges,
We've lined your sticky guides,
We've tightened your flopping jaw bolts
And we've hammered your stoker slides.

We've raised your slipping pilot,
We've greased your noisy pins,
We've planed your lateral liners,
We've excused your mechanical sins.

We've installed new cylinder packing,
We've ground in your cylinder cocks,
We've adjusted your brakes and eccentrics
And we've banished your groans and knocks.

But in spite of the work and worry,
And the pains in your running gear,
You've produced the power and service
For the NP these many years.

When they're losing dough on the main line
And the directors wish they were dead,
The W-3's from St. Paul to Duluth
Have kept us out of the red.
Webster E Wing, poem, W-3 Engine, Trains Magazine, *December 1957.*

367.14 ... never before or since has man so accurately expressed his own image, its wonder and its failings, as when he caused fire to be laid on grates beneath a boiler mounted atop rod-connected flanged driving wheels. When that fire was extinguished, we were never so warm again.
David P Morgan, Trains Magazine, *November 1965.*

367.15 Catenaries and pantographs
      May all be very well—
Appurtenances of an age
When electricity's the rage
      And coal can go to Hell;
We'll try to put the blame of Fate
When current fails to alternate,
      But railways held their highest stock
      When engines ran on igneous rock.

      Yes, Britain knew her greatest fame
      In Old King Coal's refulgent reign.

Now linear induction fills
      The next progressive need,
And diesel turbines too are planned
To hurl us to the Promised Land

At astronomic speed;
And when at length the L.M.R.
Goes wholly thermo-nuclear,
      We'll shed a sad nostalgic tear
      For those we Loved in Yesteryear.

      Great Britain played her greatest role
      When locos. ran on best steam coal.
Anon. A Lament for the Departure of The Last Officially Repaired Steam Locomotive from Crewe Works. *Noted, on the back of the celebratory leaflet "to be rendered soulfully to the tune of* The Chancellor's Song *from* Iolanthe." *The locomotive was 70013, Oliver Cromwell. 2 February 1967.*

367.16 Slowly, from Calais depôt, past the docks
A Nord *Pacific* backs; the stab and hiss
Of brake-pumps blending with the drawn-out kiss
Of sliding cross-heads on their polished blocks.

Brown, acrid smoke coils from its chimney where
Its dusky majesty it brings to rest:
Oil can in hand the driver leaves his nest
And, béret-topped, splays deftly here and there.
Henry Maxwell, poem, Flèche d'Or, *in* A Railway Rubaiyyat, *1968.*

367.17 There is a satisfying sense of fulfillment in seeing a steam locomotive down on her hands and knees, so to speak, working every inch of the way, battling beneath her pillar of smoke, blasting ever upwards.
Don Ball, caption to photograph, Portrait of the Rails, *1972.*

368. LOCOMOTIVES—Invention

368.1 The timeliness of this invention of the locomotive must be conceded. To us Americans it seems to have fallen as a political aid. We could not else have held the vast North America together which we now engage to do.
Ralph Waldo Emerson, Journals RS, *p. 249, 1849.*

368.2 No single person invented the locomotive. In strict sense the locomotive was never invented. It was a gradual growth from very small beginnings, a combination of inventions, not the handiwork of one man but evolved from the labours of many, and of some these the world knows but little. As the years passed these discoveries were blended in one harmonious whole, the early crudities removed, the defects amended, the virtues perfected, until today the locomotive stands out perhaps as the most finished example of mechanical art ever produced.
Robert Young, Timothy Hackworth and the Locomotive, *1923.*

369. LOCOMOTIVES—Manufacturers

369.1 The chart of the operations of these companies down through the years resembles nothing so much as an Alpine hiking trail.

Roland E Collons, of the big three American locomotive manufacturers, The Railway Magazine, vol. LXXX, *1937.*

369.2 All railroading is based on one thing—that 1 horse-power equals 33,000 foot-pounds per minute. And you juggle that and toss it around and set it to music and play it on the flute. And you make it come out a locomotive.

Richard Dilworth, quoted by Franklin M Reck in The Dilworth Story, *1954.*

370. LOCOMOTIVES—Midland Railway

370.1 At Derby the nice little engines were made pets of. They were housed in nice clean sheds, and were very lightly loaded. There must have been a Royal Society for the Prevention of Cruelty to Engines in existence.

D W Sanford, A Modern Locomotive History, *within* Proceedings of the Institute of Locomotive Engineers, No 190, *1946.*

370.2 Indicator diagrams in the shape of very thin bananas showed severe throttling at both inlet and exhaust, and were the sign of a thoroughly constipated front end.

E S Cox, of Class 2 locomotives, A Locomotive Panorama, *1965.*

371. LOCOMOTIVES—Musing

371.1 What was it the Engines said
Pilots touching—head to head
Facing on the single track
Half a world behind each back?
This is what the Engines said,
Unreported and unread.

Bret Harte, poem What the Engines Said, *of the locomotives at the meeting of Central Pacific and Union Pacific tracks at Promontory, Utah, 1869.*

371.2 Listen! Where Atlantic beats
Shores of snow and summer heats;
Where the Indian autumn skies
Paint the woods with wampum dyes,—
I have chased the flying sun,
Seeing all he looked upon,
Blessing all that he has blest,
Nursing in my iron breast
All his vivifying heat,
All his clouds about my crest;
And before my flying feet
Every shadow must retreat.

Ibid., verse four.

371.3 Said the Union, "Don't reflect, or
I'll run over some Director."
Said the Central, "I'm Pacific;
But, when riled, I'm quite terrific.
Yet to-day we shall not quarrel,
Just to show these folks this moral,
How two Engines—in their vision—
Once have met without collision."
That is what the Engines said,
Unreported and unread;
Spoken slightly through the nose,
With a whistle at the close.

Ibid., last verses.

371.4 A locomotive is, next to a marine engine, the most sensitive thing a man ever made; and number .007, besides being sensitive, was new. The red paint was hardly dry on his spotless bumper-bar, his headlight shone like a fireman's helmet, and his cab might have been a hardwood-finish parlor. They had run him into the roundhouse after his trial—he had said good-bye to his best friend in the shops, the overhead traveling crane—the big world was just outside; and the other locos were taking stock of him. He looked at the semi-circle of bold, unwinking headlights, heard the low purr and mutter of the steam mounting in the gauges—scornful hisses of contempt as a slack valve lifted a little—and would have given a month's oil for leave to crawl through his own driving-wheels into the brick ash-pit beneath him.

Rudyard Kipling, .007, Scribner's Magazine, August 1897.

372. LOCOMOTIVES—Names

372.1 The engines have at present neither names, numbers nor other distinctive marks and no bright brass work, so that they are certainly far more useful than ornamental.

Editorial comment, Manchester Guardian, *of the earliest locomotives of the Sheffield, Ashton-under-Lyne & Manchester Railway, 20 November 1841.*

372.2 I have heerd of one young man, a guard upon a railway, only three years opened—well does Mrs Harris know him, which indeed he is her own relation by her sister's marriage with a master sawyer—as is godfather at this present time to six-and-twenty blessed little strangers, equally unexpected, and all on 'um named after the Ingeins as was the cause.

Charles Dickens, Martin Chuzzlewit, *1844.*

372.3 Conservative, too, in a rather different sense, is the Great Western Railway in the names of eminent persons it has given to its engines—all, or practically all, are Tories...

B Sparkes, Should Locomotives be Named? The Railway Magazine, vol. XVI, *1905.*

372.4 How inexpressibly and mysteriously great some of those titles seemed to be to my boyish mind, even at that early age—Agamemnon, Hyperion, Prometheus, Ajax, Achilles, Atalanta, Memluke; they fired my imagination and filled me with strange feelings of pride and joy.
Alfred Williams, A Wiltshire Village, *1912.*

372.5 In those days most of the Cambrian engines had names, some of which were of an extremely unpronounceable character as befitted the Celtic country which the line traversed. Fortunately, unlike the broad-gauge engines of the Great Western, the Cambrian engines had numbers also, or there might have been a block on the line whilst the staff was wrestling with Welsh and classical orthography as depicted on the boilers of the engines.
E L Ahrons, The Railway Magazine, *March 1922.*

372.6 To some of the directors, however, the habit of christening engines, especially after distinguished persons or the seats of the local gentry, seemed to savour of flunkeyism and the custom was abandoned.
C P Gasquoine, The Story of the Cambrian, *1922.*

372.7 Nevertheless, the practice, in the large, was a pleasing one. And when, in the swiftly growing number of engines upon each of the more extensive roads of the land, it became necessary—or advisable—to drop it, and merely to number the locomotives sequentially, like passenger or freight cars, a distinct something was lost to the American railroad.
Edward Hungerford, The Story of the Baltimore & Ohio Railroad, vol. 1, *1927.*

372.8 When the Great Western sought to find a name
Symbolic, apt, and free from any blame,
With Swindon Engines what did they compare?
Abbeys and Castles gone beyond repair!

Equally apt the choice of L.N.E
For its Pacifics, all will sure agree;
To these fine Engines comes the appropriate gift—
Some name of Racehorse, graceful, strong and swift.

As to "Directors" who would comment dare?
By definition they're beyond compare!
"Improved Director" is, in terms, absurd!
What could "improve" that awe-inspiring word?

Britannic Heroes did the Southern choose,
But names archaic suffer oft misuse,
L.M.S. names, all jumbled, fall so flat
You'd think they simply drew them from a hat!

But if of sheer magniloquence you'd learn
To the Contractor you can safely turn;
His four wheeled "pug", a dozen tons in weight,
He boldly labels "Tamburlaine the Great".

"Doggo" (probably Kenelm Kerr), poem "What's in a Name?" Ballyhoo Review's *Christmas Annual, 1940.*

372.9 I have just seen that the Hook of Holland Continental has arrived at Liverpool Street hauled by an engine with the extraordinary name of "Bongo". Please ensure that this does not happen again.
L P Parker, telephone conversation with Terence Miller, 1948.

372.10 R.E. considered that in future railway locomotives should not generally bear names; question of naming to be reconsidered if public reaction rendered this desirable.
Railway Executive minute of 30 December 1948.

372.11 The Railway Executive has no intention of removing the existing names from locomotives, or of discontinuing the general practice of naming locomotives in suitable cases.
Railway Executive press statement, after adverse reaction to the decision of 30 December 1948, 7 March 1949.

372.12 ...so far as *George Bernard Shaw* is concerned, the answer is quite definitely "Not Pygmalion likely."
Freddie Harrison, on a suggestion for a name for a Britannia class locomotive, office memo, 29 March 1954.

372.13 No, you Christen it; I'll confirm it!
Bishop of Exeter, replying to Sir Peter Parker, who had courteously deferred to him in the naming of a locomotive on the Dart Valley Railway. Quoted in Driven by Steam, *Ian Allan, 1992.*

373. LOCOMOTIVES—Need

373.1 We are preparing for a counter-report in favour of locomotives, which I believe still will ultimately get the day, but from the present appearances nothing decisive can be said: rely upon it, locomotives shall not be cowardly given up. I will fight for them until the last. They are worthy of a conflict.
Robert Stephenson, letter, 11 March 1829.

373.2 I am verily convinced that a swift engine, upon a well-conditioned railway, will combine profit and simplicity, and will afford such facility as has not hitherto been known.
Timothy Hackworth, letter to Robert Stephenson, 1829. Quoted by Robert Carlson in The Liverpool & Manchester Railway Project 1821–1831, *1969.*

373.3 I will send the locomotive as the great Missionary over the World.
George Stephenson, quoted by L T C Rolt in George & Robert Stephenson, *1960.*

373.4 There comes a time when locomotives are more important than guns.
Attributed to Count Ludendorff, 1918.

373.5 I see no reason to suppose that these machines will ever force themselves into general use.

Attributed to the Duke of Wellington, circumstances not known.

373.6 I do declare that a steam engine is greatly to be preferred to any other first mover. . . .

John Blenkinsop, Patent 3431, 1811.

373.7 Then you've a lot to learn about trucks, little Thomas. They are silly things and must be kept in their place. After pushing them about here for a few weeks, you'll know almost as much about them as Edward. Then you'll be a Really Useful Engine.

Rev Wilbert Awdry, Thomas the Tank Engine, 1946.

374. LOCOMOTIVES—Operating

374.1 On our return home, abt 4 miles from the shipping place of the Iron, one of the small bolts that fastened the axel to the boiler broak, and let all the water out of the boiler, which prevented the engine returning untill this evening.

Richard Trevithick of his first locomotive, letter to Davies Giddy, 22 February 1804.

374.2 Locomotives must always be objectionable on a railroad for general use, where it is attempted to give them a considerable degree of speed.

Thomas Tredgold, The Steam Locomotive, 1838.

374.3 This was my first direct contact with 'motive power' work, and it was a liberal education in cause and effect, with a certain element of crime detection for good measure.

E S Cox, A Locomotive Panorama, 1965.

375. LOCOMOTIVES—Power of Achievement

375.1 It is far from my wish to promulgate to the world that the ridiculous expectations, or rather professions, of the enthusiastic speculist will be realised, and that we shall see engines travelling at the rate of twelve, sixteen, eighteen or twenty miles an hour. Nothing could do more harm toward their general adoption than the promulgation of such nonsense.

Nicholas Wood, pamphlet, no date given, quoted by Simon Garfield in The Last Journey of William Huskisson, *2002.*

375.2 May the powers of the locomotive engine continue the approximation of cities and countries until national and local prejudices give place to friendship, peace and unanimity and waft bachelors into a state of connubial bliss.

J B Roe, Toast to the Paterson & Hudson River Rail Road, 3 February 1831. Quoted in From the Hills to the Hudson, *Walter Arndt Lucas, 1944.*

375.3 Hurrah! hurrah! away we go
Without a spur or goad—
Our iron coursers snort and blow
Along an iron road.

Your noblest steeds of flesh and blood
Are soon with toil o'erdone,
But wheels impelled by fire and flood
For ever may roll on.

No load, nor length of way, fatigues
Our wild, unslumbering team;
A jaunt of a hundred thousand leagues
Is baby play for steam.

Samuel J Smith, poem Locomotives, *first three verses. Thought to be the earliest American poem on steam locomotion.* Miscellaneous Writings, *published after his death (1835), in 1836.*

375.4 The hills may lift their foreheads high,
The rivers oppose in vain—
Our smoky *motives* soon shall fly
From Mexico to Maine.

Then farewell to domestic jars,
All nullying nonsense done—
An endless chain of railroad cars
Will bind us all in one.

Ibid., verses four and five.

375.5 The goodly dames of former days
Were doomed at home to stay,
Or jog o'er dislocating ways
A dozen miles a day.

Affairs of moment only led
Their steps of course to roam,
And comfort too, was born and bred,
They might suppose, at home.

But now a feather has force enough
To send our damsels forth,
For a yard of bobbin, or thimble of snuff,
To the east, west, south or north.

Ibid., verses six to eight.

375.6 And then an aud horse brought a waggon
A' the way frae the pits to the staith,
But now it seems pretty certain,
They'll verra suin dee without baith.
For now their fine steam locomotives
A' other invention excels,
Aw've only to huik on the waggons,
And they'll bring a ship-load down their sels.

J P Robson, poem The Changes on the Tyne, *within* Songs of the Bards of the Tyne, *1849.*

375.7 Men of the East! Men at Washington! You have given the toil and even the blood of a million of your brothers and fellows for four years, and spent three thousand million dollars, to rescue one section of the Republic from barbarism and from anarchy; and your triumph makes the cost cheap.

Lend now a few thousand of men, and a hundred millions of money, to create a new Republic; to marry to the Nation of the Atlantic an equal if not greater Nation of the Pacific. Anticipate a new sectionalism, a new strife, by a triumph of the arts of Peace, that shall be even prouder and more reaching than the victories of your Arms.

Here is payment of your great debt; here is wealth unbounded; here the commerce of the world; here the completion of a Republic that is continental; but you must come and take them with the Locomotive.

Samuel Bowles, Across the Continent, *1865.*

375.8 As the "iron horse" advances towards the west, settlements spring up as if by magic along the intended route. The locomotive is the true harbinger of civilisation.

Henry Morton Stanley, 9 May 1867, republished in My Early Travels and Adventures in America and Asia, vol. I, *1895.*

375.9 Standing upon the mountain top in the red evening sunlight, which sheds a golden lustre upon earth and sky, I predict, that this country will be acknowledged in the coming future as no mean State. Time is flying, the iron horse is upon the plain, impatient to rush through the heart of the mountain towards the Pacific. Two years hence the dwellers upon the Atlantic slopes will unite hands with their brothers of the Pacific shores, and then—and then will the desert rejoice, and the wilderness be made glad.

Ibid., 23 August 1867.

375.10 What a maker of new and improved maps is the locomotive!

Benjamin Taylor, The World on Wheels, *1874.*

375.11 What speed in the engine; what priceless cargoes of time and oxygen upon the train; how fast and long we live in a little while! Let us be glad. Uneasy people sometimes wish they had been born in the days of Alexander, or Moses, or Methuselah, or somebody who looms up gigantically in the mists of history. It is better to live in the days of the Steam-engine. It has conquered more worlds than Alexander, traversed vaster wildernesses than the Israelite, and reclaimed them as it went; and behold, by the power of the Engine we live to be hundreds of years old, and never give it a thought!

Ibid.

375.12 The locomotive has not only increased the number of the King's subjects by millions, but has added more to our revenue than any tax ever invented.

The Duke of Devonshire (eighth), quoted in A Cradle of the Locomotive, *within* The Railway Magazine, vol. XI, *1902.*

376. LOCOMOTIVES—Principles

376.1 Multifarious were the schemes proposed to the Directors, for facilitating Locomotion. Communications were received from all classes of persons, each recommending an improved carriage; from professors of philosophy, down to the humblest mechanic, all were zealous in their proffers of assistance; England, America and Continental Europe were alike tributary. Every element and almost every substance were brought into requisition, and made subservient to the great work. The friction of the carriages was to be reduced so low that a silk thread would draw them, and the power to be applied was to be so vast as to render a cable asunder. Hydrogen gas and high-pressure steam—columns of water and columns of mercury—a hundred atmospheres and a perfect vacuum—machines working in a circle without fire or steam, generating power at one end of the process and giving it out at the other—carriages that conveyed, every one to its own Railway—wheels within wheels, to multiply speed without diminishing power—with every complication of balancing and countervailing forces, to the *ne plus ultra* or perpetual motion. Every scheme which the restless ingenuity or prolific imagination of man could devise was liberally offered to the Company: the difficulty was to choose and to decide.

Henry Booth, Secretary, Liverpool & Manchester Railway, An Account of the Liverpool & Manchester Railway, *1837.*

LOCOMOTIVES—Stack Design

See Locomotives—Chimney Design

377. LOCOMOTIVES—Success and Failure

377.1 We are credibly informed that there is a Steam Engine now preparing to run against any mare, horse or gelding that may be produced at the next October Meeting at Newmarket; the wagers at present are stated to be 10,000 l.; the engine is the favourite.

Concerning the second of Richard Trevithick's locomotives, later shown in London. The Times, *8 July 1808.*

377.2 RACING STEAM ENGINE.—This surprising Engine will commence to exhibit her powers of speed to the public. THIS DAY, at 11 o'clock, and will continue her experiments only a few days. Tickets of admission, 5s. each, may be had at the bar of all coffee-houses in London; and at the Orange Tree, New-road, St. Pancras.

Of the same event, The Times, *20 July 1808.*

377.3 It is the size of the boiler that denotes the strength of the engine.

George Stephenson, Parliamentary Committee hearings on the Liverpool & Manchester Bill, March 1825.

377.4 Crossher Bailey had an engine, and the engine
wouldn't go,
So they pushed the bloody engine all the way to
Nantyglo...
> *Anon.; the reference is to a leading iron-master, Cray-
> shaw Bailey, (1789–1872); the locomotive was probably a
> very early one; quoted by Derek Barrie in* A Regional His-
> tory of the Railways of Great Britain, vol. 12, *1980.*

377.5 The measure of a locomotive's success is its capacity
to boil water.
> *H A Ivatt, Locomotive Superintendent of the Great
> Northern Railway, quoted by George Dow in* British Steam
> Horses, *1950.*

377.6 Ach, anything vill for a locomotive do!
> *Charles Beyer, quoted by W J Gordon in* Our Home
> Railways, *1910.*

377.7 There is no other basic element in the design of
the steam locomotive that approaches combustion in
importance.
> *Alfred W Bruce,* The Steam Locomotive in America,
> *1952.*

377.8 Correctly proportioned and designed, well made
and maintained, worked within its capacity by keen
crews, and fed with suitable fuel and water, the steam
locomotive is one of the most elegantly simple and automatically integrated traction machines devised by man.
> *E S Cox,* Locomotive Panorama, vol. 2, *1966.*

377.9 Blowing at every joint, with tubes made up, brick
arch half down, a firebox full of ash and clinker generated by unsuitable coal, overdue for piston and valve
check, knocking and lurching at the back end due to
wear so that it was difficult to stand up, and in charge of
a crew who did not care, the second face presents itself,
just as valid as the first, and who riding such a machine
would not conclude that there must be better ways to
haul a train?
> *Ibid.*

377.10 ... the fairy wand which changed Cinderella into
a Princess scarcely equalled the genius which transformed this slut amongst engines into the most outstanding steam locomotive that ever ran upon rails.
> *H C B Rogers, of the rebuilding of Mountain type 241.101
> to a 4-8-4; Chapelon* Genius of French Steam, *1972.*

378. LOCOMOTIVES—Utility

378.1 An engine in a workshop is a useless appliance, is
costing much and earning nothing, ...
> *Clement Stretton,* The History of the Midland Rail-
> way, *1901.*

378.2 Why a certain type of locomotive may be popular
in one country and discredited in another, may be due

to the manner in which they are treated. The treatment
by most American railroads of their freight power is
simply brutal. In the British Isles the railway motive
power is petted. On Continental railways fair mileage is
demanded from locomotives well cared for.
> *Angus Sinclair,* Development of the Locomotive
> Engine, *1907.*

378.3 ... it will, I think, prove a useful tool to the traffic
people.
> *Sir Nigel Gresley, of the then new locomotive* Green
> Arrow, *in conversation with Charles S Lake, quoted in* The
> Railway Magazine, vol. 90, *1944.*

379. LOCOMOTIVES—Vulnerability

379.1 *To render Locomotives Unfit for Service:* The most
expeditious mode is to fire a cannon ball through the
boiler. This damage cannot be repaired without taking
out all the flues.
> *Herman Haupt, of experience in the American Civil
> War,* Reminiscences of Herman Haupt, *1901.*

379.2 Go for the cylinders, my boy!
> *Oliver Bulleid, locomotive designer, advice to his
> younger son who, as a Typhoon pilot in the Second Tactical
> Air Force, told his father of his attacks on trains in France,
> and of his aiming to hit the locomotive boilers to make
> them explode, 1944. Conversation with the compiler, 1993.*

379.3 The steam locomotive in 1954 is reminiscent of
some prehistoric monster unable to cope with the tide of
evolution—a hapless, harried, hunted creature rebuked
by its environment, dying off in large numbers, facing
total extinction. The simple curb of knocking down coal-
ing stages and water tanks, filling in ash pits, and cooling
backshop forges, has fenced steam off hundreds, even
thousands, of main line—forever. The very foundries
which hatched it less than a decade ago now spawn its
killer. Practically everywhere anything with a stack on
one end and a firebox under the other is fair game for
the torch, and depletion of the reserve only boosts the
bounty.
> *David P Morgan,* Trains Magazine, *April 1954.*

380. LONDON & NORTH EASTERN RAILWAY

380.1 It's quicker by rail.
> *Advertising slogan 1927–1941, both for its own posters
> and publicity material, and also for joint posters with the
> London Midland & Scottish Railway.*

380.2 The drier side.
> *Advertising slogan for East Coast resorts, 1933.*

380.3 The London & North Eastern is a distinctly fishy
railway.

Jock Scott, opening words in praise of the salmon and trout rivers within easy access of LNER stations, Salmon and Trout Rivers served by the London and North Eastern Railway, *1937.*

380.4 When the North Eastern was extended to London to connect with its Westminster offices in Cowley Street.
Anon., of the creation of the LNER, quoted by George Dow in East Coast Route, *1951.*

380.5 Loyalty to the LNER was a filial duty: at least one of the children had to be groomed in it, one of them had to carry the myths and legends into the seventies and beyond. Even if the next generation ended up living in a pressurized bubble on the moon, the LNER would not be forgotten.
Nicholas Whittaker, Platform Souls, *1995.*

381. LONDON & NORTH WESTERN RAILWAY

381.1 Just as the Great Northern gains in smartness from being the youngest born of our (great) lines, so the North-Western, because it is the oldest, is more burdened with the traditions of a bygone age when 40 miles an hour was a wonderful speed.
E Foxwell and T C Farrer, Express Trains English and Foreign, *1889.*

381.2 The tradition that 40 miles per hour is the proper speed for a gentleman still has a powerful sway over this eminently aristocratic line.
W M Acworth, The Railways of England, *1889.*

381.3 The North-Western is the only company that does all its manufacturing, from rolling many of its own rails up to fabricating artificial limbs for the injured members of its staff.
W J Gordon, Everyday Life on the Railroad, *1898.*

381.4 It is better to be a dead mackerel on the North-Western than a first-class passenger on the London, Brighton and South Coast.
E L Ahrons, The Railway Magazine, *March 1919, commenting upon the relative speeds of fish trains on the former and passenger trains on the latter.*

381.5 The London and North Western Railway is noted for Punctuality, Speed, Smooth Riding, Dustless Tracks, Safety and Comfort, and is the Oldest Established Firm in the Railway Passenger Business.
Statement on the reverse of postcards issued by the company, 1900–1914.

381.6 The Premier Line.
Advertising slogan.

381.7 The Best Permanent Way in the World.
Advertisement, 1910.

381.8 . . . no railway was more of a law unto itself than the North Western, and therein was a cause of its eccentricities as well as its qualities.
C Hamilton Ellis, British Railway History, vol. II, *1959.*

381.9 By clever use of the advertising tag *The Premier Line,* it convinced some people that it *was* the premier line, which it certainly was not, although it was a great railway in many ways.
George Dow, Railway Heraldry, *1973.*

382. LONDON & SOUTHAMPTON RAILWAY

382.1 Parsons and prawns—the one from Winchester, the other from Southampton.
Anon., noted by W M Acworth in Railways of England, *1889.*

383. LONDON & SOUTH WESTERN RAILWAY

383.1 The train stopped in a blaze of sunshine at Framlynghame Admiral, which is made up entirely of the nameboard, two platforms, and an overhead bridge, without even the usual siding. I had never known the slowest of locals stop here before, but on Sunday all things are possible to the London & South Western.
Rudyard Kipling, My Sunday at Home, *within* Idler, *March 1895.*

384. LONDON BRIGHTON & SOUTH COAST RAILWAY

384.1 One cannot but feel at times on the Brighton trains that the engines are so good that they would go faster if they were only asked to do so.
W M Acworth, The Railways of England, *1889.*

384.2 . . . truth compels me to say that I do not think that the denizens of the London Bridge offices contributed very greatly to its popularity; they were a cheerful band of railway sinners, more especially those concerned with the traffic department, and between them they could then produce more chaotic unpunctuality than could be found anywhere, except perhaps on the South Eastern.
E L Ahrons, The Railway Magazine, *January 1919.*

384.3 Had the company to deal with a longer main line, it is just conceivable that some of their 19th century trains might just be arriving at their destinations today.
Ibid.

385. LONDON, CHATHAM & DOVER RAILWAY

385.1 The locomotives were excellent, but the carriages were always poverty-stricken rabbit hutches.
E L Ahrons, The Railway Magazine, *February 1918.*

385.2 . . . a railway admired by none . . .
Michael Robbins, The Railway Age, *1962.*

386. LONDON MIDLAND & SCOTTISH RAILWAY

386.1 The Best Way.
Advertising slogan, 1920s.

386.2 I've had to make the best of it, like all of us, but the truth is that I've been shat on from such a height by these men from the Midland that I'm getting manured to it.
G Royde Smith, soon after the formation of the LMS in 1923. Quoted by Sir John Elliot in On and Off the Rails, *1982.*

386.3 The railway system had all been built by the British, and indeed was British-owned until the war. Thus, except for the fact that the carriages were bright and clean, and there was a surfeit of attendants in spotless uniform, one might have been travelling on the LMS to Derby.
Sir Stanley Hooker, of a journey by sleeping car from Buenos Aires to Cordoba, 1946, very probably on the Central Argentine Railway, Not Much of an Engineer, *1984.*

387. LONDON'S RAILWAYS

387.1 It is not enough that Parliament should spend its best endeavours on each scheme as it is presented, patiently weighing the pros and cons., and authorising or condemning each on its own merits; justice to unborn generations demands that the transit of London should be considered first of all as a whole on a basis of practical utility; London's transit network must be the arterial system of a sound body, not a heterogeneous aggregate of ill-matched limbs.
D N Dunlop, The World's Progress in Electric Traction, *within* The Railway Magazine, *vol. X, 1902.*

388. LOS ANGELES & LONG BEACH RAILROAD

388.1 Oh, sing a song of railroads,
Likewise the iron hoss,
Of all that run beneath the sun,
The L. B. is the boss;
With a thirteen cat-power engine,
That starts with a big pinch-bar,
Oh, everyone gets out and pushes,
On the G.O.A.P.R.R.

Two streaks of rust, a ballast of dust,
Amid a two-foot right of way,
We blister space at an awful pace,
Two miles and half a day;
With rails of cold rolled hairpins,
And a shoebox for a car,
Oh, everybody get out and push,
On the G.O.A.P.R.R.
Anon., of the railroad in its first form, c. 1885. Poem first appeared in the Los Angeles Times *in about 1885, without explanation of the initials G O A P R R;* RLHS Bulletin 99, *October 1958.*

389. LOUISVILLE & NASHVILLE RAILROAD

389.1 Click, click, clicketty, click!
Clicketty, clicketty, clicketty, click!
Heigh oh! Here we go
From Louisiana to O-hi-o!
Clicketty, clicketty, all day long:—
The rhyme of the rail is a rhythmical song;
It is n't a waltz, but a regular beat
Of four—full—metrical feet,
As clicketty, clicketty, on we go
From Louisiana to O-hi-o—
To Louieville, Evansville,
Nashville, Knoxville;
Louieville, Russellville,
Clarksville, Brownsville;
Louieville, Louieville, Louieville, Louieville,
Louisiana to O-hi-o!
Frank F Woodall, first verse of poem, The Song of the L&N Railroad, *c. 1885, published in* Fugitive Poems and A Christmas Story, *1913, with the instruction to read it aloud in a monotone, rapidly and without pausing—even at the end of a stanza—but beginning and ending slowly.*

390. LOVE

390.1 There were two loves in his life. His engine, And—
Text in film The General, *1927. Johnnie Gray's other (seemingly second) love was Annabelle Lee, whom he saves from Unionist soldiers during the film.*

390.2 I was brought up to hate Protestants and the Pennsylvania Railroad. After this, I've got to love them both.
Anon. Irish officer of the New York Central, of the merger into the Penn Central. Quoted by Joseph Daughen and Peter Bintzen in The Wreck of the Penn Central, *1971.*

390.3 I saw her name on the side of a train
Somewhere a long time ago;
I don't know who she was but I gave my love
To someone called Phoebe Snow.
Bruce "Utah" Phillips, song, Phoebe Snow, *1973.*

390.4 And then I awoke as the day broke
And gazed out over the plain
Thinking as how I'm better off now
Being in love with a train.
Ibid., last verse.

391. LOVE-MAKING
See also Kissing

391.1 I never see a bride going to Church with orange blossoms in her bonnet or a young couple strolling toward the kissing bridge of a summer evening that I do not involuntarily exclaim, "Heaven bless them, there goes the material to make railroads." As long as love is made in Nova Scotia and love makes children we shall be adding from twenty to thirty thousand pounds annually to our income.
Joseph Howe, in speech, 1851, quoted in Canadian National, *G R Stevens, 1960.*

391.2 It had been a wonderful trip up in the train. They had eaten the sandwiches and drunk the champagne and then, to the rhythm of the giant diesels pounding out the miles, they had made long, slow love in the narrow berth.
Ian Fleming, Goldfinger, *1959.*

391.3 After the petals and the rice, after the ribbons and lace,
We rode away together in a bridal suite
On wheels, getting to know each other better
And better, in a space of three by six
Feet, with a bunk for you and a bunk
For me, and a washbasin built for two
Martians, with retractable legs.
Who shouted I love you over the miles
Between us as the train hurtled west at unbridled speed
Toward our happily-ever-after? Who spoke
I love you as the stars stood wide-eyed
Looking down in, after the windowblind broke?
Nancy G Westerfield, poem, Pullman honeymoon, *within* Trains Magazine, *July 1973.*

391.4 She was the first girl friend who'd allowed any heavy petting, but I knew that I could never be happy with someone who got so bored in derelict railway buildings.
Nicholas Whittaker, Platform Souls, *1995.*

391.5 I think this is a very sexy train. Maybe you find a partner here. Those of you who don't have partners can . . . . find them on the Acela Express.
Dr Ruth Westheimer, sex therapist, interrupted by Michael Dukakis, Amtrak vice chairman, on the inaugural run of the Acela Express, Washington-Boston, 16 November 2000, reported in Trains Magazine, *February 2001.*

392. LUBRICATION

392.1 An oil-cup which runs out the oil faster than is needed, wastes stores, besmears everything with a coating of grease, and is likely to leave the rubbing-surfaces to suffer by running dry before it can be replenished. A cup in that condition also advertises the engineer to be incompetent.
Angus Sinclair, Locomotive Engine Running and Management, *1902.*

393. LUGGAGE

393.1 Luggage is often a perplexing point with the *paterfamilias*. One would wonder how people contrived in former days to get themselves and their wants into a stage-coach or a carriage and pair, so greatly have our wants expanded with railway times. It is not now the carrying but the paying for it that forms the difficulty; and twenty shillings is soon run away with for extra baggage by express train.
R S Surtees, Hints to Railway Travellers and Country Visitors to London, *in the* New Monthly Magazine, *1851.*

393.2 My baggage was lost; it had not come on my train; it was adrift somewhere back in the two-thousand miles that lay behind me. And, by way of comfort, the baggage-man remarked that passengers often got astray from their trunks, but the trunks mostly found them after a while. Having offered me this encouragement, he turned whistling to his affairs and left me planted in the baggage-room at Medicine Bow.
Owen Wister, The Virginian, *1902.*

393.3 The enormous weight of the trunks used by some travellers not unfrequently inflicts serious injury on the hotel and railway porters who have to handle them. Travellers are therefore urged to place their heavy articles in the smaller packages and thus minimise the evil as far as possible.
Advice to railway travelers, Baedeker's Austria-Hungary, *1905.*

393.4 When I travel alone, with my small black suitcase, I can travel anywhere and my luggage is no inconvenience, just as, when I travel alone with three incredibly heavy leather trunks, they are no inconvenience either, because they are so ludicrously disproportionate to my strength that someone will always come and help.
Lisa St Aubin de Terán, Off the Rails, *1989.*

394. LUGGAGE LABELS

394.1 'Paris!' How it thrills me when, on a night in spring, in the hustle and glare of Victoria, that label is slapped on my hat-box! Here, standing in the very heart of London,

I am by one sweep of a paste-brush transported instantly into that white-grey city across the sea.
Max Beerbohm, Ichabod, *1923.*

394.2 After money, luggage-labels are the greatest conducers to comfort in travelling.
R S Surtees, Town and Country Papers, *1929.*

395. LULLABY

395.1 Train whistle blowing makes a sleepy noise
Underneath their blankets go all the girls and boys
Heading from the station, out along the bay
All bound for Morningtown, many miles a-way

Sarah's at the engine, Tony rings the bell
John swings the lantern to show that all is well
Rocking, rolling, riding, out along the bay
All bound for Morningtown, many miles a-way

Maybe it is raining where our train will ride
But all the little travellers are snug and warm in-side
Somewhere there is sunshine, somewhere there is day
Somewhere there is Morningtown, many miles a-way
Malvina Reynolds, song, Morningtown Ride, *date not known.*

396. LUST

396.1 More Horny than Hornby.
Anon., advertising slogan of Virgin Trains, to promote its new Pendolino trains, which it claimed were "sexy." 2003. Perhaps this was a reference to the phallic look of the power cars.

397. LUTTRELL, TENNESSEE

397.1 Just a whistle stop on the Southern Railway.
Chet Atkins, of his hometown. Quoted in Atkins' obituary, The Times, *2 July 2001.*

398. LUXURY

398.1 By dictionary definition, neither Amtrak nor VIA offers "luxury." But what else do you call meals served at a table, easy chairs for conversation, a bathroom at the ready (compared to REST AREA, 20 MILES), and a bed that carries you through the night?
George H Drury, Railroad History 189, *2004.*

399. LYNTON

399.1 They are very anxious to get a railway to it, but too many people would spoil it.
Sir Daniel Gooch, diary entry 8 September 1878, Memoirs and Diary, *1972.*

400. LYNTON & BARNSTAPLE RAILWAY

400.1 A curious little affair, this miniature train, which jogs along at 16 miles an hour. It cannot go faster because of the many gradients and curves; but you are thankful for that fact, because it is taking you through scenery the like of which for extreme loveliness is unique in all England.
E P Leigh-Bennett, Woody Bay, *c. 1928.*

400.2 Oh, Little train to Lynton,
no more we see you glide
Among the glades and valleys
and by the steep hillside.

The fairest sights in Devon
were from your windows seen
The moorland's purple heather,
blue sea and woodland green.

And onward like a river
in motion winding slow
Through fairylands enchanted
thy course was wont to go.

Where still the hills and valleys
in sunshine and in rain
Will seem to wait for ever,
the coming of the train.
A Fletcher, North Devon Journal, *September 1935.*

400.3 Perchance it is not dead, but sleepeth.
Paymaster Captain Woolf, R N, inscription on card on wreath at the closure of the railway, September 1935.

Mm

401. MABLETHORPE and SUTTON-ON-SEA

401.1 The Children's Paradise.
London & North Eastern Railway poster, c. 1930.

402. MAGAZINES

402.1 For decades it bore the unmistakable stamp of its editor, Freeman Hubbard. It was a pulp magazine, and in contrast to *Trains* it was not the sort of thing you'd leave on the coffee table in the living room where your mother would see it. There were lots of adverts for trusses and such.
George Drury, of Railroad Magazine, *formerly* Railroadman's Magazine. *Letter to the compiler, 21 August 2002.*

MAIL
See Postal Service by Railway

403. MAINTENANCE

403.1 There's no glamour in maintenance.
Unidentified American maintenance engineer, in conversation with Michael Robbins, quoted in Points and Signals, *1967.*

404. MALLARD

404.1 Three cheers for the Mallard. LMS out for a duck.
Lady Wedgwood, telegram to Sir Nigel Gresley, 4 July 1938.

404.2 On 3rd July 1938 this locomotive attained a world speed record for steam traction of 126 miles per hour.
George Dow (in his capacity as Public Relations & Publicity Officer of the Eastern and North Eastern Regions of British Railways), plaques on the sides of Mallard, *1948.*

404.3 Steam loonies please note 4468 arrives today @ 1908. Please do not enquire any further as you'll drive us all round the bend.
Handwritten notice at Marylebone station, 1986. Believed to have been crafted by the station master.

405. MALLET LOCOMOTIVES

405.1 In common with the diesel locomotive, the Mallet owes its concept to the Continent but its true realization to America.
Roy W Carlson, Trains Magazine, *November 1957.*

MANAGEMENT OF RAILWAYS
Management of Railways is divided into subheadings, thus: Control, Economy, Good Fortune, Organization, Practice, Qualities and Characteristics, Skills, and Understanding.

406. MANAGEMENT OF RAILWAYS— Control

406.1 The safety of the Public also requires that upon every Railway there shall be one system of management, under one superintending authority, which should have the power of making and enforcing all regulations necessary for the protection of passengers, and for duly conducting and maintaining this new mode of communication. On this account it is necessary that the Company shall possess a complete control over their line of road, although they should thereby acquire an entire monopoly of the means of communication. But if these extensive powers are to be granted to private Companies, it becomes most important that they should be controlled so as to secure the Public as far as is possible from any abuse which might arise under this irresponsible authority.
Comment in Second Report of Select Committee on Railways, *August 1839, chaired by Lord Seymour.* Parliamentary Papers, *Transport.*

406.2 ... it will scarcely be expected that we can at once adopt any plan of operations which will not require amendment and a reasonable time to prove its worth. A few general principles, however, may be regarded as settled and necessary in its formation, amongst which are:
1. A proper division of responsibilities.
2. Sufficient power conferred to enable the same to be fully carried out, that such responsibilities may be real in their character.
3. The means of knowing whether such responsibilities are faithfully executed.
4. Great Promptness in the report of all derelictions of duty, that evils may be at once corrected.
5. Such information, to be obtained through a system of daily reports and checks that will not embarrass principal officers, nor lessen their influence with their subordinates.
6. The adoption of a system, as a whole, which will not only enable the General Superintendent to detect errors immediately, but will also point out the delinquent.
Daniel C McCallum, Superintendent's Report 25 March

1856, within Annual Report of the New York and Erie Railroad Company *for 1855.*

406.3 The railroad is entirely under your control. No military officer has any right to interfere with it. You were notified to this effect this morning. Your orders are supreme.

Major-General Halleck, to General Herman Haupt, 24 August 1862, recorded in Reminiscences of Herman Haupt, *1901.*

406.4 I, of all people, know the problems of empire.

Attributed to Edward H Harriman, when a courtier in Vienna explained why Emperor Franz Joseph of Austria-Hungary was late for an audience. Related by Robert Sobel in The Entrepreneurs, *quoted again by James A Ward in* Railroads and the Character of America, *1986.*

406.5 The evils of pluralism have never received more striking exemplification than in the history of the numerous undertakings over whose destinies he has presided.

Comment, The Railway Times, *of Sir Edward Watkin's chairmanship of six railway companies, 2 June 1894.*

406.6 Going to see the picture.

Anon. Of a summons to the boardroom of the Great Western Railway at Paddington, where hung a portrait of Charles Russell, chairman of the company. Related by Jack Simmons in The Victorian Railway, *1991.*

406.7 You run the country. I'll run the railroad.

Edward H Harriman, message to Theodore Roosevelt when President, who had expressed concern about his daughter Alice travelling on Harriman's railroad. Quoted by Maury Klein, The Life and Legend of E H Harriman, *2000.*

406.8 The third stands for one-man construction; the second stands for one-man reconstruction; the first stands only for inheritance.

Anon., Wall Street Journal, of James J Hill ("the third"), E H Harriman ("the second"), and William Vanderbilt, c. 1909. Quoted by Maury Klein, The Life and Legend of E H Harriman, *2000.*

406.9 His fingers are on pulses; he commands an army; he is part of the keystone of the economic edifice.

Of the railroad president, Fortune, *December 1932.*

406.10 Up the hill slow, down the hill fast,
Tonnage first and safety last.

Anon., of the tendency in America, early twentieth century, to run longer trains and employ fewer staff; noted by Arthur W Hecox in Railroad Magazine, *July 1939.*

406.11 Perhaps one reason why progress has been slow in improving passenger service from the standpoint of the passenger, apart from matters of train performance, is

that no system has yet been devised which will promptly inform railroad officers of inadequacies in the service. When food is improperly prepared or served, when employees do not observe the desired standards of civility, service, attention, and efficiency, when cars in service are dirty, or the temperature or ventilation is improperly controlled, there is no system of reports which goes automatically and immediately—as do reports of train delays or locomotive failures—to the executives who could correct these conditions.

John Walker Barriger, Super Railroads, *1956.*

406.12 Our railroads are too close to being the most efficient and productive insolvent industry for their presidents to be merely good businessmen.

David P Morgan, Trains Magazine, *November 1960.*

406.13 I tried to run a Railway.

Gerard Fiennes, title of book, 1967.

406.14 Even if the separation of ownership of the track and ownership and control of the moving trains had not been rejected on the grounds that it would have been difficult to manage, the aim of profit maximisation through the optional allocation of the capacity of a railway would quickly have led to integrated control of the whole railway operation.

Stewart Joy, of suggestions in early railway days that track and trains should be under separate ownership. The Train that Ran Away, *1973.*

406.15 A rail track authority offered no real solution to BR's problems, because its only effects would have been to spin off part of the BR loss under a different name, and to create enormous problems of co-ordination and integration between the owners of the track and its users.

Ibid., of suggestions made in the 1960s.

406.16 Railroads, their regulators, and their unions have become so institutionalized to such a degree that all activity has come to resemble a ritual.

Alan S Boyd, quoted in Trains Magazine, *August 1972.*

406.17 I got this train here to time today by sheer willpower. Let's see for the rest of the week what you can do.

Attributed to Richard Beeching by Gerard Fiennes, Fiennes on Rails, *1986.*

406.18 We fall flat on our interfaces.

Attributed to Sir Peter Parker, and acknowledged by him (to the compiler) as referring to railways, but not identified to a particular occasion.

406.19 What is needed is a complete overhaul of the nation's rail infrastructure by a rearrangement of its assets. This restructuring would involve splitting today's railroad assets in two parts. The roadway assets would be owned, maintained, and operated by one nationwide corporate entity, which we'll call the Roadway Company.

For a fee, any transportation company—common carriers, contract carriers, general-industry car owners, motor carriers, possibly a new RoadRailer company—could travel over Roadway Company's right of way (the industry's entire rail network) in the same manner as all types of highway carriers use the nation's Interstate Highway System, and as airlines use air corridors and airports.

Isabel H Benham, financial analyst. This was not an original thought, similar proposals having been made in the 1960s and later in Great Britain. In any case, it had been tried on the Stockton & Darlington Railway in the 1820s, and rejected as impractical. Trains Magazine, *November 1990.*

406.20 There is a remote corollary to Murphy's Law which predicts that, no matter how many trains are operating on a subdivision, they will all come together at the same place.

Frank Bryan, Trains Magazine, *November 1993.*

406.21 In addition to his own board of directors Warrington has to answer to 535 lawmakers, the White House Office of Management & Budget, the U.S. General Accounting Office, the Department of Transportation's inspector general, the Federal Railroad Administration and the congressionally-created Amtrak Reform Council . . . a shadow board of directors.

Frank Wilner, of the President of Amtrak; quoted thus from Railway Age, *April 2001, by Richard Saunders Jr,* Main Lines, *2003.*

406.22 Well, bollocks.

Sir Alistair Morton, on being asked if he was a lame duck chairman of the Strategic Rail Authority. This came at a time when it seemed railwaymen could do nothing right, notwithstanding the mess of privatization imposed by government. Perhaps he was thinking on the same lines as David P Morgan forty years before (see 522.8). BBC radio, The World at One, *26 June 2001.*

406.23 The troubles were deeper than poor management. The Jersey Central was the canary in the coal mine, and it just dropped dead.

Richard Saunders Jr, of the troubles on the Central Railroad of New Jersey, 1967, typical of the ailments of the railroad industry at that time. Merging Lines, *2001.*

406.24 It is perhaps excusable that commuters cursing the 7.33 forget to appreciate the great infrastructure of engineering and logistics beneath us. It is not excusable that Government does the same. Yet successive Cabinetsful of selfish preeners in chauffeur-driven cars have neglected or sabotaged the railways.

Libby Purves, The Times, *14 May 2002.*

406.25 It is their care in all the ages to take the buffet and cushion the shock

It is their care that the gear engages; it is their care that the switches lock;
It is their care that the wheels run truly; it is their care to embark and entrain
Tally, transport and deliver duly the Sons of Mary by land and main.

Rudyard Kipling, poem, The Sons of Martha, *quoted by Libby Purves in* The Times, *14 May 2002.*

406.26 Kipling was right: leaders and talkers and theorists forget how they depend on oily hands and long apprenticeships. The railway system was a prime victim of this insensibility.

Libby Purves, The Times, *14 May 2002.*

406.27 No Son of Martha, no engineer, would ever have countenanced a situation where the track was separated, in management and responsibility, from the wheel.

Ibid.

406.28 The investigation demonstrated the absurdity. The train belonged to HSBC Rail, was leased by West Anglia Great Northern, maintained by another company, and run on tracks owned by Railtrack, which delegated the maintenance to yet another, which in turn uses sub-contractors.

Ibid., referring to the accident at Potters Bar.

406.29 To run well, a railway depends upon concepts unknown to management schools, such as institutional memory, the discipline habit, two-way trust and devolved responsibility for risk.

Simon Jenkins, The Times, *11 July 2003.*

407. MANAGEMENT OF RAILWAYS— Economy

407.1 The cheapest way of working a railway is unquestionably by judicious centralisation.

Charles Sacré, Report to the Board of the Manchester, Sheffield & Lincolnshire Railway, 1861.

407.2 Is it in the public interest that the railroads of this country are required to make over two million reports per annum to various federal and state tribunals?

Frank Trumbull, of the nature of U.S. Government regulation; address to The National Hay Association, Railway Regulation and Locomotor Ataxia, *Cedar Point Ohio, 12 July 1916.*

407.3 Suggestions are Always in Order.

Daniel Willard, sign in his office during his career 1890–1941. Noted by David M Vrooman, in Daniel Willard and Progressive Management on the Baltimore & Ohio Railroad, *1991.*

407.4 He believed that every dollar gained by letting a road run down cost two dollars to replace.

John Moody, of E H Harriman, quoted by George Kenman in E H Harriman, *vol. 2, 1922.*

407.5 I am happy to carry more people for more money. I don't mind carrying fewer people for more money, but what you are asking me to do is to carry more people for less money, and that's the way to go bankrupt.

Sir Herbert Walker, c. 1926, quoted by Sir John Elliot, On and Off the Rails, *1982.*

407.6 If the public cannot afford to pay for services of a given quantity and quality at the appropriate level of cost, then the quantity or quality of service offered must be decreased, and the budget must be balanced with higher fares at a slightly lower level of activity. The public will not then be consuming services which it cannot afford to pay for, and which the undertaking cannot afford to supply.

British Transport policy statement, quoted by Association of American Railroads, Bureau of Railway Economics, report Nationalized Transport Operations in Great Britain, *Third Year, 1951.*

407.7 Once in an expansive moment it corrupted with gold double eagles every single member of the Nevada legislature, not just the necessary majority required for its legislative objective. The management liked to do things right.

Lucius Beebe, of the Southern Pacific, The Central Pacific and the Southern Pacific Railroads, *1963.*

407.8 Our duty is not to run a service because it is desirable: it is to run a service which will be profitable.

Sir Robert Reid, International Rail Services for the United Kingdom, *December 1989.*

407.9 The passenger problem was not irrelevant. It cast a pall over all railroading. It was the passenger train that linked the railroad to the public at large to excitement and glamour and moments of passage in people's lives. It was what set the railroads apart from other, mundane industries. As the passenger train faded, the railroad itself began to fade from the public consciousness. Watching the beautiful trains depart half empty took a toll on railroad people, as if they had given a beautiful party and nobody came. One by one in the 1960s, railroad managements that had once tended flagship trains with loving care became realistic, which meant hardnosed, which meant run the public off and get rid of the damn things. Deep in their hearts, they knew something irretrievable was slipping away. As the crowds thinned and the upholstery went threadbare and the paint began to peel, railroad morale crumbled.

Richard Saunders Jr, Merging Lines, *2001.*

407.10 More than pride was at stake. The precision on-time railroading required by the passenger train had a halo effect that carried over to the great redball fast freights. There had been a time when the streamliners rolled into the night at the appointed nanosecond with a wave of the lantern, a notch of the throttle, a blast of the horn, and a high-pitched whine of the diesel's electric motors. As the tight discipline of passenger operations waned into sloppiness, so did all the precision and discipline, and the freight operation got sloppy as well. That was just as the motor carrier was poised to roll down the new interstate highways straight from one loading dock to another with precisely the service the customer required. It was as if the very soul of the railroad was wasting away.

Ibid.

407.11 Thomas the Regulatory Engine and Alastair the superheated Engine have their buffers locked together, the coaches are unhappy and the track is in a tangle. A new character, Stephen the Grey Engine, is telling them all to stick around and try harder but Alastair has had enough. Getting this lot to budge, he says, is like hauling a heavy coal train up the Lickey Incline without a banker. Gordon the Treasury Engine is sulking. If a banker is needed, he can guess who it will have to be. All the other possibilities seem to have run out of steam.

Christopher Fildes, referring to Tom Winsor (Rail Regulator), Sir Alastair Morton (Chairman, Strategic Rail Authority), Stephen Byers (Transport Secretary), and Gordon Brown (Chancellor of the Exchequer), The Spectator, *July 2001.*

407.12 I put their reaction down to Boiling Frog Syndrome (BFS), known in the safety world as 'creeping normalcy.'

Roger Ford, of the tendency, post-privatization, to ignore the escalating cost of railway infrastructure projects. The reaction referred to was from railwaymen Chris Green and Adrian Shooter, who seemed to Ford not to be as concerned as he. Modern Railways, *August 2002.*

408. MANAGEMENT OF RAILWAYS— Good Fortune

408.1 You are a heap better off running a good road, than I am playing President.

Abraham Lincoln, to Bowen, Superintendent, Great Western of Illinois Railroad, on his way to Washington as President-elect, 30 January 1861. Quoted by Bob Withers in The President Travels by Train, *1996.*

408.2 Each Christmas Day he gave each stoker
A silver shovel and a golden poker,
He'd button-hole flowers for the ticket sorters
And rich Bath buns for the outside porters.
He'd mount the clerks on his first-class hunters
And he built little villas for the roadside shunters . . .

W S Gilbert, Thespis, *1871.*

409. MANAGEMENT OF RAILWAYS— Organization

409.1 I consider the designing of a plan for new organization one of the most important matters we have had before us for some time and we should first adopt a plan and then make officers fit into it as best we can and not make the plan to fit the officers.

Edward H Harriman, letter to J Fentress, 11 July 1889, quoted by Thomas C Cochrane in Railroad Leaders, 1845– 1890, *1953.*

409.2 Watch the details. Then the whole organization will watch the details.

Attributed to C P Huntington by Otto H Kahn, in The Railway Library, 1910, *1911.*

409.3 Small boys play trains, but grown-ups have a better game; they call it railway reorganisation.

Editorial comment, The Economist, *thought to be before 1971.*

409.4 When you reorganise, you bleed.

Stated often by Gerard Fiennes, including within an article in Modern Railways, *January 1967, but subsequently acknowledged by him, in* I Tried to Run a Railway *as having been said by a visiting speaker at the railway staff college in Woking, name unknown.*

410. MANAGEMENT OF RAILWAYS— Practice

410.1 For what *is* done in the United States, *ought* to be done in the United Kingdom.

Edward Bates Dorsey, English and American Railroads Compared, *1887.*

410.2 The old idea of management was to charge as much as you dare and give as little as you knew how.

Sir Allan Sarle, LBSCR, quoted in The Railway Magazine, vol. XVI, *1905.*

410.3 In the first place, as English railway men are ardent worshippers of the great goddess Rule-of-Thumb, and in the main disbelieve in the scientific study of their profession, it has comparatively few readers.

W M Acworth, of the then newly-published trade magazine The Railway Gazette. *Address,* Railway Professional Education, Its Objects and Limitations, *School of Commerce, University of Liverpool, 8 November 1905. Printed in* The Annals of the American Academy of Political and Social Sciences, *July 1906.*

410.4 Here was not a case of where one man dominatingly insisted that he alone was endowed with all of the functions required in successful business. The Pacific quartet recognized the value of specialization. In a general way, Huntington was intrusted with the supervision of financial affairs; Stanford of the plans for the manipulation of law and politics; Crocker was placed in charge of construction work, and Hopkins was the commandant of office details. The particular useful qualifications of each of the four were mutually appreciated and availed of. In addition to this division of overseership, all joined together as a unit in the promotion and consummation of their plans.

Gustavus Myers, of Collis P Huntington, Leland Stanford, Charles Crocker and Mark Hopkins, builders of the Central Pacific Railroad; History of the Great American Fortunes, *1910.*

410.5 Railroads are built by human beings; they are run by human beings; they are regulated by beings who are very human; and they serve human beings. Make a mistake in your treatment of the railroad question and you injure human beings. Handle it properly and you help human beings.

Ivy Lee, Introduction to Human Nature and Railroads, *1915.*

410.6 The railroad is really more remarkable for the time-honoured formulas it neglects to observe than it is for the profits it earns.

Henry Ford, of the Detroit Toledo & Ironton, owned by Ford, and which made a profit. Today and Tomorrow, *1926.*

410.7 From the ashes of yesterday's corporate plan rises the phoenix of tomorrow's cock up.

Attributed to Sir Peter Parker, Modern Railways, *July 2002.*

411. MANAGEMENT OF RAILWAYS— Qualities and Characteristics

411.1 Pooh, pooh, we don't mind principle in matters of business.

Attributed to George Hudson by Captain Watts in letter to the Railway Times, *1841 or 1842, quoted by Robert Beaumont,* The Railway King, *2002.*

411.2 Railroad companies . . . are managed nominally by those who have both head and hands full of something else, and not infrequently by those who, in addition to such an incumbrance, are also too far from the scene of action.

Comment, Hunt's Merchant's Magazine and Commercial Review, *September 1854; quoted thus by Thomas C Cochrane in* Railroad Leaders, 1845–1890, *1953.*

411.3 Intelligence and integrity are the qualities necessary to a proper management of railroads. The former must be possessed by the *public*, as well as those in charge of them.

Comment, American Railroad Journal, *16 December 1854.*

411.4 If I had time I should say a word or two for the directors and managers of our railways. They are constantly attacked—often, in my opinion, most unjustly. They are reviled often by writers in the press, but I say there is nothing in the whole world, in my opinion, to compare with the successful management of the railways of this country, whether you consider their speed, or their comfort, or their safety, or their cheapness, all that which they give to the nation; there is nothing in my opinion in the world that can excel it.

John Bright, MP, address in Rochdale, 25 September 1877.

411.5 A corporation of the first magnitude and of the character of a railroad rests its hope of successful operation very largely upon organization, discipline, and continuity of employment.

L F Loree, Railroad Freight Transportation, 1922.

411.6 The Division Superintendent is expected to know everything, and is not supposed to need any sleep.

Ibid.

411.7 We manufacture train miles, we sell ton miles.

Attributed to James J Hill by L F Loree, Railroad Freight Transportation, 1922.

411.8 Yep, shoot the bull, pass the buck, and make seven copies of everything—that's railroading.

Anon. Nickel Plate telegrapher, Indiana, to Adrian Ettlinger, late 1940s.

411.9 One of the cherished privileges of American citizenship is the right of all men, and not a few women, to vigorously damn the passenger train and to explain how the railroads could make a mint for themselves if they would only charge less for a sandwich or add a dome-lounge to the 8:11 or quit turning on the lights at midnight to punch tickets.

David P Morgan, Trains Magazine, February 1958.

411.10 If the application of fresh committees to the problems of British Railways could work their salvation, we should by now be operating one of the world's most profitable railways.

Editorial comment, Trains Illustrated, September 1960.

411.11 Almost a century ago robber barons ran the railroads; after the turn of the century one operator, E. H. Harriman, was deemed an "undesirable citizen" by Theodore Roosevelt; and as late as the 1930's "autocratic" and "inflexible" were still acceptable adjectives. Today management is merely poor; the news media, if not the Brotherhoods, no longer suspect it of cheating widows or enslaving the working man or oppressing the farmer. Management is simply mediocre, inept.

David P Morgan, of public misperceptions of railroad managers; Trains Magazine, August 1961.

411.12 I am not content to let the railroads' 15,000 executives get off the hook that easily, though. It occurs to me that they have committed a felony or worse by their stubborn refusal to annul their trains, lock up their locomotives, and otherwise desist from breaking the unwritten law of the land. For it is clear, or should be, that the railroads' dilemma is this: they are competing with the Government.

Ibid.

411.13 A railway officer does not lay down his personal and immediate responsibility when he goes home at the end of the day; it stays with him, and he may at any moment have to act on it. The work is always with him; it is a life rather than an occupation.

Michael Robbins, The Railway Age, 1962.

411.14 . . . simply a gratuitous exercise in versatility.

Lord Beeching, of his time as Chairman of British Railways Board, Extraordinary General Meeting, Redland Ltd, c. 1970, recorded in Beeching, Champion of the Railway? *Richard Hardy, 1989.*

411.15 It's the old story. They're just telling us that they're doing the impossible already.

Overheard on morning rush hour train, London, after the publication of a booklet entitled "Want to run a railway?" by the Southern Region. Noted in The Railway Magazine, vol. 109, *1963.*

411.16 . . . a confrontation of bowler hats on the ballast.

Derek Barrie, of a disagreement between officials of the Taff Vale Railway and Great Western Railway when the former tried to run a coal train over the metals of the latter, A Regional History of the Railways of Great Britain, vol. 12, *1980.*

411.17 Charm, wit, and terrorism.

Anon., of the management style soon after the creation of Norfolk Southern by merger of the Southern and Norfolk & Western Railways. Quoted by John P Kneiling in Trains Magazine, *January 1983.*

411.18 All it takes is common sense and modern equipment.

Alfred E Perlman, quoted by David P Morgan, Trains Magazine, *July 1983.*

412. MANAGEMENT OF RAILWAYS— Skills

412.1 Competent engine drivers are indispensable to good railway management.

John B Jervis, Railway Property, 1866.

412.2 It takes qualities which produce a great general and make a successful business man on a large scale to manage the intricate relations of a railway company with its several communities, with the general public, and with its employees and owners.

Chauncey Depew, to Congress of Railway Employees, Chicago, February 24 1899. Quoted in Chauncey Mitchell Depew Orator, *Welland Hayes Yeager, 1934.*

412.3 I boldly assert that there never has been a man who, in the whole course of his career, has been able to thoroughly master the details of every phase of railway working. There are no general practitioners on a railway.

W Dawson, Outdoor Assistant to Superintendent of the Line, The Great Western Railway Magazine, *19 October 1903.*

412.4 Our railway management, like our politics, is essentially amateurish.

W M Acworth, of the lack of formal training for railway managers, address, Railway Professional Education, Its Objects and Limitations, *School of Commerce, University of Liverpool, 8 November 1905. Printed in* The Annals of the American Academy of Political and Social Sciences, *July 1906.*

412.5 If the manager in charge cannot be trusted to buy mules without bringing the subject to the executive committee, a new manager should be selected. My time is worth about a mule a minute and I can't stay to hear the rest of this discussion. I vote 'Aye' on the requisition.

Edward Harriman, at a meeting on the Erie Railroad, c. 1908, at which a requisition for mules for company coal mines was being discussed. Related by George Kenman, E. H. Harriman, *vol. 1, 1922. It appears that Harriman often used the device of leaving a meeting to bring a discussion to an abrupt halt; see* The Life and Legend of E H Harriman, *Maury Klein, p. 166.*

412.6 I want to see him with his feet on his desk—thinking, thinking!

E H Harriman, of a general manager, quoted by John Droege in Freight Terminals and Trains, *1925.*

412.7 . . . he is expected to express to anybody, at any hour of the day or night, over the telephone, or at the point of the reporter's pencil, opinions on every subject relating to transport. Sometimes he knows, but it is expedient that he should not tell; sometimes he does not know, but it is inexpedient that others should know that he does not; at other times he would prefer his own time and manner of announcement.

Sir Josiah Stamp, of Chairmen and General Managers, Introduction to Railway Humour, *John Aye, 1931.*

412.8 Tch, tch! What a way to run a railroad!

Ralph Fuller, caption of cartoon, showing signalman watching two trains about to collide, Ballyhoo *magazine (U.S.), June 1932.*

412.9 The quality of men taken into the railway team from outside the industry was mixed. Some were to make an outstanding contribution: others made their mark by their very failure to make a contribution, and there were those who drifted in like shadows and so departed.

John Johnson and Robert A Long, of the influx of new managers after Nationalisation, 1948, British Railways Engineering, 1948–1980, *1981.*

412.10 It is no good being a Manager with an engineering background if you cannot get yourself out of trouble in times of breakdown and failure.

Richard Hardy, Steam in the Blood, *1971.*

412.11 None of you can be any good; you would be working in the private sector if you were.

Attributed to Margaret Thatcher by several people; speaking to British Railways managers at Woking Staff College while Secretary of State for Science and Industry, 1973. It is understood that a complaint was made, and she was subsequently reprimanded by Prime Minister Edward Heath for these words.

412.12 If you like trains, then you expect them to work properly. If you are a professional, and like trains, then you expect them to work well consistently.

Dr John Prideaux, address to Royal Society of Arts, May 1989.

412.13 We f∗∗∗ed up. We f∗∗∗ed up badly.

Richard Branson, on problems with Virgin Railways services, Radio 5 Live, 4 October 1998, reported as shown, The Times, *5 October 1998.*

413. MANAGEMENT OF RAILWAYS— Understanding

413.1 Men educated to an exact science, like civil engineering, are apt to carry its methods into everything they do, and it is no doubt true that when it came to making railroad tariffs such men undertook to construct them on mathematical principles, instead of recognizing the fact that they must in the end be regulated by the laws of supply and demand.

Charles E Perkins, Chicago Burlington & Quincy Railroad, to Joseph Nimmo, 1887. Quoted by Steven W Usselman in Regulating Railroad Innovation, *2002.*

413.2 Well, gentlemen, I suppose as regards railway management the Chairman has not yet learned as much as I have forgotten.

Mark Huish, General Manager of the London & North Western Railway, of his chairman, the Marquis of Chandos. Noted in Railway and Travel Monthly, *1911.*

413.3 . . . he had sent an order to the engineers to cut loose from their trains, run to Alexandria for wood and water and then return. As there was but a single track and no one capable of performing the Munchausen feat of picking up the engines and carrying them around the trains, the order could not be executed.

Herman Haupt, of General Halleck, an officer not well versed in railroad matters, during the Civil War; Reminiscences of General Herman Haupt, *1901.*

413.4 There are two things about which the public is most critical—one is the management of the newspaper, the other the management of the railroad. In his heart the average citizen believes that he could operate either his daily newspaper or the railroad passing through his town much better than it is being operated; he would perhaps hesitate to announce this opinion but his attitude is coldly critical, and it is to be remembered that the railroad is all out of doors—all out in the weather, everything about it exposed to the limelight and visible to anybody's naked eye. There is no human activity the operation of which is attended with so much publicity. All our earnings and expenses are published; all our charges and all our methods the subject of regulation, intelligent or otherwise.

Edward P Ripley, President, AT&SF, speech at the annual dinner of the Railway Business Association, 10 November 1909, in Railroad Library 1909, *1910.*

413.5 And in so far as he is in a position to act as an educator of the shipping public in railway matters, he is rather bound to do something in that direction. Furthermore, the time may not be so very far distant when railway managers generally will begin to conceive that the broader intelligence in railway matters is more serviceable and more desirable, even in a station agent, than the narrower; or, to speak more accurately, the two together are better than only one of them.

B C Burt, Railway Station Service, *1911.*

413.6 It frequently happens that railway administration fails to be complete and organic because the central organization fails to get into "touch" with practical conditions at the circumference of things, namely, the station.

Ibid.

413.7 A correct fate would deprive the Continent of its railways, and give them to somebody who knew about them. The continental idea of a railway is to surround a mass of machinery with forty rings of ultra-military law, and then believe they have one complete. The Americans and English are the railway peoples.

Stephen Crane, The Scotch Express, *within* Men, Women and Boats, *1921.*

413.8 The public appears to be quite willing that Charlie Chaplin should receive $1,000,000 per year for his contributions to the mirth of the movies, yet they find fault because the president of a railroad company, employing 100,000 men and representing an investment of hun-

dreds of millions of dollars devoted to public service, is paid $50,000.

W J Cunningham, American Railroads: Government Control and Reconstruction Policies, *1922.*

413.9 Except in a few notable instances, such as that of the Pennsylvania Railroad, the proportion of college men among railroad officials is small. The great majority are graduates of the school of practical experience.

Ibid.

413.10 I have never met anybody, outside the circle of those who know how it should be done, who has not a brilliant idea for the managing of railways.

Sir Frank Ree, General Manager, London & North Western Railway, quoted by George Dow in Great Central, *vol. III, 1965.*

413.11 ... I do not think it is possible to treat the organisation of a British Railway Company quite in the same way as any private industrial undertaking, however large and important. ...

Sir Alexander Butterworth, quoted in The London & North Eastern Railway, *Cecil J Allen, 1966.*

413.12 ... he had no cause to question the reasons for other companies not conforming to GWR practice.

John Johnson and Robert A Long, of Sir Allan Quartermaine, Chief Engineer of the GWR in 1947, British Railways Engineering, 1948–1980, *1981.*

413.13 Undoubtedly many of the characteristics of personality exhibited by railroaders are due in part to the process of selection which takes place when they are recruited. Until quite recently "scientific personnel management" has been almost entirely absent in the operation of American railways. In part this is due to the widespread operations of railroads, which makes central hiring prohibitive and makes it difficult to retain trained employing officers at the scattered points from which labor is to be recruited. In much larger part, perhaps, it is due to the contempt in which college training is held by most of those operating the railroads, at least in so far as college training for railroad service is concerned. Most operating supervisors on the railroads are themselves a product of the school of hard knocks.

W Fred Cottrell, The Railroader, *1940.*

413.14 To see the answer to any managerial problems of British Railways solely in terms of recruitment from the higher echelons of private industry will be as sensible as it would have been to consider the appointment of industrial barons to the command of the British Army at the time our fortunes were at their lowest in the early 1940s. There is much in common between the running of an army and a railway, for both demand an acute and forceful grip of strategies and logistics that are peculiar to their respective worlds; in both cases some field as well as theoretical training is essential to evolve an efficient commander. The general who turned the tide at Alamein was not, prior to his arrival in the Middle East, one of the very "top brass"; but he *was* a soldier, not a detergent manufacturer.

Editorial comment, Modern Railways, *February 1962.*

413.15 Together with labour relations and singing in the bath, knowledge of how the railways should be run is provided by the Almighty at the moment of birth. It is a well-known fact that the nation is divided between 27 million railway experts and 190,000 of us who earn our living on the railways.

Richard Marsh, lecture, The Economics of Indecision, *17 May 1974.*

413.16 Anyway, I had fifty-five million people to help me: everybody seemed to be an expert on railways and ready with instant advice.

Sir Peter Parker, of his assuming the chairmanship of British Rail, For Starters, *1989.*

414. MANCHESTER, England

414.1 Let Manchester, which showed London the way to make railroads, unite profits and patriotism, by risking something to open the unknown oases of Central India to British Commerce.

Samuel Sidney, Household Words, *15 March 1851.*

414.2 I am even looking forward to my journey to Manchester, supposing that there is no great rush for the place on my chosen day. The scenery as one approaches Manchester may not be beautiful, but I shall be quite happy in my corner facing the engine.

A A Milne, A Train of Thought, *within* If I May, *1920.*

415. MANCHESTER SHEFFIELD & LINCOLNSHIRE RAILWAY

415.1 I have not invited this issue of the confidence in the Manchester, Sheffield Company, or in Sir Edward Watkin personally; but if Sir Edward Watkin chooses to come here and tell a body of gentlemen such as the Great Northern Board that he cannot trust them, he invites inquiry into his own antecedents. I am merely dealing with him in his railway character—in any other character I have nothing to say to him—but I should like to know if any company in England has done such things as the Manchester, Sheffield has done—hawked itself about as buyer, as seller, as guarantor, as guarantee; bribed to shut up their traffic, bribed to open their traffic. There is not a conceivable bargain to which they have not been parties, and here they come to have justice. I only wish it to be administered to them.

Edmund Denison QC in Parliamentary Committee,

May 1871. Quoted in The History of the Great Northern Railway, *C H Grinling, 1903.*

415.2 What have the Sheffield done? They have flirted with the North Western since 1856; they then flirted with the Great Northern; they then flirted with the Midland; then they flirted with the Eastern Counties and the coal-owners. Then, in 1872, they flirted again with their old love, the London and North Western; and now, in 1873, there is a mild flirtation between Sir Edward Watkin and Mr Allport; and, like all flirts, mark my words, the Sheffield will be left without an alliance with any of them, and will entertain that feeling which flirts entertain towards all mankind when they have been left completely in the lurch, and she will move about society on her own hook, catching whom she can. This is not the less true because it creates a little mirth.

Sir Mordaunt Wells, 1873, quoted by Frederick Williams in The Midland Railway, *1878.*

416. MANCHESTER SOUTH JUNCTION & ALTRINCHAM RAILWAY

416.1 Many Short Journeys and Absolute Reliability.

Anon. long-standing nickname using the initial letters of the name, and very precisely describing its characteristics.

417. MAPS

417.1 I will ask you to look attentively for a moment at a Bradshaw's railway map, and you will see that throughout the network of rails that overspreads the land none of the meshes, so to speak, in any vital parts of the country, exceed fifteen miles across, from rail to rail, but as the eye approaches the Metropolis, or any of the commercial centres, these meshes are diminished to about one-half the area of the others.

W M McMurdo, pamphlet Rifle Volunteers for Field Service, *1869, quoted by E A Pratt in* British Railways and the Great War, *1921.*

417.2 It looks like a corridor with rooms on both sides.

Anon., at a meeting in Washington, D.C., of the map of what thus became known as the North East Corridor of railroads, USA. Related by John G Kneiling, Trains Magazine, *December 1980.*

417.3 There is something about a map that brings out the beast in a railroad. Indeed it amounts to an inexplicable lapse in an otherwise conscientious and painstaking business. A railroad will bend over backwards to refund your unused ticket to the last penny—and often on a check so impressive in its countersignatures and watermarked paper that one almost feels obliged to frame it instead of cash it. Again, a railroad will employ regiments of draftsmen working at acres of drawing tables to figure out bridge stress or driving wheel counterbalance or size of drain gutter. Accuracy is the watchword and excuses are taboo.

Except in maps—timetable maps. I think there must be rules of some sort, because the wholesale rearrangement of physical plant and nature is too universal an art to be happenstance. The standard technique is to erase all other lines except the most friendly connections; expand all states served by the company so that they occupy approximately 80 per cent of the land area of the U.S.; rub out mountains and other natural barriers; and—most important—draw all main lines with a ruler.

David P Morgan, The Mohawk that Refused to Abdicate, *1975.*

417.4　A true map of the London Underground shows the central complex as a shape suggestive of a swimming dolphin, its snout being Aldgate, its forehead Old Street, the crown of its head King's Cross, its spine Paddington, White City and Acton, its tail Ealing Broadway and its underbelly the stations of Kensington.

Barbara Vine (Ruth Rendell), King Solomon's Carpet, *1991.*

417.5　The outer configurations branch out in graceful tentacles. The seal has become a medusa, a jellyfish. Its extremities touch Middlesex and Hertfordshire, Essex and Surrey. A claw penetrates Heathrow.

Ibid.

417.6　. . . a plate of wet spaghetti . . .

Anon. official, of the map of the Chicago & North Western Railroad, quoted in The North Western, *H Roger Grant, 1996.*

418. MARKETING

418.1　Railroad men as a class, are notoriously poor salesmen. The art of persuasion does not come natural to them.

Robert E Woodruff, The Making of a Railroad Officer, *1925.*

418.2　We don't know anything about our business, about our customers, about what they need, about what we ought to be selling. Above all, we don't know anything about our costs or how to price our product.

William Brosnan Jr, Southern Railway, when seeking to employ Bob Hamilton to undertake such work, 1959. Recorded by Charles O Morgret in Brosnan, *vol. 1, 1996.*

418.3　Turbo bullshit.

Anon., (but from among the team) of proposals and other documentation written by members of Fastline Track Renewals in their successful Management Buy Out effort from British Railways during Privatisation, 1995.

419. MARRIAGE

419.1　As I went down to Louisville, some pleasure for to see,
I fell in love with a railroad man, and he in love with me.
I wouldn't marry a farmer, for he's always in the dirt,
I'd rather marry a railroader that wears them pretty blue shirts.

Anon., song, Railroad Daddy; *quoted in* Long Steel Rail *by Norm Cohen, 1990.*

419.2　So this is the sort of man you intend to marry!

Antonin Dvorak, to his daughter, of his future son-in-law, sent to the station to obtain the number of a locomotive, and who returned with the number of the tender. Found on the internet at www.rodge.force9.co.uk /faq/arts.html, but not confirmed from elsewhere (accessed 2003).

419.3　Not less decorously than a bride made ready for her groom is the Century inaugurated for departure. A strip of wedding carpet leads you down into the cathedral twilight of that long crypt. Like a bouquet of flowers her name shines in white bulbs on the observation platform. In the diner, waiters' coats are laundered like surplices.

Christopher Morley, A Ride in the Cab of the Twentieth Century Limited, *1928.*

419.4　The electric engine has a fascination and efficiency of its own, but in this ceremony one is bound to regard it mainly as the father who takes the bride up the aisle on his arm. Thirty-three miles the electric takes her to Harmon, where the bridegroom is ready, the steam engine; and (as the old rubric puts it) they are coupled together, to have and to hold.

Christopher Morley, of the practice on the New York Central to haul trains electrically out of New York City before changing to steam traction; ibid.

419.5　What Perlman and Symes announced is not so much a marriage as it is an indictment of those who carry the shotgun.

David P Morgan, of the announcement of the merger between the Pennsylvania and New York Central Railroads; Trains Magazine, *January 1958.*

420. ALBRO MARTIN

420.1　. . . the man who made deregulation academically respectable.

Anon. business historian, quoted by Albro Martin in Railroads Triumphant, *1992.*

421. MARYBOROUGH STATION

421.1　. . . a station with a town attached.

Attributed to Mark Twain by Leo J Harrison, Victorian Railways to '62, *1962.*

422. MARYLAND & PENNSYLVANIA RAILROAD

422.1 Those who rode the Maryland & Pennsylvania Railroad will not be surprised to learn that the line was not conceived as a direct route from Baltimore to New York. Indeed, many who rode the Ma & Pa doubtless concluded that it was not designed to be a direct route anywhere, and that the whole railroad was simply created as a work of art.

George W Hilton, The Ma & Pa, *1963.*

422.2 Any of the hundreds of visitors who came to see the Ma & Pa between 1935 and 1954 might easily have concluded that the whole thing came from the mind of some Velásquez or Rembrandt among model railroaders, who, having exhausted his art in HO and O gauges, came finally to the hills north of Baltimore to create his masterpiece at a scale of 12 inches to the foot.

Ibid.

423. MARYLEBONE STATION

423.1 Two days later he was at Marylebone Station, quietest and most dignified of stations, where the porters go on tiptoe, where the barrows are rubber-tyred and the trains sidle mysteriously in and out with only the faintest of toots upon their whistles so as not to disturb the signalman and there he bought a ticket to Aylesbury from a man who whispered that the cost was nine-and-six, and that a train would probably start from Number 5 platform as soon as the engine-driver had come back from the pictures and the guard had been to see his old mother in Baker Street.

A G Macdonnell, England Their England, *1933.*

424. MATHEMATICS

424.1 To an Engineer mathematical truths are like sharp instruments, they require to be handled with care and circumspection—in short they are the only truths so long as certain conditions obtain.

Robert Stephenson, letter to Rev John Bruce, 1839.

425. MEDITATION

425.1 Penitent: Father, I cannot meditate. Will you tell me what is the best church to go to?
Priest: Marylebone station, my son. It is the best place in London for meditation.

Canon Roger Lloyd, The Aesthetics of a Train, *within* The Railway Magazine, vol. LXXIV, *1934.*

426. MERGERS

426.1 The length of line under one control in 1843 was scarcely such as to obtain the most economical management, but the principle object of the companies in amalgamating since that period has been to obtain control of the districts from which they draw traffic, in order to prevent that traffic being carried by other lines. Amalgamation has thus been a matter of offensive and defensive policy than a question of economy in working the lines.

Report of the Royal Commission on Railways, 1867, quoted in Edward Cleveland-Stevens, English Railways and Their Development and Relation to the State, *1915.*

426.2 As long as railroads were purely local affairs, each locality might charter and run its own. The moment any large through traffic grew up, this was found to be a wasteful way of doing business. If they changed cars at every point of junction, the expenses were vastly increased. If they did not change cars, there was still the awkwardness of dividing responsibility, and the evil of having two separate organizations where one would do the work better.

Arthur Hadley, Railroad Transportation, *1885.*

426.3 In 1843 the Company exhibited the first symptom of that appetite for swallowing lesser railway companies which it indulged at frequent intervals during the next eighty years till the craving was finally satiated by a surfeit of Welsh coal railways forcibly administered by Parliament.

E T MacDermot, of the Great Western Railway, History of the Great Western Railway, *vol. 1, Part 1, 1927.*

426.4 Penn Central constitutes no disproof of the efficacy of consolidation; it was Civil War, not a merger.

John W Barriger, Address to Rotary Club of Houston, 13 August 1970.

427. METROPOLITAN DISTRICT RAILWAY

427.1 I am not going to tie your property which with all its unfortunate mistakes has some substance in it, to a concern like the Metropolitan District; this beautiful lady without a dower.

Sir Edward Watkin to Metropolitan Railway shareholders, 15 October 1872, quoted in The Second Railway King, *David Hodgkins, 2002.*

428. METROPOLITAN RAILWAY

428.1 The Handy Line to Anywhere.

Advertising slogan, 1910.

428.2 . . . a Squire among railways . . .

R K Kirkland, Railway World, *vol. XVII, 1956.*

428.3 Child of the First War, forgotten by the Second,
We called you Metro-land. We laid our schemes,
Lured by the lush brochure, down byways beckoned,
To build at last the cottage of our dreams,

A city clerk turned countryman again
And linked to the Metropolis by train.
> *Sir John Betjeman,* Metroland, *broadcast 1973.*

428.4 Early electric, punctual and prompt,
Off to those cuttings in the Hampstead Hills—
St John's Wood, Marlborough Road—
No longer stations, and the trains rush through.
> *Ibid.*

428.5 Sorry about the delay at Baker Street; the Met. takes priority over everything. They're a law unto themselves.
> *Circle Line driver to passengers of eastbound train delayed at Baker Street, 30 June 2000.*

429. MIDLAND RAILWAY

429.1 Euston Square holds the strings; the Midland, I regret to say, has no independent existence.
> *Edmund Denison, referring to the strength of the Euston Square Confederacy, of which the Midland Railway was a member, 1848. Quoted by Charles Grinling in* History of the Great Northern Railway, *1903.*

429.2 Here is a line with magnificent pluck and enterprise—too much sometimes for the peace of mind of its neighbours.
> *E Foxwell and T C Farrer,* Express Trains English and Foreign, *1889.*

429.3 Ours is a sort of family affair. We know, if we put our money into it, we can have it out again when we want it.
> *John Ellis, Chairman, Midland Railway, quoted in* Our Railways, *J Pendleton, 1896.*

429.4 There was no help for it; the octopus of Derby must get a tentacle into London, and to do it was well worth the money spent.
> *W J Gordon, of the Midland Railway's first access to London via the Great Northern Railway from Hitchin to King's Cross.* Everyday Life on the Railroad, *1898.*

429.5 After a somewhat long sleep, this company awoke to new energy on May 1st of this year, and put on many hundred train-miles between London and Leicester, Nottingham, Manchester and Bradford, much of it at a good, though not wonderful speed. This was partly brought about by the new competition of the Great Central.
> *W J Scott,* The Railway Magazine, *vol. V, 1899.*

429.6 The courtesy of all Midland Railway officials is proverbial . . .
> *Comment,* The Railway Magazine, *vol. XI, 1902.*

429.7 M is for Midland, with engines galore;
They have two on each train, but they hanker for more.
> *Anon., from a Railway ABC, Cecil J Allen,* The Railway Magazine, *vol. 92, 1946.*

429.8 The Midland too was utterly delicious.
Not merely for its colour, which was jolly.
That in itself might be a fact propitious
To people suffering from melancholy.
It had a luxury which even Solly
Joel would not have laughed at. I recall
Thinking that very thought. Not meretricious
But real. Cosy smoke-room of a Hall
Owned by a Peer. Can similes work out at all?
> *Terence Greenidge, verse four of* The Nationalisation of the Railways, *in* Girls and Stations, *1952. Solly Joel was a wealthy man who sponsored a cricket team which was known, in a Spoonerised version of his name, as the Jolly Souls.*

429.9 In general, however, there was no railway more agreeable than the Midland from the passenger's point of view.
> *C Hamilton Ellis,* British Railway History, *vol. II, 1959.*

429.10 Notwithstanding the magnificence justly attributed to it by more than one writer, the Midland was the railway of the ordinary man.
> *George Dow, Introduction to* Midland Railway Carriages, *1984.*

429.11 The Best Route.
> *Advertising slogan.*

429.12 Rumbling under blackened girders, Midland bound for Cricklewood,
Puffed its sulphur to the sunset where that land of laundries stood,
Rumble under, thunder over, train and tram alternate go,
Shake the floor and skudge the ledger, Charrington, Sells, Dale and Co.,
Nuts and nuggets in the window, trucks along the lines below.
> *John Betjeman, poem,* Parliament Hill Fields.

430. THE MIDNIGHT SPECIAL

430.1 Let the Midnight Special shine her light on me,
Let the Midnight Special shine her ever-loving light on me.
> *Chorus of song* Midnight Special, *anon., origin obscure, but first recorded 1926. The train was in fact the Southern Pacific's* Golden Gate, *which passed the Texas state prison at Sugarland, Texas, shining its headlight through the windows of the cells.*

431. MID-SUFFOLK LIGHT RAILWAY

431.1 . . . or the strange little Mid-Suffolk Light Railway, feeling its way eastwards from Haughley, north of Stowmarket, until it expired, unintentionally, near Laxfield.
> *Michael Robbins,* The Railway Magazine, *1951.*

432. MILFORD HAVEN

432.1 Where Fish comes from.
Great Western Railway poster, c. 1905.

433. MILK CANS

433.1 With companions of its own likeness it had been hurried, times innumerable, from its rustic surroundings, along the winding lanes of the countryside, to a little wayside railway station, there to halt for intervals of shorter or longer periods, until jostled into the freight van of a passing train, then continuing its journey with others of its kin, some having the appearance of knights in shining armour, others bearing the dull and worn effects of old service veterans.
W E Newton, LNER Magazine, October 1935.

434. MINISTER OF TRANSPORT

434.1 A job for a rather unprincipled person.
Alan Lennox-Boyd, conversation with Winston Churchill, quoted by T R Gourvish in British Railways 1948–1973, *1986.*

434.2 A Transport Secretary who wants to stay in office long enough for his own chickens to come home to roost is a novelty indeed.
Nigel Harris, of Stephen Byers, Rail *Magazine, 1 May 2002.*

435. MIXED TRAINS

435.1 The society of the cattle and hogs is not in my opinion sufficient compensation.
Anon., petitioner to Kansas Railroad Commission, 1884, concerning the slow progress of the daily mixed train on the Kansas City, Lawrence & Southern Kansas Railroad; noted (with an incorrect name for the railroad) by Clyde Freed, The Story of Railroad Passenger Fares, *1942.*

435.2 Informal railroading is the rule on the mixed train.
R W Richardson, Trains Magazine, *November 1947.*

436. MNEMONICS and ACRONYMS

436.1 Old Maids Never Wed and Have Babies Period.
Mnemonic for remembering the first stations on the Philadelphia Main Line: Overbrook, Merion, Narberth, Wynnewood, Ardmore, Haverford and Bryn Mawr. "P" stands for Paoli, at the end of the line. Noted by Nathaniel Burt, The Perennial Philadelphians, *1963.*

436.2 ycsfsoya
William Brosnan Jr, sign in Southern Railway's control center, Atlanta; an acronym for You Can't Sell Freight Sitting On Your Ass. *1965.*

437. MOBILIZATION

437.1 All the European powers had built up vast armies of conscripts. The plans for mobilizing these millions rested on railways; and railway timetables cannot be improvised. Once started, the wagons and carriages must roll remorselessly and inevitably forward to their predestined goal.
A J P Taylor, The First World War, *1963.*

438. MODEL RAILWAYS

438.1 He had an engine that he loved
With all his heart and soul,
And if he had a wish on earth
It was to keep it whole.
One day—my friends, prepare your minds;
I'm coming to the worst—
Quite suddenly a screw went mad
And then the boiler burst!
E Nesbit, first lines of verse within The Railway Children, *1906.*

438.2 A miniature railroad, unlike its big prototypes, is always successful.
Anon, Fortune Magazine, *December 1932.*

438.3 Any man with pretensions to normality knows a lot about toy trains.
Ibid.

438.4 Well, it seems like this Mrs Applebee has been married about a year, and her husband is a salesman. He is also a model-railway fan, but she never suspected it until after they were engaged. She says love must have made her blind. She says being married to a model-railway fan is a very terrible thing, and there isn't anything you can do for it except take an occasional headache tablet.
Doug Welch, short story, Mrs Union Station, *within* Saturday Evening Post, *1937.*

438.5 Always complete, never finished.
Expression describing the model railway hobby, recorded by Gilbert Thomas, Double Headed, *1963.*

438.6 Trains?!! My dear chap, electric trains are not a hobby!—They're a way of life!!!
Lynn Johnston, punch line of syndicated cartoon strip showing a scene in a hobby shop, For Better or for Worse, *c. 1980.*

438.7 Model railroading is a three-dimensional art form.
Editorial comment, Model Railroader *magazine, c. 1990.*

438.8 I admire women but I don't collect Barbie Dolls.
Hans-Udo Drees, Deutsche Bahn commercial representative, explaining his lack of interest in model railways; conversation with the compiler, 30 January 2002.

438.9 Thank God for model trains—if they didn't have model trains, they never would have gotten the idea for big trains.

Christopher Guest & Eugene Levy, screenplay of film, A Mighty Wind, *2003.*

439. MODERNISATION PLAN

(British Railways, 1955)

439.1 Once those gaunt strong engines, named after people and places one has never quite heard of—Sir Henty Thomkins, Stindon Hall—are replaced by secretive diesels; once continuous brakes in goods trains have silenced for ever the night-long mysterious bing-bong-bang from misty, moonlit yards that for generations has told millions, in their warm beds, of our ancient, endless commerce; once the fretwork of stations are replaced by pin-bright foyers, it is idle to suppose that the station hotels, the houses, and the blue-brick churches and the people in them will remain the same. A certain openness, a certain ancestral earthy communion with fire and water and lonely native hills, will have gone for ever.

Paul Jennings, The Jenguin Pennings, *1963.*

439.2 Both the diesel and the electric locomotive lack the glamour which surrounds the mighty steam engine pounding through the night with the light of its fire glowing in the faces of the crew. There is something here of real importance, and we are honestly seeking a means to avoid losing all the romance which is attached to this great iron horse.

Sir Brian Robertson, at the launch of the Modernisation Plan, quoted by Paul Jennings in The Jenguin Pennings, *1963.*

439.3 There speaks a true British voice. Sir Brian Robertson has commanded troops; he knows the strategic, practical arguments in favour of diesels. But clearly he also knows about this aspect, at once most fundamental and more true. It would be pleasant to think that what he envisages is the actual setting up of a Romance Department, with its headquarters, obviously, at Euston.

Paul Jennings, ibid. The closest that British Railways got to a Romance Department was the recalling of the Locomotive Naming Committee, specifically to propose names for the first generation of diesel and electric locomotives.

440. MONEY

440.1 Money is something to be thrown off the back platforms of trains, preferably *The Twentieth Century Limited.*

Gene Fowler, as told by Lucius Beebe in 20th Century, *1962.*

441. MONOPOLY

441.1 I certainly have a very strong feeling on the subject of these railways to be traversed by the aid of steam. I sincerely wish that all these projects could prove successful; but in the same proportion it is desirable that there should not be a perpetual monopoly established in the country.

Duke of Wellington, House of Lords debate, 3 June 1836, quoted in The Speeches of the Duke of Wellington, *1854.*

441.2 Railway companies not only have the control of all the highways of the country, but also of the entire carrying trade; and it requires very little consideration to discover that, in the absence of proper safeguards, this practically gives the railway company owning the lines of any particular district absolute power over its welfare and prosperity.

Liverpool City Council, Minutes, 1871–2, 67, quoted in Jack Simmons, The Railway in Town and Country, 1830–1914, *1986.*

441.3 Monopoly is more than a game. I don't want you kids growing up thinking you can buy the Pennsylvania Railroad for two hundred dollars. The Reading Railroad, maybe—but not the Pennsylvania.

Art Buchwald, The Fantasy World of Monopoly, *1968. http://members.aol.com/wergames/yankmono.htm. (accessed July 2005).*

442. MONORAILS

442.1 Much nonsense has been aired about the advantages of the monorail and, during recent years, money has been wasted on consultants commissioned by the uninitiated to examine the extravagant claims of its protagonists. The hard truth is that the monorail is quite incapable of meeting the heavy demands a modern railway has to fulfil, whether passenger or freight or both. The fun fair and the exhibition ground are the elements to which it is best suited and should be restricted.

George Dow, Railway Heraldry, *1973.*

443. SIR RICHARD MOON

443.1 He it was who had taken a slackly knit railway company, all too obviously the product of amalgamation, and had welded it into a supreme power of its kind, a totalitarian corporate state in nineteenth-century capitalism, which persuaded clients, and which had no gentle compunction about political dictation to its servants.

C Hamilton Ellis, of the London & North Western Railway British Railway History, vol. II, *1959.*

444. MOONING AMTRAK

444.1 The Amtrak trains will be filled to capacity with passengers to see the "moon show" between the stations of "Irvine" to the North and "San Juan Capistrano" to the South. The mooning is on the EAST side of the tracks, and most trains will slow down at this point.

Comment on website of 23rd Annual Mooning of Amtrak, scheduled for Saturday, July 12, 2003, Laguna Niguel, California.

444.2 Night Mooning is better because: it is less crowded, cooler temperature, and more authentic.

Ibid.

445. MORALS

445.1 No breath of scandal rode *The Century*. Notoriety might be on the passenger list, but strict propriety characterized its stay as guest of the New York Central.

Lucius Beebe, of the Twentieth Century Limited, 20th Century, *1962.*

446. MORECAMBE

446.1 Loosens your stumps.

London & North Western Railway poster, c. 1909.

447. MORECAMBE AND WISE

447.1 They thought of calling themselves Morecambe and Leeds, but decided it sounded too much like a cheap day return.

Obituary for Ernie Wise, who came from Leeds, The Times, *22 March 1999.*

448. DAVID P MORGAN

448.1 Concrete in print, elusive in person.

Thomas M Jacklin, Railroad History 188, *2003.*

449. SIR ALISTAIR MORTON

449.1 He likes big numbers but doesn't seem to care whether they are black or red.

Anon., quoted by Christian Wolmar in Broken Rails, *2001.*

450. MOUNTAINS

450.1 A mountain railway is a deadly sin . . .

W H Auden, of the attitude of mountaineers; Letter to Lord Byron, III.13, *1937.*

450.2 The Swiss, incidentally, not only believe that the purpose of a mountain is to have a railway built up it but also to have a hotel or restaurant on top.

George H Drury, Trains Magazine, *February 1986.*

451. MICKEY MOUSE

451.1 It seems appropriate that the birth of Mickey Mouse—a creature of mythic stature—should be shrouded in legend. Walt Disney is said to have conceived Mickey on a train, returning to Hollywood from his angry encounter with Mintz. There is no reason to suppose that this is not essentially true, but over the years this story became so polished by repetition that it began to lose its sense of reality and to take on the character of an official myth.

Christopher Finch, The Art of Walt Disney, *1973.*

452. MUGBY JUNCTION

452.1 A place replete with shadowy shapes, this Mugby Junction in the black hours of the four-and-twenty.

Charles Dickens, Mugby Junction, *published in* All The Year Round, *1866.*

453. MUSIC

453.1 It is in the texture of our popular music, however, that the railroads have left their deepest impression. Listen to the blues, the stomps, the hot music of the last fifty years, since most Americans have come to live within the sound of the railroad. Listen to this music and you'll hear all the smashing, rattling, syncopated rhythms and counter-rhythms of trains of every size and speed. Listen to boogie-woogie with its various kinds of rolling basses. Listen to hot jazz with its steady beat. Listen to the blues with those hundreds of silvery breaks in the treble clef. What you hear back of the notes is the drive and thrust and moan of a locomotive. Of course, there's the African influence, the French influence in New Orleans, the Spanish influence from Cuba to account for the character of our hot music. These cultural elements have left their mark in our music, as they have elsewhere in this hemisphere, but, in our estimation, the distinctive feeling of American hot music comes from the railroad. In the minds and hearts of the people it is the surge and thunder of the steam engine, the ripple of the wheels along the tracks, and the shrill minor-keyed whistles that have colored this new American folk music.

Alan Lomax, Editor, Folk-Song: USA, *1947.*

453.2 I like to think of Dvorak, who at about this time was Director of the New York National Conservatory, slinking off to watch the trains at the old Grand Central Terminal, as he not infrequently did. There is a passage in his *New World* symphony that peculiarly suggests some great train from afar, high-wheeling into the terminus over the complex layout of its approach roads.

C Hamilton Ellis, The Beauty of Old Trains, *1952.*

453.3 . . . Oh, th'engineer would see him sittin' in the shade,

Strummin' with the rhythm that the drivers made . . .

Chuck Berry, song, Johnny B Goode, *1957.*

453.4 Half of Jazz is railway music, and the motion and noise of the train itself has the rhythm of jazz. This is not surprising: the Jazz Age was also the Railway Age.
   Paul Theroux, The Old Patagonian Express, *1979.*

453.5 Musicians travelled by train or not at all, and the pumping tempo and the clickety-clack and the lonesome whistle crept into the songs. So did the railway towns on the route: how else could Joplin or Kansas City be justified in a lyric?
   Ibid.

454. MUSICAL FAMILIES

454.1 *Sugar Kane*: I come from this musical family. My mother's a piano teacher and my father was a conductor.
Josephine: Where did he conduct?
Sugar Kane: On the Baltimore and Ohio.
   Billy Wilder; exchange between Marilyn Monroe and Tony Curtis in the film Some Like it Hot, *1959.*

455. MUSSOLINI

455.1 The calumny is still abroad that Mussolini made the Italian trains run to time, but they were punctual long before the March on Rome.
   Bryan Morgan, The End of the Line, *1955.*

Nn

456. NAPOLEON

456.1 Tush! Don't speak to me about Napoleon! Give me men and materials, and I will do what Napoleon couldn't do—drive a railway from Liverpool to Manchester across Chat Moss!
George Stephenson, to James Cropper, Director, L&M. Date not known.

NARROW GAUGE
See Gauge

457. THE NATION

457.1 Let the country build the railroads, and the railroads will make the country.
Edward Pease, quoted on the title page of The Midland Railway, its Rise and Progress, *Frederick Williams, 1878. Also quoted by E Foxwell and T C Farrer in* Express Trains English and Foreign, *1889, as "Let the country but make the railroads..."*

457.2 It can be fairly said that in England, the country built up the railroads, but that in the United States, the railroads built up the country.
Ivy Lee, address, the London School of Economics, 7 February 1910.

457.3 Few of us can perform spectacular deeds, but we can all put our strength and intelligence to the work which lies to hand, and the railwayman who is really happy in his occupation is serving his nation as well as his company.
Robert Bell, in paper The Railways in 1927, *delivered to the LNER Debating & Lecture Society, 13 March 1928.*

457.4 For once more let it be said that the maintenance of our railways is a national question and not merely a matter of concern for our shareholders.
Ibid.

457.5 Years before, I had noticed how trains accurately represented the culture of a country: the seedy distressed country has seedy distressed railway trains, the proud efficient nation is similarly reflected in its rolling stock, as Japan is. There is hope in India because the trains are considered vastly more important than the monkey-wagons some Indians drive.
Paul Theroux, The Old Patagonian Express, *1979.*

457.6 Dining cars, I found, told the whole story (and if there were no dining cars the country was beneath consideration).
Ibid.

457.7 There are two railways in Costa Rica, each with its own terminal in San José. Their routes dramatize Costa Rica's indifference to her neighbours: they go to the coasts, not to any frontier.
Ibid.

458. NATIONALIZATION
See also Privatization

458.1 Whatever may be the rate of divisible profits on any such railway it shall be lawful for the said lords commissioners, if they shall think fit, subject to the provisions hereinafter contained, at any time after the expiration of the said term of twenty-one years, to purchase any such railway, with all its hereditaments, stock, and appurtenances, in the name and on behalf of her majesty, upon giving to the said company three calendar months' notice in writing of their intention, and upon payment of a sum equal to twenty-five years' purchase of the said annual divisible profits, estimated on the average of the three then next preceding years: ...
Clause 2, The Cheap Trains Act, (7 & 8 Vict. c. 85) making a first provision for partial nationalization of the railways in Great Britain, 1844.

458.2 State railroads are not a milch cow; they are not intended to be a source of state revenue.
Attributed to the Minister of Public Works, Prussia (name not given) in the 1870s, when Prussian railways were being nationalized. Quoted in an article in The Springfield Republican, *in which it said that the Prussian state railways later produced significant money for the state; reproduced within* The Railway Library, *1914, in which a footnote by editor Slason Thompson states that the Prussian government did not spend enough of the earnings for adequate maintenance of the railways.*

458.3 I think there is only one of two things to do; either let the railways manage the State, or let the State manage the railways.
Henry Tyler, Chief Inspecting Officer of the Railway Department, Parliamentary Papers, *1872, quoted by Geoffrey Alderman in* The Railway Interest, *1973. In* The Railway Library, *1909, edited by Slason Thompson, the words are quoted within a speech by William Acworth as*

"If the state can't control the railroads, the railroads will control the state." This was therein described as "often quoted since"; the difference in nuance will be noted. Tyler later became deputy chairman of the Great Eastern Railway. See also 522.1 and 522.4.

458.4 . . . he seriously suggested that the Federal government should work the trunk lines, and the respective state governments the branches. Even if anybody knew in every case what is a trunk and what is a branch, the result would be to create an organism about as useful for practical purposes as would be a human body in which the spinal cord was severed from the brain.

William Acworth, of a suggestion by Presidential candidate William Jennings Bryan in 1908; speech before the British Association in Dublin, 2 September 1908; The Railway Library, 1909, 1910.

458.5 The test, and the only test, to be applied to proposals for railway nationalisation is whether railways owned by the State and worked directly by Government officials would be better and more efficient than railways owned and worked by private corporations, and whether, after taking account of all the effects of the change, upon each class, each district, each interest, the net result would increase the wealth and well-being of the community, and be a permanent benefit to the public.

Sir George Gibb, paper read before the Royal Economic Society, 10 November 1908. Quoted in The Railway Library 1909, 1910.

458.6 I suppose that if railway services were as good as possible, charges as cheap as possible, profits as high as possible , and the management as perfect as it is possible for railway management to be, and these conditions were generally admitted to exist, the natural instinct to leave well alone would prevent any proposal for nationalization from obtaining a hearing.

Ibid.

458.7 State railways would not be immaculate.

Ibid.

458.8 . . . the error lay in entrusting railway questions to Parliament, and the inevitable consequence was that they received inadequate and spasmodic attention. But that is just this point that the modern advocate of railway nationalisation neglects to consider. His arguments are worthless, his case an idle one, so long as he talks at large of the evils of private railway companies, and the advantages of State management.

Edward Cleveland-Stevens, English Railways, their Development and their Relation to the State, 1915.

458.9 . . . it might even pay the State to run the railways at a loss if by developing industries, creating new ones, reviving agriculture, and placing the trader in closer contact with his markets, they stimulated a great development at home.

Winston Churchill, extract from speech in Dundee, 4 December 1918, quoted by Martin Gilbert in Winston S Churchill, The Stricken Years, vol. 4, 1975.

458.10 There is not in the principles of English institutions anything to account for the fact that in this country, in contradistinction to the rest of Europe, the whole of the railways are in private ownership. On the contrary, the customs of the Anglo-Saxons, those stern restricters of the functions of government, contain the most definite sanction for making the provision of communications a public charge.

Dixon H Davies, Solicitor, Great Central Railway; paper read before the Great Central Railway Debating Society, January 1907; reprinted in Railway News, 1914, and here found in The Railway Library 1913, 1914.

458.11 From time to time, the fingers of bureaucrats have itched to catch hold of railway management to reform and remodel it on a Whitehall pattern, but public opinion, in the light of war experience, has evinced a very decided dislike to experiment in the direction of Government ownership of industry, and it has thus happened that railways have emerged from the war's turmoil still under private ownership, but coupled with a large measure of State control.

Sir Sam Fay, Presidential address to the Institute of Transport, 2 October 1922.

458.12 From the point of view of the trader, it should be capable of demonstration that *controlled* private ownership is preferable to *uncontrolled* State ownership, and it will be upon future management that proof of this will lie.

Ibid.

458.13 We must always remember that public opinion is fickle, always alert, and prone to criticise. At some time in the future—near or distant, grievances, fancied or real, will serve to resurrect the cry for State railways. Upon management, good or bad, the appreciation or otherwise of a company's duties to the public as compared with a company's rights will depend the period of freedom from attack on private ownership.

Ibid.

458.14 After purchasing the railways, the Government should set up a body to control both road and rail transport. This public body might be run something like the London Passenger Transport Board or the Port of London Authority.

William Whitelaw, News Chronicle, 29 December 1937.

458.15 The final remnants of the competitive spirit between companies will be eliminated, and there is no reasonable prospect that dull uniformity will be offset by greater efficiency in operation. The working of the railways, indeed, may become another form of indirect taxation.

Editorial comment, The Railway Magazine, vol. 93, 1947.

458.16 But as we draw nearer and nearer, I wonder what the ordinary citizen, the man in the street, is thinking about it from the point of view of what he is to gain? There will be the same people running the railways, the same guards, the same managers, and the same rolling stock. The only thing which will disappear, to be replaced by civil servants, will be the directors, and the ordinary passenger does not meet directors of railway companies: they are always found either in the House of Commons or the House of Lords, so it will not make very much difference to the passengers.

Lord Brabazon, House of Lords debate on the 1947 Transport Bill, Hansard, 20 May 1947.

458.17 I do not say that you should take them over by nationalizing, but anyone who takes over the railways without, at the same time, taking a very close hand in road haulage, is just taking over a bankrupt affair, and that is, in logic, an absurdity and nothing else.

Ibid.

458.18 Whither?

Caption to the 1947 Christmas card of the LNER Press Relations Officer, accompanying a photograph of a train plunging into a tunnel.

458.19 Some man in Whitehall sits and tells the railway just how many slices of bread and scrape it may give us, and just how thick to cut the railway slab. It's tyranny.

Michael Innes, of food in a restaurant car from Euston to Heysham, 1948; The Journeying Boy, 1949.

458.20 It does not matter whether the nationalized railways show a deficit, though of course every possible economy should be used in their administration. What is important is that the public should have the best transport service on the roads which can only be furnished by private enterprise. It is necessary that the railways should be properly maintained. For this purpose the Road Transport should bear a levy and this it can do owing to the greater fertility of private enterprise.

Sir Winston Churchill, memorandum 7 April 1952.

458.21 Colours becoming fewer. Shades becoming
Duller. Classes of engines fewer. Kinds
Of coaches and of trucks likewise. A humming
After a singing. Melancholy finds
A way into the perforated minds
Of modern men and women, English brand.
The everlasting Marxist, who is drumming
Into our ears the blessings that a land
Receives which standardises—does he understand?

Terence Greenidge, verse seven of The Nationalisation of the Railways, *in* Girls and Stations, *1952.*

458.22 Elsewhere in the world, in nations wearied by constant struggle, one government after another resolved the conflict by nationalizing the railroads. Under the benign influence of a merciful Providence, however, nationalization was successfully resisted in these two-fisted United States, as a result of which connoisseurs have been diverted, ever since about 1840, by an unending procession of rate wars, back stabbings, raids on the public purse, bankruptcies, corruptions of legislatures, double crosses and triple crosses, virtuosities of stock manipulation, legal swindles, panics and severe depressions, larcenies, extortions, and tyrannies, together with stupidities, otiosities, effronteries, impertinences, and discourtesies customarily associated with railroad management.

Peter Lyon, To Hell in a Day Coach, 1967.

458.23 I calculate that at least half my time has been devoted to organisation, reorganisation, acquisition, denationalisation, centralisation, decentralisation, according to the requirements of the now regular political quinquennial revaluation of national transport policy.

Stanley Raymond, Chairman British Railways Board, 1965–67, quoted in The Sunday Times, *7 January 1968.*

458.24 It strikes me, however, that it is not terribly constructive to discover an illness and then prescribe death as a cure.

Donald W Wiseman, Milwaukee Road, address to Transportation Club of Portland, Oregon, quoted in Trains Magazine, *January 1976.*

459. NATURE

459.1 How temporal is transportation. Here the Illinois prairie was erasing the Big Four just as surely as the North Atlantic had smoothed over the wakes of the *Mauretania* and *Rex* and *Ile de France.*

David P Morgan, Trains Magazine, September 1989.

460. NAVIGATION

460.1 He must be mad. He won't be able to see the railway lines.

Ken Annakin and Jack Davies, scriptwriters, of a pilot who had decided to fly across the English Channel at night, in Edwardian days when pioneer pilots commonly used railways as navigation aids. Words spoken by actor Robert Morley; film, Those Magnificent Men in their Flying Machines, *1965.*

461. NAVVIES

See also Railways—Construction

461.1 Well have yon Railway Labourers to THIS ground
Withdrawn for noontide rest. They sit, they walk
Among the Ruins, but no idle talk
Is heard; to grave demeanour all are bound;
And from one voice a Hymn with tuneful sound
Hallows once more the long-deserted Quire
And thrills the old sepulchral earth, around.

Others look up, and with fixed eyes admire
That wide-spanned arch, wondering how it was raised,
To keep, so high in air, its strength and grace:
All seem to feel the spirit of the place,
And by the general reverence God is praised:
Profane Despoilers, stand ye not reproved,
While thus these simple-hearted men are moved?
　William Wordsworth, At Furness Abbey, *from* Miscellaneous Sonnets, *1845.*

461.2 Rough alike in morals and in manners, collected from the wild hills of Yorkshire and of Lancashire, coming in troops from the fens of Lincolnshire, and afterwards pouring in masses from every country in the empire; displaying an unbending vigour and an independent bearing; mostly dwelling apart from the villagers near whom they worked; with all the strong propensities of an untaught, undisciplined nature; unable to read and unwilling to be taught; impetuous, impulsive, and brute-like, regarded as the pariahs of private life, herding together like beasts of the field, owning no moral law and feeling no social tie, they increased with an increased demand, and from thousands grew to hundreds of thousands. They lived only for the present; they cared not for the past; they were indifferent to the future. They were a wandering people, who only spoke of God to wonder why he had made some so rich and some so poor; and only heard of a coming state to hope that there they might cease to be railway labourers. They were heathens in the heart of a Christian people; savages in the midst of civilisation: and it is scarcely an exaggeration to say, that a feeling something akin to that which awes the luxurious Roman when the Goth was at his gates, fell on the minds of those English citizens near whom the railway labourer pitched his tent.
　John Francis, A History of the English Railway, *1851.*

461.3 Painful it is to find that the triumphs which the human intellect has achieved should be so intimately associated with the moral degradation of so large a section of the community.
　F S Williams, Our Iron Roads, *1852.*

461.4 Verily, men earn their money like horses and spend it like asses.
　Colonel C R Savage, Union Pacific photographer, writing of the construction workers on the Union Pacific, Diary, *8 May 1869.*

461.5 You know, Sir, us chaps are just like them Israelites as you read of in the Bible; we goes from place to place, we pitches our tents here and there, and then goes on just as they did . . .
　Anon. navvy quoted by D W Barrett, Life and Work among the Navvies, *1880.*

461.6 All stout, healthy men, and as for their social standing or moral turpitude, all that is necessary to say is that nature had created them for a special purpose that people more delicately organised were unfit for.
　Edward Ordway, Reminiscences, *within* Annals of Wyoming, *vol. 5, 1927.*

461.7 Oh in eighteen hundred and forty-one
My corduroy britches I put on,
My corduroy britches I put on,
To work upon the railway, the railway,
I'm weary of the railway;
Oh poor Paddy works on the railway!
　Anon., poem, Poor Paddy works on the Railway, *date not known.*

461.8 This line, since the first sod was cut at Anley in 1869, has been a roaring trail of shanty towns. Here the old navvy, whose race will pass with the nineteenth century, has plied his pick, tipped his barrow, smashed the face of his enemy, sported his fancy waistcoat of a Sunday, lost it and everything else save his shirt, breeches and boots at crown-and-anchor on Monday night, and fallen afresh to the last great main line to the Border, every week of the five and a half years between November of 1869 and the summer of 1875.
　C Hamilton Ellis, of the Settle-Carlisle route of the Midland Railway, The Trains We Loved, *1947.*

462. NEWCASTLE & DARLINGTON JUNCTION RAILWAY

462.1 An abortion with a crooked back and a crooked snout, conceived in cupidity and begotten in fraud.
　Captain Watts, quoted in George & Robert Stephenson, *L T C Rolt, 1960.*

463. NEWCASTLE CENTRAL STATION

463.1 . . . the finest of Britain's grand stations.
　Gordon Biddle, Britain's Historic Railway Buildings, *2003.*

464. NEWPORT RAILWAY

464.1 Along the Brae of the silvery Tay,
and to Dundee straightaway,
Across the bridge of the silvery Tay,
Which was opened on the 12th of May
In the year of our Lord 1879,

Which will clear all expenses in a very short time.
Because the thrifty wives of Newport,
to Dundee will often resort,
Which will be to them profit and sport,
By bringing cheap tea, bread and jam,
And also some of Lipton's ham,
Which will make their hearts feel light and gay,
And cause them for to bless the opening day
of the Newport Railway.

William McGonagall, poem, Success to the Newport
Railway, *quoted in* LNER Magazine, *February 1937.*

465. NEWSPAPERS

465.1 As the conductor of a newspaper, published to give
the public correct information—so far as you assume to
give them any—you, of course, could not, as an honourable man, willingly publish untruths. But you might
sometimes (if as human as the average) take a little more
pains to satisfy yourselves as to the correctness of your
publications, if you had some slight feeling of personal
obligation, that you otherwise would.

*George Watrous, vice president, Hartford & Connecti-
cut Valley Railroad, letter to G H Hills, Editor,* Evening
Post, *23 November 1883; quoted by Thomas C Cochrane in*
Railroad Leaders, 1845–1890, *1953.*

465.2 One is not accustomed to search the columns of the
daily newspapers for reliable facts and figures relating to
railways.

Cecil J Allen, British Locomotive Practice and Performance, *in* The Railway Magazine, *vol. LII, 1923.*

465.3 For my part, since being railroaded, I find that I live
with 'Chaos'. With twenty thousand trains running a day,
railwaymen are acquainted with grief, and human fallibility is daily in the headlines. We have an open day every
day. But, more often than not, any minor incident is Rail
Chaos. There seems little in between no news bliss and
Rail Chaos—except possibly Fares Shock.

Sir Peter Parker, Foreword to Great Commercial Disasters *by Stephen Winkworth, 1980.*

465.4 Bad news made the best news, and so got the fullest
treatment. When the railways were working normally
and efficiently, not much credit came from pointing that
out; their well-doing was passed over in silence.

Jack Simmons, The Railway in Town and Country,
1830–1914, 1986.

465.5 I don't want any more of that train and truck stuff.
Nobody's interested in it except you and a few of your
friends.

Anon. section chief, the Washington Post, *to Don Phillips, transportation correspondent;* Trains Magazine, *April
2004.*

466. SIR CHARLES NEWTON

466.1 Our Principal. No Nazi although a G.E.R. man.
Able to make and take a joke. Can be very firm but has
a soft spot in his heart for all his masters, prefects and
boys. His gift to us of The Stables, The Billiard Room and
The Lounge came as no surprise to those who have witnessed his successful efforts to improve L.N.E.R Stations.

*"Keyhole Charlie," Who's Who at the Hoo, (wartime
HQ of the LNER), within* Ballyhoo Review, *19 December
1939.*

467. NEW YORK & NEW ENGLAND RAILROAD

467.1 The strangest, most loose-jointed, shambling,
anomalous congeries of rails, with the most melodious
name in all the six states.

Alvin F Harlow, Steelways of New England, *1946.*

468. NEW YORK CENTRAL RAILROAD

468.1 The Water Level Route—You Can Sleep.

*Advertising slogan, drawing attention to the more tor-
tuous route taken by its great rival, the Pennsylvania Railroad, through the Allegheny Mountains, on routes between
New York and Chicago, 1930s.*

468.2 No need to remind you that this corporate entity,
New York Central, is much more than an individual railroad. It is an influence. It is a seat on the board of the
Association of American Railroads, a market for countless railroad suppliers, a conversation-stopper in Broad
Street, Philadelphia, and a most important slot on the
Big Board of Wall Street. It is a paycheck to 90,000 men
and women, an investment to more than 41,000 stockholders, the name on the yard engines that switch the
sidings of countless on-line industries, the operator of
the world's most famous train.

David P Morgan, open letter from the Editor of Trains
Magazine *to Robert R Young, who had just won a long and
bitter battle for control of the NYC. The reference to Broad
Street, Philadelphia is to the head office of the Central's
chief competitor, the Pennsylvania Railroad; the last item
referred to is the 20th Century Limited.* Trains Magazine,
September 1954.

469. NEW YORK CHICAGO & ST LOUIS RAILROAD

469.1 The great New York and St Louis double track,
nickel plated railroad of which we have heard so much
talk for the last six months, struck Norwalk very suddenly last Saturday; or rather a small portion of it came
along in the shape of an advance guard or engineering
corps.

Comment in the Norwalk Chronicle, *March 10, 1881, acknowledged to F R Loomis, Editor, as the origin of the nickname of the railroad.*

470. NEW YORK NEW HAVEN & HARTFORD RAILROAD

470.1 Its two good trains are really good and one of them, the Merchants' or "the Five o'Clock" as we say in Boston, is part of our communal life—no Boston banker, broker, college girl, or mink coat would dream of using any other train.

Bernard de Voto, Harpers Magazine, *April 1947.*

471. NICKNAMES

471.1 Some latitude is allowable, perhaps, to halfpenny papers, in the use of nicknames, but for a railway to adopt its gutter title, is not what we expect from a railway company.

Editorial comment, The Railway Magazine, vol. XIX, *1906, on the American-owned Baker Street & Waterloo Railway's having adopted the name* Bakerloo.

472. O S NOCK

472.1 . . . he who has been described as the Edgar Wallace of the railway world.

Rixon Bucknall, in a manuscript note on the front endpaper of his copy of The South Eastern and Chatham Railway, *written by O S Nock.*

472.2 If the Railway Technical Centre is located at Derby, The Railway Writing Centre must surely be at Batheaston where O. S. Nock produces books at a rate which would earn him the title "Stakhanovite Writer" in the Soviet Union.

Roger Ford, reviewing O S Nock's 107th book, Modern Railways, *August 1980.*

472.3 There is ample evidence that he read considerably less than he wrote.

Kevin P Jones, Steam Index, *on http://www.steamindex .com/library/nock.htm (accessed February 2005).*

473. NORFOLK & WESTERN RAILWAY

473.1 Carrier of Fuel Satisfaction.

Advertising slogan, 1930s.

473.2 Precision Transportation.

Advertising slogan, probably first used in the 1940s.

473.3 An objective ride over its main lines instills the railroadman with a feeling of well-being—a sense that everything is in "apple-pie order" and as it should be.

Robert G Lewis, The Handbook of American Railroads, *1951.*

473.4 Roanoke had a way of obtaining from orthodoxy what others strove for in gadgetry.

David P Morgan, of the locomotive design team of the N&W, Trains Magazine, *June 1956.*

473.5 They run a tight, disciplined property, only spending a dollar where it will return $1.40 or more, and disbursing the balance to the owners so unstintingly that the dividend record is unbroken since 1901 and the mortgage maturities of 1989 and 1996 are already provided for, thank you.

David P Morgan, Trains Magazine, *September 1959.*

473.6 The only people I knew who worked for the Norfolk & Western always seemed so loyal and careerminded. Are they still like that?

Jean Stapleton, actress, who worked for the N&W in the 1940s, interviewed by Norfolk & Western Magazine, *15 September 1973.*

473.7 We haven't formed a holding company and do not have any plans to diversify. We plan to concentrate all of our energy on the rail business.

John P Fishwick, N&W, report to the New York Society of Security Analysts, 21 April 1975, reported in Trains Magazine, *August 1975.*

473.8 A conveyor belt for coal masquerading as a railroad.

Anon., quoted by Albro Martin, Railroads Triumphant, *1992.*

474. NORTH BERWICK

474.1 North Berwick is occasionally designated the "Scarborough of Scotland" but it is only so in so far as Scarborough might be called "The North Berwick of England".

North British Railway, Tourist Guide, *1914.*

475. NORTH BRITISH RAILWAY

475.1 From west to east, along the Central Rift, Soft from North British lums the reek did drift.

Anon., quoted in The North British Railway, *C Hamilton Ellis, 1955.*

476. NORTH EASTERN RAILWAY

476.1 No company has more powerful engines or better drivers; all that is wanting is stimulus.

E Foxwell and T C Farrer, Express Trains English and Foreign, *1889.*

477. NORTH LONDON RAILWAY

477.1 The North London Railway started life as a sort of handmaiden to the London and North Western Railway, and has been more or less one of that family ever since.

E L Ahrons, The North London Railway, *within* The Railway Magazine, vol. LIV, *1924.*

477.2 There was a certain quality of apparent immortality about North London passenger stock; it could be found on hire to the LNWR at Blackpool or to the Great Western at Henley, and coaches survived for years after the North London had disposed of them to the Burry Port and Gwendraeth Valley, Mawddy, Whittingham Asylum, Isle of Wight, Isle of Wight Central, Cleobury Mortimer & Ditton Priors Light, Kent & East Sussex, East Kent, Easingwold and Neath & Brecon Railways.
Michael Robbins, The North London Railway, 1953.

477.3 From Canonbury, Dalston and Mildmay Park
The old North London shoots in a train
To the long black platform, gaslit and dark,
Oh Highbury Station once and again.
John Betjeman, poem The Scandinavian Meeting-House in Highbury Quadrant.

478. NORTH MIDLAND RAILWAY

478.1 Minerals, not men.
George Stephenson; attributed to him by E L Ahrons in The Railway Magazine, September 1921.

479. NORTH STAFFORDSHIRE RAILWAY

479.1 . . . th'owd North Stafford's getting up i'th'world. It'll be a five per cent. line yet. Then thou' mun sell out.
Arnold Bennett, Anna of the Five Towns, 1902.

480. NORTHERN PACIFIC RAILROAD

480.1 The Route of the Great BIG Baked Potato.
Advertising slogan, 1912.

480.2 Main Street of the Northwest.
Advertising slogan.

480.3 We think this is the way to run a railroad. And this is the way we run the Northern Pacific.
Walter Gustafson, manager of advertising and publicity, NP, on advertisements for the company. Quoted by David P Morgan, Trains Magazine, December 1985.

481. NOTICE OF CLOSURE

481.1 Here's your death warrant.
T E B Clarke; guard to porter, handing him Notice for display at Titfield station. Opening words of film The Titfield Thunderbolt, 1952.

482. NOTTINGHAM, England

482.1 Through cuttings deep to Nottingham
Precariously we wound;
The swallowing tunnel made the train
Seem London's Underground.
John Betjeman, poem, Great Central Railway Sheffield Victoria to Banbury.

Oo

483. OBAN, Scotland

483.1 The Charing Cross of the North.
Poster, Caledonian Railway, 1909.

484. OBSERVATION CARS
See also Dome Cars

484.1 There are, shamefully enough in a degenerate age, transcontinental Pullman trains without observation cars.
Lucius Beebe, Highball: a Pageant of Trains, *1945.*

485. OFFICIALDOM

485.1 . . . in the teeth of official apathy.
John Betjeman, of the achievement of the preservation society that saved the Talyllyn Railway. Foreword to Talyllyn Adventure, *1971.*

486. *THE OFFICIAL GUIDE*
See also Bradshaw's Guide, *Timetables*

486.1 It was a thick-bound volume always carefully consulted by the sales boys before going off on their travels. Richard found it more to his taste than fiction. It was called *The Official Guide of the Railways and Steam Navigation Lines of the United States, Canada, Mexico and Cuba*. It was filled with timetables and the rather violently simplified maps of railroad companies, in which the route of the company under consideration is shown as strong and direct as possible while all the others are very spiderweb.
Christopher Morley, Human Being, *1934.*

486.2 Wanderlust on pulp.
David P Morgan, Trains Magazine, *December 1956.*

487. OIL

487.1 If you want to strike oil, drill on the right of way of the Texas & Pacific Railway.
Saying among Texas wildcatters, quoted by Charles M Mizell, Jr, in Trains Magazine, *October 1972.*

488. OPERATING

488.1 Tracks and trains are, in themselves, real railroad problems. Yet, large as they are, they are overshadowed, always, by that of their proper usage; of their correlation, if you prefer to put it that way. The railroad operating executive must have good right of way, including terminals and all the rest that goes with it; he must be provided with efficient engines and cars, in generous plenty; all these are his tools. But a real master workman regards his tools as but accessories to the real end to be accomplished. Over tracks, trains must be operated—here is the very soul and essence of the railroad. But it is in the manner of operation that one finds the real efficiency with which business is, or is not, being conducted. On this thing, it stands, and stands well; or falls, and falls miserably.
Edward Hungerford, The Story of the Baltimore & Ohio Railroad, *vol. 2, 1928.*

488.2 The English have a disconcerting habit of coupling up a train's locomotive a handful of minutes before departure.
David P Morgan, Trains Magazine, *July 1975.*

488.3 Railways do not run on "if", "maybe", "possibly", or even "probably"; they run on certainties.
John Heppell, MP, Commons debate, second reading, Railways Bill, *Hansard, 2 February 1993.*

489. THE ORANGE BLOSSOM SPECIAL

489.1 Well, I'm going down to Florida
and get some sand in my shoes.
Or maybe Californy
and get some sand in my shoes
I'll ride that Orange Blossom Special
And lose these New York blues.
Ervin T Rouse, song, Orange Blossom Special, *1938.*

489.2 Hey talk about a-ramblin'
She's the fastest train on the line
talk about a-travellin'
She's the fastest train on the line
It's the Orange Blossom Special
Rolling down the Seaboard Line.
Ibid.

489.3 . . . its happy name sounded like some concoction blended of gin and clickety-clack.
David P Morgan, Trains Magazine, *May 1966.*

490. THE ORIENT EXPRESS

490.1 The Great Trains are going out all over Europe, one by one, but still, three times a week, the Orient Express thunders superbly over the 1400 miles of glittering steel track between Istanbul and Paris.

Ian Fleming, borrowing the words of Sir Edward Grey, when Foreign Secretary, "The lamps are going out all over Europe. . .", perhaps unaware that Grey had been a railwayman, as a director of the North Eastern Railway. From Russia with Love, *1957.*

490.2 Turkey was retreating from Europe, but the Orient Express still remained the umbilical cord which attached her . . .
Paul Morand, quoted by E H Cookridge in Orient Express, *1978.*

490.3 You can never be overdressed.
Dress Code of the Venice-Simplon-Orient Express, as advised in brochure, 2002.

491. OXFORD STATION

491.1 That old bell, presage of a train, had just sounded through Oxford station; and the undergraduates who were waiting there, gay figures in tweed or flannel, moved to the margin of the platform and gazed idly up the line. Young and careless, in the glow of the afternoon sunshine, they struck a sharp note of incongruity with the worn boards that they stood on, with the fading signals and grey eternal walls of that antique station, which, familiar to them and insignificant, does yet whisper to the tourist the last enchantments of the Middle Age.
Max Beerbohm, Zuleika Dobson, *1911.*

Pp

492. PADDINGTON STATION

492.1 In fact, it obtained its Act of Parliament on the understanding that it was to effect its junction with the London and Birmingham in a field between the Canal and the Harrow Road at the Western side of Kensal Green Cemetery; and it was only when the scheme of a joint station was abandoned that Paddington was thought of.
> W J Gordon, of the Great Western Railway's entry into London, Everyday Life on the Railroad, 1898.

492.2 There is a healthy, well-to-do look about Paddington which no other station possesses. It is the most English of railway stations.
> Ibid.

492.3 Do you remember the five-thirty from Paddington? What a dear old train it was!
> Siegfried Sassoon, Memoirs of a Fox-Hunting Man, 1928.

492.4 Here a leisured pace prevails and you get only the best people—cultured men accustomed to mingling with basset hounds and women in tailored suits who look like horses.
> P G Wodehouse, Uncle Fred in Springtime, 1939.

492.5 The two-forty-five express—Paddington to Market Blandings, first stop Oxford—stood at its platform with that air of well-bred reserve which is characteristic of Paddington trains, and Pongo Twistleton and Lord Ickenham stood beside it, waiting for Polly Pott.
> Ibid.

492.6 Mr and Mrs Brown first met Paddington on a railway platform. In fact, that was how he came to have such an unusual name for a bear, for Paddington was the name of the station.
> Michael Bond, A Bear Called Paddington, 1958.

492.7 'A bear? On Paddington station?' Mrs Brown looked at her husband in amazement. 'Don't be silly, Henry. There can't be!'
> Ibid.

493. THE PAN AMERICAN

493.1 I have heard your stories about your fast trains
But now I'll tell you about one all the southern folks have seen.

She's the beauty of the Southland, listen to that whistle scream.
It's the *Pan American* on her way to New Orleans.
She leaves Cincinnati headin' down that Dixie Line
When she passes the Nashville tower, you can hear that whistle whine.
Stick your head out of the window and feel that Southern Breeze.
You're on the *Pan American* on her way to New Orleans.
> Hank Williams, song, Pan American.

494. PAOLI

494.1 Along that green embowered track
My heart throws off its pedlar's pack
In memory commuting back
Now swiftly and now slowly—
Ah! lucky people, you, in sooth
Who ride that caravan of youth
The Local to Paoli!
> Christopher Morley, The Paoli Local, within Travels in Philadelphia, 1920.

494.2 The saying that was good enough for Queen Mary and Mr Browning is good enough for me. When I die, you will find the words PAOLI LOCAL indelibled on my heart.
> Ibid. See also 515.2.

494.3 When the Corsican patriot's bicentennial comes along, in 1925, I hope there will be a grand reunion of all the old travelers along that line. The railroad will run specially decorated trains and distribute souvenirs among commuters of more than forty years' standing. The campus of Haverford College will be the scene of a mass-meeting. There will be reminiscent addresses by those who recall when the tracks ran along Railroad avenue at Haverford and up through Preston. An express agent will be barbecued, and there will be dancing and song and passing of the mead cup until far into the night.
> Christopher Morley, The Paoli Local, within Travels in Philadelphia, 1920.

495. PARLIAMENT
See also Government, Politicians

495.1 ... but experience had convinced him that it was very difficult to please that House on the subject of railway legislation. He had always found a disposition to exist to quarrel with that which existed; to demand a change; and then immediately to quarrel with the alteration.

Henry Labouchere, reported, as was then common practice, in the third person; House of Commons debate, Hansard, 28 March 1848.

495.2 What had the railways done that they should be treated in this arbitrary manner?

Joseph Henley, past President of the Board of Trade, on a proposed Parliamentary Tribunal with extra-Parliamentary powers, 1873, noted in Edward Cleveland-Stevens, English Railways, their Development and their Relation to the State, 1915.

495.3 Whatever the future may bring, a strong, continuous, certain and comprehensive policy of State control must be evolved, and the outstanding lesson to be drawn from the history of English railways is the danger of entrusting control to the Legislature. The central problem, whether the railways remain in private hands or be taken over by the State, is the creation of a permanent Board of Control, and one that is as far removed as is possible from the interference of Parliament.

Edward Cleveland-Stevens, English Railways, their Development and their Relation to the State, 1915.

496. PARLIAMENTARY TRAINS

496.1 On and after the several days hereinafter specified all passenger railway companies which shall have been incorporated by any Act of the present session, or which shall be hereafter incorporated, or which by any Act of the present or any future session have obtained or shall obtain directly or indirectly any extension or amendment of the powers conferred on them respectively by their previous Acts, or have been or shall be authorized to do any act unauthorized by the provisions of any previous Acts, shall, by means of one train at the least to travel along their railway from one end to the other of each trunk, branch, or junction line belonging to or leased by them, so long as they shall continue to carry other passengers over such trunk, branch, or junction line, once at the least each way on every week day, except Christmas Day and Good Friday (such exception not to extend to Scotland), provide for the conveyance of third-class passengers to and from the terminal and other ordinary passenger station of the railway, under the obligations contained in their several Acts of Parliament, and with the immunities by law to carriers of passengers by railway; . . . . .

Clause 6, The Cheap Trains Act, (7 & 8 Vict. c. 85), requiring the operation of what became known as "Parliamentary" trains, 1844.

496.2 In a word, a parliamentary train collects,—besides mechanics in search of work, sailors going to join a ship,

and soldiers on furlough,—all whose necessities or tastes lead them to travel economically, among which last class are to be found a good many Quakers.

Samuel Sidney, Rides on Railways, 1851.

496.3 We cannot say that we exactly admire the taste of the three baronets whom a railway superintendent found in one third-class carriage, but we must own that to those to whom economy is really an object, there is much worse than travelling by the Parliamentary.

Ibid.

497. PARLOR CARS

497.1 The parlor car is one of the last refuges of tranquility in our craven society.

Richard Bissell, Holiday magazine, early 1960s, quoted by George H Drury in Trains Magazine, October 1996.

PASSENGERS

Entries on passengers are divided into subheadings, thus: Actions, Character, and Handling.

498. PASSENGERS—Actions

498.1 In second and third-class carriages avoid sitting near the doorway; the constant opening and shutting of the door subject those near it to cold and draughts, and the incessant passing to and fro, with the usual accompaniments of shoulder-grasping and toe-crushing, is more than sufficient to test the best of tempers.

Anon. advice, The Railway Traveller's Handy Book, 1862.

498.2 It also frequently happens that one of the passengers, a "navvy," for instance, is ever and anon thrusting his head out the window, thus favouring you with an uninterrupted view of his substantial but by no means picturesque proportions, and at the same time beating time on your shins with his hob-nailed and iron-shod high-lows.

Ibid.

498.3 With regard to conversation, the English are notoriously deficient in this art. Generally speaking, the occupants of a railway carriage perform the whole of the journey in silence; but if one passenger be more loquaciously inclined than the rest, he is soon silenced by abrupt or tart replies, or by a species of grunt expressive of dissent or dissatisfaction.

Ibid.

498.4 If you want to find out what cowards the majority of men are, all you have to do is rob a passenger train.

O Henry, (William Sydney Porter), Holding up a Train, 1916, within Short Lines, ed. Rob Johnson, 1996.

498.5 To each of the passengers his seat was his temporary home, and most of the passengers were slatternly housekeepers.

Sinclair Lewis, Main Street, *1922.*

498.6 In the train, however fast it travelled, the passengers were compulsorily at rest; useless between the walls of glass to feel emotion, useless to try to follow any activity except of the mind, and that activity could be followed without fear of interruption.

Graham Greene, Stamboul Train, *1932. (Orient Express in U.S.)*

498.7 Complacently you weigh your chances of a foreign countess, the secret emissary of a Certain Power, her corsage stuffed with documents of the first political importance. Will anyone mistake you for No. 37, whose real name no one knows, and who is practically always in a train, being 'whirled' somewhere? You have an intoxicating vision of drugged liqueurs, rifled dispatch-cases, lights suddenly extinguished, and door-handles turning slowly under the bright eye of an automatic. . . .

Peter Fleming, One's Company, *1934.*

498.8 I have made the test through a dozen years, but I made it again—with the same results: on these Boston-New York trains, as one walks through, there are more people reading books than on any other trains in the United States. It must be said also that there are more feet stuck out in the aisle, more people who glance up in disgust as you wish to put the aisle to other purposes.

Rollo Walter Brown, I Travel by Train, *1939, quoted by H Roger Grant,* We Took the Train, *1990.*

498.9 The only type of passenger whom I find it hard to stomach is the man who pares his nails with a penknife.

John Betjeman, BBC broadcast, quoted in the Great Western Railway Magazine, *September 1940.*

498.10 At the station all the passengers were talking at once. They were telling the fat director, the stationmaster and the guard what a bad railway it was.

Rev W Awdry, Thomas the Tank Engine, *1946.*

498.11 In Europe and America people travel in a train fully aware that it belongs either to a state or a company and that their ticket grants them only temporary occupation and certain restricted rights. In Russia people take them over. They move in with their luggage, bundles and children as if for permanent occupation. I saw empty trains, newly washed and shining, pull in with great dignity to their allotted wayside shrines. Within a few minutes they would be full of the devout spreading out themselves, their food and their possessions with the look of rapture on their faces. Before long the train would resemble a second-rate boarding-house or a whole village on wheels.

Laurens van der Post, Journey into Russia, *1964.*

498.12 I have even seen long-distance trains with clothes lines and washing hanging out in the crowded compartments, dropping an odd drop of water or two on the heads of people who were eating, sleeping, or playing chess below.

Ibid.

498.13 Experience had shown me that there was always a German in Second Class, slumbering on his pack-frame and spitting orange pips out of the window.

Paul Theroux, The Old Patagonian Express, *1979.*

498.14 On entering an Underground train it is customary to shake hands with every passenger.

Anon., from a proposed spoof book of advice to foreigners in London; used in an advertisement for New Statesman, *1997.*

498.15 If you must use your cellphones, please keep your voices at the lowest possible decibel level. Unless you are giving stock tips. Then speak up so we can all hear.

Conductor on Long Island Railroad, noted by Nicholas Wapshott, The Times, *14 June 2002.*

499. PASSENGERS—Character

499.1 On the banks of a canal navigators and loiterers infest the towing-paths and create a nuisance, but all descriptions of travellers on a railroad may rather be compared to a flock of pigeons or swallows, that confine their flight to the regions of the air, and leave neither track nor trace behind.

Sir George Head, A Home Tour through the Manufacturing Districts and other parts of England, Scotland and Ireland, *1835.*

499.2 Within the car, there was the usual interior life of the railroad, offering little to the observation of other passengers, but full of novelty for this pair of strangely enfranchised prisoners. It was novelty enough, indeed that there were fifty human beings in close relation with them, under open long and narrow roof, and drawn onward by the same mighty influence that had taken their two selves into its grasp. It seemed marvellous how all these people could remain so quietly in their seats, while so much noisy strength was at work on their behalf.

Nathaniel Hawthorne, The House of the Seven Gables, *1851.*

499.3 It must be remarked that the character of the Parliamentary varies very much according to the station from which it starts. The London trains being the worst, having a large proportion of what are vulgarly called "swells out of luck." In a rural district the gathering of smock-frocks and rosy-faced lasses, the rumbling of carts, and the size, number, and shape of the trunks and parcels, afford a very agreeable and comical scene on a frosty, moonlight, winter's morning, about Christmas time, when visiting commences, or at Whitsuntide.

Samuel Sidney, of Parliamentary trains; Rides on Railways, 1851.

499.4 No man who has a taste for studying the phases of life and character should fail to travel at least once by the Parliamentary.

Ibid.

499.5 The real sufferers are those poor fellows the rich.

Comment in The Daily Telegraph, *January 1875, concerning the abolition of second class on the Midland Railway. Quoted in* The Midland Railway, *Frederick Williams, 1878.*

499.6 It was an occasion, I felt—the prospect of a large party—to look out at the station for others, possible friends and even possible enemies, who might be going. Such premonitions, it was true, bred fears when they failed to breed hopes, though it was to be added that there were sometimes, in the case, rather happy ambiguities. One was glowered at, in the compartment, by people who on the morrow, after breakfast, were to prove charming; one was spoken to first by people whose sociability was subsequently to show as bleak; and one built with confidence on others who were never to reappear at all—who were only going to Birmingham.

Henry James, The Sacred Fount, *1901.*

499.7 The average excursionist is not ordinarily a very grateful sort of individual. As a rule he takes all he can get, wants as much for his money as possible, and reserves to himself the right to grumble freely.

Comment, Great Western Railway Magazine, *March 1905.*

499.8 Every one has observed the beautiful system and efficiency of service performed at the large stations in cities where multitudes of people are safely and comfortably handled almost every hour of the day. All the more may one be pardoned if he makes mention of the little trials which the agent at a smaller station undergoes in dealing with people who, unlike the city folk, do not ride every day, travel only occasionally, and are lacking somewhat in the intelligence which is acquired by an experienced traveler. At such a station many grown people manifest a childish simplicity which is at once annoying and amusing.

B C Burt, Railway Station Service, *1911.*

499.9 Often the would-be traveler knows merely the name and the state and county in which his destination may be located; he has never thought to make a little study as to how to get there. And the agent, when he should be taking care of a dozen other passengers, finds himself totally engrossed by the necessity of having to do for the seemingly witless passenger the thinking and

investigating which said passenger should have tried to do for himself.

Ibid.

499.10 The most trying passenger to deal with is probably the man whose fondness for stimulants, or perhaps sedatives, causes him to keep his intellectual condition such a dazed or crepuscular one that he can only with difficulty make known even the name of the place he wishes to reach by train.

Ibid.

499.11 The shipper here is also that which is shipped. What is loaded and unloaded, loads and unloads itself; the charges collected for the services performed must, in order to insure for collection, be collected before transportation occurs.

Passenger traffic as distinguished from freight traffic, ibid.

499.12 Where we came out strong in the carriage company at large, was in our superior familiarity with the route. We knew all the points of interest to be looked out for. "We are going to Cornwall." "We always go there." "We'll show you when it comes." By such delicate expressions of superiority we managed to conceal our contempt for the poor creatures who 'were only going to Bristol', or some degraded person who had to 'change at Didcot'.

M Vivien Hughes, A London Child of the Seventies, *1934.*

499.13 Why not? One more jackass on this trip won't make any difference!

Ralph Budd, President, Chicago Burlington & Quincy Railroad, when asked at the last minute (and at the height of a technical problem) to carry a donkey on the inaugural run of the Zephyr, *from Denver to Chicago, 26 May 1934. Quoted in* Trains Magazine, *September 1952.*

499.14 In two points only does 'The Coronation Scot' overstep the bounds of travelling comfort. At 80 miles an hour it is impossible to resist the window-closing fellow-passenger and—even at 90—which I suspect is touched by the 'Queens' and 'Princesses' more often than the guide admits, her progress is so quiet and so steady that, unless you are very rude indeed, you cannot foil the conversational advances of that other human nuisance, the Chatty Stranger. To be honest, however, I must admit that my fellow-passenger up to Glasgow was so tactful in restraint and so amusing when the ice was broken that I was glad she spoke to me.

Naomi Royde-Smith, The Times LMS Centenary Number, *20 September 1938, quoted in* The Coronation Scot, *Edward Talbot, 2002.*

499.15 Since we are men and women rather than moles, we dislike travelling in the dark. Some of us made the best of it, in the war's first weeks, by noticing that we could find unexpected interest in the conversation of our unseen fellow-travellers; after sitting on their laps in mistake for empty seats, we formed temporary friendships, instead of maintaining the dour Non-recognition Treaty of the peacetime English railway-compartment.

Evan John, Time Table for Victory, *c. 1946.*

499.16 A passenger is the most precious and perishable quantity the railways carry—the most interesting, the most articulate and the least remunerative.

Donald Gordon, Canadian National Railway, Trains Magazine, *March 1956.*

499.17 I don't travel on planes.
I travel on trains.
Once in a while, on trains,
I see people who travel on planes.
Every once in a while, I'm surrounded
By people whose planes have been grounded.
I'm enthralled by their air-minded snobbery,
Their exclusive hobnobbery.
And I'll swear to, before any notary,
The cliches of their coterie.
They feel that they have to explain
How they happen to be on a train,
For even in Drawing Room A
They seem to feel declasse.
So they sit with portentous faces,
And clutch their attache cases.
As the Scotches they rapidly drain
That they couldn't have got on the plane,
They grumble and fume about how
They'd have been in Miami by now.

Ogden Nash, poem (part), The Unwinged Ones.

499.18 British railways are not like other railways. Fundamentally, they are mine tramways. Trevithick and Stephenson both came from a background of mines and mining. Coal, iron, *things,* never mind about people.

Things have a much better time of it. A crowd of commuters stands, wet and shivering, on the platform, when there is a click in the loudspeakers, then an announcement: Grerksqxx krkk the hai naw nee do lunn runna box ditty dirty binny day. While the more experienced travellers are working out that this means: "The 8.14 to London is running approximately 30 minutes late," (they always say approximately, never just about) a train of huge identical containers thunders through, blam, blam, blam, at 90 mph.

Paul Jennings, The Times Saturday Review, *5 February 1972.*

499.19 The sort of person who will talk to you on a train is almost by definition the sort of person you don't want to talk to on a train.

Attributed to Bill Bryson by George H Drury, Trains Magazine, *December 1997.*

500. PASSENGERS—Handling

500.1 Carry him safely, dismiss him soon: he will thank you for nothing else.

John Ruskin, The Seven Lamps of Architecture, *1849.*

500.2 The most important feature to be observed at all times is to satisfy and please passengers.

Pullman Company, instructions to employees, c. 1880.

500.3 You are not to consider your own personal feelings when you are dealing with these people. You are the road's while you are on duty; your reply is the road's; and the road's first law is courtesy.

William Van Horne, to a trainman on the Canadian Pacific Railway who had engaged in a dispute with an irritable passenger. Quoted by Pierre Berton in The Last Spike, *1972.*

500.4 The passenger is a patron; he ought to be treated in such a manner that his patronage will be continued; it is for the interest of the railway company that he shall be courted, surrounded with conveniences and placated by attentions.

Charles Paine, Elements of Railroading, *1895.*

500.5 Which does NP want, dignity or passengers?

Hazen J Titus, Northern Pacific Railroad, who faced management resistance to his suggestion that the railroad should advertise by using an image of a large baked potato to entice hungry passengers. The potato won; see 480.1 under Northern Pacific Railway. Quoted by Michael Zega in Travel by Train, *2002.*

500.6 Pack them in nice and tight, they'll keep each other warm.

Anon. old "Brighton" line man, to John Elliot, about passengers on the London Brighton & South Coast Railway, which had train heating on very few trains. Journal of Transport History, *May 1960.*

500.7 The Passenger Is Always Right, Except When He Doesn't Have His Fare.

Doug Welch, slogan of the fictional Chicago, Omaha, Salt Lake & Pacific Railway, short story Mrs Union Station, *in the* Saturday Evening Post, *1937.*

500.8 I want to emphasize that it was my job to ensure that the wheels rolled and not to concern myself with what was transported.

Walter Mannl, railway yard worker at Auschwitz during the Second World War, quoted by Alfred C Mierzejewski, Railroad History No 186, *Spring 2002.*

500.9 If special inducements such as reduced fares, budget priced meals, nursing service and tolerant porters

could attract the infantile trade, the management figured it would remove a ponderable menace to the tranquility and repose on its more adult trains. *The Scout*, as the rolling nursery was named, remained in service for well over a decade, a mobile nightmare of diaper changes and the inconvenient feeding habits of the very young and, although it was a sort of negative blessing, was appreciated by travelers in search of abated tumults aboard *The Chief, Super Chief* and *California Limited*.
Lucius Beebe, of a train introduced by the Santa Fe in the 1930s, The Trains We Rode, *1965.*

500.10 I prefer football fans over racegoers any day.
Les Benson, Northern Spirit ticket collector, 22 May 1999.

501. PASSES

501.1 Dear Sir,
Says Tom to John, "here's your old rotten wheelbarrow. I've broke it, usin' on it. I wish you would mend it, 'case I shall want to borrow it this arternoon.
Acting on this as a precedent, I say "Here's your old 'chalked hat.' I wish you would take it, and send me a new one, 'case I shall want to use it by the first of March." Yours truly.
Abraham Lincoln, letter to R P Morgan, requesting a replacement free pass, known by railroaders at the time as a "chalked hat," 13 February 1856. Quoted in Lincoln and the Railroads, *John W Starr Jr, 1927.*

501.2 Sheriffs are given passes for somewhat the same reason that clergymen get them. The latter are encouraged to raise the standard of character throughout the State and the former to lay hands upon those whose standard of character is so low as to make treatment of a penal nature necessary. The railroads feel that the more the parson and the sheriff can be encouraged to travel the more safe life and property will be on their lines.
Iowa Railroad Commissioners, report, 1882; quoted by Clyde H Freed, The Story of Railroad Passenger Fares, *1942.*

502. PASSING TRAINS

502.1 Hereafter, when trains moving in opposite directions are approaching each other on separate lines, conductors and engineers will be required to bring their respective trains to dead halt before the point of meeting, and be very careful not to proceed till each train has passed the other.
Anon. railroad rule, quoted in Railway Adventures and Anecdotes, *Richard Pike, 1888.*

502.2 When two trains approach each other at a crossing, they shall both come to a full stop and neither shall start up again until the other has gone.

Old Kansas state law, quoted in A Treasury of Railroad Folklore, *Botkin & Harlow, 1953.*

503. PENN CENTRAL

503.1 You can always rely on the Americans to do the right thing after they've done everything else. The fact that Penn Central could fall apart plus a generational change in management engineered a revolution.
Henry Livingston, banker, quoted by Nicholas Faith, The Economist, *24 August 1985.*

503.2 Worms-in-love.
Anon. Of the Penn Central logo of the letters P and C intertwined. Noted by Robert F Holzweiss in Railroad History, *188, Spring 2003.*

504. PENNSYLVANIA LIMITED

504.1 The most beautiful train in the world.
Lucius Beebe, referring to an epithet applied to this train c. 1900; 20th Century, *1962.*

505. PENNSYLVANIA RAILROAD

505.1 The Standard Railroad of the World.
Advertising slogan. In its centenary year, when the PRR made a loss it apparently admitted it had lost title to this slogan.

505.2 Finest Roadbed and Heaviest Rail in America—You can sleep restfully.
Advertising slogan derived to respond to that of the New York Central, 1930s.

505.3 The moment that this Company forgets that its duty is to be at the head of the list of carrying companies of the United States and ceases to have the ambition to become the first in the world, that moment do I wish to pass from its management.
George B Roberts, President 1880–97, quoted in Trains Magazine, *March 1970.*

505.4 I always like to see what kind of locomotive is going to draw my train; especially on the Pennsylvania Railroad, where, in their struggle for motive power as efficient as the New York Central's, they have indefinitely multiplied their types of engine.
Owen Wister, Stanwick's Business, *1904, quoted by Rob Johnson in* Short Rails, *1996.*

505.5 Do not think of the Pennsylvania Railroad as a business enterprise. Think of it as a nation.
Comment, Fortune Magazine, *May 1936.*

505.6 The roster of engines of the Pennsylvania over the past 70 years contains literally hundreds of classes, of all types; yet the line of succession from one class into

the next is as clear and distinct as the hereditary characteristics of a long line of carefully-bred racehorses.
Roland E Collons, The Railway Magazine, *vol. LXXX, 1937.*

505.7 This short era of Public Benevolence lasted only until 1852. It was followed by a thirty-year long and very different era of Hard-Boiled Railroading.
Nathaniel Burt, The Perennial Philadelphians, *1963.*

505.8 When you've seen one you've seen them all.
Anon. commentator, recorded as a typical reaction to the varnish trains of the Pennsylvania Railroad, no doubt as a direct result of its dedication to standardization. Lucius Beebe, The Trains We Rode, *1965.*

505.9 Of all the railroads in the United States, the only one whose annals can appropriately be chronicled in terms of its presidents as English history is by the reigns of its monarchs is the Pennsylvania, and Broad Street was their palace and shabby-genteel seat of power. The power, of course, had nothing either shabby or genteel about it.
Lucius Beebe, The Trains We Rode, *1965.*

505.10 . . . The Standard Railroad of the World did many nonstandard things.
David P Morgan, Trains Magazine, *July 1970.*

505.11 What Cole Porter wrote was finally true of trains: anything goes.
David P Morgan, of the collapse of the Pennsylvania Railroad. Trains Magazine, *April 1988.*

506. PENNSYLVANIA STATION (New York)

506.1 The station, as he entered it, was murmurous with the immense and distant sound of time. Great, slant beams of moted light fell ponderously athwart the station's floor, and the calm voice of time hovered along the walls and ceiling of that mighty room, distilled out of the voices and movements of the people who swarmed beneath. It had the murmur of a distant sea, the langorous lapse and flow of waters on a beach. It was elemental, detached, indifferent to the lives of men. They contributed to it as drops of rain contribute to a river that draws its flood and movement majestically from great depths, out of purple hills at evening.
Thomas Wolfe, You Can't Go Home Again, *1934.*

506.2 The Pennsylvania Station is indefinitely scheduled to be torn down. May the President and Board of Directors be blasted by the curse of all those who value American architectural monuments. May they go down in history as infamous.
Nathaniel Burt, Appendix in second printing, The Perennial Philadelphians, *1963.*

506.3 Any city gets what it admires, will pay for, and, ultimately, deserves. Even when we had Penn Station, we

couldn't afford to keep it clean. We want and deserve tin-can architecture in a tin-horn culture. And we will probably be judged not by the monuments we build but by those we have destroyed.
Of the destruction of Pennsylvania Station, The New York Times, *Editorial, 30 October 1963.*

506.4 A rich and powerful city, noted for its resources of brains, imagination and money, could not rise to the occasion. The final indictment is of the values of our society.
Ibid.

506.5 Pennsylvania Station was an oversize entrance to a subterranean world that has grown less and less pleasant to contemplate.
Albro Martin, Railroads Triumphant, *1992.*

507. ARTHUR PEPPERCORN

507.1 'Pep' as he was usually called seemed to be loved by all, and his contributions kept closely to his prepared brief. Not for him were any excursions into the mental stratosphere, but if he himself was not a very live wire he had the inestimable ability of letting his assistants get on with it. . . .
E S Cox, A Locomotive Panorama, *1965.*

508. PERAMBULATION

508.1 March Up and Down.
Applicability of permanent way work in March, Cambridgeshire; Railtrack notice, 1994.

509. ALFRED E PERLMAN

509.1 . . . this deceptively mild, immensely self-confident man who turned Vanderbilt's paradise lost into a lean, aggressive plant.
David P Morgan, Trains Magazine, *July 1983.*

510. PERMANENT WAY MEN

510.1 They were working with that extreme deliberation which seems to characterise permanent way men wherever there are railways and which makes the work appear to be like some game of chess where each move, if it be only one step forward, must first be carefully pondered.
L T C Rolt, Railway Adventure, *1953.*

511. PERRY BARR, England

511.1 The Goodwood of Greyhound Racing.
London Midland & Scottish Railway poster, 1928.

512. PERSHORE STATION

512.1 The train at Pershore Station was waiting that
     Sunday night
Gas light on the platform, in my carriage electric light,

Gas light on frosty evergreens, electric on Empire wood,
The Victorian world and the present in a moment's
   neighbourhood.
   John Betjeman, poem, Pershore Station, *or* A Liverish
Journey First Class.

513 PERTH STATION

513.1 The station cannot well be enlarged any more, for it
is a Sabbath day's journey from one end of it to the other
already.
   W M Acworth, The Railways of Scotland, *1890.*

514. PETERBOROUGH, England

514.1 Our main object is to shorten the distance between
London and Yorkshire. If a line through Peterborough is
best for the general public, Peterborough shall have it,
but, if not, it shall pass outside the town.
   *Edmund Denison, at a public meeting in Peterborough
prior to the construction of the Great Northern Railway,
1 August 1844.*

514.2 The last station, Peterborough, presents an instance
of a city without population, without manufacturers,
without trade, without a good inn, or even a copy of the
Times, except at the railway station; a city which would
have gone on slumbering without a go-a-head principle
of any kind, and which has nevertheless, by the accident
of situation, had railway greatness thrust upon it in a
most extraordinary manner.
   *Samuel Sidney, of the three routes from Peterborough to
London,* Rides on Railways, *1851.*

514.3 There is, therefore, the best of consolation on being
landed in this dull inhospitable city, that it is the easiest
possible thing to leave it.
   Ibid.

515. PHILADELPHIA

515.1 It is one of the few places in the country where it
doesn't matter on what side of the tracks you are. These
are very superior tracks . . .
   *John Gunther, of the Main Line area of Philadelphia,
built around the Philadelphia-Harrisburg line of the Pennsylvania Railroad,* Inside USA, *1947. The first use of the
expression "the wrong side of the tracks" has not been
found. Lucius Beebe stated in* Hear the Train Blow *that the
expression came into the language when, at the command
of Wyatt Earp, Abilene's "love stores" were moved to the
south side of the Kansas Pacific tracks.*

515.2 Nothing was so holy as the local to Paoli.
   *Christopher Morley, of the suburban service on the
Main Line in Philadelphia, within poem* Elegy in a railroad station, *published in* The Saturday Review, *reprinted
in* Trains Magazine, *April 1955.*

516. PHOTOGRAPHERS AND PHOTOGRAPHY

516.1 A not unimportant official at Derby is the Com-
pany's photographer in ordinary.
   W M Acworth, of a Midland Railway employee, The
Railways of England, *1899.*

516.2 To catch the *Kaysee Flyer* on the Mopac as her big
Pacific leans to the curve coming out of Webster Groves
and stop the side motion with a shutter speed of ¹⁄₆₆₀ is to
experience what the big game hunter knows as a bull elephant looms in the sight line of his Ross rifle.
   Lucius Beebe, Highliners, *1940.*

516.3 The unforgettable images created by O Winston
Link on the Norfolk & Western in the 1950s have become
such a celebrated and durable part of the railroad photography tradition that they are in danger of becoming
mere icons.
   Kevin P Keefe, Trains Magazine, *October 1995.*

516.4 Take only pictures, leave only footprints.
   Anon., railroad photographers' rule, quoted in Trains
Magazine, *March 1997.*

517. FRANK PICK

517.1 The man who built London Transport.
   Christian Barman, title of Biography, 1979.

518. PILOTS

518.1 Admirable indeed were the contours struck by the
Northern Pacific, the Katy, and the Richmond, Freder-
icksburg & Potomac, with flanged-up and perforated
pilots declaiming any allegiance to the commonplace.
And what about the plowlike arcs on some Rock Island
and Rio Grande power? What say you to the stamped-
out rig gracing certain Illinois Central hogs, with great
ridges like boxcar ends? Monstrous steel castings, regular
Alcatraz architecture, dominated a few Wabash passenger engines, and big Rock Island racers too.
   F H Howard, Trains Magazine, *January 1956.*

519. PLANT LIFE

519.1 The chemists tell us that the air of cities and their
neighborhood is richer in available nitrogen (in shape of
ammonia or nitric acid) than the air of the country, by
reason of the outpourings of so many chimney-tops, and
the attendant processes of combustion. May not the
cinders and the fine ash and the gases evolved from a
great highway of engines always puffing and smoking in
the lower strata of the atmosphere contribute somewhat,
and that not inconsiderably, to the plants found along
lines of such highway?
   Donald G Mitchell, Rural Studies, *1867.*

520. POETRY ON RAILWAYS

520.1 Had I the knack of rhyme, I would write a sonnet-sequence of the journey to Newhaven or Dover—a sonnet for every station one does not stop at. I await that poet who shall worthily celebrate the iron road. There is one who describes, with accuracy and gusto, the insides of engines; but he will not do at all. I look for another, who shall show us the heart of the passenger, the exhilaration of travelling by day, the exhilaration and romance and self-importance of travelling by night.

Max Beerbohm, Ichabod, *1923*.

520.2 It could be said with considerable justification that railroad poetry is largely emotion governed by intellect. One short poem will convey more of the atmosphere of railroading than will many pages of carefully written prose. The measured click of rails, the throbbing of the air pump, the mighty roar of the stack, all lend themselves to the poet's fine sense of rhythm and balance. But more: the inherent fascination of railroading, whether it be a brakeman walking cat-like along the treacherous footboard of a swaying box car, the flickering "markers" of a passenger train leaving a country station in the still of the night, or the noise and activity of a vast freight yard are all blended and united in the versifier's pen. To be sure, this is not done blatantly, for often it is implied or suggested rather than stated outright, and yet its effect is far-reaching. Like a puff of smoke, a good poem expands and coils lazily in the mind and then slowly dissipates into the imagination.

Frank P Donovan Jr, The Railroad in Literature, *1940*.

521. POLITICIANS

See also Government, Parliament

521.1 In the history of no country, has there been such a barefaced sacrifice of the public interests for the benefit of private associations, who, without any efficient restraint or restriction, have been suffered to monopolize, and for their own selfish purposes to employ the means of communication of a great industrial nation.

Anon., The Influence of English Railway Legislation on Trade and Industry, *1844*.

521.2 We have heard that this gentleman intends at the next election to turn his talents from the engineering line to the Parliamentary; that is, he intends to obtain a seat in Parliament if he can. This is a silly ambition.

Editorial comment, of Joseph Locke, distinguished railway civil engineer, Herapath's Journal and Railway Magazine, *28 September 1846*.

521.3 . . . a politician generally makes bad work at anything but tricks in his own game.

John Brooks, Superintendent, Michigan Central, letter to John Forbes, President, MCR, 1 May 1851; quoted by Thomas C Cochrane in Railroad Leaders, 1845–1890, *1953*.

521.4 It sounds odd to talk of a Railroad President as being popular, as if he was a comedian or a politician.

Frederick J Kimball, President, Norfolk & Western Railroad, letter to Everett Gray, 12 September 1889; quoted by Thomas C Cochrane in Railroad Leaders, 1845–1890, *1953*.

521.5 You know, politicians in opposition often make criticisms and say things which, if they bore the responsibility of office, they wouldn't say. You may have read some pretty strong things I said about you. I want you to forget them entirely, if you will. I know nothing about the railway business; and you have spent your lifetime in it. I want your help in mastering my job, as freely as you care to give it to me.

William C Kennedy, Minister of Railways and Canals, Canada, 1921–1923, to D B Hanna, quoted by D B Hanna, Trains of Recollection, *1924*.

521.6 Lord Walkden:Just as they do in the various departments today, they will carry on the day-to-day business; but they will be entirely without interference from politicians.

A Noble Lord: Oh!

Lord Walkden: It will only be the Minister. The Minister belongs to Parliament, and no-one would suggest that he is going to descend to "pruggling" about amongst the day-to-day working of this vast business. He will be above all that.

Exchange about railwaymen and politicians, Lords debate on the 1947 Transport Bill, Hansard, *20 May 1947*.

521.7 Next to getting houses built, railroad passenger service is, perhaps, the biggest single political issue; and politicians rise or fall on their record in this regard.

William H Schmidt Jr, Associate Editor, Railway Age, *in article* It Could Happen Here. *Of the nationalization of British Railways,* Trains Magazine, *July 1950*.

521.8 . . . it is seldom indeed that any politician gets his facts right where railways are concerned.

George Dow, Great Central, vol. II, *1962*.

522. POLITICS

522.1 Politics would corrupt the railways. And the railways would corrupt politics.

Attributed to an Italian Commission of the mid-1870s, apparently looking into the possibility of state ownership of railways. Quoted by W M Acworth at an address delivered to the Leeds Luncheon Club, 16 November 1914. Of this attribution, Acworth (later Sir William) said "The Italians since then have put the question to the test of experience, and have abundantly proved that the commission was right." The Railway Library, 1914, *1915*.

522.2 You cannot turn in any direction in American politics without discovering the railway power. It is the power behind the throne.

Richard T Ely, 1890, quoted in The Wreck of the Penn Central, *Joseph Daughan and Peter Bintzen, 1971.*

522.3 The time has come to take the railroads out of politics. Dog in the manger politicians who cannot bear to recognise merit or honesty on the part of the railroad managers, and who do not want to admit of the possibility of efficiency in transportation, are now obsolete.

Editorial Comment, New York Journal of Commerce, *7 August 1920; quoted by P Harvey Middleton in* Railways and Public Opinion, *1941.*

522.4 Politics corrupts the railway management and the railway management corrupts politics.

Attributed to Sir Eric Geddes, February 1920, by Keith Grieves, Sir Eric Geddes, *1989.*

522.5 We have no politics at Paddington.

Viscount Churchill, Chairman, Great Western Railway, in an address to a former director, Stanley Baldwin, visiting the board at Paddington while Prime Minister, April 1925. Four years later Harold Macmillan joined the board of the company. Great Western Railway Magazine, *May 1925.*

522.6 In the days when railways in this country competed with one another, they seemed to have marked distinguishing characteristics; and what was more, they differed not only in external appearance but also in their political affiliations. In Lancashire, the London & North Western Railway was generally thought to be an Anglican and Conservative institution; the Midland, on the other hand, represented Nonconformity and Liberalism. Above all, the Great Western stood for Conservatism, if not High Toryism, in the regions that it and its associates had made their own; and in the high Imperialist period its engines proclaimed that Roberts, Kitchener, and Baden-Powell were its heroes.

Michael Robbins, Railways and Politics in East Cornwall, *within* The Railway Magazine, vol. 104, *1958.*

522.7 Because the railroad manager is so much at home with practicalities, he is frustrated to discover that politically 1+1 seldom equals 2.

David P Morgan, Trains Magazine, *December 1959.*

522.8 Like people, railroads need to be loved. Few are. And they don't know why. Railroads can haul anything, anywhere, any time, at a lower true cost than can anybody else. They maintain their own right of way, pay their taxes promptly, abide by most of the rigid regulation, and yet somehow wind up in the political deep freeze.

David P Morgan, Trains Magazine, *June 1961.*

PORTERS
See Station Staff

523. POST

523.1 The strangest people and institutions choose to send me letters. The railways, for example, keep sending me messages boasting about how much money they have made, which I think dreadfully bad taste.

Robertson Davies, The Papers of Samuel Marchbanks, *1985.*

524. POSTAL SERVICE BY RAILWAY

524.1 The mail train is preferred by many persons travelling on business, for inasmuch as it is employed in the postal service, its movements are necessarily regulated by the most scrupulous punctuality.

Anon., Comment, The Railway Traveller's Handy Book, *1862.*

524.2 This is the Night Mail crossing the Border,
Bringing the cheque and the postal order,
Letters for the rich, letters for the poor,
The shop at the corner, the girl next door,
Pulling up Beattock, a steady climb:
The gradient's against her but she's on time.

W H Auden, poem Night Mail, *written for the Post Office film of that name.*

524.3 Past cotton-grass and moorland border,
Shovelling white steam over her shoulder,
Snorting noisily, she passes
Silent miles of wind-bent grasses.

Ibid.

524.4 There's nary a slip 'twixt the pouch and the hitch.

Anon. caption to photograph, emphasizing the reliability of apparatus for picking up mail at speed. Trains Magazine, *April 1941.*

524.5 There's a whisper down the line at 11.39
When the Night Mail's ready to depart . . .

T S Eliot, poem Skimbleshanks: the Railway Cat.

524.6 This is the Night Mail, sluggish and sad:
Better a crawl than a dangerous spad.
Mail that is junk, mail unexpected
Till afternoon, since it's seldom collected.

Peter Pindar, poem, You have mail, *with apologies to W H Auden. The reference to "spad" is to "signal passed at danger," an item of railway argot in the public domain;* Sunday Telegraph, *6 October 2002.*

524.7 Railtrack's replacement by Network Rail's like
A one-legged man, having stolen your bike
And taken your money, now asking for more
To repair it, while you're left worse off than before.

Ibid.

524.8 Now as the Night Mail jolts on its wheels
Spinning on autumn leaves on the rails' steel,

Long-delayed passengers slowly approach
Detrained and decanted in road-weary coach.
Muffled loudspeakers on dark platforms say
Stuff about maintenance causing delay.
   Ibid.

525. POSTERS
See also Advertising

525.1 The big railway termini in America have no posters,
but are in themselves fine architectural schemes. The big
termini in this country, especially recent ones, like Victoria, have no architectural scheme, but plenty of posters. One can imagine the English director saying, "It
does not matter about the shape of our stations if we
plaster them with these", and then, more touchingly, "If
we go to the Royal Academy for the posters, all will
indeed be well."
   *Charles Reilly, reflecting the use of Royal Academicians
by the LMS; Some Architectural Problems of Today, 1924.*

525.2 The image is everything.
   *Ian Wright, auctioneer, of the collectible value of old
railway posters, at Sheffield Railwayana auction sale,
14 September 2002.*

526. POVERTY

526.1 I have even gone through tunnels in those vile open
standing-up cars, called by an irreverent public "pig-
boxes", and seemingly provided by railway directors as a
cutting reproach on, and stern punishment for, poverty.
   *George A Sala, Poetry on Railways, in Household
Words, 2 June 1855.*

526.2 In many respects and especially Railroad matters, it
is better to show moderate poverty than great success.
   *James C Clarke, President, Illinois Central, letter to
Stuyvesant Fish; quoted by Thomas C Cochrane in Railroad Leaders, 1845–1890, 1953.*

526.3 American engineers have often shown that poverty is the mother of invention. For example they used
wooden cross-ties as a temporary substitute, being too
poor to buy stone blocks, and so made good roads
because they were not rich enough to make bad ones.
   *Ashbel Welch, Presidential Address to the American
Society of Civil Engineers, quoted by James Vance Jr in The
North American Railroad, 1995.*

527. PRESERVATION

527.1 With further reference to the closing down of the
rly service on the Mansfield line we would like to say that
the nearest approach to what you require as a souvenir is
a L.N.E.R. single ticket to Mansfield which would cost
you 10d.
   The most ancient thing as a memento we could offer

would of course be the clerk & he gives us to understand
he would have no objections to being regarded as a curiosity or even being preserved.
   *Chalked notice at Kirkby station, 2 January 1956, shown
in* Trains Illustrated, *March 1956.*

527.2 Either it can be preserved in exactly its original state
and so become a dead thing, or it can be used in which
case both the nature of that use and the thing itself must
change as all living things are bound to do.
   L T C Rolt, Introduction to Talyllyn Adventure, *1971.*

527.3 When I acquire pieces of railroad hardware, I'm not
stealing them; I'm saving them for posterity. Other people steal. I just don't want them to go into someone's
hands who won't appreciate them.
   *David Lustig, in a satirical article about the activities
and attitudes of some railfans,* Trains Magazine, *February
1993.*

528. PRESTON STATION (England)

528.1 The glory of the refreshment rooms has to some
extent departed.
   J L Lawrence, Notable Railway Stations, *within* The
Railway Magazine, *vol. XII, 1903.*

529. PRINCETON, New Jersey

529.1 The village of Princeton, as you may know, is not on
the main railroad line. One waits at the Junction for a
dumpy little train which ambles back and forth from the
main line to the village.
   William McAdoo, Crowded Years, *1931.*

530. PRIVATE CARRIAGES

530.1 He and Clegg owned and lived in a private car decorated in opulent but insufferable magnificence, a consequence of the two trying to squeeze every decorative element of the Venetian Renaissance into the narrow
confines of a single car.
   Of Lucius Beebe; Carl Condit, Railroad History No 142,
1980.

531. PRIVATE ENTERPRISE

531.1 The English railway system, as everyone knows, is
the result of private enterprise.
   —Clarke, The Board of Trade, *in* Household Words,
3 March 1855.

532. PRIVATIZATION
See also Nationalization

532.1 Railway privatisation will be the Waterloo of this
government. Please never mention the railways to me
again.

Attributed to Lady Thatcher, noted by Jon Shaw in
Competition, Regulation and the Privatisation of British
Rail, 2000, *and repeated by Christian Wolmar in* Broken
Rails, *2001.*

532.2 The architects of Privatisation were interested in
financial, not mechanical, engineering.
   Ian Hargreaves, The Secret History of Rail, *BBC Tele-*
vision, 1 December 2001.

532.3 The Bill is a lawyer's paradise but will prove to be a
passenger's hell.
   Nick Harvey, MP, Commons debate, second reading,
Railways Bill, *Hansard, 2 February 1993.*

532.4 It might have been preferable to have had a Bill of
one clause which simply said, "The Secretary of State can
do what he likes, how he likes, when he likes, and where
he likes." It would have had more or less the same effect.
   Ibid.

532.5 I say this to my hon. Friends, if the size of their
majorities is bigger than the number of rail users in their
constituencies, they should not have too much to worry
about.
   Robert Adley, MP, of fellow Conservative party mem-
bers who were inclined to support the government and vote
for privatization. Commons debate, second reading, Rail-
ways Bill, Hansard, *2 February 1993.*

532.6 Even the poll tax had its enthusiasts. This poll tax
on wheels has no enthusiasts whatever.
   Attributed to Robert Adley, MP. These words were spo-
ken by Keith Hill, MP, Commons debate, second reading,
Railways Bill, Hansard, *2 February 1993, but he assures the*
compiler that the expression "poll tax on wheels" origi-
nated with Robert Adley, very probably in conversation
rather than in recorded debate.

532.7 We want the country to be able to regard the railways not only with affection but with pride.
   Earl of Caithness, opening the debate on second reading
of the Railways Bill, Hansard, *15 June 1993.*

532.8 The government will probably have to give it away.
There is the prospect of extra capital gains, asset-
stripping and a reasonably stable profit flow in the
short term. You have a duty to go for it, but I wouldn't
want to hang around with it.
   Anon. banker, quoted by Lord Marsh, Lords debate on
second reading of the Railways Bill, Hansard, *15 June 1993.*

532.9 Rail privatisation will bring extra efficiency, free-up
investment, expand choice, enhance services that the
customer wants.
   Privatisation should be judged by the result.
   Dr Brian Mawhinney, Department of Transport Press
Release, *17 October 1994.*

532.10 Broken railways, broken rail.
   Christian Wolmar, Broken Rails, *2001.*

532.11 Privatise in haste—repent at leisure.
   Frank Paterson, book review in Friends of the National
Railway Museum Review, *Spring 2002.*

532.12 If the intention of privatisation was to inject some
new managers into train operating, there has been precious little evidence of it.
   Terry Gourvish, British Rail 1974–1997, *2002.*

532.13 How can you have a Railtrack in which the profits
go up if they do less maintenance?
   William Withuhn, conversation with the compiler,
19 August 2003.

533. PROCLIVITIES

533.1 At our Board Meeting yesterday there was a long
discussion as to our present and future prospects, one
section of our Board having strong Midland proclivities,
another section, and, I am happy to say, the largest, having Great Western proclivities.
   J C Wall, General Traffic Manager, Bristol & Exeter
Railway, letter to James Grierson, Great Western Railway,
23 September 1875.

534. PROFESSIONAL OPINIONS

534.1 Do not discompose yourself, my dear Sir; if you
express your manly, firm, decided opinion, you have
done your part as their adviser. And if it happened to be
read some day in the newspapers—'Whereas the Liverpool and Manchester Railway has been strangled by
ropes,'—we shall not accuse you of guilt in being accessory either before or after the fact.
   Timothy Hackworth, concerning rumors that the Liver-
pool & Manchester might use stationary engines. Letter to
Robert Stephenson, March 1829.

535. PROFITS

535.1 The road has earned money for us and would earn
more were it not for the fact that an Act of Congress
limits the return on our investment to 6 per cent. We
are limited in our service by laws conceived in part by
ill-informed theorists who cannot understand the real
function of profits, and in part by those who see in regulated business the inevitable necessity for banker finance.
   Henry Ford, of the Detroit Toledo & Ironton Railroad;
the Act referred to was the Transportation Act of 1920.
Today & Tomorrow, *1926.*

535.2 Tremendous profits are something that we don't
envisage.
   Richard Beeching, in answer to a question about profits
from enhanced inter-city passenger traffic after the publica-

tion of the Beeching Report. Press conference, 1963, BBC Sound Archive.

535.3 There remains one big element of doubt, and that concerns railway "profitability". In the world of private railway enterprise, in which I was brought up on the Southern Railway, there was always the strong discipline of the *necessity* to earn a reasonable dividend or there would be no new capital to keep the railway alive. That discipline, which is invaluable in itself to energetic management, can still be brought to bear on our railways, but *only if the Government of today and of tomorrow creates equal conditions under which the various forms of public transport—rail, road and air—can operate.* This is much easier said than done when about half the transport system is publicly owned and half is in private hands. Such a situation becomes, inevitably, highly political, and only political decisions can solve it, one way or the other.

Sir John Elliot, Modern Railways, April 1965.

535.4 *Profit* is not a four-letter word, but *loss* is.

George Shimrak, Penn Central, quoted in Trains Magazine, August 1975.

535.5 The last time the New Haven Railroad earned a profit was when it carried cement to build the Connecticut Turnpike.

Attributed to Frank Wilner, Association of American Railroads, 1993, by Stephen B Goddard in Getting There, 1994.

535.6 They must have cooked the books—nobody carries passengers at a profit.

Anon. European railway president, on the announced profit of £57 million announced by British Railways Inter-City, 1989; quoted in The InterCity Story, 1994.

536. PROGRESS

536.1 The principal but generally unrecognized adverse result of curtailed earnings following over-regulation has been the inevitable reduction of the rate at which technological improvement could be integrated into railway plant and equipment. Deferred maintenance, which arises from the same cause, at least provides concrete evidence of its presence to serve as a warning. Deferred *development* is not similarly manifest. The most insidious characteristic of retarded progress is the absence of positive proof of its existence.

John Walker Barriger, Super Railroads, 1956.

536.2 George M Pullman intended that the sleeping cars bearing his name be built by his company, heated by steam, manned by black porters, and contracted to private railroads. Pullman's intentions have progressively been annulled by an antitrust suit instigated by the Budd Company, by electric train heat, by equal-opportunity employment, by women's liberation, and by Amtrak.

Humanists, liberals, moralists and sociologists acclaim civilization's advances.

David P Morgan, Trains Magazine, August 1977.

537. PROHIBITIVE AND RESTRICTIVE SIGNS

See also Trespass

537.1 All boys and inexperienced men must not grind on the rest side of this stone.

Next to a grinding wheel, in Derby Works, Midland Railway.

537.2 These closets are intended for the convenience of passengers only, workmen, cabmen, fishporters and idlers are not permitted to use them.

Cheshire Lines Committee notice.

537.3 Please Do Not Shoot Buffalo From The Train.

Sign in Rio Grande Southern Business Car Nomad. Illustrated in Narrow Gauge in the Rockies, 1958.

537.4 Passengers are requested not to throw rocks, as there may be fishermen below.

Sign above Toltec Gorge, Denver & Rio Grande Railroad, recalled by Archie Robertson, Slow Train to Yesterday, 1945, quoted by H Roger Grant, We Took the Train, 1990.

537.5 Only Great Western Railway horses may drink at this trough.

In goods depot, Birmingham. It prompted the comment that at least the Great Western credited other companies' horses with the ability to read.

537.6 Engines must proceed with caution when approaching Cemetery Junction.

Sign on Manchester Ship Canal Railway.

537.7 Engines Must Condense.

Instruction to enginemen (to turn on condensing gear) at entrances to tunnels on the line from King's Cross to Moorgate, mid 1920s.

537.8 Another characteristic of the Underground Railways which must strike every traveller is how remarkably prolific they are in notices and warnings. Passengers are requested to "Beware of Pickpockets," to "Stand clear of the gates,", not to attempt to travel without a ticket, only to cough behind their hand, to live only in Ruislip, and finally not to spit.

John Aye, who might have added "Mind the Gap"; Railway Humour, 1931.

537.9 Mind The Gap

Warning sign on sharply curved London Underground platforms.

537.10 Obstructing the doors causes delay and can be dangerous.

Michael Robbins; sign on the inner face of the doors of London Underground trains, devised after most careful thought.

537.11 Load limited to One loaded Brute
Redundant sign, dating from 1970s, seen at Doncaster station, 13 November 2002.

538. PROMISCUITY

538.1 There was an old man at a Station,
Who made a promiscuous oration.
Attributed to Edward Lear, but not found in published works of his limericks.

539. PROMOTION OF STAFF

539.1 It is urgently requested that every person, whether on or off duty, shall conduct himself in a steady, sober and creditable manner, and that on Sundays or other holy days, when he is not required on duty, that he will attend a place of worship, *as it will be the means of promotion when vacancies occur.*
Rule Book, Taff Vale Railway, quoted by John Aye, Railway Humour, 1931.

540. PROSTHETICS

540.1 My leg is working O.K. I have worn it every day since I put it on last April. I am running a locomotive every day.
W J Angier, locomotive engineer, letter of endorsement in advertisement for artificial limbs with rubber hands and feet, available by mail order from A A Marks of New York City, 7 December 1891, appearing in Railway Age *and reproduced in* Railroad History 189, 2004.

541. PROVINCIAL TOWNS

541.1 Beside those spires so spick and span
Against an unencumbered sky
The old Great Western ran
When someone different was I.
John Betjeman, poem, Distant View of a Provincial Town.

542. THE PUBLIC

542.1 The people are in favour of building a new road and do what they can to promote it. After it is once built and fixed then the policy of the people is usually in opposition.
James C Clarke, President, Illinois Central, letter to Stuyvesant Fish, 13 September 1883; quoted by Thomas C Cochrane in Railroad Leaders, 1845–1890, 1953.

542.2 The public be d——d. I don't take any stock in that twaddle about working for the good of anybody but our-selves. I continue to run the 'limited express'—well, because I want to.
William Henry Vanderbilt, in response to a question from a reporter from The Chicago Herald; *reported in that newspaper on Monday 9 October 1882, and in the* Fort Wayne Daily Gazette *of 10 October 1882. See appendix B for a discussion on this entry.*

542.3 We must be conservative and keep the public with us.
William H Vanderbilt, President, Lake Shore & Michigan Southern, letter, no date, to H Ledyard; quoted by Thomas C Cochrane in Railroad Leaders, 1845–1890, 1953. *In this book Cochrane also quotes another letter from Vanderbilt to Ledyard in which he says that "we don't want to get the public excited."*

542.4 The Public Be Pleased.
Slogan, Hudson & Manhattan Railroad, coined c. 1909 by William G McAdoo, President.

542.5 We believe in the public be pleased policy as opposed to the public be damned policy; we believe that the railroad is best which serves the public best; that decent treatment of the public evokes decent treatment from the public; that recognition by the corporation of the just rights of the people results in recognition by the people of the just rights of the corporation. A square deal for the people and a square deal for the corporation! The latter is as essential as the former, and they are not incompatible.
William McAdoo, at the opening of the downtown tunnels, Hudson & Manhattan Railroad, July 1909. Related in Crowded Years, 1931.

542.6 And remember that you can never injure the railway without injuring yourselves.
James J Hill, public speeches in Tacoma and Seattle, 9/10 November 1909, published in local newspapers and reprinted by Slason Thompson in The Railway Library 1909, 1910.

542.7 My natural disposition in discussing railroads and the public is to growl, while, if I understand your officers' wishes, I am here expected to "purr."
Edward P Ripley, President, AT&SF, opening remarks of speech at annual dinner of the Railway Business Association, 10 November 1909; published in The Railway Library 1909, 1910.

542.8 In spite of the fact that the railway property is owned in this country by the small investor, and in spite of the fact that the return on capital has always been a modest one, the railways have figured in the mind of the masses as a great, bloated capitalist organisation, an object not of sympathy, but rather of suspicion and jealousy. The roads were something that the public felt they had a right to, just as they had the right to breathe the air.

Sir Eric Geddes, inaugural Presidential address of the Institute of Transport, 22 March 1920, reported in Modern Transport, *27 March 1920.*

542.9 The railroads, despite all of our fancy advertising and our proud heritage, are not absolutely essential to the American public today. The public can get along very nicely without us, thank you. They might even do it if we should be so unfortunate as to have another war. What the public wants from us today is what they want from Ford and General Motors, what they want from Bethlehem and United States Steel, what they want from the corner grocery and the dime store, and that is value received for their money.

William Brosnan Jr, speech before New England Railroad Club, 1955, recorded by Charles O Morgret in Brosnan, *vol. 1, 1996.*

543. PUBLIC RELATIONS

543.1 Every effort is worthwhile to convince the members of the public that the railroad which serves a particular community is a home institution, a neighbor of other citizens of the community, and operated for the benefit of the public primarily. The railroads have suffered in public esteem from a rather widespread misconception that they are an example of absentee landlords. Railroad men, on the contrary, are substantial and interested citizens of the communities in which they live, and their welfare cannot be disassociated from the welfare of the community generally.

Association of American Railroads, Transportation in America, *1947.*

543.2 There are no short cuts to satisfactory public relations, no wands to wave, no magic words to utter. Public relations of the right sort is the product of doing the job right, and telling the public about it in the right way.

Ibid.

543.3 We must take the public into our confidence. When properly informed the judgment of the public is almost always sound. I am strongly persuaded that much of the public misunderstanding of railway affairs, much of the criticism we hear of railway management, much of the adverse legislation that has been enacted, and many of the ills which now beset the railway industry are due to the failure on the part of railway management to inform the public fully concerning railway achievements and railway problems.

Charles H Markham, President, Illinois Central Railroad, 1919, quoted in Main Line of Mid America, *Carlton Corliss, 1950.*

543.4 Answers are frequently needed at once or in a matter of minutes. Most of the replies have to be given over the telephone and it requires experience, knowledge and tact to know what to say and how to say it.

George Dow, internal report The London Midland Public Relations & Publicity Department, *British Railways, January 1955.*

543.5 If an accident, or a strike or something else unpleasant has happened it is the responsibility of the department to ascertain the facts with the minimum of delay and disclose them at once unless, for some very sound reason, it would be impolitic to do so. If this is not done the newspapers will get their information elsewhere, perhaps from an unreliable or tainted source.

Ibid.

543.6 Our Face is Our Bottom.

Richard Beeching, of the fact that most of the responsibility for the public perception of the railways lies with the lowest grades of staff such as porters, booking clerks and guards; quoted by Terry Gourvish in British Rail 1974–1997, *2002.*

543.7 Our train has crossed this huge city from one end to another. Let's reflect on that.

Public address announcement on Moscow Metro reported thus in The Times, *5 January 2004.*

544. PULLMAN TRAVEL

544.1 Let George do it.

Anon. referring to the universal nickname of Pullman attendants, who attended to the passengers' every wish; probably from the name of the founder of the company, George Mortimer Pullman. Eric Partridge, in his Dictionary of Catchphrases *does not make the connection with Pullman. One wonders if the expression led to the adoption of "George" as the nickname for automatic aircraft pilots.*

544.2 Elegant as a hotel bar and rigid as a church pew.

Samuel Hopkins Adams, spoken by a character, of Pullman sleepers, in Success, *1921.*

544.3 The moment he entered the Pullman he was transported instantly from the vast allness of general humanity in the station into the familiar geography of his home town.

Thomas Wolfe, You Can't Go Home Again, *1934.*

544.4 The closest the average man ever gets to true luxury in America is in a Pullman sleeping car.

Attributed to Franklin D Roosevelt, Albro Martin, Railroads Triumphant, *1992.*

544.5 Green aisles of Pullmans soothe me like trees Woven in old tapestries . . .

William Rose Benet, quoted by Lucius Beebe in High Iron, *1938.*

544.6 I see gorgeous drawings of the train of tomorrow, which is to have frequency modulation I don't want and pretty hostesses to amuse the children I don't travel with, but the Pullman Palace Car goes on spoiling my shirt, tiring me out with badly designed cushions, and bouncing so much I can't work on my proofs.

Bernard de Voto, Harper's, *April 1947.*

544.7 Let me take you in a private car from Chicago to the West, a car specially designed by George Mortimer Pullman. The luxury of it will soothe and startle you; sofas and chairs of damask of the most violent patterns but infinitely comfortable; dark, grinning servants to wait on you; fresh iced celery from Kalamazoo. Rainbow trout from the Rocky Mountains, and outside the wide windows of your drawing room you shall see the world passing by . . .

Noel Coward, Quadrille, *1952.*

544.8 Bond stepped on to the train and turned down the drab olive green corridor. The carpet was thick. There was the usual American train-smell of old cigar-smoke. A notice said 'Need a second pillow? For any extra comfort ring for your Pullman Attendant. His name is,' then a printed card, slipped in: 'Samuel D. Baldwin'.

Ian Fleming, Live and Let Die, *1954.*

544.9 These cars will bear the name of Pullman. They will be serviced with linen and other supplies in the same manner as all other Pullmans. There will be a berth for every man and bedding will be changed nightly by the porter. The kind of service on which we have built our reputation will be the same in every possible respect.

Pullman Company bulletin to staff, of troop sleeping cars built during the Second World War; quoted by H Roger Grant, We Took the Train, *1990.*

544.10 That George M. Pullman, and to a somewhat lesser degree, his imitators and competitors, contributed materially to the national well-being and taste for comfort is an inescapable conclusion to any survey of the part the palace cars played in the noontide of their riding. That they may have been overdecorated is only the sterile judgement of an era that produced nothing half so fine in any field of transport, decor or the amenities of living. Or, it may be added, a dimension to the language.

Lucius Beebe, Mr Pullman's Elegant Palace Car, *1961.*

544.11 Our staff take every possible care and precaution in the service of refreshments and the Company cannot be held responsible for minor acts of spillage, etc., which may occur on account of excessive movement of the train.

Cautionary words, in small print, on the menu of the Brighton Belle, 1950s, quoted by George Behrend in Pullman in Europe, *1962.*

544.12 You know that a smart, clean, turn-out and unfailing courtesy will rebound to your benefit and satisfaction. Everything possible is done to enable our staff to appear smartly turned-out and I count on all of you to maintain this.

F D M Harding, Managing Director, Pullman Car Co. (Great Britain), notice to staff in kitchen cars, 1950s.

544.13 What joy unequalled, once, to be inside
A Pullman in the morning in that train;
And, now, what paradise to feel again
The sudden jerk and see the platform glide

Past the side windows, hear the muffled bark
Of banking engine thrusting up the grade
To Grosvenor Road and see the white cascade
Of steam that pours away upon an arc;

And hear once more the creakings and complaints
Of costly Pullman panellings the while,
Until the wheel-beats break into a smile
And *Rosalind* is free of her constraints!

Henry Maxwell, poem, Boat Express, *in* A Railway Rubaiyyat, *1968.*

544.14 The Pullmans, with their windows wide-ablaze
And table-lamps with silken-shaded sheen,
Like shop-fronts in some Eve-of-Christmas scene,
Project a glowing beam upon the haze.

Henry Maxwell, poem, Fog at Folkestone, *verse eighteen, in* A Railway Rubaiyyat, *1968.*

544.15 First-class passengers expected the absolute best from the sleeping cars, dining cars, club cars, and observation cars carrying the livery of the largest hotel corporation in the world, the Pullman Company.

John R Stilgoe, Metropolitan Corridor, *1983.*

544.16 The peripatetic Pullman sleeper played a role in the development of American society that no one could possibly put in quantitative terms. Even the most casual observer of human nature, however, can see the implications of putting thousands of industrious, ambitious people in close proximity for many hours at a time. The table shared in the dining car, the midnight conversations in the Pullman smoking room between business travelers, who would not or could not sleep, the running commentary on the mines and mills that flashed past the windows of the observation car, all intensified the exchange of ideas and the making of deals that went on in cities and towns, great and small, throughout an America on the move.

Albro Martin, Railroads Triumphant, *1992.*

545. PUNCTUALITY

545.1 To Uxbridge by Great Western railway, sad want of punctuality at starting, and when off, rate only fourteen miles an hour. How have the mighty promises of Brunel fallen.

William Lucas, entry for 9 July 1838, A Quaker Journal, *1934.*

545.2 I hate to be made to wait for a steam-engine, and for a steam-engine never to wait for me.

Anon., Blackwood's Magazine, *April 1839.*

545.3 The railways are introducing their rigid mechanism into the habits of the people. Though there are complaints of the want of punctuality on individual lines, yet is punctuality to them the very soul of safety.

Comment, The Economist, *30 August 1851.*

545.4 The rail, like time and tide, will wait for no man.

Ibid.

545.5 Have not men improved somewhat in punctuality since the railroad was invented? Do they not talk and think faster in the depot than they did in the stage-office?

Henry David Thoreau, Walden, *1854.*

545.6 The startings and arrivals of the cars are now the epochs in the village day. They go and come with such regularity and precision, and their whistle can be heard so far, that the farmers set their clocks by them, and thus one well-conducted institution regulates a whole country.

Ibid.

545.7 It is a fact well known to railway officials of all grades, but not appreciated by the public, that it is much more difficult to keep time with trains stopping at many stations, than with express trains. Yet the explanation is very easy. People who travel by stopping trains are of totally different habits from those who travel by express. The greater part of the passengers of the former belong to the agricultural classes; and it is just like this,—the farmer is not accustomed to being hurried, and he won't be hurried, neither are the farmer's wife, the farmer's sons, the farmer's daughters; the tradesman in farming towns and villages is like the farmer, he won't be hurried, and it is the same with the tradesman's wife, his sons and his daughters; neither will the workman, nor the workman's wife, nor his sons, nor his daughters, be hurried and flurried. In fact, no one is ever in a hurry, except in large towns and cities. The country was not made for people who are always in a hurry, therefore it is impossible to get country people in and out of trains with the same speed as railway servants can manage with town and city people, and the effect of all this is, a stopping train is, much more frequently than "fast" or express trains, unpunctual, and it is especially so when it exceeds very much its ordinary amount of traffic. It is then sure to be very late, solely from circumstances beyond the control of the company, but of course the company gets all the discredit of the delay.

Sir Cusack Roney, in his appropriately named Rambles on Railways, *probably the most circumlocutory of all railway books, 1868.*

545.8 The locomotive is an accomplished educator. It teaches everybody that virtue of princes we call punctuality. It waits for nobody. It demonstrates what a useful creature a minute is in the economy of things.

Benjamin Taylor, The World on Wheels, *1874.*

545.9 Take my advice and start on your journey earlier, and if you miss one train you will get the next. I cannot afford to wait till 9 o'clock in the morning. I start long before that hour on my work.

Archibald Scott, General Manager of the London & South Western Railway, in reply to a complaining passenger, June 1883. Recorded by G P Neele in Railway Reminiscences, *1904.*

545.10 I loaf aroun' the depo' jest to see the Pullman scoot,
An' to see the people scamper w'en they hear the engine toot;
But w'at makes the most impression on my som'w'at active brain,
Is the careless men who get there jest in time to miss the train.

Sam Walter Foss, poem, The Men Who Miss the Train, *1892.*

545.11 If any train must be delayed, let it be the one that is already late rather than the one that is on time.

Charles Delano Hine, advising train despatchers, Letters from an Old Railway Official, *1904.*

545.12 'Well,' said the guard, as I stepped forth on to the deserted platform of Loughranny, 'that owld Limited Mail's th'unpunctualist thrain in Ireland! If you're a minute late she's already gone from you, and maybe if you were early you might be half an hour waiting for her!'

Edith Oenone Somerville and Martin Ross, "Poisson d'Avril," Further Experiences of an Irish R M, *1908.*

545.13 The train to Birmingham was in ironical mood, for it ran into New Street to the very minute of the timetable.

Arnold Bennett, Catching the Train, *1912.*

545.14 Punctuality is one of the few things about which there is doubt. A passenger knows beforehand how much his ticket will cost and what sort of accommodation he will get, and does not trouble himself with anxious thoughts on these matters. But the fear of unpunctuality, of missing a connection, or arriving late for an appointment, is a constant source of annoyance, and a railway which relieves its customers of these apprehensions does perhaps more to win their approbation than could be done in any other way.

Lord Monkswell, The Railways of Great Britain, *1913.*

545.15 There was a young man of the Tyne
Put his head on the South-Eastern line;
But he died of ennui,
For the 5.53
Didn't come till a quarter-past nine.
Anon., of the reputation of the South Eastern Railway.
The Complete Limerick Book, Langford Reed, 1924. An
updated version, concerning the South Eastern's successor
the Southern Railway, appeared in H A V Bulleid's Bulleid
of the Southern, 1977.

545.16 They may talk of Columbus sailing,
Across the Atlantical sea,
But sure he never went railing,
From Ennis as far as Kilkee.
You run for the train in the morning,
The excursion train starting at eight,
You're there when the clock gives the warning,
And there for an hour you'll wait.
And as you're sitting in the train,
You'll hear the guard sing this refrain;
"Are ye right there Michael, are ye right?
Do ye think that we'll be there before the night?
Ye couldn't say for sartain, ye were so late in startin',
But we might now Michael, so we might."
Percy French, first verse of poem, Are ye right there,
Michael?

545.17 Long distance trains have a perfectly miraculous
ability to make up time in the last 50 miles or so of a run.
A C Kalmbach, editorial comment, Trains Magazine,
November 1941.

545.18 Well, suh, we is not only on time, but we is 15 minutes ahead of time, except of course that we is an hour
late!
Anon. Pullman porter, the morning after America had
put clocks forward one hour, early in the Second World
War. Reported in Trains Magazine, March 1942.

545.19 Folks around these parts get the time o' day
From the Atchison, Topeka And the Santa Fe.
Johnny Mercer, from song in the film The Harvey Girls,
1946.

545.20 No suh! The *Century* is never late; she ain't exactly
on time tonight, but she's never late.
Anon. redcap porter at Syracuse, N.Y., to Willard V
Anderson, of the New York Central's Twentieth Century
Limited, 1946, recounted in Trains Magazine, November
1948.

545.21 That's just like the Long Island, always late—even
in getting into bankruptcy court!
Anon. commuter, Long Island Railroad, 2 March 1949,
reported in Trains Magazine, May 1949.

545.22 I had a genius for missing trains in those days.
Charles Chaplin, of his early life in England, My Auto-
biography, 1964.

545.23 King Edward was not amused. Indeed, he was
vastly irritated, for he heartily disliked unpunctuality.
George Dow, of the late arrival of a Royal Train at
Marylebone; opening words of Great Central, vol. III,
1965.

545.24 Better Never—Than Late?
Headline of one of a series of self-mocking British Rail-
ways advertisements featuring comedian Tony Hancock,
during the Beeching era, 1960s.

545.25 The politeness of Kings, the discourtesy of
railways.
Gerard Fiennes, paper The Future of British Railways,
c. 1968.

545.26 Of course, I date myself by even mentioning a
microscopic 6-minute deficit in an 18-hour overnight
schedule. No-one, I'm sure, took either Amtrak or BN to
task for it, assuming anyone besides myself even noticed
it. America in our time isn't much on punctuality, and
hasn't been since the go-when-you-please automobile
replaced the we-leave-on-schedule-whether-you're-on-
board-or-not train. Airlines count any arrival of less than
15 minutes late as on time. In 1971 only a B&O or Milwaukee Road veteran would even understand what the
discussion is about. Alas.
David P Morgan, Trains Magazine, January 1972.

545.27 You're gonna schedule that thing right down to
the day?
Anon. railroad vice president operations, when dis-
cussing freight train schedules with John G Kneiling,
and recalled by him in Trains Magazine, November
1990.

545.28 It was a sort of special train, put on by Virgin.
Pretty damn special as it turned out, for it arrived with-
out delay.
Matthew Parris, of a train bearing the Prime Minister,
The Times, 17 May 2001.

545.29 I always dread arriving at Paddington and seeing a
notice that "FGW is operating normally".
Richard Hall, of train operator First Great Western,
email to the compiler, 27 October 2002.

545.30 We're on time. Are you?
Metra billboard beside suburban line in Chicago, facing
Interstate highway 94, clearly directed at the motorist in
traffic jams; seen August 2003.

546. PUSH-AND-PULL WORKING

546.1 Hitherto we have been led to understand that push-and-pull working in this country was restricted by ancient statute . . .

Editorial Comment, Trains Illustrated, *December 1961.*

547. PYROTECHNICS

547.1 Ah, but they could stand it, those 0-4-2s. They were grand litttle engines. I wish you could have seen them tackling the 1 in 70 grades of our Dalmellington road. I shall never forget the night of Saturday, 16th November, 1918, just after the armistice of the great war. No. 231 was on the 7 p.m. from Ayr, Willie Craig driving and Tom Rennie firing, and a great packed train of 191 tons tare. With a tenderful of splendid coal, 231 was going up like an engine possessed, exhaust ringing out clear in the frosty air and a great column of sparks shooting to the moonlit sky, while all down the curving line behind us you could see the little fires springing up. No trouble to fire—easy on coal and water, and *steam*—she could have made steam to run the *Queen Mary*!

We were in high heart that night, for the war was over and we thought that a great peace had come to the world. It'll come yet, and maybe on that glad night the ghost of little 231 will return to the Doon Valley, and go soaring up in the moonlight, with her exhaust echoing from the silent hills, her high column of sparks leaping to the stars, and the gallant little heart of her singing with a great joy.

D L Smith, More Glasgow & South Western Nights' Entertainments, *within* The Railway Magazine, *vol. LXXXVI, 1940.*

547.2 The sparks from the express became visible, like hordes of scarlet beetles tempted into the air by night; they fell and smouldered by the track, touched leaves and twigs and cabbage stalks, and turned to soot.

Graham Greene, Stamboul Train, *1932. (Orient Express in U.S.).*

548. QUALITY

548.1 Quality is remembered long after price is forgotten.
Attributed to George M Pullman by Nigel Harris.

548.2 We want to go one better than the Pullman.
Sir Nigel Gresley, of the intended quality of the Coronation *train introduced in 1937, quoted by A J Mullay,* Streamlined Steam, *1994.*

548.3 We have quality and low prices. Which do you want?
Andy Sprague, cartoon, showing "steam engineer" behind a counter, talking to a customer. Date not known; used by Roger Waller in an address to the Institution of Mechanical Engineers, 15 March 2004.

549. RAILCARS

549.1 The "first-class" train we shall dismiss with a polite gesture of disdain and deprecation, inasmuch as it is a gasoline car possessing a set of rubber-tired wheels and a set of flanged wheels which may be retracted for highway service, trips to the filling station, etc. With due charity for this unhappy hermaphroditic travesty in the sight of all true sons of the order of the holy smokebox, let us simply say that she *is* economical, and unobtrusively quiet.

> *Thomas U Johnson, of the passenger equipment of the Coudersport & Port Allegheny Railroad in Pennsylvania;* Trains Magazine, *July 1944.*

RAILFANS
See Enthusiasm and Enthusiasts

550. RAILS

550.1 Even at that early period, he was in the habit of regarding the road and the locomotive as one machine, speaking of the Rail and the Wheel as "Man and Wife."

> *Samuel Smiles, of George Stephenson's view in 1816.* Life of George Stephenson, *1881.*

550.2 In order still further to remedy the evil arising from the joint of the Rail-Road, I propose to weld the ends of the Bars together as they are laid down, so as to form a considerable length of Iron Rail in one place.

> *John Birkinshaw, patent specification, 1820.*

550.3 Probably no other technological development has done so much to increase the capacity of the railroads and to reduce their operating costs as this substitution of steel for iron rails.

> *George H Burgess and Miles C Kennedy,* Centennial History of the Pennsylvania Railroad, *1949.*

550.4 This is not to depreciate the great part played by the Westinghouse Air Brake, which made it practicable to control longer and heavier trains, or the substitution of steel for iron in locomotives and cars, or any of the other improvements which came along from time to time; but without the improvement of the track structure, the benefit of these other improvements would have been decidedly limited.

> *Ibid.*

550.5 When romance and color were being passed around the railroad scene many decades ago, the locomotive walked off with an alarmingly large and disproportionate share of the glory while rail was left with practically none. Rail remains dead and dull to the same lay mind that is excited by thundering 2-10-4's, that is amused by a semaphore arm's muscle action, and is enchanted by the marble monuments that are certain union passenger stations.

> *David P Morgan,* Trains Magazine, *February 1950.*

551. RAILTRACK

551.1 The heart of the railway.

> *Advertising slogan, 1997.*

551.2 Railtrack is not about taking a nostalgic journey back to a golden age; it is about travelling into the future on a mode of transport that is more relevant to the needs of the 21st century than any other.

> *Railtrack advertising brochure, 1997.*

551.3 Railtrack was a bastardised construct at the heart of a dysfunctional railway, neither truly in the private nor in the public domain; but due to luck and happenstance it took a couple of years to be found out.

> *Christian Wolmar,* Broken Rails, *2001.*

551.4 This most cooperative of industries, traditionally operated on military lines because of the need for safety, routine and integration, was now maintained by a ragbag of individuals, some excellent, some hopeless, but few with any real understanding of their role in keeping the trains running.

> *Ibid.*

551.5 Network Rail knew they had taken on a problem child—but they had no idea they needed an anti-social behaviour order too.

> *Tom Winsor, Railway Regulator, at a meeting of an all-party group of Members of Parliament; on the successor to Railtrack. Recalled by Nigel Harris to the compiler, 19 January 2004.*

552. RAILWAY & LOCOMOTIVE HISTORICAL SOCIETY

552.1 What a group of individualists we 2000 plus members of the Society are! We are variously known and unknown, retiring and forthright, scholarly and unschooled, provincial and sophisticated, mild and opinionated, yet we hold in common this inexplicable emotion about flanged wheels and rails, semaphores and stations, the precise date that Pittsburg, Shawmut & Northern was incorporated, and the height above water of Hell Gate.

> *David P Morgan, Tribute to Charles P Fisher,* RLHS Bulletin 125, *October 1971.*

553. RAILWAY BUSINESS

553.1 You find a dying railway: you say to it, live, blossom anew with scrip; and it lives, and blossoms into umbrageous flowery scrip, to enrich with golden apples, surpassing those of the Hesperides, the hungry souls of men. Diviner miracle what God ever did?

> *Thomas Carlyle,* Latter-Day Pamphlets, *1850.*

553.2 The more business it does, the more money it loses, and the greatest favor that could be conferred upon it—if public wants permitted—would be the privilege of quitting business until the end of the war!

> *John P King, president, Georgia Railroad, of the effect of the Civil War upon his line, 1864. Quoted by Professor John Stover,* American Railroads, *1968.*

553.3 Stick to the original line; keep down capital, and let competing schemes do their worst.

> *Lord Westbury, quoted at a GNR shareholders' meeting, 1872.*

553.4 Building railways from nowhere to nowhere is not legitimate business.

> *Attributed to Cornelius Vanderbilt, without date, by Robert Keith Middlemas in* The Master Builders, *1963.*

553.5 It has become the tritest of commonplaces that the present position of our railways, from the proprietary point of view, is highly unsatisfactory. With degrees of phrasing, the unpalatable fact has formed the burden of many authoritative deliverances during the past few weeks, and it were idle to disguise the apprehension that the future is pregnant with grave reasons for disquietude.

> *G B Bayley,* The Position of Railway Proprietors, *within* The Railway Magazine, vol. IX, *1901.*

553.6 It is almost axiomatic that the passenger carrying business of most steam railroads is not entirely self-sustaining.

> *John A Droege,* Passenger Trains and Terminals, *1916.*

553.7 A railroad sells transportation, but sells something else along with it. It sells "service." The price of a railroad ticket covers both.

> *Howard Elliott,* What a Railroad Ticket Covers, *quoted by John A Droege,* Passenger Trains and Terminals, *1916.*

553.8 Over there they produce train-miles as we do ton-miles.

> *David P Morgan, of passenger railway operation in Great Britain.* Trains Magazine, *January 1966.*

553.9 The poor health of the railroad industry could hardly be denied when one noted that the 1965 rail mileage was less than that of 1904, although capitalized at more than twice as much; that railroad passenger traffic in 1965 was below the level of 1902 and yet received an average fare that was only a penny per mile higher; that railroad employment in 1965 was below the level of 1890, with the average annual wage up nearly fourteenfold. The golden age that American railroads had known in the generation before the First World War had, by the sixties, turned a sickly yellow.

> *Professor John Stover,* American Railroads, *1968.*

553.10 The pendulum has swung, from stultifying regulations under which railroading could never change, to an obsession with quarterly results under which railroading can never plan ahead.

> *Mark Hemphill,* Trains Magazine, *October 2004.*

554. RAILWAY MANIA

554.1 Railway Shares! Railway Shares!
Hunted by Stags and Bulls and Bears—
Hunted by women—hunted by men—
Speaking and writing—voice and pen—
Claiming and coaxing—prayers and snares—
All the world mad about Railway Shares!

> *Anon. poem in* Illustrated London News, *1845.*

554.2 The railway gambling of 1844 and 1845 probably saved the country from a depression of profits and interest, and a rise of all public and private securities, which would have engendered still wilder speculations, and when the effects came afterwards to be complicated by scarcity of food, would have ended in a still more formidable crisis than was experienced in the years immediately following.

> *John Stuart Mill,* Political Economy, *1848.*

554.3 Folly and knavery were for a time in the ascendant. The sharpers of society were let loose, and jobbers and schemers became more and more plentiful. They threw out railway schemes as lures to catch the unwary. They fed mania with a constant succession of new projects. The railway papers became loaded with their advertisements. The post-office was scarcely able to distribute the multitude of prospectuses and circulars which they issued.

> *Samuel Smiles,* Life of George Stephenson, *1881.*

554.4 But there came the day when the names of Watt and Stephenson waxed great in the land, and these slow citizens caught the railway frenzy. They took it, however, in their own fashion. They never dreamed of connecting themselves with other towns and a larger world, but of aggrandizement by means of a railway that should run from Tregarrick to nowhere in particular, and bring the intervening wealth to their doors.

> *Sir Arthur Quiller-Couch,* Cuckoo Valley Railway, *1944.*

555. RAILWAYMEN

See also Enginemen, Station Staff, Working on the Railway

555.1 Here he is, in velveteen or in a policeman's dress, scaling cabs, storming carriages, finding lost articles by a sort of instinct, binding up lost umbrellas and walking sticks, wheeling trucks, counselling old ladies, with a wonderful interest in their affairs—mostly very complicated—and sticking labels on all sorts of articles. I look around—there he is, in a station master's uniform, directing and overseeing, with the head of a general, and with the courteous manners of a gentleman; and then there is the handsome figure of the guard, who inspires confidence in timid passengers. I glide out of the station, and there he is again with his flags in his hand at his post in the open country, at the level crossing, at the cutting, at the tunnel mouth, and at every station on the road until our destination is reached.

Charles Dickens, at the ninth Anniversary Festival of the Railway Benevolent Society, 5 June 1867, recorded in The Speeches of Charles Dickens, *edited by Bernard Darwin.*

555.2 Well, one of'n went on the line, and he became a station-master; an 'nother, he went on the line, and he became a ganger; and t'other, he were a-runned over by a train, and so, as us may say, they was all connected with the railway.

Cecil Torr, Small Talk at Wreyland, *1870.*

555.3 The man with the big opportunity today is the man in the ranks.

James J Hill, quoted within his obituary, New York Times, *30 May 1916, repeated in* The Railway Library, *1915, 1916.*

555.4 There is no business in the world in which a man is judged more on the results he achieves than in the railroad business. When you report to your superior on any job, he is interested solely in knowing whether or not you have turned the trick. It is a case of you did or you didn't, not why you did or why you didn't.

Elisha Lea, Asst. General Manager, Pennsylvania Railroad, address before Delaware College, c. 1915, reproduced in The Railway Library, *1915, 1916.*

555.5 A feature of the Lancashire and Yorkshire landscape of this period was the extremely high percentage of red-headed individuals amongst the traffic staff. Just as every London and South Western employee today has to wear a red tie, so it would appear that many of the corresponding Lancashire and Yorkshire functionaries of those days had to be provided with red hair.

E L Ahrons, of the Lancashire & Yorkshire Railway, The Railway Magazine, *September 1917.*

555.6 From pre-natal days I was destined for the railway service, as an oyster to its shell.

Joseph Tatlow, Fifty Years of Railway Life, *1920.*

555.7 The very name of the Railway Executive Committee would hardly be recognised by the multitude; yet this body, which controlled the railways with such astonishing skill, did much to ensure our ultimate victory in the field. Its members paid the price. Two general managers of railways died through overwork in the war, and the constant tension left its unmistakable mark on those who have survived. They received very little thanks or recognition and none at all from the Government, while the bounteous showers from the fountain of honour left them almost unsprinkled.

Editorial Comment, special railway number, The Times, *15 August 1921.*

555.8 Railwaymen are not a perfectionist crowd. They never set themselves up for paragons. But take them by and large, railwaymen are as worthy a segment of society as any other.

D B Hanna, Trains of Recollection, *1924.*

555.9 Ambition that was all self-sacrifice never braced a continent with steel. The money changers do not open the purse of Fortunatus to Simon Pure altruists. Railways must be carried to completion on the financial engines that are available to mundane men.

Ibid.

555.10 Let it not be forgotten that ingenuity begins where the rules end.

John Droege, of railwaymen, Freight Trains and Terminals, *1925.*

555.11 The successful railroad man must be large in every way; big in thought, brain and conception—an organizer and an executive; and broadminded. He must be a keen analyst—able to separate the wheat from the chaff and to use the kernel in the heart of the nut.

Ibid.

555.12 Eight times that spring, after all the fury, wildness, and debauch of night, he rode back at morning towards the city in a world of waking men: they were for the most part railroad men—engineers, firemen, brakemen, switchmen, and train conductors, on their way to work. And their homely, seamed and pungent comradeship filled him with the health of morning and with joy.

Thomas Wolfe, Of Time and the River, *1935.*

555.13 A man is in degrees and by degree a railroader. Only when his habits are reduced to those typical of the norm, and his values are squared with those of his intimates, is he one of the craft. On many issues he may differ vitally with his fellows and still be accepted; but on some he must conform or be an outcast. To the degree that he participates in the world outside of railroading and retains attitudes, habits, and values derived from other occupational groups, he fails to qualify as a real railroader.

W Fred Cottrell, The Railroader, *1940.*

555.14 Railwaymen are trained to emergency.

Evan John, Time Table for Victory, *c. 1946.*

555.15 There never has been and there never will be a ceiling on improvement in railroad operations and services. Railroad management and railroad employees never rest on past accomplishments; they are never satisfied with what has been done, but instead are always looking for ways and means of doing better.

William T Faricy, President, Association of American Railroads, quoted in Trains Magazine, *May 1949.*

555.16 Railroad men are cut of the same cloth everywhere.

Brian Fawcett, Railways of the Andes, *1963.*

555.17 The Reds and the Greens.

Anon. During the Penn Central merger, late 1960s, of former employees of the Pennsylvania Railroad ("Reds") and New York Central ("Greens"), referring to the colours used on passenger rolling stock and freight rolling stock respectively. Also used in the mid 1980s of employees during the merger of the Norfolk & Western ("Reds," passenger rolling stock) and Southern ("Greens," locomotives) into Norfolk Southern.

555.18 There's railwaymen, and there's others.

Michael Cook, in conversation with Nigel Harris, c. 1973.

555.19 They could charge me money to let me work as a railroader and I'd still do it.

John W Barriger III, quoted by Charlton Ogburn, Railroads, *1977.*

555.20 Oh, it's Tommy this an' Tommy that,
And chuck 'im out the brute,
but it's Mr Bloody Railway
when the Beeb begins to shoot.

Simon Jenkins, after Rudyard Kipling, concerning a common attitude to railwaymen in times of trouble. The Times, *22 November 2000.*

555.21 The trainmaster at Niagara Falls was an old-fashioned railroader: he was hired by the railroad before he finished high school, spent his life on the railroad, saw his job as running trains, and did it exactly as he had been taught to do it in the years around World War 1. He had no managerial skill, and did not inspire those who worked for him. He did not frighten them: he just avoided them. He had no concept of how to please a cus-

tomer or even how to deal with a customer. His clerks were left alone to take the customers' wrath and to deal with them as best they could.

Richard Saunders Jr, of conditions on the New York Central in 1958; Merging Lines, 2001.

555.22 This was an age-old rivalry on every railroad. The traditional attitude of operating people was that real men ran trains and sissies wined and dined customers. The attitude of the traffic people was that they worked their butts off to solicit business that paid everyone's salaries, and then the operating people lost it or bonged it up or otherwise screwed up.

Ibid.

555.23 Don't call them contractors, they're railwaymen.

Robin Gisby, speech, referring to those men whose operations had been sold to contracting companies during the privatization process; 26 September 2003, reported in Modern Railways, *November 2003.*

555.24 I knew the part of Alf Garnett before I played it: I'd worked as a porter at Euston station.

Warren Mitchell, who played Alf Garnett in television sitcom Till Death Us Do Part, *in U.K. (translated to U.S. as* All in the Family*); comment in BBC radio program* Loose Ends, *6 December 2003.*

556. RAILWAY RACES AND RIVALRY

556.1 The L. & N. W Co. have expressed their intention to reach Aberdeen before us. This of course we cannot permit . . .

Patrick Stirling, Locomotive Superintendent, Great Northern Railway, internal memorandum, 14 August 1888.

556.2 There'll be no speeding on this railway. But don't let the other guy win either.

Lord Shaughnessey, President CPR, admonishing a driver who had been reported for racing his train with one of the Canada Atlantic, 1904. Quoted in CP Rail News, *14 November 1984.*

556.3 Do you think we could go faster than the LMS?

Sir Nigel Gresley, in conversation with Norman Newsome, June 1938, and leading to the successful speed record attempt by Mallard *the following month.*

RAILWAYS

Railways are divided into sub-headings, thus: Achievements and benefits, Anticipation, Appreciation, Character, Completion, Construction, Disdain, and Inspiration.

557. RAILWAYS—Achievements and Benefits

557.1 And the said Huntingdon Beaumont hath used new and extraordinary invencions and practises for the spedy and easy conveyance of the said coales, and especially by breaking the soyle for layinge of rayles to carry the same uppon with great ease and expedicion the length of two myles or thereaboutes, and by drawinge of certen carrylaggs laden with coales uppon the same rayles whereby a hundred and twenty rooke of coales in a weeke att the least being worth twenty poundes are conveyed away uttered and solde more that otherwise there could bee by the ordinarie and customarie course of the gettinge and carryinge away of the said coales.

Francis Strelley, Proceedings before Court of Chancery, (Jas. I, S27/4) 1615. Carrylaggs were carriages. Quoted by Richard S Smith, England's First Rails, *in* Renaissance and Modern Studies, IV, *1960.*

557.2 The manner of the carriage is by laying rails of timber, from the colliery down to the river, exactly straight and parallel, and bulky carts are made with four rowlets fitting these rails, whereby the carriage is so easy that one horse will draw four or five chaldron of coals and is an immense benefit to the coal merchants.

Roger North, The Life of Francis North, Baron Guildford, *1676.*

557.3 The success of the Darlington railway experiment, and the admirable way in which the locomotive-engine does all, and more than all, than was expected of it, seems to have spread far and wide the conviction of the immense benefits to be derived from the construction of new railways.

Comment, The Times, *2 December 1825.*

557.4 . . . is probably connected with changes in the country far greater than have hitherto been witnessed from any other recent improvement.

Comment, Liverpool Mercury, *5 October 1827.*

557.5 Railroad iron is a magician's rod in its power to evoke the sleeping energies of land and water.

Ralph Waldo Emerson, Nature, *1840s, quoted by John R Stilgoe in* Metropolitan Corridor, *1983.*

557.6 Yes, truly, nothing, next to religion, is of so much importance as a ready communication.

Edward Oxley, director of the Great North of England Railway, in speech at opening of the line between Darlington and York to passenger traffic, 30 March 1841.

557.7 Railway travelling is a delightful improvement of human life. Man is becoming a bird; he can fly longer and quicker than a Solan goose. The early Scotsman scratches himself in the morning mists of the North, and has his porridge in Piccadilly before the setting sun.

Attributed to Reverend Sydney Smith, 1842, by Alan Jenkins in The Stock Exchange Story, *1973.*

557.8 Railroads make the country transparent.

Ralph Waldo Emerson, Journal U, *page 135, 1844.*

557.9 All Hail to this utilitarian age
Of mighty works, which mighty minds engage!

Gigantic fabrics cover all the land.
Steam power the tedious labours of the hand
Supplants—intelligence from pole to pole
Triumphant flies by iron and by coal.
The fields our fathers misemployed in grain,
By stupid rustics sown, shall soon sustain
Unnumbered families, close packed together,
In dwellings which keep out the wind and weather.

Basil Montague, Railroad Eclogues, *1846.*

557.10 There were railway patterns in its draper's shops, and railway journals in the windows of its newsmen. There were hotels, office-houses, lodging-houses, boarding-houses; railway plans, maps, views, wrappers, bottles, sandwich-boxes, and timetables; railway hackney coach and cab stands; railway omnibuses, railway streets and buildings, railway hangers-on and parasites, and flatterers out of all calculation. There was even railway time observed in clocks, as if the sun itself had given in.

Charles Dickens, of Stagg's Gardens, Dombey & Son, *1847.*

557.11 Railroads are to civilization what mathematics were to the mind. Their immense promise made the whole world nervous with hope & fear, & they leave society as they found it. The man gets out of the railroad car at the end of 500 miles in every respect the same as he got in.

Ralph Waldo Emerson, Journal CD, *page 90, 1847.*

557.12 A railway is a more beneficial means of setting aside a tyranny than a revolution.

Comment in The Economist, *23 August 1851.*

557.13 Truly Queen Victoria can set the railway conquests of her reign against the glories of the war victories of Queen Anne and her grandfather, King George.

Samuel Sidney, Rides on Railways, *1851.*

557.14 My impression is, that our wonderfully increased, and still increasing, facilities of locomotion are destined to bring us round again to the nomadic state.

Nathaniel Hawthorne, The House of the Seven Gables, *1851.*

557.15 Railways having brought London within reach of everyone, thousands will visit it who have never thought of doing so before. Railways have done more; they have annihilated the prejudice against public conveyances, and abolished the old stiff-necked line of demarcation between posting and coaching company. They have opened out the world to everyone.

R S Surtees, Hints to Railway Travellers and Country Visitors to London, *in the* New Monthly Magazine, *1851.*

557.16 The rail is an English art; it is carrying English arts, English men, and English opinions all over the world. It is civilising Egypt. If it be capable of teaching

mildness to Mussulman tyranny, and placing restraint on the power of the Pachas, what other power is there in either Europe or Asia it will not in time subdue? To the progress of railways, and to the knowledge, arts, and skill they will diffuse, we look for the political regeneration of Europe, rather than to the political constitutions of Young Germany, or the insurrections of Young Italy.
Ibid.

557.17 All the lesser towns through which or near to which a railway passes, have virtually changed face; they confront the outside world no longer upon their embowered street or quiet common, but at the "station."
Donald G Mitchell, Rural Studies, *1867.*

557.18 The village cannot ignore the railway: it is the common carrier; it is the bond of the town with civilization; it lays its fingers upon the lap of a hundred quiet valleys, and steals away their tranquillity like a ravisher.
Ibid.

557.19 . . . the railway is everything. It is the first necessity of life, and gives the only hope of wealth. It is the backbone of existence from whence spring, and by which are protected, all the vital organs and functions of the community. It is the right arm of civilization for the people, and the discoverer of the fertility of the land.
Anthony Trollope, North America, *1862.*

557.20 It is essentially a modern invention, alike in its inception and in its execution. It is altogether a product of the nineteenth century, and the greatest product that that greatest of all the centuries has bequeathed to posterity.
J S Jeans, of the railway system, Jubilee Memorial of the Railway System, *1875.*

557.21 But the dollar is more mighty than the sword, the musical ring of the locomotive-bell more potent than the rattle of the drum, the thrust of the piston-rod more irresistible and permanent in its effects on civilization than that of the lance; and the throb of the engine in its peaceful conquest, is more frequent than was the sob of the widow and orphan in New Mexico in its warlike conquests of bygone days.
John George Hyde, Our Indian Summer in the Far West, *1880.*

557.22 The iron rail proved a magician's road.
Samuel Smiles, Preface to the Life of George Stephenson, *Centenary edition, 1881.*

557.23 The railway as an agent of civilization is not inferior to the art of printing.
Attributed to the Chicago Times, *upon the introduction of the Pennsylvania Railroad's Broadway Limited, 1887; in a bookseller's catalog on the internet, at www.dogbert .abebooks.com/servlet/BookSearchPL?cmid= hp-search-form, February 2004.*

557.24 The railway, in its present phase of development, enters so intimately into the social life of the community in its every detail, and has become so potent a factor in its every movement and operation, whether of business or of pleasure, that it must clearly be material for every individual to know something of the great agency which does so much for his happiness and welfare, and to realise clearly what he has a right to expect from it.
George Findlay, The Working and Management of an English Railway, *1889.*

557.25 "The poet scoffed, and deemed the end was come
To things romantic," said the iron horse;
"And road of steel would scar a beauteous face
Of lovely nature, and would end perforce
In death to all things worth the singing of."
But, spite of him, the engine pushed its way
Through desert, forest, over hill and dale,
And is—well, what it is today.

I'll sing a moment what the poet scoffed—
Sing of the victory of the road of steel,
Which spreads its arms embracing through the land,
And binds us brothers in the common weal.
The panting steed, which daily speeds its course,
Makes light of time, and bridges over space,
Brings friend to friend, wafts greetings near and far,
And makes one brotherhood the human race.
A C Chauncy, The Poet and the Railway, *within* The Railway Magazine, *vol. III, 1898.*

557.26 Railways have rendered more services, and have received less gratitude, than any other institution in the country.
John Bright, quoted as a subheading to Pertinent Paragraphs *within* The Railway Magazine, *1898–1908.*

557.27 The Railway is my right hand and the telegraph my voice.
Attributed to Cecil Rhodes by The Railway Magazine, *1963.*

557.28 A nebulous dream was a reality: an iron ribbon crossed Canada from sea to sea. Often following the footsteps of early explorers, nearly 3,000 miles of steel rail pushed across vast prairies, cleft lofty mountain passes, twisted through canyons, and bridged a thousand streams. Here on Nov. 7, 1885, a plain iron spike welded east to west.
Inscription on signpost at Craigellachie, British Columbia, where the transcontinental Canadian Pacific Railway was completed.

557.29 Thanks to Stephenson, George Hudson, and the many other Georges, who invested their talents and valuable money in the undertaking, railways have brought wealth and salubrity to everyone's door. It is no longer the class distribution that used to exist, this place for that set, that for another; but a sort of grand quadrille of gaiety in which people change places continually, and whirl about until they finally settle down, finding out beauties that none can see but themselves.

R S Surtees, Plain or Ringlets? *1860.*

557.30 The grand, the crowning benefit of all, however, were railways.

Ibid.

557.31 We build railroads, we know not for what or for whom, but one thing is certain, that we who build will receive the very smallest share of benefit.

Ralph Waldo Emerson, The Young American, *within* Complete Works, vol. 1, *1903.*

557.32 Irish peasants used to go in thousands to England and Scotland to work in the harvesting of the grain crops and thereby earned enough money to pay the rent of their small holdings. Steel rails and Consolidation locomotives stopped the cultivation of so many wheat fields in the British Isles, that the help of the Irish worker was no longer needed, and the suffering and discontent arising therefrom led to the vigorous agitation for home rule in Ireland.

Angus Sinclair, Development of the Locomotive Engine, *1907. See also 557.37.*

557.33 The railroad was more than a means of transportation to Gopher Prairie. It was a new god: a monster of steel limbs, oak ribs, flesh of gravel, and a stupendous hunger for freight; a deity created by man that he might keep himself respectful to Property, as elsewhere he had elevated and served as tribal gods the mines, cotton-mills, motor-factories, colleges, army.

The East remembered generations when there had been no railroad, and had no awe of it; but here the railroads had been before time was. The towns had been staked out on barren prairie as convenient points for future train-halts; and back in 1869 and 1870 there had been much profit, much opportunity to found aristocratic families, in the possession of advance knowledge as to where the towns would rise.

Sinclair Lewis, Main Street, *1922.*

557.34 If a town was in disfavor, the railroad could ignore it, cut it off from commerce and slay it. To Gopher Prairie the tracks were eternal verities, and boards of railroad directors an omnipotence.

Ibid.

557.35 ... the Intercolonial must not be treated as a commercial railway; but as a sort of douceur to induce the maritime provinces to remain confederated with Canada.

D B Hanna, Trains of Recollection, *1924.*

557.36 The winning of the west has been ascribed to the early trappers, explorers, stockmen and settlers, but the real winning of the west belongs to the railroads.

Edward Gillette, c. 1925, attributed by Stewart H Holbrook in The Story of American Railroading, *1947.*

557.37 It has only been as a consequence of this introduction of railways that the English farmer is regularly, and ordinarily, exposed to competition with the wheat growers of the most fertile regions of the West. The development of the railway system in America has done much to deprive the landed classes in England of their natural protection which was afforded by distance and difficulty of transport.

Professor W Cunningham, quoted in Railways, *W V Wood and Sir Josiah Stamp, 1928.*

557.38 They bind the vast, far-flung regions together; they give easy access to every part of the national domain. Their economic influence is enormous—incalculable in its potential force. With their help commerce spreads; local concerns grow into national enterprises; and the products of a town, sold everywhere in the United States, may become better known than the town itself.

William McAdoo, Crowded Years, *1931.*

557.39 At the same time our system of transportation has the effect of segregating industries and activities. Various sections of the country have turned to a specialization which would be impossible without the railroads. Pittsburgh is a specialist in iron and steel; Minneapolis produces flour; Danbury makes hats. Florida and California grow oranges; the Northwest grows wheat; the South grows cotton. Connecticut makes clocks and tools; the North Carolina towns specialize in cigarettes and other tobacco products, while Detroit is a huge automobile factory. The food you eat, the clothes you wear, and the furniture in your house constitute an assemblage of articles that originated in at least a hundred different places.

Ibid.

557.40 The railroad cannot shut down when times are bad.

Anon., Fortune Magazine, *December 1932.*

557.41 The urbanism of the railroader is heightened by his mobility. Railroads were the first truly modern means of transportation, and as such they have had much to do with the creation of "streamlined" morality and personality. The coming of the railroad has intruded new forces into the geographical community. It serves as a stream not only bearing with it the culture of other places but also carrying a channel of its own. In some cases the significance of the stream is not only what it bears but its own action, carving away local landmarks and setting up eddies, which threatens the foundations of local morality.

W Fred Cottrell, The Railroader, *1940.*

557.42 Take that mark of civilization away, and we are a depressed area.

Taxi driver in Sedbergh, of the closing of the Lune Valley line, quoted in the Manchester Guardian, *25 September 1953.*

557.43 The national interest requires that the future of the railways should be assured. As carriers of passengers over long distances and of suburban passengers to and from work, and as carriers of bulk freight the railways are essential and will continue to be for as long as can be foreseen.

Comment, British Railways Modernisation Plan, *1955.*

557.44 Even if railroading ended tonight at 11:59, you and I would have been the fortunate ones—those who experienced the great drama, who were close by when man's genius produced something at once useful and beautiful.

David P Morgan, Trains Magazine, *May 1957.*

557.45 The railway's obligation to carry every load that is offered to it is something that has become very firmly fixed in the thinking of railwaymen and of the public. It arose naturally in the conditions of near-monopoly that the railways once enjoyed; tradition and a decent anxiety to meet all kinds of social needs make many railwaymen unhappy about giving up any of the services they have been rendering for many years.

Michael Robbins, The Railway Age, *1962.*

557.46 In the first half of the nineteenth century, public transport, if not yet an essential element in London's life, was becoming a very important adjunct to it. Later, in the railway phase, it became a prerequisite of continued outward growth.

T C Barker, A History of London Transport, vol. 1, *1963.*

557.47 The railways need to be seen not only as mechanical contrivances but as instruments of social change, in the context of the Victorian age.

Jack Simmons, The Railway in Town and Country, *1830–1914, 1986.*

557.48 The network of steel rails that criss-crossed the nation irrevocably altered the face of the land, bringing wealth and opportunity to many, but hardship and bitterness to others. As the great wheels clattered across the landscape, the dust they raised settled slowly on the pages of American literature, providing major themes for our writers for almost a century.

Norm Cohen, Long Steel Rail, *1990.*

557.49 The most successful British industrial invention of all time.

Michael Binyon, book review in The Times, *9 October 1997.*

557.50 ... the utter centrality of the railway system to the maintenance of decent daily life in a crowded, compli-

cated island. It is up there with the NHS and the police: an essential shared source.

Libby Purves, The Times, *21 November 2000.*

557.51 They are a vital, shared resource without which the kingdom is disunited. There is no point in spewing out wish-lists and election manifestos disguised as Urban and Rural White Papers unless you put some proper thought into what joins it all up.

Ibid.

557.52 Railways are a growth industry. Their most sustained attempts to drive away their customers have not succeeded.

Christopher Fildes, The Spectator, *July 2001.*

558. RAILWAYS—Anticipation

558.1 It is scarcely possible to contemplate an institution from which would result a greater quantity of harmony, peace, and comfort to persons living in the country, than would naturally result from the introduction of railroads.

Dr James Anderson, Recreations in Agriculture, *c. 1800.*

558.2 I consider it in every point of view, so exclusively an object of national concern that I shall give no encouragement to private speculations until it is ascertained that Congress will not be disposed to pay any attention to it. Should it, however, be destined to remain unnoticed by the General Government, I must confess that I shall feel much regret, not so much from personal as from public considerations. I am anxious and ambitious that my native country should have the honour of being the first to introduce an improvement of such immense importance to society at large, and should feel the utmost reluctance at being compelled to resort to foreigners in the first instance.

John Stevens, of the construction of railroads in America; pamphlet Documents Tending to Prove the Superior Advantage of Railways and Steam Carriages over Canal Navigation, *1812.*

558.3 I do verily believe that carriages propelled by steam will come into general use, and travel at the rate of 300 miles a day. But one step in a generation is all we can hope for. If the present generation shall adopt canals, the next may try the railway with horses, and the third generation use the steam carriage.

Oliver Evans, c. 1813, quoted by Stewart Holbrook in The Story of American Railroads, *1947.*

558.4 So many and so important are the advantages which these States would derive from the general adoption of the proposed railways, that they ought, in my humble opinion, to become an object of primary attention to the national government.

Ibid.

558.5 Come give me a spade, let it never be said that we have not made a beginning.

George Stephenson, during the survey of the Stockton & Darlington Railway, 1821. Quoted by M Heavisides, The History of the First Public Railway, *1912.*

558.6 The Memorial of Benjamin Dearborn of Boston, respectfully represents,—that he has devised in theory, a mode of propelling wheel-carriages, in a manner probably unknown in any country; and has perfectly satisfied his own mind, of the practicability of conveying Mails and Passengers with such celerity, as has never before been accomplished, and in complete security from Robberies on the highway.

For obtaining these results he relies on carriages propelled by steam on level rail-roads; and contemplates that they be furnished with accommodations for Passengers to take their meals and their rest during the passage, as in a Packet; that they be sufficiently high for persons to walk in them without stooping, and so capacious as to accommodate twenty, thirty, or more Passengers and their baggage.

Benjamin Dearborn, Memorial to Congress, *presented by Senator Harrison Gray Otis on 12 February 1819. Quoted in* Railroad History, *Bulletin 132, Spring 1975.*

558.7 The inequalities of the earth's surface will require levels of various elevations in the rail-road; and your Memorialist has devised means, which he believes will be completely effectual, for lifting the carriages by the inherent power of its machinery, from one level to another; as also for the passage of carriages *by* each other on the same road; and he feels confident; that whenever such an establishment shall be advanced to its most improved state, the carriage will move with a rapidity at least equal to a mile in three minutes.

Ibid.

558.8 Don't be surprised if I should tell thee there seems to us after careful examination no difficulty of laying a railroad from London to Edinburgh on which waggons would travel and take the mail at the rate of 20 miles per hour, when this is accomplished steam vessels may be laid aside! We went along a road upon one of these engines conveying about 50 tons at a rate of 7 or 8 miles per hour, and if the same power had been applied to speed which was applied to drag the waggons we would have gone 50 miles per hour—previous to seeing this locomotive engine I was at a loss to conceive how the engine could draw such a weight, without having a rack to work into the same or something like legs—but in this engine there is no such thing.

Edward Pease, letter to Thomas Richardson, 10 October 1821. Quoted in The Origins of Railway Enterprise, *Maurice Kirby, 1993, in which he describes the "truly visionary terms" of Pease's statement, made after a meeting*

with George Stephenson. There seems little doubt, however, that Pease is reflecting George Stephenson's vision (see 558.26).

558.9 What can be more palpably absurd and ridiculous than the prospect held out of locomotives travelling twice as fast as stage-coaches! We would as soon expect the people of Woolwich to suffer themselves to be fired off upon one of Congreve's ricochet rockets, as trust themselves to the mercy of such a machine going at such a rate.

Anon. commentator, of the proposed Woolwich Railway, Quarterly Review, *c. 1825. Quoted by Samuel Smiles, in* Life of George Stephenson, *1881.*

558.10 . . . I don't like the look of these tramroads; there's mischief in them.

The Duke of Bridgewater, prominent canal owner, at the time of the promotion of the Liverpool & Manchester Railway, quoted in Railway Adventures and Ancedotes, *1888.*

558.11 The importance, to a commercial state, of a safe and cheap mode of transit, for merchandise from one part of the country to another, will be readily acknowledged. This was the plea, upon the first introduction of Canals: it was for public advantage; and although the new mode of conveyance interfered with existing and inferior modes, and was opposed to the feelings & prejudices of landholders, the great principle of public good prevailed, and experience has justified the decision.

Paragraph 2, Prospectus, Liverpool & Manchester Railway, *29 October 1824, quoted by Henry Booth in* An Account of the Liverpool & Manchester Railway, *1831.*

558.12 It is upon the same principle that Rail-roads are now proposed to be established; as a means of conveyance manifestly superior to existing modes: possessing, moreover, this recommendation in addition to what could have been claimed in favour of Canals, namely, that the Rail-road scheme holds out to the public not only a cheaper, but a far more expeditious conveyance than any yet established.

Ibid., paragraph 3.

558.13 All I ask you is, not to crush it in its infancy. Let not this country have the disgrace of putting a stop to that which, if cherished, may ultimately prove of the greatest advantage to our trade and commerce; and which, if we do not adopt it, will be adopted by our rivals.

William George Adam, (Leading Counsel to organizing committee of Liverpool & Manchester Railway) Report of Parliamentary Committee on Liverpool and Manchester Railway Bill, *May, 1825.*

558.14 Mr Creevey was appointed a member of the Committee to deal with the Bill of the Liverpool & Manchester Railway Company, to which, it would appear, he applied himself in no judicial frame of mind. He acted openly in the interests of his friends. Lords Derby and Sefton, who, like most territorial magnates at that time, viewed the designs of railway engineers with the utmost apprehension and abhorrence.

Sir Herbert Maxwell, of the first attempt to secure a bill for the L&MR; as editor, The Creevey Papers, *1903.*

558.15 We had to fight this battle against an almost universal prejudice to start with—interested shareholders and perfidious Whigs, several of whom affected to oppose us upon *conscientious* scruples.

Thomas Creevey, MP, in contrast with Sir Herbert Maxwell's account above; Journal, *1 June 1825.*

558.16 Coaches to run without horses—mercy on us! In a short time, instead of travelling along in a comfortable coach with a set of beautiful tits moving gracefully along the road, we may expect to hear of a monster of a thing rolling furiously along, frizzing, steaming and clattering, leaving a trail of smoke behind it like one of Whiston's comets, and woe to them what come within its vortex.

Anon., letter to a local Staffordshire newspaper before the building of the North Staffordshire Railway, c. 1830, quoted in The North Staffordshire Railway, *Manifold, 1952.*

558.17 It is in this point of view that Railways will be found invaluable—increased facility and security of transport, increased despatch, and diminished expense will keep the balance even with our foreign rivals, if not turn it in our favour, and will bring conveniences and comforts of life cheapened more readily to the door of every consumer in these kingdoms. In short, the principle of a Railway is that of commerce itself—it multiplies the enjoyment of Mankind by increasing the facilities and diminishing the labour by which the means of those enjoyments are produced and distributed throughout the world.

William Huskisson, MP, part of speech prepared for delivery on the evening on 15 September 1830, the opening day of the Liverpool & Manchester Railway, found in his pocket after he was killed on that day. Quoted by Simon Garfield in The Last Journey of William Huskisson, *2002.*

558.18 Objects of exalted power and grandeur elevate the mind that seriously dwells on them, and impart to it greater compass and strength. Alpine scenery and an embattled ocean deepen contemplation, and give their own sublimity to the conceptions of beholders. The same will be true of our system of Rail-roads. Its vastness and magnificence will prove communicable, and add to the standard of intellect of our country.

Charles Caldwell, Thoughts on the Moral and Other

Indirect Influences of Rail-roads, New England Magazine, *April 1832.*

558.19 Now first we stand and understand,
And sunder false from true,
And handle boldly with the hand,
And see and shape and do.

Dash back that ocean with a pier
Strow yonder mountain flat,
A railway there, a tunnel here,
Mix me this Zone with that!

Alfred, Lord Tennyson, Mechanophilus, *published 1892 but written c. 1833.*

558.20 These things may appear very strange today; yet in the womb of the future there slumbers the seed of great development in railways, the results of which it is, as yet, quite beyond our powers to foresee.

Friedrich Harkort, probably c. 1833 quoted by Edwin A Pratt in The Rise of Rail Power in War and Conquest, *1915.*

558.21 We denounce the mania as destructive of the country in a thousand particulars—the whole face of the Kingdom is to be tattooed with these odious deformities; huge mounds are to intersect our beautiful valleys; the noise and stench of locomotive steam-engines are to disturb the quietude of the peasant, the farmer, the bleating of sheep, and the grunting of pigs to keep up one continual uproar through the night along the lines of these most dangerous and disfiguring abominations.

Anon., in John Bull, *15 November 1835.*

558.22 Let the great world spin for ever down the ringing grooves of change.

Alfred Lord Tennyson, in a reference which is undoubtedly to railways, although at first glance (apparently in poor light and in a large crowd) Tennyson thought that rails were grooved. Locksley Hall *1842.*

558.23 Is then no nook of English ground secure
From rash assault? Schemes of retirement sown
In youth, and 'mid the world kept pure
As when their earliest flowers of hope were blown,
Must perish; how can they this blight endure?

William Wordsworth, poem, concerning the Windermere and Kendal Railway, October 1844.

558.24 The railways will do as much for mankind as the monasteries did.

Benjamin Disraeli, Sybil, *1845.*

558.25 We who lived before railways, and survive out of the ancient world, are like Father Noah and his family out of the Ark. The children will gather around and say to us patriarchs, "Tell us, grandpapa, about the old world". And we shall mumble our old stories; and we shall drop off one by one; and there will be fewer of us, and these very old and feeble. There will be but ten pre-railroadites left: then three—then two—then one—then O!

W M Thackeray, De Juventute, *1860.*

558.26 I will do something in coming time which will astonish all England.

George Stephenson, quoted in George Stephenson, The Father of Railways, *Thomas Summerside, 1878.*

558.27 I meant to make the mail run between London and Edinburgh by the locomotive before I died, and I have done it.

George Stephenson, when asked in later life what he meant by "astonish all England." Quoted in Railway Adventures and Anecdotes, *Richard Pike, 1888.*

558.28 I venture to tell you that I think you will live to see the day when railways will supersede almost all other methods of conveyance in this country—when mail coaches will go by railway and railroads will become the great highways for the king and all his subjects. The time is coming when it will be cheaper for a working man to travel on a railway than to walk on foot. I know that there are great and almost insurmountable difficulties to be encountered; but what I have said will come to pass as sure as you now hear me.

George Stephenson, after dinner conversation during construction of Stockton & Darlington Railway. Quoted by Samuel Smiles, The Life of George Stephenson, *Centenary Edition, 1881.*

558.29 Men shall take supper in London and breakfast in Edinburgh.

George Stephenson, quoted in Railway Adventures and Anecdotes, *Richard Pike, 1888.*

558.30 Bright Summer spreads his various hue
O'er nestling vales and mountains steep,
Glad birds are singing in the blue,
In joyous chorus bleat the sheep.
But men are walking to and fro,
Are riding, driving, far and near,
And nobody as yet can go
By train to Buttermere.

J K Stephen, Poetic Lamentation on the Insufficiency of Steam Locomotion in the Lake District, *(written with Wordsworth in mind), first verse, 1891.*

558.31 Presumptuous nature! do not rate
Unduly high thy humble lot,
Nor vainly strive to emulate
The fame of Stephenson and Watt.
The beauties which thy lavish pride
Has scattered through the smiling land
Are little worth till sanctified
By man's completing hand.

Ibid., last verse.

558.32 Build railways, and population will follow as the hounds do the fox.

Attributed to L R Robinson by A W Pearse, The Railway Library, *1910, 1911.*

558.33 Until this great work is completed, our Dominion is little more than a 'geographical expression'.

Sir John Macdonald, of the Canadian Pacific Railway, in conversation with Sir Stafford Northcote, quoted in The National Dream, *Pierre Berton, 1970.*

558.34 The war has shown how much could be effected in the way of rapid development of our transportation system behind the lines, especially by the laying down of hundreds, if not thousands, of miles of light railways. If this were done in this country, especially in connection with an improvement of our roads, it would achieve great results in the opening up of the resources of the country and the spreading of the population into healthier areas.

David Lloyd George, in Parliament, 6 December 1918, quoted by Edwin A Pratt in British Railways in the Great War, *1921.*

558.35 Even Ilfracombe had only as yet talked of a railroad. Lynton had not even dreamed of such a thing. It would have seemed to us nothing short of sacrilege.

A G Bradley, of North Devon in 1864. The line to Ilfracombe was opened ten years later. Lynton started to look for railway proposals in the 1880s and the narrow gauge line to Lynton was opened in 1897. Exmoor Memories, *1926.*

558.36 Railways will sew up the Italian boot.

Massimo d'Azeglio, quoted in The Economic History of Modern Italy, *Shephard Clough, 1963.*

559. RAILWAYS—Appreciation

559.1 Of each wonderful plan
E'er invented by man,
That which nearest perfection approaches,
Is a road made of iron,
Which horses ne'er tire on,
And travelled by steam, in steam coaches.

G W P Custis, operetta, The Rail Road, *c. 1829.*

559.2 A common sheet of paper is enough for love, but a foolscap extra can alone contain a railroad and my ecstasies.

Fanny Kemble, actress, in a letter after a footplate ride with George Stephenson, August 1830.

559.3 Meanwhile, the genius of the age, like a mighty river of the new world, flows onward, full, rapid, and irresistible. The spirit of the times must needs manifest itself in the progress of events, and the movement is too impetuous to be stayed, were it wise to attempt it. Like the "Rocket" of fire and steam, or its prototype of war and desolation—whether the harbinger of peace and the arts, or the Engine of hostile attack and devastation—though it be a futile attempt to oppose so mighty an impulse, it may not be unworthy our ambition, to guide its progress and direct its course.

Henry Booth, last paragraph of An Account of the Liverpool & Manchester Railway, *1831.*

559.4 ... the monarchical governments of Europe duly appreciate the vast importance of railroads as a means of controlling the people; but if we are not mistaken, railroads are the precursors of liberty and equality to the people everywhere.

J Kimball Minor, American Railroad Journal, *August 1843.*

559.5 Let us clearly understand our position. We have arrived at a new epoch in the history of the world. A new element of civilization has been developed. As was the invention of letters, as was the printing-press, so is the railway in the affairs of mankind. It is a revolution among nations. A moral revolution as affecting the diffusion of knowledge, the interchange of social relations, the perpetuation of peace, the extension of commerce; and a revolution in all the relations of property.

Comment, Westminster Review, *December 1845, quoted by John B Jervis in* Railway Property, *1866.*

559.6 No poetry in Railways? Foolish thought
Of a dull brain, to no fine music wrought.
Charles Mackay, poem Railways, *1846.*

559.7 It must be borne in mind here, that the boon conferred upon the public is not limited to written correspondence. Viewed in reference to the postal facilities they afford, the railways are the great public instructors and educators of the day.

Robert Stephenson, Address as President of the Institution of Civil Engineers, 8 January 1856, quoted by Sir Cusack Roney in Rambles on Railways, *1868.*

559.8 Contrast the size of *The Times*, in 1830, and in 1856. Do you suppose that the huge mass of paper, which you are permitted to forward by tonight's post, would have been conveyed upon the same terms, if the means of conveyance had remained limited to the mails and its four horses? Look at the immense mass of parliamentary reports and documents, now distributed, every session, amongst all the constituencies of the Empire, at almost a nominal charge. To what do the public owe the valuable information embodied in these documents, but to the railways? except as parcels, by waggons or by canal boats, they could never have been conveyed prior to the existence of the railway system, and if they could never have

been distributed, we may rely upon it they would never have been printed.

Ibid.

559.9 Whilst most people in our country have, in fact, repelled the railroads from their localities, the people of the United States have invited them to their very streets and doors.

Sir Morton Peto, The Resources and Prospects of America Ascertained during a visit to the States in the Autumn of 1865, *1866.*

559.10 We, who accept with as little gratitude as wonder the marvellous results of our railway system—who complain of the slightest delay or inconvenience, and in whose estimate the slightest casual annoyance outweighs a thousand continuous and unspeakable benefits—find it difficult, even if we try, to realise the enormous difficulties of the first beginning and the early progress.

Samuel Smiles, Memoir of the Late Henry Booth, *1869.*

559.11 Colorado without railroads, is comparatively worthless.

John Evans, former Governor of Colorado, to William Jackson Palmer, builder of the Denver & Rio Grande, c. 1870. Quoted by Robert G Athearn, Rebel of the Rockies, *1962.*

559.12 Somebody told you the railroad was a bad thing. That was a lie. It may do a bit of harm here and there, to this and to that; and so does the sun in heaven. But the railway's a good thing.

George Eliot, Middlemarch, *1871.*

559.13 Hurrah for the rail! for the stout iron rail,
A boon to both country and town,
From the very first day that the permanent way
And the far-famed fish-point was laid down.
'Tis destined, you'll find, to befriend all mankind,
To strew blessings all over the world;
Man's science, they say, gave it birth one fine day,
And the flag of King Steam was unfurled.
Ned Farmer, quoted in Our Railways, *J Pendleton, 1896.*

559.14 If the people of Milo did well to put up a statue in gold to the man that invented wheels, so should we also put one up in Portland stone or plaster to the man who invented rails, whose property is not only to increase the speed and ease of travel, but also to bring on slumber as can no drug: not even poppies gathered under a waning moon.

Hilaire Belloc, The Path to Rome, *1902.*

559.15 I will stake my reputation as a railway man that the country could not concentrate men and materials half so fast as the railways could deal with them; but the management of railways must be left in the hands of railway men.

Sir Charles Owens, c. 1908, in reply to a question from the Secretary of State for War, as to the preparedness of the railways for national emergency, as quoted by Edwin A Pratt in The Rise of Rail Power in War and Conquest, *1915.*

559.16 Your railways have reached a higher standard in international comparison than your farmers or your government and under greater difficulties.

Sir William Acworth, of American railroads, 1910, quoted by Albro Martin in Enterprise Denied, *1971.*

559.17 I can remember when we were almighty glad to pay 5 cents a mile to ride in rickety little jim crow cars over a roadbed as rough as a Virginia corduroy road, 25 miles an hour, and the liver shaken out of us before we had travelled half-way to our destinations. Now we complain because we are compelled to pay 2 cents a mile to ride in cars more elegant than most of our homes, over roadbeds as smooth as a billiard table, speeding along at 50 or 60 miles an hour.

Anon. Nebraska farmer, originally published in The Midwest, *date not known, reproduced in* The Railway Library, *1913, 1914.*

559.18 There never will be a country where you will not find complaints against the railways. That is part of the reason the railways are there, so that you can complain about them. But, taking the whole figures, comparing them with what the railways are doing today and what they did in 1913—a phenomenal year—I say they are playing their share in the reconstruction of this country in a way that deserves credit and not blame.

Sir Eric Geddes, House of Commons, 24 March 1920, quoted by Edwin Pratt in British Railways in the Great War, *1921.*

559.19 In the new world of science and finance, and particularly of democratic and semi-socialized forms of government in which we now live, transport occupies a more important place than ever, and the position of railways in the economic structure of all countries requires to be re-examined without prejudice from the past, but losing none of its lessons.

Sir Josiah Stamp, Railways, *1928.*

559.20 I like to see it lap the Miles—
And lick the Valleys up—
And stop to feed itself at Tanks—
And then—prodigious step

Around a pile of Mountains—
And supercilious peer
In Shanties—by the sides of Roads—
And then a Quarry pare...

Emily Dickinson, poem, I like to see it lap the Miles, *1929.*

559.21 Before the discovery of steam as motive power, mankind did not expect much help from nature. Its immutability was taken for granted. The earth had served, through untold centuries, merely as a food-producer.

The invention of the steam engine, with all its implications, opened up long vistas of industrial progress. Mechanical contrivances acquired a splendid dignity. The minds of men went exploring among cog-wheels and pistons, seeking new combinations that would take the place of hand and muscle.

William McAdoo, Crowded Years, *1931.*

559.22 The railways were England's gift to the world.

G M Trevelyan, Illustrated English Social History, *1942. See also 76.4.*

559.23 We, the general public, pay our railways the highest possible compliment in taking them very much for granted.

Bernard Darwin, War on the Line, *1946.*

559.24 There is a lot of goodwill here. The trouble is that it is mostly spectator goodwill; it is ceasing to carry over to the cash customer.

Bernard de Voto, Harpers Magazine, *April 1947.*

559.25 The Railways will never in the lifetime of the youngest of us be allowed to decay.

British Railways publication Careers in the Railway Service, *early 1950s.*

559.26 One is the obvious glamour of the express—the distances, the speed, the exciting sounds, the fires in the night, the call to escape from a present place and way of life, the sense of momentary suspension between two worlds in a third and independent one. The other appeal is that of the rural branch-line train, so perfectly integrated with the country which it serves and yet so self-contained a thing.

Bryan Morgan, The End of the Line, *1955.*

559.27 Who says railroading's lost its romance?

David P Morgan, Trains Magazine, *October 1955.*

559.28 The German attitude to railways is, of course, a very serious one. That they are the arteries of national economy far more than in most countries is shown by a dozen phenomena—the learned publications, the social status and high pay of the railwaymen, the fact that bomb-damage has not been used for the closing of a single line, the centrally-sited stations, the teaching of railway lore in schools, the transport of lorries by rail and so on, right down to the fact that when a German has a job to do he starts by building a 600 mm line.

Ibid.

559.29 Every form of transportation has contributed to the Epic of America, but the railways have played the most important role in it. Their place is secure in history but not in the national economy. Paradoxically, at a time when mass production and mass merchandising are accelerating the nation's material growth through the integration of technological progress into manufacturing and distribution, the future status of the principal mass producers of transportation is in jeopardy. This is inimical to the public interest, because the railways provide the only functionally and geographically complete and flexible transportation service. That is, of all the forms of transportation, it is the railways alone which haul anything, anywhere, and at any time.

John Walker Barriger, Super Railroads, 1956.

559.30 In the late 1950's a party came back from the U.S.A. with the irresistible logic that the Yanks would be out of the inter-city business by 1970 and so should we. Since then, with the perversity which characterises our nation and with the obstinacy which so often saves it, we have developed our express passenger services in speed, frequency, punctuality and comfort, and match them with the French, the Japanese, the Dutch, the Germans, any nation you name, they are the best in the world.

Gerard Fiennes, Paper to Leeds Branch of The Institute of Transport, 9 January 1968.

559.31 Railways and ships are still the only civilized means of travel.

Rogers E M Whitaker, All Aboard with E. M. Frimbo, 1974.

559.32 It was part of your life. Railways went through the back of your spine like Blackpool went through rock.

Jim Howarth, driver, within The Ballad of John Axon, 1958.

559.33 A Railway always does me good.

Lady Charlotte Schrieber, quoted in Lady Charlotte, Revel Guest and Angela John, 1989.

559.34 One cannot understand the resurgence of the modern railroad without understanding Big John.

Richard Saunders Jr, of the Southern Railway's high capacity grain cars and the cathartic effect of the case fought with the ICC over the reduced charges which they permitted; Merging Lines, 2001.

559.35 . . . the true worth of the day filtered down—a person could travel thirty miles in less time than it took to eat a meal.

Simon Garfield, of the opening of the Liverpool & Manchester Railway, 15 September 1830, The Last Journey of William Huskisson, 2002.

560. RAILWAYS—Character

560.1 The declivity is in general so gentle as to be almost imperceptible: the road, sometimes conveyed in a straight line, sometimes winding around the sides of precipices, is a picturesque object, and the cars filled with coals or iron, and gliding along occasionally without horses, impress the traveller, who is unaccustomed to such spectacles, with pleasing astonishment.

William Coxe, of a tramroad, A Historical Tour through Monmouthshire, 1801.

560.2 A railway and its carriage form together one machine; it is only from the extent of one of its parts that they are separate in idea, and no excellence in the one will compensate for an imperfection in the other.

Henry R Palmer, Description of a Railway on a New Principle, 1824. (This pamphlet was written to promote a raised monorail with pannier vehicles of a type that generally preceded the Lartigue monorail by very many years.)

560.3 On the very line of this railway, I have built a comfortable house; it enjoys a pleasing view of the country. Now judge, my friend, of my mortification, whilst I am sitting comfortably at breakfast with my family, enjoying the purity of the summer air, in a moment my dwelling, once consecrated to peace and retirement, is filled with dense smoke or foetid gas; my homely, though cleanly, table covered with dirt; and the features of my wife and family almost obscured by a polluted atmosphere. Nothing is heard but the clanking iron, the blasphemous song, or the appalling curses of the directors of these infernal machines.

"Ebenezer," The Leeds Intelligencer, 13 January 1825.

560.4 One has dim foresight of hitherto uncomputed mechanical advantages who rides on the rail-road and moreover a practical confirmation of the ideal philosophy that Matter is phenomenal whilst men & trees & barns whiz by you as fast as the leaves of a dictionary. As our teakettle hissed along through a field of mayflowers, we could judge of the sensations of a swallow who skims by trees & bushes with about the same speed. The very permanence of matter seems compromised and oaks, fields, hills hitherto esteemed symbols of stability do absolutely dance by you. The countryman called it 'Hell in harness.'

Ralph Waldo Emerson, Journal, 10 June 1834.

560.5 "I con-sider," said Mr Weller, "that the rail is unconstitootional and an inwaser o' priwileges. . . ."

Charles Dickens, Master Humphrey's Clock, 1840.

560.6 As to the *h*onour and dignity o' travellin', vere can that be vithout a coachman; and wot's the rail to sich coachmen and guards as is sometimes forced to go by it, but a outrage and a insult?

Ibid.

560.7 The railroad is in all its relations a matter of earnest business, to be got through as soon as possible. It transmutes a man from a traveller into a living parcel.

John Ruskin, The Seven Lamps of Architecture, 1849.

560.8 The Iron Civiliser.

Thomas C Keefer, Philosophy of Railroads and other Essays, *1849.*

560.9 Railways will remain safe in the midst of panic; and though times of pressure, severe, hazardous, ruinous pressure, have been felt in this country, and unfortunately must be felt again, yet it will only prove them to be part and parcel of the genuine sources of wealth and avenues for labour, in which this country lives and moves and has its being.

Anon., quoted by John Francis, A History of the English Railway, *1851.*

560.10 In fact, civility to all, gentle and simple, is the rule introduced by the English railway system; every porter with a number on his coat is, for the time, the passenger's servant.

Samuel Sidney, Rides on Railways, *1851.*

560.11 Our railways are dearer than the foreign, so is our government,—we make both ourselves . . .

Ibid.

560.12 . . . compare the speed, the universal civility, attention, and honesty, that distinguish our railway travelling, and you cannot fail to come to the conclusion that for a commercial people to whom time is of value, ours is the best article, and if we had not been a lawyer-ridden people we might also have had the cheapest article.

Ibid.

560.13 The Railway Ham, Beef, and German Sausage Warehouse I was prepared for. The Railway Pie Shop, I have purchased pastry from. The Railway Hat and Travelling Cap Depot, I knew to be an establishment which in the nature of things must come. The Railway Haircutting saloon, I have been operated upon in; the Railway Ironmongery, Nail and Tool Warehouse; the Railway Bakery; the Railway Oyster Rooms and General Shell Fish Shop; the Railway Medical Hall; and the Railway Hosiery and Travelling Outfitting Establishment; all these I don't complain of. In the same way I know that the cabmen must and will have their beer-shops, on the cellar-flaps of which they can smoke their pipes among the watermen's buckets, and dance the double-shuffle. The railway porters must also have *their* houses of call; and at such places of refreshment I am prepared to find the Railway Double Stout at a gigantic threepence in your own jugs. I don't complain of this; neither do I complain of J. Wigzell having absorbed two houses on each side of him into The Railway Hotel (late Norwich Castle), and setting up an illuminated clock, and a vane at the top of a pole like a little golden Locomotive. But what I do complain of, and what I am distressed at, is, the state of mind—the moral condition—into which the neighbourhood has got. It is unsettled, dissipated, wan-

dering (I believe nomadic is the crack word for that sort of thing just at present), and don't know its own mind for an hour.

Charles Dickens, Household Words, *11 November 1854.*

560.14 The railway is no respecter of persons, living or dead.

Frederick Miller, St Pancras Past and Present, *1874.*

560.15 The railroad system of the United States, with all its excellences and all its defects, is thoroughly characteristic of the American people. It grew up untrammelled by any theory as to how it ought to grow; and developed with mushroom rapidity, without reference to government or political systems.

Charles Francis Adams, Railroads; their Origins and Problems, *1878.*

560.16 There is no subject of easier declamation, or at which the impassioned orator can fire his guns with a louder report on a small charge of powder, than the railway system of the country.

Congressman Long, House of Representatives debate on the bill to establish a Board of Commissioners of Interstate Commerce, 1884; quoted by P Harvey Middleton, Railways and Public Opinion, *1941.*

560.17 But for the purposes of debate I do not think it is profitable to call the railway system either an angel of beneficence or a monster of iniquity. We all know that it has developed our country, turned the wilderness to a garden, peopled our great territorial domain into flourishing states, belted the Republic together, cheapened transportation, and, best of all, laid at the feet of the poorest citizen, by means of mutual and prosperous exchange, the products, whether of shop or farm, of every section of the Union. On the other hand we all know it has been the raw material, if that term shall offend no sensitive ear, of wild and ruinous speculation, and that the red blood of its stocks has often been watered until in too many cases it has only swashed a bloated corpse, offensive to the nose and damaging to health.

Ibid.

560.18 Railroad property, being of a permanent character, seems to be the only species of property which the demagogues, politicians and communists can attack at any and all times, they knowing full well that the owners of it cannot remove it out of their reach.

James C Clarke, President, Illinois Central, letter to Hon Lorenzo S Coffin, 2 January 1885, quoted by Thomas C Cochrane in Railroad Leaders, 1845–1890, *1953.*

560.19 A railroad differs from many other business enterprises, in the existence of a large permanent investment, which can be used for one narrowly defined purpose, and for no other. The capital, once invested, must remain. It is worth little for any other purpose than the one in question. A railroad cannot contract its capital merely because it does not pay; nor can it be paralleled at short notice when it happens to pay remarkably well. In these respects it differs quite sharply from a bank or store; and, to a certain extent, from a factory. The different lines of business—bank, store, factory, railroad—form a series, at one end of which we have an elastic business capital, which can be readily expanded or contracted, while at the other end we have a large permanent investment of "fixed" capital, which cannot thus adapt itself to the wants of trade. This is why it is so often said that the ordinary laws of political economy do not operate in the case of railroads.

Arthur Twining Hadley, Railroad Transportation, *1885.*

560.20 Pioneer railways are not like works at home. The lines are single, with crossing places every five, ten, or twenty miles; ballast is not always used, the lines on prairies being laid for long stretches on the earth formation; rivers, chasms, canons and cataracts are crossed by timber trestle bridges.

Sir Edward Watkin, of early railways in Canada, Canada and the States, *1887.*

560.21 All these lines across the Continent have very similar features. They each have prairies to pass, with long straight lines and horizons which seem ever vanishing and never reached; mountain ranges of vast altitudes to cross, alkaline lands, hitherto uncultivable, hot sulphur springs, prairie-dogs, gophyrs, and other animals not usually seen. The buffalo has retired from the neighbourhood of these iron-roads and of the "fire-wagons," as the Indians call the locomotives.

Ibid.

560.22 No enterprise is so seductive as a railroad for the influence it exerts, the power it gives, and the hope of gain it offers.

Comment, Poor's Manual of Railroads, *1900 edition.*

560.23 I do not say that the railways of Great Britain have no sins of omission and commission to answer for, but I forgive and forget them all when travelling over a foreign railway.

Attributed to E T's Diaries, *by* The Railway Magazine, *which used the quotation as a subheading to* Pertinent Paragraphs, *February 1908–February 1910.*

560.24 There is nothing, unless perhaps, the English language, more terrible than the workings of an English railway line.

Rudyard Kipling, The Day's Work, *1898.*

560.25 A railway should bend itself to the population and not leave the towns.

Attributed to George Hudson by a staff member of The Times, *in* The Complete Railway Traveller, *part I,* The Times, *21 November 1905.*

560.26 To suppose that railways, by which man thus shows himself master of his world, are—or could be—ugly and devoid of beauty betrays that regard for the obvious which is the very essence of Philistinism.

Anon. staff member of The Times, *in* The Complete Railway Traveller, *part III,* The Times, *5 December 1905.*

560.27 It is natural and necessary that the railway should adapt itself to the general conditions of its environment; and this to the end not merely of exploiting it, but also of making it a suitable and valuable return for what has been taken from it.

B C Burt, Railway Station Service, *1911.*

560.28 The capacity of a modern railroad is measured by the capacity of its terminals rather than by that of its main line tracks.

Edward Hungerford, The Modern Railroad, *1911.*

560.29 The railroad system of the United States is a great piece of commercial machinery, essential to every one in this complicated modern civilization; without this piece of machinery, there could not be the volume of business—agricultural, manufacturing and commercial—that there now is. The magnitude of these transactions is so great that this piece of commercial machinery must be kept in the very best order, and its capacity must be increased all the time. Without this, much business could not exist; on the other hand, without the business, this piece of commercial machinery would be idle and rust.

Howard Elliott, The Truth About the Railroads, *1913.*

560.30 People have too long been led to believe that the interests of the railroads are diametrically opposed to all other business interests; they should be made to understand that the maintenance of the transportation system in an unimpaired state is of vital importance to the economic fabric of the nation, that the railway service should continue to develop and expand, and that existing obstacles to a healthy growth of transportation enterprise should be promptly removed.

T W Van Metre, Failures and Possibilities in Railroad Regulation, *within* War Adjustments in Railroad Regulation, *published as* The Annals of the American Academy of Political and Social Science, vol. LXXVI, *March 1918.*

560.31 Historically and etymologically a railway is merely a road on which rails are laid.

W M Acworth, The Elements of Railway Economics, *1924.*

560.32 A railway is a republic and a monarchy. It is a republic because there is no pre-emption of high offices for any favoured class among its servants. It is a monarchy in the virtual dictatorship of its President.

D B Hanna, Trains of Recollection, *1924.*

560.33 "This bird says he's the train dispatcher. Says this train's got to move."

Dutch turned his hard-boiled visage toward the telegrapher.

"Oh yeah? Well, this is a murder, see?"

The dispatcher was unimpressed.

"This is a railroad," he snapped.

Rodger Denbie, exchange between policemen and a railroad man, Death on the Limited, *1933.*

560.34 Railways must conform to the first law of life: they must develop or die.

Evan John, Time Table for Victory, *c. 1946.*

560.35 The most characteristic symbol of a railroad is not, as many may believe, the dramatic steam locomotive or the streamlined Diesel, but rather a flanged wheel on a steel rail.

Ernest E Norris, President, Southern Railway System, Trains Magazine, *October 1945.*

560.36 This wouldn't be a railroad without passenger service.

Anon. operating executive, Rutland Railroad, when abandonment of passenger service was considered (and later carried out) because of a damaging strike, 1953. Quoted by Dwight R Ladd in Cost Data for the Management of Railroad Passenger Service, *1957.*

560.37 The germ of railroading is this: you can pull more than you can carry.

Comment, Trains Magazine, *September 1952.*

560.38 Railways are living organisms of such vast scope and complexity that events and developments, which are imperceptible singly, become striking when accumulated over years. Present conditions undermining the financial vitality of many railroads constitute a warning which informed and responsible citizens should not disregard, even though the general public may not become aware of the problem until it has grown too serious to be corrected by moderate means.

John Walker Barriger, Super Railroads, *1956.*

560.39 Because railroading is big, so it produces big men, who can speak of the almost incredible fact of moving 14,000 tons in a single train with the ease of selling a shirt in Sears Roebuck.

David P Morgan, Trains Magazine, *May 1957.*

560.40 There always exists that tiny, yet sly and insidious danger that railroads and their train-watchers will lose faith in themselves. The incomparable value of the flanged wheel riding the steel rail is a phrase tossed about so much that is could become mechanical and meaningless amidst more topical headlines like car shortages and rate hikes.

David P Morgan, Trains Magazine, *October 1957.*

560.41 You can, with diligence, make of a railway a very perfect *machine*. But it takes little touches like that to give it a *soul*.

David Smith, of a driver's devotion to keeping his locomotive clean while his fireman was away ill, unable to do so himself. Tales of the Glasgow & South Western Railway, *c. 1960.*

560.42 It is a curious thing that people think it a compliment to say that railways are at their best when they are like a conveyor belt or a liner. Railways are in fact at their best when they are like railways, which, I believe, they have never been. Certainly George Stephenson and Brunel and Hudson and the railway maniacs made them as like a horse and cart as they could, doing custom-tailored transport into every hole and corner. They never ran a train in their lives. And to run trains is the distilled essence of a railway; trains which remain during the hours of service continuously coupled.

Gerard Fiennes, paper Co-operation in Transport, *no date.*

560.43 This was railroading as God and George Stephenson intended it.

Anon., seen in a postwar American railroad book, not since found again, but never forgotten.

560.44 Firstly, railways are a world of their own; they are segregated from the rest of the nation, and yet they serve it. They are self-contained, definable, understandable even by attentive amateurs and therefore welcoming to escapists; yet they are ubiquitous, infinitely diverse, complex within their own limits and wrapped in their own mystique. They have their own language, their own telephone network, their eating houses, factories and estates; they have their own slums, palaces, mausoleums and rustic beauty; they offer majesty and meanness, laughter, wonder and tears.

David St John Thomas, Double Headed, *1963.*

560.45 In the course of the bitter battles over government control, the indispensability of railroads to American economic growth was elevated to the status of an axiomatic truth. Critics and defenders of railroad management argued about the appropriate basis for setting rates, the size of a fair profit, the necessity of various types of discrimination, the effectiveness of competition, and the wisdom of private pools, but they rarely debated the indispensability of railroads. Quite the contrary, the invocation of the "axiom of indispensability" was usually the first step in the argument of every disputant.

Robert William Fogel, Railroads and American Economic Growth, *1964.*

560.46 There is yet, praise be, an essential efficacy about the flanged wheel on the steel rail which commends itself to the mind as well as to the heart. There is yet in that mechanical formula an extraordinary combination of speed, capacity, safety, economy and flexibility which, when employed in concert, elude the skills of the competition.

David P Morgan, Trains Magazine, *November 1970.*

560.47 It was an institution, an image of man, a tradition, a code of honor, a source of poetry, a nursery of boyhood desires, a sublimest of toys, and the most solemn machine—next to the funeral hearse—that marks the epochs in man's life.

Jacques Barzun, quoted by Joseph Daughen and Peter Bintzen, The Wreck of the Penn Central, *1971.*

560.48 There has always been a sense of massive continuity about railways. Their great appetite for capital investment, the rugged longevity of their equipment, the bold confidence of their early building, and the statutory structure within which they function have all contributed to an aura of enduring inevitability. Indeed, is there not a grandly assertive ring in the very use of the term 'permanent way'?

John Johnson and Robert A Long, British Railways Engineering, 1948–1980, *1981.*

560.49 Railways had a polymorphic polygenesis.

Prof. Liviu Sofonea, paper The Wooden Mining Truck from Transylvania: Historical, Technical, Epistemological, Museological, Axiological Aspects, *First Early Railways Conference, Durham, 12 September 1998.*

560.50 The similarities between different railway systems are more important than their differences.

Christian Wolmar, Rail Magazine, *1 May 2002.*

560.51 Any railway, working properly, is a marvel of civilised co-operation.

Libby Purves, The Times, *14 May 2002.*

560.52 ... we might have to agree that what truly makes a railroad wonderful is what it does, not what it is.

Mark W Hemphill, editorial comment, Trains Magazine, *April 2004.*

561. RAILWAYS—Completion

561.1 The pale face has completed a mighty work. He has overcome the most imposing natural barriers; he has pierced the valleys of the Delaware, Susquehanna, Chemung, Alleghany and levelled the hills which were roamed by my ancestors. Now their descendant marvels at the doings of the mighty pale face. They cannot but be pleased to see him accomplish his great destiny; to see him fly from hill to valley; and ride upon the wings of lightning.

Peter Wilson, a native Cayugan, banquet 15 May 1851, upon the opening of the New York & Erie Railroad to Dunkirk. Quoted in Men of Erie, *Edward Hungerford, 1946.*

561.2 Done.

W N Shilling, Western Union telegrapher, message to announce the driving of the last spike in the transcontinental railroad at Promontory, Utah, 10 May 1869.

562. RAILWAYS—Construction

562.1 ... some short time before this conveyance a new method was invented for carrying coals to the river in large machines called waggons made to run on frames of timber fixed in the ground for that purpose and since called a waggon-way which frames must of necessity lye very near, if not altogether upon a level from the colliery to the river and therefore wherever there are any hills or vales between the colliery and the river and the same cannot be avoided, it is necessary in order to the laying of such waggon ways, then to make cutts through the hills or level the same, and raise or fill up the vales so that such waggon may lye upon a level as near as possible.

Anon. Counsel's opinion, 1763, on a conveyance (mentioned herein) of 1672. Quoted by Charles E Lee in The Evolution of Railways, *1943.*

562.2 The principal object in constructing a rail-road is to form hard, smooth and durable surfaces for the wheels of the carriages to run upon.

Thomas Tredgold, opening words of A Practical Treatise on Railroads and Carriages, *1825.*

562.3 The construction of canals is considered here a very simple operation, but the proper adjustment of a railway, on a system which admits of no derangement, is deemed the highest production of art and science.

George Featherstonehaugh, writing from Scarborough, letter to General Theodore Sill, Albany, N.Y., 18 December 1826.

562.4 I consider this among the most important acts of my life, second only to my signing of the Declaration of Independence, if second even to that.

Charles Carroll, then the sole survivor of the signatories, at the laying of the foundation stone of the Baltimore & Ohio Railroad, 4 July 1828; quoted by Edward Hungerford in The Story of the Baltimore & Ohio Railroad, *1927.*

562.5 Why, gentlemen, if this sort of thing be permitted to go on you will in a very few years *destroy the nobility!*

Sir Astley Cooper, to Robert Stephenson, of the construction of the London & Birmingham Railway, recalled in later years by Robert Stephenson and quoted in Railway Adventures and Anecdotes, *1888.*

562.6 You know that my system has always been in Main Lines, to keep the Line in a low country, so as to allow of Branches diverging into high country if such should exist.

George Stephenson, letter to Michael Longridge, 7 August 1838.

562.7 The railroads in America are not so well made as in England, and are therefore more dangerous; but it must be remembered that at present nothing is made in America but to last a certain time; they go to the exact expense considered necessary and no further; they know that in twenty years they will be better able to spend twenty dollars than one now. The great object is to obtain quick returns for the outlay, and except in a few instances, durability or permanency is not thought of.

Frederick Marryat, A Diary in America, *second series, vol. 1, 1839.*

562.8 The twanging of horns, the grating noise of the iron borers and the heavy and incessant explosions stunning the ear on all sides like the roar of artillery might have induced the traveller to believe himself in the neighbourhood of a sharp engagement.

Contemporary account of the construction of the Glasgow Paisley & Greenock Railway, 1839, quoted by N W Webster in Joseph Locke, *1970.*

562.9 They build not merely roads of earth and stone, as of old, but they build iron roads; and, nor content with horses of flesh, they are building horses of iron, such as never faint or lose their breath.

Anon., c. 1840, quoted by Albro Martin, Railroads Triumphant, *1992.*

562.10 All this without the aid of one dollar from a New York City stockholder, or even the sympathy of its citizens.

Eleazar Lord, of the construction of the New York & Erie Railroad. Quoted by Edward Hungerford in Men of Erie, *1946.*

562.11 Fear haunts the building railroad but it will be American power & beauty, when it is done.

Ralph Waldo Emerson, Journal U, *p.33, 1843.*

562.12 'Shut, shut the door,' as, Pope, disgusted, said
'Tie up the knocker—say I'm sick, I'm dead.'
Say what you please, good James, but pray keep out,
At all events, the speculating rout;
Those railroad bores, who, papers in each hand,
Request permission to cut up our land—
Request permission! I should rather say,
Who, leave unmasked, invade our lands, survey,
Trample, and trespass, and break people's legs,
By stumbling over their confounded pegs.

Basil Montague, Railroad Eclogues, *1846.*

562.13 But I have observed, fellow citizens, that railroad directors and railroad projectors are no enthusiastic lovers of landscape beauty; a handsome field or lawn, beautiful copses, and all the gorgeousness of forest scenery, pass for little in their eyes. Their business is cut and slash, to level or deface a finely rounded field, and fill up beautifully winding valleys. They are quite utilitarian

in their creed and in their practice. Their business is to make a good road. They look upon a well-constructed embankment as an agreeable work of art; they behold with delight a long, deep cut through hard pan and rock, such as we have just passed; and if they can find a fair reason to run a tunnel under a deep mountain, they are half in raptures.

Daniel Webster, speech at opening of the Northern Railroad from Franklin to Grafton, 28 August 1847; Writings and Speeches of Daniel Webster, *vol. IV, 1903.*

562.14 We are bringing into this world *something*—no child of mine—which if it prove better than an absolute abortion will only have one leg or be otherwise incapable of sustaining itself. However I shall do my best for it and be glad if in my time I hear the first *whistle*!

Lord Dalhousie, of the construction of the earliest railways in India; letter to George Carr Glyn, 1850, quoted by Robert Keith Middlemas in The Master Builders, *1963.*

562.15 We are beginning to find that he who buildeth a railroad west of the Mississippi must also find a population and build up business.

Charles Russell Lowell, letter to Judge Charles Mason, 20 October 1859, quoted by Richard C Overton in Burlington West, *1941.*

562.16 The *beau ideal* of a railway is one that comes about a mile from one's own house and passes through a neighbour's land.

The Earl of Powis, c. 1860, quoted by C P Gascoine, The Story of the Cambrian, *1922.*

562.17 We have drawn the elephant. Now let us see if we can harness him up.

Of the signature into law of the Pacific Railroad Bill, July 1862. Attributed to Collis P Huntington in Southern Pacific *by Neill C Wilson and Frank Taylor, 1952, and by Cerinda Evans in* Collis Potter Huntington, *1954; attributed to Theodore Judah in* American Railroads *by Professor John Stover, 1968.*

562.18 A "Grand Anvil Chorus", in triple time, three strokes to a spike.

Anon. reporter, quoted loosely by Robert Hedin in The Great Machines, *1996.*

562.19 Hear ye, Dakotas! When the Great Father at Washington sent us his chief soldier to ask for a path through our hunting grounds, a way for his iron road to the mountains and the western sea, we were told that they wished merely to pass through our country, not to tarry among us, but to seek gold in the far west.

Chief Red Cloud, quoted in The Great Machines, *Robert Hedin, 1996.*

562.20 Ten thousand track layers will bring in enough whiskey to kill every Indian within a hundred miles.

Attributed to General William Sherman by Lucius Beebe, concerning the dangers of construction of the Union Pacific Railroad, Trains Magazine, *June 1952.*

562.21 Wonderful, indeed is the rapidity with which the rolling hills are cleft, roads graded, ties laid down, and the rails secured in their places by the railroad-makers. The regularity and order of the operations are as perfect as could be devised. Within the short space of five weeks a most surprising change has happened. From Harker to Junction City the bottom lands are studded with ranches lately built, and the voice of the ploughman is heard where but two short months ago the war whoop of the savage Indian echoed on the green woods of the Smoky. Pedestrians bound for the west line the road. Pilgrims with their little all, their wives, children, and earthly substance, flock in numbers, and answer to your queries as to wither bound, with "To the west." The energy of the tracklayers seems to be infectious; everybody seems imbued with railroad haste and vigour.

Henry Morton Stanley, 9 May 1867, republished in My Early Travels and Adventures in America and Asia, vol. I, *1895.*

562.22 ... this system of England, where George Stephenson's thumb, pressed upon a plan, is an imprimatur which gives it currency and makes it authority; while here in the United States no man's imprimatur is better than another's. Each is trying to surpass his neighbor. There is rivalry here out of which grows improvement. In England it is imitation—in America it is invention.

Franz von Gerstner, quoted by John H B Latrobe in The Baltimore & Ohio Railroad: Personal Recollections, *1868.*

562.23 The time is coming, and fast too, when in the sense that it is now understood, there will be no West.

Arthur Ferguson, of the rapid progress of survey parties opening up the route to the West, diary, *21 July 1868.*

562.24 Its work all day for damn sure pay
on the Atchison Topeka and Santa Fe.

Anon., chant of track workers, c. 1870, noted by Keith L Bryant Jr, History of the Atchison Topeka and Santa Fe Railway, *1974.*

562.25 Now, my lads, you can't hinder the railroad: it will be made whether you like it or not.

George Eliot, Middlemarch, *1871.*

562.26 You Enterprised a Railroad through the valley—you blasted its rocks away, heaped thousands of tons of shale into its lovely stream. The valley is gone, and the gods with it; and now every fool in Buxton can be at Bakewell in half an hour, and every fool in Bakewell at Buxton; which you think a lucrative process of exchange —you Fools Everywhere.

John Ruskin, Fors Clavigera, *1871.*

562.27 Workin' all day without sugar in me tay,
Hammerin' rocks on the old railway,
The months roll by and I don't get no pay,
Drill, ye tarriers, drill.
　　　Drill, ye tarriers, drill
　　　Drill, ye tarriers, drill

Workin' in the tunnel, shovelin' out the dirt,
I worked so hard that I wore out me shirt,
The tunnel caved in and we all got badly hurt,
Drill, ye tarriers, drill.

Attributed to Thomas Casey, song, Drill, ye tarriers, drill, *first published 1888. The etymology of* tarrier *is uncertain.*

562.28 When I think how the railroad has been pushed through this unwatered wilderness and haunt of savage tribes; how, at each stage of the construction, roaring, impromptu cities of gold and lust and death sprang up and then died away again, and are now but wayside stations in the desert; how in these uncouth places pigtailed pirates worked side by side with border ruffians and broken men from Europe, talking together in a mixed dialect, mostly oaths, gambling, drinking, quarrelling, and murdering like wolves; how the plumed hereditary lord of all America heard in this last fastness the scream of the 'Bad Medicine Wagon' charioting his foes; and then when I go on to remember that all this epical turmoil was conducted by gentlemen in frock coats, and with a view to nothing more extraordinary than a fortune and a subsequent visit to Paris, it seems to me, I own, as if this railway were the one typical achievement of the age in which we live; as if it brought together into one plot all the ends of the world and all the degrees of social rank, and offered the busiest, the most extended, and the most varying subject for an enduring literary work. If it be romance, if it be contrast, if it be heroism that we require, what was Troy town to this?

Robert Louis Stevenson, of the transcontinental railroad, Across the Plains, *1892.*

562.29 They were flung out over boundless wastes inhabited by wild beasts and red Indians, much as the gossamer throws its delicate lines upon the summer air.

Editorial comment, on the construction of American railroads, in special edition of The Times *(London) devoted to American railways, 28 June 1912. Quoted in* The Railway Library, *1912, 1913.*

562.30 They were built in faith that the population would come and that the cities would arise—a faith fully justified by events, but resting at the time upon evidence so slight that the British investor whose money mainly paid for construction would have tightened his pursestrings had he known the real state of the case.

Ibid.

562.31 The builders seem to have had but one desire—to push further and ever further into the empty continent. Their building was hurried and flimsy. The single-track lines just held together under very moderate traffic; and American railroads suffer to this day the consequences of that initial insufficiency. Enormous capital has been sunk in the conversion of the unsubstantial lines into the solid and heavily metalled railroads that now serve a vast and busy population. Immense further expenditure is called for to provide equipment to deal with ever-increasing business, and the money will beyond doubt be found. Yet American railroad managers have not even now escaped from the odium incurred by poor lines compelled to extract an exiguous income from a sparse population of struggling pioneers.

Ibid.

562.32 A railway is never completely built. The construction accounts are never finished. What is called the closing of the period of construction, is nothing more than the transfer of responsibility for such further construction work as may be necessary, to those officials who are in charge of the general operation of the property.

Henry C Adams, American Railway Accounting, 1918.

562.33 Verily, the Union Pacific of the Pacific Railway was built from one end.

Edwin L Sabin, Building the Pacific Railway, 1919.

562.34 In some ways Belgium led Europe in railway building. She was ahead of all the Continent in ordered construction, and ahead of England in that she had a railway policy when England was fumbling for a policy which she never found.

J H Clapham, Economic Development of France and Germany, 1921, quoted in Railway Economics, K G Fenelon, 1932.

562.35 John Henry sez to his cap'n:
"Send me a twelve-poun' hammer aroun',
A twelve-poun' hammer wid a fo-foot handle,
An' I beat yo' steam drill down,
An' I beat yo' steam drill down.

Anon., quoted from The American Songbag, 1927, by Frank P Donovan Jr in The Railroad in Literature, 1940.

562.36 200 miles of engineering impossibilities.

William Van Horne, of the route of the Canadian Pacific Railway east of Fort William. Quoted in The National Dream, vol. 2, Pierre Berton, 1972.

562.37 The solidity of English railway construction gradually evinced itself in the highly substantial workmanship of the first twelve miles of the American road.

Edward Hungerford, The Story of the Baltimore & Ohio Railroad, vol. 1, 1927.

563. RAILWAYS—Disdain

563.1 You are welcome to the use of the schoolhouse to debate all proper questions in, but such things as railroads and telegraph are impossibilities and rank infidelity. There is nothing in the Word of God about them. If God had designed that His intelligent creatures should travel at the frightful speed of fifteen miles an hour by steam, He would have clearly foretold it through His holy prophets. It is a device of Satan to lead immortal souls down to hell.

Anon., letter from a school board in Lancaster, Ohio, 1828, recounted by Elisha Lea, in address before Delaware College, c. 1915, reproduced in The Railway Library, 1915, 1916.

563.2 Rail-ways, Mr Speaker, may be well enough in old countries, but will never be a thing for so young a country as this. When you can make rivers run backward, it will be time enough to make a railway.

Mr Cogswell, Representative for Ipswich, Massachusetts, House debate, c. 1829; Quoted by Alvin F Harlow, Steelways of New England, 1946.

563.3 Poor Huskisson is dead, or must die before tomorrow. He has been killed by a steam carriage. The folly of seven hundred people going fifteen miles an hour, in six carriages, exceeds belief. But they have paid a dear price.

Lord Brougham, of the accident on the opening of the Liverpool & Manchester Railway, 15 September 1830. Quoted in Notes on Railway Accidents, *Charles Francis Adams, 1879.*

563.4 Proud were ye, Mountains, when in times of old,
Your patriot sons, to stem invasive war,
Intrenched your brows; ye gloried in each scar:
Now, for your shame, a Power, the Thirst of Gold,
That rules o'er Britain like a baneful star,
Wills that your peace, your beauty, shall be sold,
And clear way made for her triumphal car
Through the beloved retreats your arms enfold!
Hear YE that Whistle? As her long-linked Train
Swept onwards, did the vision cross your view?
Yes, ye were startled;—and, in balance true,
Weighing the mischief with the promised gain,
Mountains and Vales, and Floods I call on you,
To share the passion of a just disdain.

William Wordsworth, Shorter Poems.

563.5 Gentlemen, as an individual, I hate your railways; I detest them altogether; I wish the concoctors of the Cheltenham and Oxford, and the concoctors of every other scheme, including the solicitors and engineers, were at rest in Paradise.

A Cheltenham MP, c. 1840, quoted in Railway Adventures and Anecdotes, 1888.

563.6 All railways are public frauds and private robberies.
Col Charles Sibthorp, MP for Lincoln, c. 1844, quoted by Michael Robbins, The Railway Age, *1962.*

563.7 In the same carriage with me there sat an ancient gentleman (I feel no delicacy in alluding to him, for I know he is not in the room, having got out far short of Birmingham), who expressed himself most mournfully as to the ruinous effects and rapid spread of railways, and was most pathetic upon the virtues of the slow-going old stage coaches. Now I, entertaining some lingering kindness for the road, made shift to express my concurrence with the old gentleman's opinion, without any great compromise of my own. Well, we got on tolerably comfortably together; and when the engine, with a frightful screech dived into the darkness, like some strange aquatic monster, the old gentleman said this would never do, and I agreed with him. When it parted from each successive station with a shock and a shriek, as if it had had a double tooth drawn, the old gentleman shook his head, and I shook mine. When he burst forth against such new-fangled notions, and said that no good could come from them, I did not contest the point. But I invariably found that when the speed of the engine was abated, or there was the slightest prolongation of our stay at any station, the old gentleman was up in arms, and his watch was instantly out of his pocket, denouncing the slowness of our progress.
Charles Dickens, Speech *at Conversazione of the Polytechnic Institution, Birmingham, 28 February, 1844. Published in* Speeches of Charles Dickens, *ed. K J Fielding, 1988.*

563.8 A plague on railroads!—they're a vile device,
For propagation of all sorts of vice.
Basil Montague, Railroad Eclogues, *1846.*

563.9 A railroad is like a lie—you have to keep building it to make it stand. A railroad is a ravenous destroyer of towns, unless those towns are put at the end of it and a sea beyond, so that you can't go further and find another terminus. And it is shaky trusting them, even then, for there is no telling what may be done with trestlework.
Mark Twain, letter to San Francisco Alta California, *26 May 1867. www.twainquotes.com (accessed 2002).*

563.10 Forget six counties overhung with smoke,
Forget the snorting steam and piston stroke,
Forget the spreading of the hideous town;
Think rather of the pack-horse on the down,
And dream of London, small, white, and clean.
William Morris, Prologue to The Earthly Paradise, *1868.*

563.11 I am sorry to find the intelligent people of the north country gone quite mad on the subject of railways.
Lord Eldon, quoted by Henry Pease, at the Stockton &

Darlington Railway Jubilee celebration, 1875, recorded by Gordon Biddle in Britain's Historic Railway Buildings, *2003.*

563.12 RAILROAD, n. The chief of many mechanical devices enabling us to get away from where we are to where we are no better off. For this purpose the railroad is held in highest favor by the optimist, for it permits him to make the transit with great expedition.
Ambrose Bierce, The Devil's Dictionary, *1911.*

563.13 No governmental action was taken, and the introduction and building of railways—most powerful of all instruments affecting the development of the nation—thereafter proceeded under private auspices, without systematic experimentation, control, uniformity of construction, or harmonious relationship as parts of one continental system designed and operated as the chief mechanical servant of the public. Instead—for two generations or more—the railways of the country were usually brought into being without sufficient regard for their need, coöperation or location; were built with more thought of cheapness than of efficiency and safety; sometimes became political machines that ruled cities or dominated whole commonwealths; occasionally came into the hands of unscrupulous men who systematically used them to filch money from others; and at last the entire fabric—drunk with a fallacious belief in its own power—fell into such a state of moral collapse that its members conspired with outsiders and among themselves, to injure communities, to destroy private enterprises, and to favor one man at the expense of another.
Seymour Dunbar, of the early development of railroads in the United States; A History of Travel in America, *1915.*

563.14 There has been a great deal of critical comment about exceptional cases of railway administration, but if you will put it all together you will find it relates to less than 10% of the mileage in the country, and that it has very much exceeded in volume and sound the praise bestowed upon the other 90%.
Frank Trumbull, address to The National Hay Association, *concerning Railway Regulation, Cedar Point Ohio, 12 July 1916.*

563.15 This railway system of ours is a very poor bag of physical assets. The permanent way is badly worn. The rolling stock is in a state of great dilapidation. The railways are a disgrace to the country. The railway stations and their equipment are a disgrace to the country. We are talking about the values of these things and I am saying that they are a pretty poor bag of physical assets. It cannot be supposed that the Stock Exchange is seriously undervaluing it, at present terms.
Hugh Dalton, (Chancellor of the Exchequer), House of Commons, 17 December 1946.

563.16 The railroad stems direct from Hell. It is the Devil's own invention, compounded of fire, smoke, soot, and dirt, spreading its infernal poison throughout the fair countryside. It will set fire to houses along its slimy tracks. It will throw burning brands into the ripe fields of honest husbandmen and destroy his crops. It will leave the land despoiled, ruined, a desert where only sable buzzards shall wing their loathsome way to feed upon the carrion accomplished by the iron monster of the locomotive engine. No, sir, let us hear no more of the railroad.

Anon. Quoted in The Story of American Railroads, *Stewart H Holbrook, 1947.*

563.17 . . .let us not try to disguise the unpleasant truth that the reputation of our railways has gone down in recent years; unjustly, for the most part, but the fact is there. We seem to have many critics but few friends and some quite dramatic proof is needed that the railways have not lost the will or the ability to play a progressive part in the social and industrial life of the nation.

A J White, (Asst General Manager of Eastern Region, British Railways), British Railways Magazine, *March 1955.*

563.18 I often thought about the line to Alexandria when, in the middle of the Second World War, I was stationed in the Pentagon as a minute part of the Transportation Corps. While the Army was wondering how to get the thousands of people in the Pentagon to and from work, men were lifting the double track of the Alexandria line which lay inches away from the Pentagon on its way south. Is this not the history of transportation in this country?

Rogers E M Whitaker, All Aboard with E M Frimbo, *1974.*

563.19 And now, after a century's interruption of technology, horses had repossessed the route. I had seen this all over Britain—defunct viaducts, abandoned cuttings, former railway stations, ruined railway bridges—and I thought of all the lost hopes and all the wasted effort. Then, small dismantled England seemed simple and underdeveloped—and too mean to save herself—deceived by her own frugality.

Paul Theroux, The Kingdom by the Sea, *1983.*

563.20 There is nothing romantic about the railways.

Sir Richard Marsh, on his retirement. Quoted in The Railway Station, *J Richards and J M MacKenzie, 1986.*

563.21 I saw a pile of steel plates, curiously shaped, with bent corners and holes in the middle and each stamped with a number. There must be, somewhere, in some dusty-windowed railroad office peopled by fat, grumpy middle-aged men in rumpled white shirts, a venerable municipal-green file cabinet, and in the back of the cabinet's bottom drawer must be a dog-eared manila folder, its edges soft as felt, and in the folder must be a sheet of flimsy among many other sheets, smudged with fourth-generation carbon, and on the sheet must be, I suppose, an order of some sort, an invoice, with a list of numbers. And one of them would be the number of this particular steel plate. And maybe you could track it down. Maybe if you looked in enough drawers and folders you would locate this plate by paperwork alone, predict that it be found in a burlap bag 2227 yards east of Alderson, W. Va.

Michael Kernan, Washington Post Magazine, *3 July 1988. Quoted in* Trains Magazine, *December 1988.*

563.22 So much effort is put into denial. Railways may have a future—but not as railways. Stations must look like airports, trains must look like aeroplanes, drivers must look like pilots (no walking down the platform with a billycan of tea). Even the ticket staff are redesigned as beaming hostesses and speak with robotic phrases that pass for "customer care".

Nicholas Whittaker, Platform Souls, *1995.*

563.23 Railways are not a priority.

Attributed to Tony Blair in cabinet meeting, 1997.

564. RAILWAYS—Inspiration

564.1 We owe all our railways to the collieries in the North.

Capt. J M Laws, General Manager of Manchester & Leeds Railway, evidence to the Gauge Commissioners, 1845.

564.2 Lay down your rails, ye nations near and far—
Yoke your full trains to Steam's triumphal car;
Link town to town; unite in iron bands
The long-estranged and oft embattled lands.

Charles Mackay, Railways, *1846.*

564.3 May God continue the unity of our country as this Railroad unites the two great oceans of the world.

Engraving on the Golden Spike driven at Promontory, Utah, to signify the completion of the transcontinental railway, 10 May 1869.

564.4 Someone has blundered in Egypt. Had Palmerston built a railroad from Cairo to Khartoum, there would not now be a rebel in the Soudan to annoy Gladstone.

Alexander Hogg, The Railroad as an Element in Education; *address to the International Congress on Education, New Orleans, 1887.*

565. RATES

565.1 It would puzzle wiser men than Railroad freight agents, however, to state upon the witness stand, and in brief answers, all of the considerations and influences which make the basis of the rates for transportation by rail.

Charles Perkins, vice president, Chicago Burlington &

Quincy, letter to John Forbes, 24 September 1879, quoted by Thomas C Cochrane in Railroad Leaders, 1845–1890, *1953.*

565.2 What the traffic will bear.

Attributed to George B Blanchard (giving evidence in Albany) by Edward P Ripley, with the comment 'a most excellent answer, but a most unfortunate one—for it has passed into history meaning "all the traffic will bear," which is a very different thing.' Quoted by Slason Thompson in The Railway Library 1909, *1910.*

565.3 The phrase "What the traffic will bear," has been crucified for many transgressions, but its real result has been to annihilate distance, to decentralize markets and population, and to develop prosperity.

Ivy Lee, address, the London School of Economics, 7 February 1910.

565.4 It was very early in the history of railroads perceived that if these agencies of commerce were to accomplish the greatest practicable good, the charges for the transportation of different articles of freight could not be apportioned among such articles by reference to the cost of transporting them severally; for this, if the apportionment of cost were possible, would restrict within very narrow limits the commerce in articles whose bulk or weight was large as compared with their value.

Anon., Report of Interstate Commerce Commission, *(stated to be vol. I, 1888, p. 303, but not found therein); quoted by Dixon H Davies, in* The Railway Library, 1913, *1914. Probably quoted in support of the principle of charging what the traffic will bear.*

565.5 We may well fear that like causes will produce like results. Western railroads are compelled to sell their services at such prices as may be fixed by the people they serve. Slaves in the South served their masters on similar terms, but the law dealt more kindly with the slave. His master was required to support him.

T B Blackstone, Chicago & Alton Railway, Annual Report 1888, *of the activities of railroad commissions in fixing rates. Quoted by P Harvey Middleton in* Railways and Public Opinion, *1941.*

565.6 There's a cuss they call Per Diem now that keeps me
on the jump,
He never lets me stop and rest. 'Tis always up and hump.
In the good old times I had a chance to sleep a month
or so
Every summer on a siding, waiting for the wheat to grow,
Or they'd fill me up with cotton, and I'd have a little quiet
Till the broker man that owned it found a customer to
buy it.
But now those peaceful days are gone. 'Tis always, "Get
thee hence!
What are you doing standing there, a-wasting twenty
cents?"

It's hustle all day long, and then at night, unless I'm
ditched,
At eleven fifty-nine p. m. I'm certain to be switched.

Anon., attributed to Gondola Hopperbottom Flatbox (probably a pseudonym), and written as an added verse to Strickland Gillilan's famous poem Song of the Freight Car *(part reproduced under 742.2, 742.3); quoted in* The Railway Library 1911, *1912.*

565.7 The most sensitive spot in the great business of the country is the railroad rate. This rate must be raised or lowered, not in obedience to a rigid statutory law, but in obedience to the varying conditions of trade and commerce.

Martin A Knapp, evidence to Senate Committee on Interstate Commerce, 29 January 1908.

565.8 There was never any better definition than that which was given many years ago by somebody and which has been used as a by-word and a reproach ever since, namely, 'what the traffic will bear.'

Edward P Ripley, president, AT&SF, 1910, ICC Hearings, *1910.*

565.9 That does not mean all the traffic will bear, it does not mean all that can be extorted or squeezed out of it, but what the traffic will bear having regard to the freest possible movement of commodities, the least possible burden on the producer and the consumer. The middleman can take care of himself.

Ibid.

565.10 The purpose of a railroad is to give good service at fair rates and to earn a dividend on its capital.

Ivy Lee, address to Transportation Club, Indianapolis, 11 May 1914, published in Human Nature and Railroads, *1915.*

566. READING ON THE TRAIN

566.1 I never travel without my diary. One should always have something sensational to read in the train.

Oscar Wilde, The Importance of Being Earnest, *1895.*

566.2 Give me three hours of absolute leisure, with nothing to do but read, and I instantly become almost incapable of the act. So it is always on railway journeys, and so it was that evening.

Arnold Bennett, The Grim Smile of the Five Towns, *1907.*

566.3 But do you suppose that I could continue with Wordsworth in the train? I could not. I stared out of the windows; I calculated the speed of the train by my watch; I thought of my future and my past; I drew forth my hopes, examined them, polished them, and put them back again; I forgave myself for my sins; and I dreamed of the exciting conquest of a beautiful and brilliant woman that I should one day achieve. In short, I did everything that men habitually do under such circumstances.

Ibid.

567. READING RAILROAD

567.1 In the old days nice young Philadelphia girls were permitted to travel unescorted on Mr. Roberts' railroad but not the Reading.

Attributed to Nathaniel Burt. Mr Roberts' Railroad was the Pennsylvania; George B Roberts was the fifth president of the PRR, from 1880 to 1897. Lucius Beebe, The Trains We Rode, *1965.*

REGULATION
See Government

568. SIR ROBERT REID

568.1 A hearty welcome to the new Chairman of the British Railways Board, "Neutron Bob" Reid—the one that takes out the people while leaving the buildings standing.

Anon., in reference to the neutron bomb of the same characteristic. Modern Railways, *November 1983.*

569. REMOTENESS

569.1 In 1914 Cornwall remained, as it had been in 1859, the remotest part of England. But in the course of half a century railways had changed the meaning of remoteness itself.

Jack Simmons, The Express Train and Other Railway Studies, *1994.*

570. RESPONSIBILITY

570.1 The Buffer Stops Here.

Sign on the desk of the Head of the National Railway Museum, 1992–94.

571. RETURN TICKETS

571.1 "Two return tickets, please."
"Where for?"
"Back here, of course."

Exchange recorded in Railway Humour, *John Aye, 1931. It was used in an episode of* To the Manor Born, *BBC television series.*

572. RISK

572.1 Periculum Privatum Utilitas Publica.

(*At private risk for public service.*)

Rev Daniel Peacock; suggested and adopted as the motto of the Stockton & Darlington Railway, and incorporated in the company seal, 1821.

573. RIVETS

573.1 What are all those buttons?

Raymond Loewy, industrial designer, asking about the rivets used to hold the casing of the first Pennsylvania Railroad class GG1 electric locomotive, 1934. Seen in Trains Magazine, *but date not noted.*

574 ROAD COMPETITION
See also Cars and Trains

574.1 I beseeche you to take order with Sir Thomas that we maie have libertie to bring coales down the railes by wagen, for our cariadges onely, and we will bring them down by raile ourselves, for Strelley cartway is fowle as few cariadges can passe.

Anon.; letter concerning the coal traffic from the estates of Sir Francis Willoughby of Wollaton, and the possibility of sending it over the rails of Sir Thomas Beaumont, 1610. Extract quoted by C F Dendy Marshall in A History of British Railways down to the Year 1830, *1938.*

574.2 That railways are inadequate appears
Indubitable now. For sixty years
Their comfort grew until the *train de luxe*
Arrived, arousing in conducted Cook's,
And other wholesale, tourists, an envious smart,
For here they recognised the perfect art
And science of land travel. Now we sing
A greater era, hail a happier Spring.
The motor-car reveals ineptitude
in railway-trains; and travellers conclude
The railway is archaic: strictly true,
Although the reason sounds as false as new:-
Railways are democratic, vulgar, laic;
And who can doubt Democracy's archaic?

John Davidson, poem, The Testament of Sir Simon Simplex concerning Automobilism, *1908.*

574.3 Not till the motor was the contrast plain,
Because the separate classes of the train
Deceived us with a choice of company;
And, when he liked, the tame celebrity,
The genius, man of wealth, aristocrat,
By means of tips through any journey sat
In cornered state; or with sufficient pelf
Could purchase a compartment for himself.
He rather would have deemed himself a snob
Than that the train could turn him into Mob,
Till automotion's privacy and pride
Exposed the grossness of the railway ride;
For 'twas freedom of the motor-car
That showed how tyrannous the railways are.

Ibid.

574.4 In the old days it was assumed that the motor transport vehicle would be the deadly enemy to the railway companies. Time has, however, shown that on the contrary the motor vehicle is a most valued ally to the railways both in the United Kingdom and abroad, for service in outlying districts, and for collection and delivery work in large towns.

Caption to photograph of lorry and locomotive, The Railway Magazine, *vol. XXXIX, 1916.*

574.5 Once aboard the lorry and the girl is mine.

Caption to Punch *cartoon, 8 October 1919, showing girl ("The Goods") accepting a lift from a lorry driver while a locomotive is left idle through strike action. Its origins are in the expression "Once aboard the lugger and the girl is mine," as noted in Eric Partridge's* Dictionary of Catchphrases, *1974.*

574.6 For goods traffic the motor lorry is an excellent jackal for the railway lion.

Pertinent Paragraphs, The Railway Magazine, *vol. XLV, 1919.*

574.7 We have not fully appraised the evolution from the ox-cart to the motor age. The automobile and motor truck have made greater inroads on railway revenues than the electric lines with their intimate appeal to the local community. There will never be a backward step in motor transportation. But we shall do better if we find a plan to coordinate this service with the railways rather than encourage destructive competition.

President Warren G Harding, quoted by John Droege, Railroad Terminals and Trains, *1925.*

574.8 In the new order of things the destructive war between road and rail would cease. It does not follow that there would be no duplication of road and rail services; public tastes and preferences must be taken into account. It would, however, be a duty of the Transport Board to build up a complete co-ordinated system and see that the most efficient means of transport are available for appropriate classes of traffic.

Herbert Morrison, putting the case for nationalization, British Transport at Britain's Service, *pamphlet, August 1938.*

574.9 No vision of a Sunday drivers' Utopia, four lanes wide with no traffic, can lure me away from taking a train ride, sensing the greatness of the railroad business and being a part of its activity.

Al Kalmbach, Railroad Panorama, *1944.*

574.10 For many years now, not, of course, because of any fault of the railway companies, but because of the advance of the internal combustion engine, transport by road has eaten into the guts of the railway companies, and it will go on.

Lord Brabazon, House of Lords debate on the 1947 Transport Bill, Hansard, 20 May 1947.

574.11 The bus, the truck, and the family auto made a lot of headway because of the negative, defensive "we-wuz-here-first" railroad approach to problems which required their concerted, affirmative, aggressive action.

R W F Schmidt, Trains Magazine, *February 1954.*

574.12 The fact that the railroads had it within their power to speed and clean up many services long before they did, particularly on runs which are still logically theirs because of inherent disadvantages of competitive surface or air transportation, must forever go down as the blackest of marks in the book.

Ibid.

574.13 We have said before that New York must learn to live with the automobile. But it must keep it under control. Every encouragement to ride the railroads, to ride the subway, to improve such riding, to extend the area served by rail, improves the traffic outlook. No sound future can be planned that doesn't recognize the railroad, surface or underground, as our real salvation from a worsening metropolitan traffic crisis.

Editorial Comment, New York Times, *supporting a recent study conducted by that newspaper into traffic problems in New York City, and quoted in* Trains Magazine, *August 1954.*

574.14 Trucks ride noisily over highways paid for and paid for by you and me, bearing the riches of the West over the mountains, while railroads maintain their expensive rights of way and then pay taxes on them.

Nathaniel Burt, The Perennial Philadelphians, *1963.*

574.15 It might pay to run the railways at a loss in order to prevent the incidence of an even greater cost which would arise elsewhere if the railways were closed.

Richard Beeching, The Reshaping of British Railways, *1963.*

574.16 It's quite apparent that it will overall be favourable, because where we push freight off rail, on to the roads, it will go on to the roads in the lightly populated, lightly trafficked areas of the country, where the roads are, at the moment, in general, very much underused.

Richard Beeching, of the effect of his railway closure proposals upon road congestion. Press Conference, 1963, BBC Sound Archive.

574.17 On the other hand, where we draw freight from roads to the railways, it will be from trunk roads on to the trunk routes of the railways, so that we shall be drawing substantial quantities from roads which are at the moment, tending towards, or are already actually congested.

Ibid.

574.18 The motor-car has become the symbol of hurry in a manner railways never were. Speed during a railway journey was expected and enjoyed; but also, by most passengers, was the leisurely and decorous procedure at the beginning and end.

Gilbert Thomas, Double Headed, *1963.*

574.19 We are blacktopping ourselves to death. Los Angeles now gives sixty percent of its land to highways, streets and parking lots. This is land removed from productive use and from the tax rolls. The cloverleaf has become our national flower.

Stuart Saunders, Pennsylvania Railroad, quoted by Peter Lyon, To Hell in a Day Coach, *1967.*

574.20 The British people had never admitted that the invention of the motor car and truck changed the railways from exploiters to exploited. By the time the nation had achieved the ability to bleed the railways white, the red corpuscles had already been killed by the speed and convenience of the internal combustion engine.

Stewart Joy, The Train that Ran Away, *1973.*

574.21 You know, we used to cuss the railroads. We used to say, 'Boy, we'll fix them; we'll ship by truck.' Well, we fixed them alright; we fixed them right out of business and now we'd like to have them back.

Harold Stark, whose manufacturing plant was on a branch line closed as a result of the liquidation of the Chicago Rock Island & Pacific; Quad City Times, *15 September 1980, quoted in* Trains Magazine, *January 1981.*

574.22 Co-ordination had little place in British transport policy until the latter part of the 1960s, and then only in local passenger movements.

John Johnson and Robert A Long, British Railways Engineering, 1948–1980, *1981.*

574.23 *Do not delay* this truck. It pays *taxes* to support railroads & people on welfare.

Anon., bumper sticker seen on a truck in South Dakota, 13 January 1983; reported by Robert Fryml in Trains Magazine, *May 1983.*

574.24 The railroad, which created the heavy industry that created the cheap car, also introduced the modern era, in which the trade-off between life, convenience, and speed has been considered a fair bargain.

Albro Martin, Railroads Triumphant, *1992.*

575. ROCKET

575.1 George Stephenson's "Rocket" had a glorious youth, a humiliating middle life and an honored and eternal old age.

Ada Louise Barrett, George Stephenson, *1948.*

576. ROSTERS (of Locomotives)

576.1 The jet age has yet to make the world so small that the locomotive historian can trace every engine from its builder serial to its ultimate disposition. Some engines vanish.

David P Morgan, caption to photograph, Trains Magazine, *October 1962.*

576.2 John Henry Cardinal Newman argued that knowledge is its own justification. There are those who go further, holding that relevance is a positive tarnish on a fact. From either point of view, how perfect is a steam roster! It won't get rid of slack, redefine the working day, or do anything else useful. All it can do is settle some arguments, preserve the record, and remind us how sensible we were to love steam locomotives so much.

George W Hilton, Trains Magazine, *February 1971.*

577. ROYAL GORGE

577.1 Nowhere else does man come closer to realization of the Infinite.

Anon., Denver & Rio Grande guidebook, quoted by David P Morgan in Trains Magazine, *April 1960.*

578. ROYAL TRAINS

578.1 Make it as like a yacht as possible!

King Edward VII, on being consulted upon the design of new carriages for the royal train to be built by the London & North Western Railway, c. 1902. Quoted by C Hamilton Ellis in Nineteenth Century Railway Carriages, *1949.*

579. RUGBY STATION

579.1 Rugby Station is so hideous as to be positively distinguished. It is in fact hardly a building at all; a conglomeration of shapeless erections would really be a more accurate phrase.

Canon Roger Lloyd, The Fascination of Railways, *1957.*

579.2 And quite where Rugby Central is
Does only Rugby know.

John Betjeman, poem, Great Central Railway Sheffield Victoria to Banbury.

580. RULES

580.1 Employees of the Company disapproving of these rules, or other regulations of the road, or not disposed to aid the Superintendent in carrying them out are requested not to remain in the employ of the road.

Notice in rulebook, Portland Saco and Portsmouth Railroad, 1861, reproduced in RLHS Bulletin 50, *October 1939.*

580.2 The General Superintendent, in preparing these rules, would appeal to the manliness and gentlemanly feeling of everyone in the Company's service under his direction.

H T Peake, South Carolina Railroad, Preface to Rulebook, c. 1860. Quoted by Robert C Black in Railroads of the Confederacy, *1952.*

580.3 They closed the door, but in a few moments came out, and the conductor told the brakeman to go and flag off the train which was coming about three miles behind. The brakeman replied, "I dare not—the Indians are all around here." To which the conductor replied, "D—n the Indians. Go and flag off that train, or by G-d she'll be into us."

Henry Morton Stanley, 8 August 1867, republished in My Early Travels and Adventures in America and Asia, *vol. I, 1895. Repeated in a slightly different form (without reference to Henry Morton Stanley and naming the conductor as William Kinney), by Edwin L Sabin,* Building the Pacific Railway, *1919.*

580.4 Every rule in the time-table has its history written in suffering and dollars.

James O Fagan, Confessions of a Railroad Signalman, *1908.*

580.5 The railroad didn't believe in lengthy formal training. They offered a two-week class that covered the book of rules, a three-hundred-page document with a dual purpose—to keep trains from running into one another and to prevent any situation in which the company might get sued.

Linda Niemann, of the Southern Pacific, Boomer, *1990.*

580.6 Rules of the road that you had to learn were mixed in with rules that you had to ignore in order to get the work done.

Ibid.

581. RUTLAND RAILROAD

581.1 So until the arrival of the modernization that the Rutland must someday have or die, the line will remain one for the canvas and the negative—fragrantly geared to the old and fine economy of manual block and 4-6-0, Stephenson valve motion and stations untouched by Loewy.

Wake Hoagland, (David P Morgan), Trains Magazine, *April 1950.*

582. S1 LOCOMOTIVE
(Pennsylvania Railroad)

582.1 She made a wonderful exhibit out at Flushing. They should have left her there.
>   *Anon. PRR executive, of the S1 exhibited at the New York World's Fair, 1939. Noted in* Trains Magazine, *May 1965.*

583. SACKS

583.1 The holes in the sacks, which are eaten by rats, which are not eaten by cats, are darned by twelve women, who are employed by the company.
>   *W M Acworth, of activity at the Midland Railway sack warehouse, Trent. There were "three or four hundred thousand" sacks, eight cats, and an unknown number of rats.* The Railways of England, *1899.*

584. SAFETY
See also Accidents, Derailment

584.1 And one of the great errors into which you have fallen, and into which the officers you have appointed have fallen, is this; in supposing that, with all your desire to watch the public safety, you can have anything like the desire that we have, anything approaching the deep feeling of interest that the companies have, I mean with reference to their pockets, in doing every thing which their particular circumstances require, and which their particular means enable them to do; because you must have reference to the quantity of traffic upon the line. I am quite sure that nothing on your part can approach the feeling of interest that the companies feel even in that particular department of providing for the public safety. Their all, their daily bread, depends entirely upon their not only carrying the public without accident, but upon their carrying them without their being afraid of accident; and a great deal of mischief has been done to the companies, not by what is really public feeling against us, for I doubt its existence, but by that which is called public feeling, by the great attention which has been drawn to our accidents by the newspapers.
>   *I K Brunel, Evidence before the Select Committee on Railways, para. 599, Parliamentary Papers, 22 March 1841.*

584.2 We have been, up to this point, very careless of our railway regulations. The first person of rank who is killed will put everything in order, and produce a code of the most careful rules. I hope it will not be one of the bench of bishops; but should it be so destined, let the burnt bishop—the unwilling Latimer—remember that, however painful gradual concoction by fire may be, his death will produce unspeakable benefit to the public. Even Sodor and Man will be better than nothing.
>   *Sydney Smith, letter to* Morning Chronicle, *1842.*

584.3 The road must be run safe first, and fast afterward.
>   *New York & Erie Railroad, Rulebook, 1854.*

584.4 ... grave and serious dangers may arise from too great a reliance on mechanical appliances as substitutes for manual labour.
>   *Sir Daniel Gooch, response to report on safety, 1872, quoted by E T MacDermot in* History of the Great Western Railway, vol. II, *1931.*

584.5 I say, my man, would you mind leaning toward the center on the curves?
>   *Caption to cartoon, probably 1870s, spoken by a nervous Englishman to a fat miner on a train winding its way through the mountains; quoted by Oscar Lewis in* The Big Four, *1938.*

584.6 It must, however, be borne in mind, that in endeavouring to guard against every danger, one can 'buy gold too dear'; for if every possible known precaution is to be taken, regardless of cost, it may not pay to work a railway at all.
>   *John Wolfe Barry, Railway Appliances, 1876.*

584.7 Improvements in railway safety appliances will be retarded unless a proper subordination of that which is mechanically possible to that which is financially rational, is recognised, and unless the money that is available for the purpose is spent in the best possible manner.
>   *Ibid.*

584.8 Though quite incomprehensible, there is at the same time something superb in such an exhibition of stolid conservatism. It is British. It is, however, open to but one description of argument, the *ultima ratio* of railroad logic. So long as luck averted the loss of life in railroad disasters, no occasion would ever have been seen for disturbing time-honored precautions or antiquated appliances.
>   *Charles Francis Adams, of the refusal by the management of the Great Western Railway to accept the need for continuous brakes or train communication after the accident at Shipton-on-Cherwell, near Oxford,* Notes on Railway Accidents, *1879.*

584.9 A practically irresistible force crashing through the busy hive of modern civilization at a wild rate of speed, going hither and thither, across highways and by-ways and along a path which is in itself a thoroughfare,—such an agency cannot be expected to work incessantly and yet never come into contact with the human frame.

Charles Francis Adams, Notes on Railroad Accidents, *1879.*

584.10 . . . to walk at will on any part of a railroad track is looked upon as a sort of prescriptive and inalienable right of every member of the community, irrespective of age, sex, color, or previous condition of servitude.
Ibid.

584.11 In this exacting age the traveling public are much more disposed to find fault with systems that do not provide against fatalities resulting from human fallibility than to commend perfection of appliances which annually save more lives than would be lost in a sanguinary war.

Angus Sinclair, Locomotive Engine Running and Management, *1902.*

584.12 The fad for main track derails at interlocking plants seems nearly to have ditched itself. We are realising that it is not necessary to kill an engineman who runs past a signal.

Charles Delano Hine, Letters from an Old Railway Official, *1904.*

584.13 The railway company is expected to provide not only transportation, but an immunity from risk far beyond what is obtained under other ordinary conditions of everyday life.

C B Byles, Signal Engineer, Lancashire & Yorkshire Railway, Great Western Railway Magazine, *February 1911.*

584.14 And yet the Nation goes placidly along discussing eugenics and health laws without number, and hardly a passing thought is given to the 5,284 trespassers killed on our railways in the single year ending June 30, 1912, and the continued killing day after day, year in and year out, that needs only a Nation's attention to eliminate.

Marcus A Dow, General Safety Agent, New York Central, The Outlook, *27 September 1913, reproduced in* The Railway Library, *1913, 1914.*

584.15 Railway travelling is safer than walking, riding, driving, than going up and down stairs . . . and even safer than eating, because it is a fact that more people choke themselves in England than are killed on all the railways of the United Kingdom.

Sir Edward Watkin, source unknown, quoted in The Story of the Cambrian, *C P Gasquoine, 1922.*

584.16 It appears that a few drunken people, as well as certain others of an experimental turn of mind, had taken it into their heads to step out of trains when in motion and had provided employment for coroners and juries, who had held the railway companies responsible for the actions of these erratic passengers. As a result the Great Western Railway Company had ordered that the doors of all carriages should be kept locked between stations.

Hesketh Pearson, The Smith of Smiths, *1932.*

584.17 Never get into a place you can't get out of in a hurry.

Attributed to O M Hansen, signal foreman, Boston & Maine Railroad, c. 1949.

584.18 This technician's mackintosh could be dangerous.

Comment, concerning the dangers of loose clothing for permanent way workers, in film Safety on the Track, *made by British Transport Films, 1954.*

584.19 What's more, loose clothes might get caught up by passing trains, which would be very unpleasant.
Ibid.

584.20 Cap, the dime just made it safe.

Anon. union official, to Bill Brosnan, of the walkie-talkie radio just issued to Southern Railway men. Their objections to it as unsafe were assuaged by an extra dime an hour. Told by Charles O Morgret, Brosnan, *1966.*

584.21 Almost the only advantage that railroad passenger service now has to offer in competition with motor and air travel is an incomparably better safety record. This factor is of negligible value.

Howard Hosmer, ICC Report, 1958, quoted in Railroad History No 184, *Spring 2001.*

584.22 Say, lad, will you fetch me a bucket of red oil for a red tail lamp?

Anon. railwayman, The Ballad of John Axon, *broadcast 2 July 1958.*

584.23 To annihilate the unprofitable, but extremely safe, railway system by increasing the lethal propensities of our ridiculous and costly road system will transfer the price to be paid in money to an account which will be paid in blood.

Leonard Fisher, Mayor of Chard, late April 1963, quoted by Lord Stonham in Lords debate on the Beeching proposals, Hansard, *vol. 249, 2 May 1963.*

584.24 Safety—then punctuality.

Vic Perkin, advice to Nigel Harris when training as a volunteer railwayman, 1973.

584.25 And no one was ready to defend the dismal safety record resulting from flimsy track, blind curves, unprotected road crossings, unfenced tracks, or wobbly trestles. Yet financiers, not technicians, controlled America's business, and they found expediency and profit in wooden railroading. These were men conditioned to high risk; running fast over a shaky track seemed no more perilous than selling short in the rough-and-tumble arena of the Wall Street trader. No believer in Darwin would doubt their logic.

John H White Jr, of early American railroads built extensively with wood. Chapter 6 of Material Culture of the Wooden Age, *1981.*

584.26 Get the money wrong and you lose your job; get safety wrong and you go to jail.

Anon., warning to managers of British Rail as the privatization process started, mid 1990s. Recalled by Graham Nicholas.

584.27 Attention all passengers: We regret to inform you that the next train you board from any platform in Britain may misread a badly-sited signal run a red light and kill you. In the old days British Rail would apologise for any inconvenience this may cause. But now nobody apologises because none of the people who make millions from our privatised rail system and none of the politicians entrusted with guaranteeing your safety is prepared to take the blame. Sorry!

Front page display, in the form of a dot-matrix departure sign, following the derailment at Ladbroke Grove, The Mirror, *20 October 1999.*

584.28 An industry that was already on its knees is being kicked to the floor. As a result, hundreds more people will die.

Simon Jenkins, of the decision to prosecute railwaymen as a result of the accident at Hatfield, October 2000; The Times, *11 July 2003.*

585. SAFETY VALVES

585.1 It so happened that this functionary while in the discharge of his duties was much annoyed by the escape of steam from the safety valve, and, not having made himself complete master of the principles underlying the use of steam as a source of power, he took advantage of the temporary absence of the engineer in charge to effect a radical remedy to this cause of annoyance. He not only fastened down the valve lever, but further made the thing perfectly sure by sitting upon it. The consequences were hardly less disastrous to the *Best Friend* than to the chattel fireman. Neither were of much further practical use.

Charles Francis Adams, of a boiler explosion which befell the Best Friend of Charleston, *within* Railroads: Their Origins and Problems, *1878.*

586. ST PANCRAS STATION (London)

586.1 It is too beautiful!

Attributed to Sir Gilbert Scott, architect of the station building and hotel, by John Betjeman in Ghastly Good Taste, *1933.*

586.2 The harmony that is apparent in all parts of this great work is a practical proof, if any were wanted, of the advantages that accrue from the united working of architects and engineers.

Comment, Building News 16, *1869.*

586.3 This monument of confectionery is a fair specimen of the result of the competition among architects for the approval of judges whom they know to be incompetent.

John T Emmet, The State of English Architecture, *April 1872, quoted in* Culture and Society in Britain 1850–1890, *1986.*

586.4 We may regret the plainness of the Great Northern station, but it is better it should remain as it is, rather than that it should be disfigured with incongruous medievalism like the station of the Midland Railway.

James Fergusson, quoted by Jack Simmons in St Pancras Station, *1968.*

586.5 St George for England!
St Pancras for Scotland!

Exchange in musical comedy, Funny Face, *1928. A similar chant, King George for England, King's Cross for Scotland, current mid-twentieth century, has not been traced to such a source.*

586.6 He found himself repeating, with the solemn relish of one who achieves nonsense, "Pancrastination is the thief of time."

J B Priestley, of Adam Stewart, who has time to kill before his train leaves St Pancras, in Adam in Moonshine, *1927.*

586.7 St Pancras, the most canonical of all our stations, seemed to rebuke his levity. Indignant puffs of smoke and steam, sudden red glares of anger, ascended to the great arched roof. The locomotives grunted and wheezed like outraged sacristans. The thin high voices of the newsboys ran together into a protesting chorus of virgins and elders.

Ibid.

586.8 St Pancras works: leave it alone.

Paul Hodgkinson, letter to The Times, *12 September 1966.*

586.9 C'est magnifique—mais ce n'est pas la gare!

Brian Haresnape paraphrasing General Pierre Bosquet; caption to photograph of ticket hall at St Pancras in 1876, Modern Railways, *January 1977.*

586.10 I suggest London Agincourt: "Travel like an arrow to the Heart of France."

David Webb. Suggestion in a debate on a new name for St Pancras station in its new role as terminus for Eurostar services to France and elsewhere; in response to comments that Frenchmen are unhappy to arrive at the first such terminus at Waterloo. Letter to The Times *17 March 2004.*

587. ST PAUL MINNEAPOLIS & MANITOBA RAILROAD

587.1 ... a snug property, with very little, if any, waste mileages.

Charles Perkins, Chicago Burlington & Northern, letter to H D Minot, summer 1885, quoted by Julius Grodinsky in Trans-Continental Railway Strategy 1869–1893, *1962.*

588. SALUDA

588.1 With short explosive thunders, ripping the lilac night, 36 began to climb Saluda. She bucked helplessly like a goat, her wheels spun furiously on the rails, Tom Cline stared seriously down into the milky boiling creek, and waited. She slipped, spun, held, ploughed slowly up, like a straining mule, into the dark. Content, he leaned far out the cab and looked: the starlight glimmered faintly on the rails. He ate a thick sandwich of cold buttered fried meat, tearing it raggedly and glueily staining it under his big black fingers. There was a smell of dogwood and laurel in the cool slow passage of the world.

Thomas Wolfe, Look Homeward, Angel, *1929. Saluda is the steepest grade on a Class One railroad in America, at 4.5%.*

589. SCHOOLING

589.1 Naturally I chose Colorado to go to school as it was near the Denver & Rio Grande Western narrow gauge lines.

Donald Duke, Introduction to American Narrow Gauge, *1978.*

590. SCIENCE

590.1 But I confess *Light* has at length *shone* from the *North*.

William James, acknowledging the superiority of malleable iron rails as demonstrated on the Bedlington; letter 22 June 1821.

590.2 May the triumph of science prove the blessing of the people.

Banner displayed at the opening of the Leicester & Swannington Railway, 17 July 1832. Quoted in The History of the Midland Railway, *Clement Stretton, 1901.*

591. SCOTLAND

591.1 There <u>are</u> two ways about it, and they are both LMS.

London Midland & Scottish Railway poster advertising services from Euston and St Pancras to Scotland, c. 1925.

591.2 You tak' the high road and I'll tak' the railroad, and I'll be in Scotland 'afore ye.

British Railways poster, c. 1965.

592. SEASON TICKETS & SEASON TICKET HOLDERS

592.1 Persons wishing to commute for transportation at the present reduced price can enter into such an agreement by calling on an agent of the company.

Advertisement, Paterson & Hudson River Railroad, 28 November 1843, quoted by Jerome Beatty Jr, Show Me the Way to go Home, *1959.*

592.2 A season ticket is, after all, but a small bait to tempt a man to live in a malodorous swamp.

A P Crouch, Silvertown, *1900.*

592.3 The simple, but imperative, requirements of the commuter, with the newspaper under one arm and a package of garden seed under the other, is satisfied when he has an ample concourse through which he can race to a train, whose schedule he knows to a second. He buys a commutation ticket once or twice a month; he has no baggage to carry or check; and, if the railroad will furnish him the means of coming home to dinner on time comfortably, he will be pretty well pleased.

Samuel O Dunn, Scribner's Magazine, *October 1912.*

592.4 The 2:15 train is a good one to take, for it affords an interesting opportunity to observe those who may be called sub-commuters: the people who come in town in the morning, like honest working folk, but get back to the country after lunch. These, of course, are only half-breed commuters. They are the silver-chevron suburbanites, deserving not the true golden stripes of those who moil all day.

Christopher Morley, The Paoli Local, *within* Travels in Philadelphia, *1920.*

592.5 They are mere cherubim and seraphim, not archangels. Stern and grizzled veterans, who go home on the Hjw6:05 ("H" Will not run New Year's, Memorial, Independence, Thanksgiving and Christmas Days; "j" will not run Saturdays June to Sept. 27, both inclusive; "w" No baggage service), speak of them scornfully as "Sam Brown belt commuters."

Ibid.

592.6 The Main Line commuters, it is true, seem to stroll trainward like a breed apart, with the air of leisurely conquest and assurance. They have the bearing of veterans who have conquered the devils of transportation and hold them in leash.

Christopher Morley, Broad Street Station, *within* Travels in Philadelphia, *1920.*

592.7 The short-distance daily rider on the railroad was a new creature in the 'sixties; by the eighties, however, he was to become a recognized fixture upon it; a citizen forever to be patronised and never, never to be scorned.

Edward Hungerford, of the growth of commuter traffic

from the 1860s, The Story of the Baltimore & Ohio, *vol. 2, 1928.*

592.8 If anyone thinks that commuting is not travelling, let him try 5,400 trips between Fairview Manor and town—at his own expense, of course. I think he will find out very quickly that it is not just a question of being dragged back and forth like a tennis-court roller.
Edward Streeter, Daily—Except Sundays, *1938.*

592.9 *Every commuter should have a regular train.* This is essential to an ordered life. Select one that will get you to the office earlier than necessary. Having done so make up your mind that you are going to miss it regularly. Nothing is more demoralizing than to *catch* it one morning and *miss* it the next. It upsets your digestion and your stenographer. By having a regular train which you *never* catch, however, you will be leading an orderly existence and an independent life at the same time.
Ibid.

592.10 A word of practical advice in case you ever want to venture out on the platform when a Rock Island commutation train pulls in just before working hours: *Don't!* I tried it once and it was only by the grace of God that I escaped being knocked down and trampled on. Your commuter may be a gentleman, scholar and churchgoer, but when the 8.15 is late and the iron gates of the Rock Island's suburban cars are opened he's an animal let out of a cage. His one object is beating the time-clock, and never let anyone tell you otherwise. I know.
Frank P Donovan Jr, Trains Magazine, *August 1948.*

592.11 In every railroad terminal there is a station bar whose function is to see that the commuter arrives home in a state ranging from slightly moistened to downright sodden.
Caskie Stinnett, Will Not Run Feb. 22nd., *1949.*

592.12 Standing as it does, at a point where one world ends and another begins, the temptation to use it as a sort of emotional punctuation—the application of a period to a harrowing day—becomes almost irresistible. For this reason, the station bar must be integrated into the life of the commuter, and this is not easy.
Ibid.

592.13 Commuters can be regarded as a group only for the period of time when they are on the train, or when they are bunched together at regular intervals along some platform, hoping that the engineer will stop at exactly the same location as yesterday, so that they will be quick enough to get on board and grab a seat up front, over the wheels or not over the wheels, on the shady side, next to the window or next to the aisle, with somebody or far away from somebody—or maybe just a plain, ordinary seat.
Jerome Beatty Jr, Show Me the Way to go Home, *1959.*

592.14 They represent a cross-section of the creative brains that converge on the metropolitan area. Commuters are executives, artists, proprietors, vice-presidents, stockholders, directors, gourmets, consultants, authors, publishers, importers, toy manufacturers, and college professors.
Ibid.

592.15 The peculiar fact is that the railroads have little, if any, conception of all this. It usually hits them with a stunning blow when they find that some newspaperman or commentator, who is screaming about a lapse in efficiency, turns out to be a commuter. Generally railroad officials don't associate their commuter trains either with people, or with individuals.
Ibid.

592.16 Mr Brown goes off to Town on the 8.21,
But he comes home each evening
And he's ready with his gun.
Jimmy Perry, song, Who do you think you are kidding, Mr Hitler? *for BBC television show* Dad's Army, *1968.*

592.17 The commuter—*l'homme moyen de notre epoque.* The anti-hero of our age. More than the soldier, the nuclear physicist, the political prisoner or the starving child, he indicates where we've gone wrong.
Roger Green, Notes from Overground, *1984.*

592.18 There was a young lady of Staines,
Who loved to be ravaged in trains.
She was given a Season
for no other reason,
By a man with more urges than brains.
Anon. verse quoted by Gerard Fiennes in Fiennes on Rails, *1986.*

593. SECOND CLASS
See also Class Distinction, First Class, Third Class

593.1 . . . an unfriendly act, calculated to do damage to railway property.
Sir Edward Watkin, of the abolition of second class by the Midland Railway, 1 January 1875. Quoted by George Dow in Great Central, *vol. II, 1962.*

594. SELSEY

594.1 A Healthy Rural Village.
Poster, London, Brighton & South Coast Railway, 1911.

595. SETTLE-CARLISLE RAILWAY

595.1 The works are extremely substantial; in fact, there is not a more perfect line of railway in the world.
John Ellis, Chairman, Midland Railway, half-yearly shareholders' meeting, 22 February 1876.

595.2 There's little reason for the trains to stop. The area is populated by more sheep than people. If the east coast route is the NYC and the west coast route is the Pennsylvania, the Settle & Carlisle is the Erie.

George H Drury, Trains Magazine, *drawing an analogy with the principal routes between New York and Chicago, December 1997.*

596. SHAREHOLDERS

See also Investment in Railways

596.1 Tell me, oh, tell me dearest Albert, have *you* any railway shares?

Anon., Queen Victoria to Prince Albert, Punch *cartoon, 1845.*

596.2 Every man of the present day is a holder of shares in a railway; that is, he has got some pieces of paper called scrip, entitling him to a certain proportionate part of a blue, red or yellow line drawn across a map, and designated a railway. If the coloured scratch runs from south to north, it is generally called a Trunk-line; if it 'turns about and wheels about' in all directions, leading to nowhere on its own account, but interfering with every railway that does, ten to one but it is a Grand Junction; and if it lies at full length along the shore, it is a Coast Line.

Quoted from Cruikshank's Table Book *by F S Williams* in Our Iron Roads, *1852.*

596.3 Here are the shareholders diffused throughout the whole kingdom, in towns and country houses; knowing nothing of each other, and too remote to co-operate were they acquainted. Very few of them see a railway journal; not many a daily one; and scarcely any know much of railway politics. Necessarily a fluctuating body, only a small number are familiar with the company's history—its acts, engagements, policy, management. A great proportion are incompetent to judge of the questions that come before them, and lack decision to act out such judgements as they may form—executors who do not like to take steps involving much responsibility; trustees fearful of interfering with the property under their care, lest possible loss should entail a lawsuit; widows who have never in their lives acted for themselves in any affair of moment; maiden ladies, alike nervous and innocent of all business knowledge; clergymen whose daily discipline has been little calculated to make them acute men of the world; retired tradesmen whose retail transactions have given them small ability for grasping large considerations; servants possessed of accumulated savings and cramped notions; with sundry others of like helpless character—all of them rendered more or less conservative by ignorance or timidity, and proportionately inclined to support those in authority.

Herbert Spencer, Railway Morals and Railway Policy, *1855.*

596.4 The speech of the Chairman of the Great Western Railway line was one long homily on the virtue of patience and the consolation of hope. The shareholders were always exhorted to forget the past, be content with the present, and hope for the future.

"A vivacious shareholder," quoted as speaking at a meeting sixty years before (thus referring to a meeting c. 1871), Great Western Railway Magazine, *January 1932.*

596.5 The railway shareholder is certainly not as other men. There is something difficult to define, yet something distinctive about him: an air of solid prosperity that is superior to foppery of dress and is not concealed by shabbiness of apparel; a dignified confidence and a comfortable gravity that stamp him a man in possession of wealth. It is possible that he has a large holding; that he is accustomed to bank notes and the crackle of scrip, and that he invariably carries a cheque-book. Such a man cannot do otherwise than command respect, especially in this country where the worship of money runs other forms of devotion a very close race.

J Pendleton, Our Railways, *1896.*

597. SHAVING ON A TRAIN

597.1 I once tried it, but I shall not do so again, unless I have a good supply of sticking plaster, or wish to pose as a German student.

J S Stuart, A Railway Journey Forty Years Ago, *within* The Railway Magazine, *vol. V, 1899.*

597.2 The legendary smoothness of *The Century's* passage over the Water Level Route encouraged shaves among even the most timid who normally viewed a straight razor at eighty miles an hour with distrust.

Lucius Beebe, 20th Century, *1962.*

598. SHAY LOCOMOTIVES

598.1 They were powerful locomotives but their top speed was about 8 mph—uphill or down!

M I Dunn, of Shays on the Chesapeake & Ohio Railway, 1910. Trains Magazine, *August 1965.*

599. SHEFFIELD BUXTON & LIVERPOOL RAILWAY

599.1 A veritable High Peak Line, evidently conceived in pique.

Sir Edward Watkin, of a railway proposed by the London & North Western Railway in 1866 in reprisal for action against its interests by the Manchester Sheffield & Lincolnshire Railway, 1866. The SB&L was not built. Quoted by George Dow in Great Central, *vol. II, 1962.*

600. SHILDON, England

600.1 Shildon, if not the birthplace of the locomotive, has been its nursery.

John Dixon, quoted in Railway Centenary Supplement *to* The Locomotive Magazine, *30 June 1925.*

601. SHIPPING

601.1 If British Railways ruin coastal shipping how shall we ever again take a British Army off the Beaches at Dunkirk?

Anon. Royal Navy Admiral, "roaring" at Gerard Fiennes. Quoted in paper "What is wrong with British Railways?" *Gerard Fiennes, undated.*

602. PHILIP SHIRLEY

602.1 He was like an emetic: the first dose works wonders, but problems arise when the medicine is taken repeatedly.

Terry Gourvish, of probably the most unpopular member of the British Railways Board under Richard Beeching. British Railways 1948–1973, *1986.*

603. SHORTLINES

See also Branch Lines

603.1 The Shortline is so delicately integrated to the region it serves or traverses, touches so closely the lives of its countryside, that it ranks in importance of function with those of the banker, the leading merchant and the parish man of the cloth.

Lucius Beebe, Mixed Train Daily, *1947.*

604. SHUNTING and SWITCHING

604.1 The greater part of the shunting is performed by six-coupled tank engines, but four-coupled tender engines can also be seen doing this work, although they appear to do it under protest.

H Schloesser and W E Napier, Railway Goods Depots, *within* The Railway Magazine, vol. VIII, *1901.*

604.2 Now when you get to these big towns if you look out of the window, you'll notice some things called sidings, on which a quite inexplicable thing called shunting takes place. This of course is just an excuse on the part of the railway company to provide homes for old engines. You see there comes a time when an engine when it ceases to be virile and hearty, and it just becomes long-funnelled and tiresome, so it's put on one of these sidings and given a lot of trucks to play with.

Reginald Gardiner, monologue, Trains, *1934.*

604.3 And if you happen to have a bedroom anywhere near the siding you'll find that you're woken up very early in the morning by the trucks taking umbrage. And it

goes like this: Huh-huh-huh-huh-huh-huh-be-big-ping-ting-per-pong-tung-ding-ding-bom-bom-ting-ting-ting-pong-pung-pum to the far end. Then of course there's another old gentleman the other end who decides to do the same thing, and biffs them all back again. And in this way the railway company is carried on.

Ibid.

604.4 He runs a few yards by the side of the wagon, thrusts his brakestick into its anatomy, and thus prevents it from being indiscreet.

Bernard Darwin, of shunters (brakemen) at Feltham hump yard, War on the Line, *1946.*

604.5 Couldn't switch a giraffe out of a herd of sheep.

Anon., concerning switching crews on the Long Island Rail Road, noted by Gene Collora, Trains Magazine, *August 2001.*

605. SIGNAL BOXES

605.1 Bright lights overhead; the glare of large electric lamps outside; the flashing of hand-signals, red, white and green, on every hand below; the snort of passing engines, their deafening whistles, far and near, and of every variety; whiffs of nauseous smoke; shouting from the shunters below; and as a constant accompaniment to all that, the ringing of bells, the snap and clang, the rattle and wrench of numberless lever handles.

James Scott, Railway Romance and other Essays, *1913.*

605.2 This high house contains many levers, standing in thick, shining ranks. It perfectly resembles an organ in some great church, if it were not that these rows of numbered and indexed handles typify something more acutely human than does a keyboard. It requires four men to play this organ-like thing, and the strains never cease. Night and day, day and night, these four men are walking to and fro, from this lever to that lever, and under their hands the great machine raises its great hymn of a world at work, the fall and rise of signals and the clicking swing of switches.

Stephen Crane, The Scotch Express, *within* Men, Women and Boats, *1921.*

605.3 The old manual interlock, in its day
Of dominating the railroad diamond, stood
Neatly nautical here as a pilot-house
Over cross-currents of steel. Like the watch
At his wheel, full body thrust was what
The towerman gave to throw the irons
Controlling the click and lock of the rails:
From the mammoth surge of train
Just under the windows, the engineer waved.
Then the knit of linkage to be undone
Again, the diamond reset with a straining
Of levers. Quiet settled. The Hamilton

Ticked. The towerman idled, savored a smoke.
What his eye saw, and remembered, is now
Wholly lights on a board. Steps up
To the tower reach rotted wood,
Rusty irons, a window unwedged by weather,
Creaking darkness, the long trick over.
Nancy G Westerfield, verse, The Interlocking Tower,
within Trains Magazine, *April 1995.*

606. SIGNALMEN

606.1 The sun shines brightly, though it is a cold sun, this
piercing day; and when the Birmingham tunnel disgorges
us into the frosty air we find the pointsman housed in no
mere box, but in a resplendent pavilion, all bejewelled
with dazzling icicles, at least a yard long. A radiant pointsman he should be, we think, invested by fairies with a
dress of rainbow hues, and going round and round in
some gorgeously playful manner on a gold and silver
pivot. But, he has changed neither his stout great coat, nor
his stiff hat, nor his stiff attitude of watch; and as (like the
ghostly dagger of Macbeth) he marshals us the way that
we were going, we observe him to be a mortal with a red
face—red, in part from a seasonable joviality of spirit, and
in part from frost and wind—with the encrusted snow
dropping silently off his outstretched arm.
Charles Dickens, Household Words, *21 January 1854.*

606.2 The signalman in charge of the traffic should be
placed so that he can easily be approached or spoken to
by the station-master, or other authorised persons; but it
should be rendered difficult for anyone to go into the signal cabin, except in open view; and the cabin should be
glazed all round, so that not only can the signalman see
outwards, but the station-master can see inwards, and
detect any gossiping or idling.
John Wolfe Barry, Railway Appliances, *1887.*

606.3 Someone speaks to him, and the sentence opens
with "tonk . . . tonk . . . tink, tink, tink," and continues in
that strain.
Dell Leigh, On The Line, *1928.*

606.4 The lovely lights of evening, the grey smudge of
Edinburgh Castle against a background of deepening
blue and crimson are nothing to him. He does not see
them. All that concerns him, and deeply, is the thought
that the 6.3 Dalmeny is half a minute late.
Ibid.

607. SIGNALS

607.1 A Signal Ball will be seen at the entrance to Reading
Station when the Line is right for the train to go in. If the
Ball is not visible the Train must not pass it.
Within Regulations for the Engines working the
Trains on and after the 30th March 1840, *Great Western*

Railway, quoted in History of the Great Western Railway,
E T MacDermot, 1927.

607.2 The fan or arm of the auxiliary signal at Ched-
dington, Leighton, Roade, Blisworth and Weedon are
repainting to a yellow colour which will be more discernible than green, the ground colour is obliged to be nearly
red, the signal will, however, continue to be shown for
the assistance of the drivers, although it is not intended
they should stop thereat but come on as heretofore as far
as the stationery post, when the policeman will tell them
why the train has been stopped.
*Notice, London & North Western Railway, 24 February
1846, quoted by G P Neele in* Railway Reminiscences, *1904.*

607.3 It is to be hoped that the drivers understood it.
Ibid. G P Neele's comment on the above.

607.4 Unknown languages in the air, conspiring in red,
green and white characters.
Charles Dickens, Mugby Junction, *published in* All the
Year Round, *1866.*

607.5 It is astonishing to see the blind faith the English
engine-driver places in his block signals.
Edward Bates Dorsey, English and American Railroads
Compared, *1887.*

607.6 There is something about signals very fascinating to
the inventive faculty.
*Anon., Catalog of Interlocking and Railroad Signaling
Appliances, Johnson Railroad Signaling Co., 1889. Quoted
by Steven W Usselman in* Regulating Railroad Innovation, *2002.*

607.7 . . . some of the lights are almost equalled in bril-
liance by the glow worms on the embankments.
*J W Walker, of the new signal lamps that burned for a
week at a time,* Railway Notes, No. 23, *November 1910.*

607.8 Oft when I feel my engine swerve,
As o'er strange rails we fare,
I strain my eyes around the curve
For what awaits us there.
When swift and free she carries me
Through yards unknown at night,
I look along the line to see
That all the lamps are white.
Cy Warman, poem, Will the Lights be White?, *verse 1,
reflecting the former practice of using white lights for the
All Clear.*

607.9 Practically the only purpose a railroad has is to give
train service to the public and its industries; and whatever will facilitate and expedite train movements to the
best advantage to secure this purpose with a reasonable
expenditure of capital will work to the best interests of
the public generally.
Everett Edgar King, Railway Signalling, *1921.*

607.10 The signal's down! Oh, can we see the train?
From childhood days, I hear that plea again!
No 'Upper quadrants' then, the signals *fell*
The coming benefaction to foretell.
   Henry Maxwell, poem, Ave Atque Vale, in A Railway
Rubaiyyat, 1968.

608. THE SILVER PHANTOM

608.1 It lay, a quarter of a mile of silver carriages, quietly
in the dusk of the underground station.
   Ian Fleming, Live and Let Die, 1954.

609. SINGLE LINE WORKING

609.1 It is now universally admitted that when a sufficient
object is to be attained, arrangements may easily be made
by which a short piece of single line can be worked without any appreciable inconvenience . . .
   I K Brunel, justifying the single track of the Saltash
Bridge on an otherwise double line, Report to Board of
Trade, 1852.

610. SIZE

610.1 Multum in parvo.
(*Much in little.*)
   Adopted by A B MacLeod as the motto of the Romney
Hythe & Dymchurch Railway, and incorporated in the
heraldic device, c. 1946.

611. SKEGNESS, England

611.1 Skegness is so bracing.
   Advertisement, Great Northern Railway, 1908.

612. SLAVES

612.1 Lincoln freed the slaves, and the Pullman Company
hired 'em.
   George Anderson Smock, recalled by George Henry
Smock, a grandson, in Those Pullman Blues, *David D*
Perata, 1996.

613. SLEEPING ON TRAINS

613.1 A sleeper is one who sleeps. A sleeper is that in
which the sleeper sleeps. A sleeper is that on which the
sleeper which carries the sleeper while he sleeps runs.
Therefore, while the sleeper sleeps in the sleeper the
sleeper carries the sleeper over the sleeper under the
sleeper, until the sleeper which carries the sleeper jumps
off the sleeper and wakes the sleeper in the sleeper by
striking the sleeper under the sleeper, and there is no
longer any sleeper sleeping in the sleeper of the sleeper.
   Doncaster schoolboy, asked to write an essay on a
sleeper. Quoted in The Railway Magazine, vol. III, *1898.*

613.2 It has been found to be within the powers of human
ingenuity to avoid an examination of tickets at four
o'clock in the morning, and on one line it is reported
that the porters have received instructions to announce
the names of the stations in a considerate whisper.
   W J Gordon, Everyday Life on the Railroad, *1898.*

613.3 When Phoebe Snow went East one night
Upon the road of Anthracite,
She'd dined on lobster ere she left,
And so of peaceful sleep bereft
She squirmed and tossed the whole night through,
As restless sleepers always do;
And ere her "tummy" quit a hurtin',
She'd poked her "tootsies" thro the curtain.
Then sweetest dreams came to her while
Soft breezes fanned them in the aisle.

Next morn the porter found them there
And tucked them in with zealous care;
Said he, "I begs yo pardon, Miss,
De Lackawanna can't stand dis,
We ain't objectin to deir style,
You shore got Trilby beat a mile,
But dem dat travels on dis line
Is jest de most pertickler kine,
An' throo de aisle dey'd 'fuse to go,
Blockaded by TWO FEET OF SNOW!"
   K M Chapman; postcard legend, parodying the popular
Phoebe Snow advertisements on the Delaware Lacka-
wanna & Western Railroad, during the advertising cam-
paign featuring her and her spotless clothing, 1907.

613.4 Please take off your boots before retiring.
   Request printed on Pullman company tickets, as
recorded by Stewart H Holbrook, The Story of American
Railroading, *1947.*

613.5 Sleeping-cars are for the strange beings who love
not the act of travelling. Them I should spurn even if I
could not sleep a wink in an ordinary compartment.
I would liefer forfeit sleep than the consciousness of
travelling.
   Max Beerbohm, Ichabod, *1923.*

613.6 All through the night the vibration of the train
keeps one-third of me awake, while the other two parts
of me profoundly slumber. Whenever the train stops,
and the vibration ceases, the one-third of me falls asleep,
and the other two parts stir.
   Ibid.

613.7 The long line of coaches stand with shrouded windows all down the platform. The train has rather a furtive
air, as if it is about to creep away upon some unlawful occasion. A faint feeling of adventure grips you, as if you too
were about to do something naughty, and therefore nice.
   Dell Leigh, On The Line, *1928.*

613.8 QUIET is requested for the benefit of those who have retired.
Sign seen universally in Pullman sleeping cars, United States of America.

613.9 Sleep like a kitten.
Lionel C Probert, legend on posters that launched Chessie, the cat featured in numerous advertisements for sleeping car and other services; Chesapeake & Ohio Railway, 1933.

613.10 Despite these precautions, thousands of Americans kept to the end their resolve never to go to bed on a railroad train. Other thousands, forced to that extremity by some emergency, lay broad awake until dawn, the ladies removing only their hats and gloves, and keeping foot-long hatpins, bought for the occasion, close at hand.
Oscar Lewis, of the patrols of railway train crews in Pullman sleepers; The Big Four, 1938.

613.11 The up-to-the-minute single bedroom, most in demand of all Pullman sleeping arrangements, is unquestionably a vast improvement over the green carpeted ladder, the contortionist poses for trouser removal, and hopeless feeling of irrevocable isolation characteristic of the upper berth of tradition. But the traveler who has not groped blindly for his morning boots, has not encountered the curiosa of masculine undergarments on view in the matutinal men's wash room and finally emerged into the breakfast car, scarred with razor nicks staunched with fragments of toilet paper, but unquestionably victorious over a number of the major problems of civilization, has not seen life.
Lucius Beebe, Trains in Transition, 1941.

613.12 The nun, I suspected, felt quite at home in a train, since I imagined monastic life to be a little like existence in a sleeping-car: the embryonic warmth and intimacy, the isolation, the feeling of walking about in your own mind.
Peter Ustinov, God and the State Railways, within The Frontiers of the Sea, 1966.

613.13 I washed, brushed my teeth, and dressed. Afterward I still felt as though I hadn't washed or brushed my teeth, or ever undressed. That's sleeping cars.
Ibid.

613.14 Oh it's very nice when you have found your little den
With your name written upon the door.
And the berth is neat with a newly folded sheet
And there's not a speck of dust on the floor.
There is every sort of light—you can make it dark or bright;
There's a button that you turn to make a breeze.

There's a funny little basin you're suppose to wash your face in
And a crank to shut the window if you sneeze.
Then the guard looks in politely and will ask you very brightly
'Do you like your morning tea weak or strong?'
T S Eliot, poem, Skimbleshanks the Railway Cat. Complete Poems and Plays, 1969.

613.15 I don't like this, being carried sideways
Through the night. I feel wrong and helpless—like
A timber broadside in a fast stream.

Such a way of moving may suit
That odd snake the sidewinder
In Arizona: but not me in Perthshire.

I feel at rightangles to everything,
a crossgrain in existence—it scrapes
the top of my head, my footsoles.

To forget outside is no help either—
then I become a blockage
in the long gut of the train.

I try to think I am a through-the-looking glass
mountaineer bivouacked
on a ledge five feet high.

It's no good. I go sidelong.
I rock sideways . . . . I draw in my feet
to let Aviemore pass.
Norman MacCaig, poem, Sleeping Compartment.

613.16 Those Great Western cars, though elegant and spacious, were not popular with travellers to the West, who objected strongly to sleeping in close-packed rows like recumbent cod on a fishmonger's slab . . .
C Hamilton Ellis, of broad-gauge sleeping carriages with four berths arranged side-by-side, Nineteenth Century Railway Carriages, 1949.

613.17 One cannot explain why an ornate dormitory with 16 men and women struggling out of their clothes behind green repp buttoned curtains was tolerated by a society which delicately banned beds from illustrations to furniture catalogues, but it was so.
C Hamilton Ellis, of the Pullman style of sleeping carriages used by the Midland Railway. Ibid.

613.18 He turned off his own light and lay thinking of Solitaire and listening to the steady gallop of the wheels beneath his head and the comfortable small noises in the room, the gentle rattles and squeaks and murmurs in the coachwork that bring sleep so quickly on a train at night-time.
Ian Fleming, Live and Let Die, 1954.

613.19 The train howled on through the night. Bond sat and watched the hurrying moonlit landscape and concentrated on keeping awake.

Everything conspired to make him sleep—the hasty metal gallop of the wheels, the hypnotic swoop of the silver telegraph wires, the occasional melancholy, reassuring moan of the steam whistle clearing their way, the drowsy metallic clatter of the couplings at each end of the corridor, the lullaby creak of the woodwork in the little room. Even the deep violet glimmer of the night-light above the door seemed to say, 'I will watch for you. Nothing can happen while I am burning. Close your eyes and sleep, sleep.'

Ian Fleming, From Russia with Love, *1957.*

613.20 The sheets were crisp and the wheels were round.
David P Morgan, Trains Magazine, *January 1972.*

613.21 Sleep the night and gain a day.
Headline of advertisements for sleeping car services, British Rail, 1972.

613.22 Company International of Wagons Beds.
Dr Jack Hollick, translating the title of the Compagnie Internationale des Wagons-Lits, conversation with the compiler, 1975.

613.23 The romance associated with the sleeping car derives from its extreme privacy, combining the best features of a cupboard with forward movement.
Paul Theroux, The Great Railway Bazaar, *1975.*

613.24 And I hope we get a rough stop at Newcastle just as he is using his potty.
Sleeping Car attendant identified as Joe, concerning an awkward passenger, quoted by Gerard Fiennes, Fiennes on Rails, *1986.*

613.25 A sleeping car named desire.
Magnus Linklater, title of article on the threatened ending of the sleeping car service between London and Fort William, The Times, *15 June, 1995.*

614. SLIP CARRIAGES

614.1 It is as well to be cautious in getting into last carriages, lest they should be slips.
W J Gordon, Our Home Railways, *1910.*

615. SMELLS

615.1 This morning at nine o'clock I took passage on a railroad car (from Boston) for Providence. Five or six other cars were attached to the locomotive, and uglier boxes I do not wish to travel in. They were made to stow away some thirty human beings, who sit cheek by jowl as best they can. Two poor fellows who were not much in the habit of making their toilet, squeezed me into a corner, while the hot sun drew from their garments a villainous compound of smells made up of salt fish, tar and molasses.
Samuel Breck, Journal, *22 July 1825.*

615.2 A gust of that characteristic railroad restaurant odor drifts outward from the dining room—a warm, soupy blend of browned chicken-skin and crisp roll-crust.
Christopher Morley, Broad Street Station, *within* Travels in Philadelphia, *1920.*

615.3 He could feel, taste, smell, and see everything with an instant still intensity, the animate fixation of a vision seen instantly, fixed forever in the mind of him who sees it, and sense the clumped dusty autumn masses of the trees that bordered the tracks upon the left, and smell the thick exciting hot tarred caulking of the tracks, the dry warmth and good worn wooden smell of the powerful railway ties, and see the dull rusty red, the gaping emptiness and joy of a freight car, its rough floor whitened with soft siltings of thick flour, drawn in upon a spur of rusty track behind a warehouse of raw concrete blocks, and see with sudden desolation, the warehouse flung down rawly, newly, there among the hot, humid, spermy, nameless, thick-leaved field-growth of the South.
Thomas Wolfe, Of Time and the River, *1935.*

615.4 And to these familiar sounds, filled with their exultant prophecies of flight, the voyage, the morning, and the shining cities—to all the sharp, thrilling odors of the trains—the smell of cinders, acrid smoke, of musty, rusty freight cars, the clean pine-board of crated produce, and the smells of fresh stored food—oranges, coffee, tangerines and bacon, ham and flour and beef—there would be added now, with an unforgettable magic and familiarity, all the strange sounds and smells of the coming circus.
Thomas Wolfe, Circus at Dawn, *1935.*

615.5 The smell of a street car at this hour of day is also good—a dynamic smell of motors, wood work, rattan seats, worn brass, and steel-bright flanges. It is a smell of drowsy, warm excitement, and a nameless beating of the heart; it speaks of going somewhere.
Thomas Wolfe, The Web and the Rock, *1939.*

615.6 It is not so much of a station, as stations come and go, and it is surrounded by an atmosphere of leather, wool, roasting coffee, and dank sea water.
Rollo Walter Brown, of South Station, Boston, Mass.; I Travel by Train, *1939, quoted by H Roger Grant,* We Took the Train, *1990.*

615.7 In the hot station there is an all-pervading smell, a compound of empty loose-lid milk-churns, horses, Welsh coal and oil-lamps, the London and South Western smell.
C. Hamilton Ellis, The South Western Railway, *1956.*

615.8 At Paddington are more milk churns, horses, Welsh coal smoke and oil lamps, yet the Great Western smell is different. These things have never been satisfactorily analysed, and probably never will be, though it may be remarked that on the Great Western there was also more gas.

Ibid.

615.9 The City and South reeked like a changing-room;
Its orange engines and rolling-stock,
Its narrow platforms, undulating tracks,
Seemed even then historic. Next in age,
The Central London, with its cut-glass shades
On draughty stations, had an ozone smell—
Not seaweed-scented ozone from the sea
But something chemical from Birmingham.

John Betjeman, Summoned by Bells, *1960.*

615.10 Once free of the city the big cars hurried along at exhilarating speeds, swaying and "nosing" from side to side on the often uneven track. Windows flung open against the warmth of a summer's day scooped up the rich odors of the countryside, sometimes mingled with the ozone smell generated by the electric traction motors or the pungent odor of grinding brake shoes as the car slowed for a stop.

William D Middleton, of interurban cars, The Interurban Era, *1961.*

615.11 Why does that tremendous smell of stale bread come from the grilles under Platform 9 at Waterloo?

Paul Jennings, The Observer, *29 October 1962.*

615.12 Ventilation had always been a problem with the Woodhead tunnels, which possessed an odour all of their own, an odour one could almost taste as well as smell, which somebody once said could be approached by drinking cheap port!

George Dow, Great Central, vol. II, *1962.*

615.13 There has always been the smell of great wealth— the most compelling of all American perfumes—about the Southern Pacific, blended with the scent of magnolia in the Deep South.

Lucius Beebe, The Central Pacific and the Southern Pacific Railroads, *1963.*

615.14 The new-mown hay adds a fine bouquet
To the breath of the Western breeze,
And there still is room for the rich perfume
From a grove of orange trees.

Now mix these well with the pungent smell
Of smoke from a Mallet's stack,
And there, my friend, is the perfect blend
I found on U. P. track.

H L Kelso, poem, On a Western Breeze, *within* Great Poems from Railroad Magazine, *1968; published earlier in that magazine at unstated date.*

615.15 The sun always seemed to shine on that crowded platform. It smelled of shellfish, newsprint, disinfectant —and the lime-encrusted urinal at the far end.

Alan Delgado, of the station at Sutton Bridge, The Annual Outing and Other Excursions, *1977.*

615.16 Every vintage has its bouquet, however, and the B&M vintage carriage had a bouquet all its own, compounded of faded upholstery, shag, beer and coal dust, but with an additional aroma acquired on each trip past the Bargoed coke ovens.

Derek Barrie, on Brecon & Merthyr Railway carriages, A Regional History of the Railways of Great Britain, vol. 12, *1980.*

616. SMOKE

616.1 That the furnace of every Steam Engine to be erected or built by the said Company of Proprietors, under or by virtue of the Powers of this Act, shall be constructed on the Principle of consuming its own Smoke.

Provision of the Liverpool & Manchester Railway Act, 1826.

616.2 The smoke is very inconsiderable, but sparks of fire are abroad in some quantity.

Thomas Creevey, letter to Miss Elizabeth Ord, 14 November 1829.

616.3 Ancient maiden lady impiously remarks,
That there must be danger from so many sparks,
Roguish looking fellow turning to the stranger,
Says, its his opinion, she is out of all danger.

John Godfrey Saxe, song, Rhyme of the Rail, *(one of three known titles), c. 1840.*

616.4 . . . a whirlwind of bright sparks, which showered about us like a storm of fiery snow.

Charles Dickens, of a trip on the Boston & Lowell, which at the time used wood-burning locomotives—presumably without adequate spark-arrestor chimneys. American Notes, *1842.*

616.5 Black Smoke is Waste.

Marked on the front coal plate of all Norfolk & Western Railway steam locomotive tenders.

616.6 It is my contention that a diet of good, oleaginous tunnel smoke from the stack of an engine with weeping flues will ward off the pangs of mountain sickness and all the ills that the human nose and throat are subject to.

Brian Fawcett, of oil-fired locomotives, Railways of the Andes, *1963.*

617. SMOKING

617.1 No smoking will be allowed in any of the First Class Carriages, even with the general consent of the Passengers present, as the annoyance would be experienced in a still greater degree by those who may occupy the same coach on the succeeding journey.

Bylaw of Liverpool & Manchester Railway, 1837.

617.2 And there are still public places where even daring and desperate women do not venture to smoke. A duchess might smoke in a restaurant-car of a train, but she would never smoke on the top of an omnibus. Still, evolution proceeds.

Arnold Bennett, Things That Have Interested Me, *1921.*

617.3 The smoker was always up at the front and usually had leather seats and a strong smell. The floor was stained with spittle, the air was blue. The population, distinct from that in the rest of the train, consisted of unshaven men in work clothes with tobacco-stained mouths, or sports wearing tight-fitting models, the zoot-suits of yesterday, with loud stripes, bright shirts, ties and stick-pins and hats with feathers in them. A game of seven-up usually was in progress over a suitcase between three drummers and a brakeman dead-heading, or riding home. In one corner the news-butcher guarded his pile of unsold magazines.

Archie Robertson, Slow Train to Yesterday, *1945.*

618. SOMERSET, England

618.1 Smiling Somerset.

Title of holiday booklet, Great Western Railway, 1931.

619. SOMERSET & DORSET JOINT RAILWAY

619.1 The Somerset and Dorset used to be
My favourite. But that was long ago,
Even before England and Germany
Selected one another for The Foe.
Came death into the world, and all our woe,
With loss of—well, I used to revel in
The royal blue, the royal gold, to me
Medley as dear as a symphony's dear din.
I always seemed to see them neat as a new pin.

Terence Greenidge, verse three of The Nationalisation of the Railways, *in* Girls and Stations, *1952.*

619.2 Forget motor cars. Get rid of anxiety. And here, to the rhythms of the Somerset and Dorset Joint Railway dream again that ambitious Victorian dream, which caused this long railway still to be running through deepest, quietest, flattest, remotest least spoilt Somerset.

John Betjeman, commentary for television film on the S&DJR, 1963.

619.3 Old companions, that's what you were
Sharing the vale across time
Like chalk and cheese: the river Stour
By the Somerset and Dorset branch line.

Tim Jones, verse, Friend of the Stour, *winner of poetry competition in* Dorset, *August 2002.*

619.4 The river meanders, the lines converge
Wildlife is used to the noise:

A slow chattering build-up; from tunnels emerge,
Shuddering, fire-breathing, mechanical toys.

The Stoker is hot, the farm-girl is not,
Each passenger cradling a dream:
Of Whitsun Weddings, or 'life's lot'
Of Sturminster market, cheese, milk and cream.

Ibid., verses 7 & 8.

620. SOUNDS
See also Whistles

620.1 The sound of the engines, on the Stockton & Darlington railroad, may be distinctly heard on a still day at the Dinsdale Hotel, like the flapping of mighty wings, as they pass along; and the line being in many parts circuitous, the puffs of smoke appear here and there among the trees in a thickly wooded country, enabling the spectator to mark the progress of the trains and trace their direction.

Sir George Head, A Home Tour through the Manufacturing Districts and other parts of England, Scotland and Ireland, *1835.*

620.2 As train after train of rolling wagons approached, a black speck first appeared in the distance, gradually and by slow degrees extending its dimensions; meanwhile the sound, like the roaring of the sea, became as a heavy gust of winds, and then, as the carriages receded, grew again less and less audible, till it expired in a low gentle murmur.

Ibid.

620.3 The whistle of the locomotive penetrates my woods summer and winter, sounding like the scream of a hawk sailing over some farmer's yard, informing me that many restless city merchants are arriving within the circle of the town, or adventurous country traders from the other side.

Henry David Thoreau, Walden, *1854.*

620.4 Clicketty, clicketty, clucketty, cluck!
Clicketty, cluck!
From Gallatin up to Bowling Green
A mile-a-minute is often seen;
Into the darkness, into the night,
Clippetty, clip, with all our might—
Gemini Crimini!—terrible gait!
And an awful, terrible, horrible fate,
If BOOM-AH! KERBANG-WHANG! We're into the DITCH!
But heavens, no! 'Twas only a switch,
And God be praised! whatever it be,
There's a wife at home and a prayer for me,
As clippetty, clippetty, on we go,
Slam! Bang! to O-hi-o—
To Louieville, Evansville,
Nashville, Knoxville;

Louieville, Russellville,
Clarksville, Brownsville;
Louieville, Louieville, Louieville, Louieville,
Louisiana to O-hi-o!

Frank F Woodall, fifth verse of poem, The Song of the
L&N Railroad, *c. 1885, published in* Fugitive Poems and A
Christmas Story, *1913. The sounds in the ninth line refer to
old target switches, of which the L&N had several when this
poem was written. This verse refers to a particular one at
Memphis Junction, about three miles south of Bowling
Green. See also Louisville & Nashville Railroad and Speed.*

620.5 They hang us now in Shrewsbury jail:
The whistles blow forlorn,
And trains all night groan on the rail
To men that die at morn.

A E Housman, poem, A Shropshire Lad, IX. *1896.*

620.6 It was a long train, moving slowly, methodically,
with a measured coughing of its locomotive and a rhythmic cadence of its trucks over the interstices of the rails.

Frank Norris, The Octopus, *1901.*

620.7 A composite wheel in company with a rail anything
more than 85 or 90 lbs. to the yard will yield a semitone
of such persistency as to become the *bete noir* of the
neurotic.

Robert Weatherburn, Wheels and Wear, *within* The
Railway Magazine, *vol. XIV, 1904.*

620.8 America is the noisiest country that ever existed.
One is waked up not by the singing of the nightingale,
but by the steam whistle.

Oscar Wilde, Impressions of America, *1906.*

620.9 It had doubtless not been merely absurd, as the wild
winter proceeded, to find one's self so enamoured of the
very name of the South that one was ready to take it in
any small atmospheric instalment and to feel the echo of
its voice in the yell of any engine that appeared not to
drag one either directly North or directly West.

Henry James, The American Scene, *1907.*

620.10 The low rich purr of a Great Western express is
not the worst background for conversation, and the journey passed pleasantly enough.

E M Forster, Howards End, *1910.*

620.11 The working engines each had a voice of its own,
so that it was easy to tell by ear which of them was passing with its load of trucks. *Fox* informed the world that
there was "such a hurry, such a hurry". *Hunslet,* a tank
engine that arrived much later on the scene, was particularly clear in her enunciation, informing all the world
of her huffy temper, though I never heard she was ill to
deal with as a worker—"I'm in a huff, I'm in a huff!" she
puffed in her way along the line. *Progress,* who laboured
at the Wood Green end, proclaimed continually the

name of the chief engineer—"Mr Claringbull, Mr Claringbull" she shouted with a strong accent on the last syl-
lable. *Ferret* seldom left the Enfield portion of roadmak-
ing, perhaps because everything was "such a heavy load,
such a heavy load."

Henrietta Cresswell, Winchmore Hill, *1912.*

620.12 All these little dramas were enacted to a merry
undertone of constant sound: the clear chime of bells,
the murmur and throb of hissing steam, the rumble of
baggage trucks, the slither of thousands of feet.

Christopher Morley, of an arriving train; Broad Street
Station, *within* Travels in Philadelphia, *1920.*

620.13 In town, she listened from bed to the express
whistling in the cut a mile north. Uuuuuu!—faint, nervous, distrait, horn of the free night riders journeying to
the tall towns where were laughter and banners and the
sound of bells—Uuuuu! Uuuuu!—the world going by—
Uuuuuu!—fainter, more wistful, gone.

Sinclair Lewis, Main Street, *1922.*

620.14 *Clunk! Clank!* over the switch-points the sullen
monster glides,
To the stately lift of the rising rods that shimmer along its
sides,
And coupler locks into coupler; the plaints of the air-
valves cease:
A green star glows in the open block as the shuddering
brakes release.

Henry Herbert Knibbs, poem, Right of Way *in* Songs of
the Trail, *1920.*

620.15 The noise of a great terminus is no jar to me. It is
music. I prick up my ears to it, and paw the platform.
Dear to me as the bugle-note to any war-horse, as the
first twittering of the birds in the hedgerows to the light-
sleeping vagabond, that cry of 'Take your seats please!'
or—better still—'En voiture!' or 'Partenza!'

Max Beerbohm, Ichabod, *1923.*

620.16 Chemins de Fer de l'Ouest is perforated on the
white antimacassars. Familiar and strange inscription! I
murmur its impressive iambs over and over again. They
become the refrain to which the train vibrates on its way.

Ibid.

620.17 From end to end of the platform a fog, that was
coloured like sulphur and smelt of it, welled steadily out
of the black tunnel-mouths into the glass-domed cavity
of the Midland Station. It was as though some superhu-
man chemist were generating noxious gases and passing
them into a sealed retort, in which, like imprisoned ver-
min, the crowd of which Clare formed part, ran hither
and thither in a sort of dazed and doomed automatism,
coughing and gasping as they hurried about their incon-
ceivable business, harassed by the shrieks of engines, the
melancholy echoes of shunters' horns, the detonations

of distant fog-signals, the rumbling bourdon made by thousands of other human feet crossing the wooden bridge that spanned the station.

Francis Brett Young, Portrait of Clare, *1927.*

620.18 Now the rains are falling,
Hear the trains a-calling,
Hooeee!
My Mama done told me,
Hear that lonesome whistle,
Blowing 'cross the trestle,
Hooeee!
My Mama done told me,
Clickety clack,
Is echoing back,
The blues in the night.

Johnny Mercer, song, Blues in the Night, *date not known.*

620.19 . . . the sudden slamming racket of the elevated trains . . .

Thomas Wolfe, of the character of New York City, You Can't Go Home Again, *1934.*

620.20 He found it difficult to go to sleep again at once. For one thing, he missed the motion of the train. If it *was* a station outside it was curiously quiet. By contrast, the noises on the train seemed unusually loud. He could hear Ratchett moving about next door—a click as he pulled down the washbasin, the sound of a tap running, a splashing noise, then another click as the basin shut to again. Footsteps passed up the corridor outside, the shuffling footsteps of someone in bedroom slippers.

Agatha Christie, of Hercule Poirot's sleeplessness, Murder on the Orient Express, *1934.*

620.21 The vision nears and deepens once again. Faintly, mournfully, infinitely far away, the cry of a great train is heard, as it wails back across America: Whoo-hoo-oo-hoo-oo-hoo-oo. (The cry fades back, away, into the moon-drenched scenery of America: there is faint thunder of wheels pounding at the river's edge. The scene nears and deepens with terrific instancy—the train is now heard plainly: it is the great *Pacific Nine* stroking the night with the pistoned velocity of its full speed): chucka-lucka, chucka-lucka, chucka-lucka, chucka-lucka, chucka-lucka. (The scene nears and deepens through the night: *Pacific Nine,* like a lighted thunderbolt, is smashing westward through Nebraska.)

PACIFIC NINE: Ho-Idaho! Ho-Idaho ho-ho-ho-ho-ho-ho-ho! Chucka-lucka, chucka-lucka, chucka-lucka, chucka-lucka, chucka-lucka: with fierce bull-bellows, hoarse with pride, she laughs her jolly laughter.)

Thomas Wolfe, A Prologue to America, *1978.*

620.22 The hotshot grunted to a stop. The desert silence throbbed at the temples. Then the air pump hammered as it fought for pressure to release the clamped shoes. Air sneezed at last and the reluctant brakes slackened. Drawheads mumbled as the engine stamped hard in reverse. The tail-lights on the caboose bobbed and crawled back toward the siding.

Harry Bedwell, Smart Boomer, *originally published in the* Saturday Evening Post, *1941; within* Short Lines, *ed. Rob Johnson, 1996.*

620.23 The rails go westward in the dark,
Brother, have you seen starlight on the rails?
Have you heard the thunder of the fast express?

Thomas Wolfe, quoted in A Treasury of Railroad Folklore, *B A Botkin and Alvin Harlow, 1953.*

620.24 Down by the river's edge, in darkness now, he heard the bell, the whistle, and the pounding wheel of the night express coming into town, there to pause for half an hour and then resume its northward journey. It swept away from them, leaving the lonely thunder of its echoes in the hills and the flame-flare of its open firebox for a moment, and then just heavy wheels and rumbling cars as the great train pounded on the rails across the river—and, finally, nothing but the silence it had left behind. Then, farther off and almost lost in the traffic of the town, he heard again and for the last time its wailing cry, and it brought to him once more, as it had done forever in his childhood, its wild and secret exultation, its pain of going, and its triumphant promise of morning, new lands, and a shining city.

Ibid.

620.25 I was born in Dixie in a boomer's shack,
Just a little shanty by the railroad track;
The humming of the drivers was my lullaby,
And a freight train whistle taught me how to cry.
  I got the freight train blues,
  Lordy, Lordy, Lordy (I) got 'em in the bottom of
    my ramblin' shoes;
  And when the whistle blows I gotta go,
  Oh Lordy, guess I'm never gonna lose the freight
    train blues.

John Lair, song, Freight Train Blues, *c. 1934.*

620.26 The sigh of midnight trains in empty stations
Silk stockings thrown aside, dance invitations . . .

Holt Marvell (Eric Maschwitz), song, These Foolish Things, *fourth and fifth lines of last verse as originally written, 1936.*

620.27 One thousand five hundred and three people slept in this station-shelter the other evening. 1,503! 1,650 of whom seemed to be snoring.

Anon., news for those sleeping in the tube station at Swiss Cottage to shelter from air-raids; Bulletin No 3, The Swiss Cottager, de profundis, *September 1940. Quoted in* Lines on the Underground, *compiled by Dorothy Meade and Tatiana Woolf, 1994.*

620.28 Ponkshumbleeze.

Margaret Moore (née Dow), at the age of about eighteen months, of a Gresley three-cylinder goods engine, leaking on one cylinder, under wartime maintenance restrictions. Hitchin, 1943.

620.29 It's at night
In the still moments of the night,
That you hear the pulse-beat of America.

Anon., advertisement, Chesapeake & Ohio Railroad, 1943.

620.30 Listen . . . From across the sleeping countryside
Comes the steady, rhythmic rumble of the trains,
The great, husky trains of America.
They've talked to you since childhood.
They've told you, in the lonely silence of the night,
Of far-off places of romance and adventure.
But their rhythm's faster now,
And their whirring wheels speak different words.
Their cargo now is fighting tools for men of freedom
And through the night their rumble says—
We'll get 'em there . . . We'll get 'em there . . .
We'll get 'em there!
Listen in the night, listen to the rumble of the trains in
the night
And you'll hear the strong, surging pulse-beat of
America!
Ibid.

620.31 Not long ago your poet wrote
Of shimmering steel and creosote,
Expanding rails, perspiring crews and air conditioned
diners,
But now he sits behind the bay,
The old pot-bellied roars away,
While tightening wires rehearse a lay in weird harmonic
minors.

I like the melodies sublime
Of railroads in the wintertime,
The deep exhaust, the whistle's shriek, the joints' recoil
emphatic.
I like the icy winds that roar
Around my shack and semaphore,
The frosted panes, the rattling door and undertones
erratic.

Charles D Dulin, poem, first two verses, Frosted Steel, in Railroad Magazine, 1945.

620.32 As she hove in sight far up the track
She was working steam, with her brake shoes slack,
She whistled once at the whistling post,
Then she flittered by like a frightened ghost.

He could hear the roar of the big six wheel
As the drivers pounded on the polished steel,
And the screech of the flanges on the rail
As she beat it west up the desert trail.

Anon. poem The Gila Monster Route, Hobo Song of the South West; *place of publication not known, contributed by David Elliot.*

620.33 Where the express had raced, the local ambled; where the express had played something like Schumann's *Arabesque* on the railjoints, the local played a gentle pavane.

C Hamilton Ellis, The Trains We Loved, *1947.*

620.34 Only a loud catastrophic roar told them, even, when they were in a tunnel. But by now speed had begun to slacken; from the sound of the train, more and more often constricted deep in cuttings between and under walls, they must be entering London: no other city's built-up density could be so strongly felt . . . Euston.

Elizabeth Bowen, The Heat of the Day, *1949.*

620.35 . . . a rolling stack blast like a helper mallet singing alto.

W H Hutchinson, of the sound of a Shay locomotive. Trains Magazine, *August 1951.*

620.36 There are lots of sounds that reach us in our
retrospective dreams,
Such as semaphores a-clankin' and the crashing
couplers' din,
But there's none that's quite as thrilling as the wild,
insistent screams
Of a mammoth locomotive as she calls her flagman in.

Charles D Dulin, poem, first verse, Whistle Thrills, *within* Sage & High Iron, *1951.*

620.37 The noise is rhythmic, constant, deafening: the pound of a harness of main and side rods, crossheads and valve motion; the monotone of the stoker screw; the asthmatic whine of the injector; the exhalation of brass valves; the rattle of the deck plates.

David P Morgan, Trains Magazine, *May 1953.*

620.38 When you hear an A crack with the reverse notched up just right, why you wonder what anybody would want to dieselize for.

Clarence E Pond, of Norfolk & Western class A 2-6-6-4 locomotives, quoted in Trains Magazine, *November 1954.*

620.39 For generations Britons have lain awake at night and listened to the soothing music of shunted trucks, the slow, quiet breathing of distant engines, and derived a deep contentment from the thought that others are out there working while they lie warmly in bed. This innocent pleasure is to end. For the future, the silence of death will lie like a pall over the swift and sanitary scene.

H F Ellis, Full Diesel Ahead, *about the BR Modernisation Plan, within* Punch, *2 February 1955.*

620.40 Unthinkable is the din of the London and North Western at Euston.

C Hamilton Ellis, The South Western Railway, *1956.*

620.41 . . . a mighty cascade of noise, a mile long . . .
Peter Allen, of American freight trains, On the Old Lines, *1957.*

620.42 . . . when they are standing still, they have a curious and irritating habit of making what a small child once called Very Rude Noises.
Canon Roger Lloyd, of the vacuum pump on Great Western Railway locomotives, The Fascination of Railways, *1957.*

620.43 Many a night I have run her on a manifest or time freight, stepping through the country with 100 to 150 loads behind her; and the whole world seemed to be asleep. She would be cracking at the stack like a machine gun, the headlight boring a hole in the night for us, the signals coming up green. I would hang out the window and listen to her talk in a language only she and I understood.
D Stewart Baals, locomotive engineer, C&O, of locomotive 2727 when she was retired to a museum. Trains Magazine, *December 1957.*

620.44 A high-pitched screaming came from the traction motors and gears, and the steady thump and hiss of the trolley wheel overhead was faintly heard. The wheels beat a measured rhythm over staggered rail joints, and now and then, to the clank of loose fitting switch points and frogs, the car lurched through the turnouts that led to spurs or sidings. Occasionally the air compressor beneath the car cut in with its characteristic *lung-a-lung-a-lung*. The conductor's signal cord, suspended from the ceiling, flip-flopped back and forth, and there was a muffled creaking from the car's ornate woodwork.
William D Middleton, of the interurban car, The Interurban Era, *1961.*

620.45 Thus you bunch the couplers for the reverse. As the engine comes to a grinding stop you twirl the screw reverse counter-clockwise into back gear and look out of the cab for the back-up signal. When it comes, almost at once, you blow three toots on the whistle, put the automatic brake handle back into *Running Position* and crack her open. The bunched couplers allow of a smooth start; so that by the time you are clattering over the switch the train is moving at its usual running speed, the exhaust shouting 'Chaupichaca, Chaupichaca!' and the pistons answering with a contented, 'Fosdick, Fosdick, Postlethwaite and Fosdick!'—the marching song of the mountain engine when all is going right.
Brian Fawcett, of working a train through a zig-zag on the Central Railway of Peru, Railways of the Andes, *1963.*

620.46 . . . the sound from the front end was 99 per cent U.S. stack talk and 1 per cent French pip-squeak whistle.
David P Morgan, of a journey in Southern France behind an Alco-built 141R locomotive. Trains Magazine, *July 1966.*

620.47 Nothing like the sound of steam, to keep love warm, to keep love moist, to bring it to the throat.
Harold Pinter, Monologue, *BBC Sound Archive, no date given.*

620.48 The welkin rang under the echoing overall roofs of the great termini with the beat of Westinghouse pumps on both sides at Victoria, at London Bridge, and with the hush and thump of the same apparatus—in Great Eastern style—at Liverpool Street. The continuous brake apparatus was indeed a more noticeable contributor of sound impressions than the exhaust notes of the motive power itself. At Paddington the explosive plomp of vacuum pumps of Swindon make was noticeable as shunting engines moved; at Broad Street the tall chimneys of North London Adams 4-4-0 tanks made organ pipe resonance for the ejectors of the vacuum brakes; steam brakes, less sensitive in control, could jam the wheels to stop a locomotive with a roaring slide. Four-wheel coaches, still to be the rule for many years of passenger service on the North London and much in evidence, with six-wheelers elsewhere, especially on suburban operations, made a clatter over points and crossings and rail joints unknown to modern rail fans.
Charles Klapper, of London termini in Victorian days, London's Lost Railways, *1976.*

620.49 The sounds of a train climbing through the countryside, for instance, can be likened to a symphony in three movements, played without a break: first, pianissimo, the birdsong and a distant whistle emphasise the silence out of which the train is heard approaching, perhaps with a brief and abrupt change of tempo when the wheels slip; the train comes closer at a steady and now slower tempo, reaches a crescendo as it passes by, then climbs away into the distance, now pianissimo again, with maybe a long, lonely whistle as a coda.
Peter Handford, Sounds of Railways and their Recording, *1980.*

620.50 But do you know, once I walked out of the place, I could hear once more the thud of ticket daters, the nonstereo call of the train announcer for the *Pan-American*, and the high-born sigh of a Pennsy K4 at the bumper post.
David P Morgan, of a visit to the derelict Louisville Union station, Trains Magazine, *September 1989.*

620.51 You could sit on the office steps at my hometown roundhouse and tell as surely as you had been standing in the cab beside him when the engineer of the class J 4-8-4 on the eastbound *Pelican*, half a mile distant by then and going away fast, made his last cut-off adjustment at speed.
Ed King, Trains Magazine, *October 1993.*

620.52 Equally impressive was the sound of a hotshot grooving the diamond at 60 mph or a shade more, as

when N&W's auto parts and multilevel DS-11 would hit the Ann Arbor crossing at Milan, Mich. First was the ear-shattering *crash* of the three units hitting the diamond, then the kadang-kadang-kadang-kadang-kadang of the auto cars and the 86-foot high-cubes with their trucks inset about 10 feet towards the middle of the car.
Ibid.

620.53 . . . some of the other very interesting sounds of railroading—the *ka-whop-shhhhhh* of a Union Switch pneumatic switch machine, for instance.
Ibid.

620.54 Alcos *burble* [or, in the case of those about-to-die 539's, *wheeze*]; Baldwins *chortle*; EMD's *chant*; FM's *drum*; and GE's *chug*.
J David Ingles, Trains Magazine, October 1993.

620.55 "Turbo" implies the whine of a jet or at least the whistle of a turbocharger. No, the noise from under the floor is *kdlkdlkdlkdl*. Pure diesel.
George H Drury, of Turbo Express DMUs operated by Thames Trains. Trains Magazine, December 1997.

620.56 yangyangyangyangyangyangyang.
Ray Towell, of the sound of an idling General Motors class 59 diesel, England, to the compiler, late 1990s.

620.57 You could set music to that.
Jim Smith, traction inspector, to the compiler, on the footplate of Green Arrow and referring to the sound of the locomotive as it accelerated on a climb east from Carlisle, 8 November 1998.

620.58 . . . for a railroad fan, the total train experience involves quite a bit more than photography, and for total train experience, it's hard to beat a place like Cajon Pass. Try pacing a freight out of San Berdoo, up Cajon Boulevard on a moon-lit summer evening, windows rolled down, warm breeze rushing past . . . it's a total sound experience. Camp at Sullivan's Curve. Hang out up near Summit where you can see two or three trains at once. Listen simultaneously to the whine of dynamics on one track above you and the throaty burbling of a matched set of G.E.s heading uphill on the track below. As a railfan, these are the experiences that make a place a "must go." Unless you're getting paid to produce it, as a railfan, the photography is secondary in Mecca.
Brian Jennison, in a website newsgroup, April, 2003.

621. SOUTH EASTERN RAILWAY

621.1 It is audaciously unpunctual.
E Foxwell and T C Farrer, Express Trains English and Foreign, 1889.

621.2 Both railways must have maintained a clerical staff for the special purpose of writing letters of complaint to

each other, which of course were worded very politely—but what they said of each other in private wasn't.
E L Ahrons, The Railway Magazine, June 1917. The other railway was the London Brighton & South Coast.

622. THE SOUTHERN BELLE

622.1 The most luxurious train in the World.
Advertisement, London Brighton & South Coast Railway, 1909.

623. SOUTHERN PACIFIC RAILROAD

623.1 Fast as the Fastest. Finer than the Finest.
Advertising slogan, 1902.

623.2 The friendly Southern Pacific.
Advertising slogan, 1949–50.

623.3 While its politics and economic warfare were conducted on a scale of truculence and ferocity that would have frightened the warring barons of medieval Italy, the Southern Pacific's love affair with its passengers, as long as the management was so inclined, was on a positively idyllic scale.
Lucius Beebe, The Trains We Rode, 1965.

623.4 . . . a railroad legacy left by four shopkeepers, 14,000 Chinese, 2,000 whites, 6,000 horses.
Anon., caption to photograph, Trains Magazine, August 1958.

623.5 Nothing is as for real as the Southern Pacific Railroad. It has a reputation as a cloud gatherer of western railroads. Its economics—perhaps more before the depression than after—were the economics of the State of California. And then some.
Bruce MacGregor, Portrait of a Silver Lady, 1977.

624. SOUTHERN RAILWAY (England)

624.1 For Holidays I always go Southern 'cos it's the Sunshine Line!
Anon., caption on poster, from a photograph by Charles Brown of a Southern locomotive, Fireman W Woof of Nine Elms, and Ronald Witt, then a small child, 1925.

624.2 Live in Surrey
Free from Worry
Poster slogan, 1926.

624.3 Live in Kent
And be content
Poster slogan, 1926.

624.4 As for the Southern, with its trim green bourgeois stations, all exactly alike, it is really just an extension of the Underground, not a real railway at all.
Paul Jennings, The Observer, 30 October 1949.

625. SOUTHPORT, England

625.1 Southport for mild winters.
Advertisement, Lancashire & Yorkshire Railway, c. 1921.

626. SOUTHWOLD, England

626.1 Southwold was one of those coastal villages which had become remote with the closure of its railway. It was now emptier and more rural than it had been twenty years before.
Paul Theroux, The Kingdom by the Sea, *1983.*

627. SPEED

See also Trains—Express, Travel—Speed

627.1 It seemed, indeed, to fly, presenting one of the most sublime spectacles of human ingenuity and human daring the world has ever beheld.
Liverpool Mercury, *9 October 1829, commenting upon* Sans Pareil, *one of the locomotives competing in the Rainhill trials.*

627.2 As accuracy upon this subject was my great object, I held my watch in my hand at starting, and all the time; and as it has a second hand, I knew I could not be deceived; and it so turned out there was not the difference of a second between the coachee or conductor and myself. But, observe, during these five miles, the machine was occasionally made to put itself out or *go it*; and then we went at the rate of 23 miles an hour, and just with the same ease as to motion or absence of friction as the other reduced pace.
Thomas Creevey, of a trip on the Liverpool & Manchester, and probably the earliest account of train timing; letter to Miss Elizabeth Ord, 14 November 1829; The Creevey Papers, *1903.*

627.3 It is really flying, and it is impossible to divest yourself the notion of instant death to all, upon the least accident happening. It gave me a headache which has not left me yet.
Ibid.

627.4 You cannot conceive what that sensation of cutting the air was; the motion as smooth as possible too. I could either have read or written; and as it was I stood up, and with my bonnet off "drank the air before me." The wind, which was strong, or perhaps the force of our own thrusting against it, absolutely weighed my eyelids down.
Fanny Kemble, actress, letter August 1830, after a footplate ride with George Stephenson.

627.5 When I closed my eyes this sensation of flying was quite delightful, and strange beyond description; yet strange as it was, I had a perfect sense of security and not the slightest fear.
Ibid.

627.6 I can only judge of the speed by putting my head out to spit, which I did, and overtook it so quick, that it hit me smack in the face.
Davy Crockett, quoted by Professor John Stover in American Railroads, *1968.*

627.7 The public will always prefer that conveyance which is the most perfect, and speed within reasonable limits is a material ingredient in perfection in travelling.
I K Brunel, letter to directors of the Great Western Railway, 15 August 1838.

627.8 Now we pass through a deep cutting—now a tunnel! Now trees flying past! A pretty country—a canal—across a valley; rushing at the rate of thirty-three miles an hour on an embankment high above the surrounding country—Father holding his watch in hand marking speed by the mileposts.
Elizabeth King, in an old journal, 24 May 1839. Quoted in Lord Kelvin's Early Home, *1909.*

627.9 Spanked along the road to Liverpool. It is quite a just remark that the Devil, if he travelled, would go by train.
Lord Shaftesbury, Journal, *quoted in* The Life and Work of the Seventh Earl of Shaftesbury, *1888.*

627.10 It is true I have said the locomotive engine *might* be able to travel a hundred miles an hour; but I always put a qualification to this, namely, as to what speed would best suit the public. The public may, however, be unreasonable; and fifty or sixty miles an hour *is* an unreasonable speed. Long before railway travelling became general, I said to my friends that there was no limit to the speed of the locomotive *provided the works could be made to stand.* But there are limits to the strength of iron, whether it be manufactured into rails or locomotives; and there is a point at which both rails and tyres must break. Every increase of speed, by increasing the strain upon the road and the rolling stock, brings us nearer to that point.
George Stephenson, Evidence to Parliamentary Select Committee on Railways, 1841.

627.11 Not quite so fast next time, Mr Conductor, if you please.
Prince Albert, after a journey between Paddington and Windsor, quoted in the Morning Post, *February 1842.*

627.12 On it whirls headlong, dives through the woods again, emerges in the light, clatters over frail arches, rumbles upon the heavy ground, shoots beneath a wooden bridge which intercepts the light for a second like a wink, suddenly awakens all the slumbering echoes in the main street of a large town, and dashes on haphazard, pell-mell, neck-or-nothing, down the middle of the road. There—with mechanics working at their trades, and people leaning from their doors and windows, and boys flying kites and playing marbles, and men smoking, and women talking, and children crawling, and pigs burrowing, and unaccustomed horses plunging and rearing, close to the very rails—there—on, on, on—tears the mad dragon of an engine with its train of cars; scattering in all directions a shower of burning sparks from its wood fire; screeching, hissing, yelling, panting; until at last the thirsty monster stops beneath a covered way to drink, the people cluster round, and have time to breathe again.

Charles Dickens, American Notes, *1842.*

627.13 From the fire and the water we drive out the steam
With a rush and a roar and the speed of a dream;
And the car without horses, the car without wings,
Roars onward and flies
On its grey iron edge, . . .

Elizabeth Barrett Browning, poem, A Rhapsody of Life's Progress, *VII.*

627.14 . . . when it is remembered how violently the body is thrown from side to side, and every now and then pitched in an opposite direction, and this incessantly for perhaps three or four hours, it is not to be wondered at that the delicate and ailing who have been subjected to such a process should feel *shaky* for some days afterwards.

Anon., The Railway Traveller's Handy Book, *1862.*

627.15 Some o' the sparks are no' doon yet!

Driver Norman Kellock, the day after a fast run on the Glasgow & South Western Railway.

627.16 Celerity of communications can only take place by increased celerity of trains.

Attributed to Frederick Hill (brother of Sir Rowland Hill) by Sir Cusack Roney, Rambles on Railways, *1868.*

627.17 Clicketty, clicketty, clicketty, click!
Look at your watch a third of a minute;
How many rail-clicks are there in it?—
One- two- three- four;
Twenty, or thirty, or forty or more
If there be sixty, then you may know,
That sixty miles an hour we go
At the rate of a mile a minute we go,
From Louisiana to O-hi-o—
To Louieville, Evansville,
Nashville, Knoxville;
Louieville, Russellville,
Clarksville, Brownsville;
Louieville, Louieville, Louieville, Louieville,
Louisiana to O-hi-o!

Frank F Woodall, second verse of poem, The Song of the L&N Railroad, *c. 1885, published in* Fugitive Poems and A Christmas Story, *1913. The calculation suggested in lines 2, 6, and 7 confirms that the Louisville & Nashville at this time used the then standard 30 foot length of rail. See also* Louisville & Nashville Railroad *and* Sounds.

627.18 The race against time which followed was grand and terrible. The engine seemed to be not so much running as coursing with great lion-like bounds along the track, and the spectacle from the locomotive as it rose and fell in its ceaseless rapid motion, while houses, fields and woods rushed by, was wonderful and glorious, almost worth the risk to enjoy!

William Pittenger, of the dash towards Chattanooga in the Andrews Raid, Daring and Suffering, *1887.*

627.19 What with third-class cushions and room for our knees, hat-racks and foot-warmers, smooth steel rails and the easier carriage springs that followed, with bogie trucks and differential axles, we are whirled along now so unexcitedly that we hardly realise the pace.

E Foxwell, Express Trains, English and Foreign, *1889.*

627.20 Faster than fairies, faster than witches,
Bridges and houses, hedges and ditches;
And charging along like troops in a battle,
All through the meadows the horses and cattle:
All of the sights of the hill and the plain
Fly as thick as driving rain;
And ever again in the wink of an eye,
Painted stations whistle by.

Robert Louis Stevenson, poem, From a Railway Carriage.

627.21 "We are going well," said he, looking out of the window, and glancing at his watch. "Our rate at present is fifty-three and a half miles an hour."
"I have not observed the quarter-mile posts," said I.
"Nor have I. But the telegraph posts upon this line are sixty yards apart, and the calculation is a simple one."

Sir Arthur Conan Doyle, of the Great Western main line; Silver Blaze, *within* Memoirs of Sherlock Holmes, *1892.*

627.22 Far away, from among the Kentish woods there rose a thin spray of smoke. A minute later a carriage and engine could be seen flying along the open curve which leads to the station. We had hardly time to take our place behind a pile of luggage when it passed with a rattle and a roar, beating a blast of hot air into our faces.
"There he goes", said Holmes as we watched the carriage swing and rock over the points.

Sir Arthur Conan Doyle, of a special train chartered by Professor Moriarty on the London, Chatham & Dover Railway, seen from Canterbury station. The Final Problem, *within* Memoirs of Sherlock Holmes, *1892.*

627.23 You can depend on me, Mr Casey. I'll sure keep her hot.

Sim Webb, fireman to Casey Jones, when it was realized that they had to run fast to make time, 30 April 1900. Recalled in interview by Sim Webb, 1936. Recorded in Railroad Magazine *and repeated in* Railroad Avenue, *Freeman Hubbard, 1945.*

627.24 Oh, Sim! The old girl's got her high-heeled slippers on tonight!

Luther "Casey" Jones, to his fireman Sim Webb during their fateful run to Vaughan, Tenn., 30 April 1900. The "old girl" was Illinois Central 4-6-0 No. 382. Recalled in interview by Sim Webb, 1936. Recorded in Railroad Magazine *and repeated in* Railroad Avenue, *Freeman Hubbard, 1945.*

627.25 And occasionally the black valley space between was traced, violated by a great train rushing south to London or north to Scotland. The trains roared by like projectiles level on the darkness, fuming and burning, making the valley clang with their passage. They were gone, and the lights of the towns and villages glittered in silence.

D H Lawrence, Sons and Lovers, *1913.*

627.26 Faster schedules do not earn more money—they burn it—but the public wants fast trains.

John Droege, Passenger Terminals and Trains, *1916.*

627.27 It was something in the nature of a triumphal expression conducted at thrilling speed. Perhaps there was a curve of infinite grace, a sudden hollow explosive effect made by the passing of a signal box that was close to the track, and then the deadly lunge to shave the edge of a long platform.

Stephen Crane, The Scotch Express, *within* Men, Women and Boats, *1921.*

627.28 Trade per Hesperias cito nos, via ferrea, partes, Hoc iubet urbs; Felix hoc iubet ipse Polus.

"Traveller by G W R"; Letter in The Times, *13 August 1928; it translates as: Bear us quickly through the Western parts, O Railway; / The town demands this; Felix Pole himself demands it. Noted by Felix Pole in* His Book, *1954.*

627.29 And even while the *Silver Fox* was thus getting over the country in the fastest run ever known, the *Queen Mary* also may have been making history by breaking the record for the eastward crossing of the Atlantic. Everybody will agree that both of these Atalantas deserve a whole bunch of blue ribbons to tie up whatever may be the equivalent of their bonny brown hair. But even in these days of 'speed-mania' there must be a good number of old sobersides left who are glad to know that both ship and train (as no doubt Milanion was glad to know about Atalanta after they had been married a few years) are built for comfort as well as for speed.

Anon., of the 113 mph achieved on a routine run of the Coronation *on 27 August 1936, hauled by A4 Silver Fox. The assumption about RMS Queen Mary obtaining the Blue Riband at the very same time was correct. RMS Queen Mary had broken the transatlantic record, westbound, with a speed of 30.14 knots on a voyage between 20 and 24 August 1936, and upon returning eastbound, 26 to 30 August, she* broke the eastbound record with a speed of 30.63 knots. Both records stood until the following year, when taken by the Normandie. *LNER Magazine, October 1936.*

627.30 In this country record locomotive runs are usually done by stealth.

Anon., LNER Magazine, December 1936.

627.31 A party of Scottish journalists yesterday carried out what was described as the swiftest raid ever made into England, as far as Newcastle. Incidentally, they were sent back to Edinburgh as quickly as they came.

Comment on the Scottish press run of the LNER Coronation train the previous day, The Scotsman, *3 July 1937.*

627.32 The exhaust was humming with a continuous roar like that of an aeroplane engine. The white mileposts flashed past and the speedometer needle shot up through the "90's" into the "100's" to 110-111-112-113-114 miles per hour, but beyond it—No!

Robert Riddles, concerning the high speed run of LMS locomotive Coronation, *29 June 1937, in speech to Junior Institution of Engineers, 12 December 1947.*

627.33 We were doing 60 to 70 m.p.h. when we spotted the platform signal. The crockery in the dining car crashed. Down to 52 m.p.h. through the curve, the engine riding like the great lady she is; there wasn't a thing we could do but hold on and let her take it. Take it she did; past a sea of pallid faces on the platform, we ground to a dead stand, safe and sound, and still on the rails.

Ibid.

627.34 Lima built 'em in 'forty-three . . .
Now we wheel 'em down Espee.
Rapid exhaust! Hear the "varnish" roar!
Big steam hog! Big four-eight-four!

Howard W Bull, a fireman on Southern Pacific, of new locomotives used on Rapid Exhaust passenger runs on that line. Trains Magazine, *November 1944.*

627.35 There is also a branch line from Whitchurch to Chester, with a very few trains each way a day, all of which emulate an elderly and asthmatic tortoise.

Canon Roger Lloyd, The Art and Mystery of the Railway, *within* The Railway Magazine, vol. 94, *1948.*

627.36 Charlie, send someone to tell that engineer there's no need to get us to Denver at this rate of speed. Eighty miles an hour is good enough for me.

Harry S Truman, on realizing that his train was running at 105 mph, en route from Independence, Mo.; as recorded by Margaret Truman in Harry S Truman, *1973.*

627.37 We are a remarkably happy-go-lucky nation. Where people like the Germans would experiment with meticulous care in working up to record speeds on rails, we are inclined to make the experiment without any inquiry as to the possible cost, and then, when it is all over, blandly to inquire whether we had been doing anything risky. Scarcely any of the highest British speed records have been made without some element which has caused the hair of the knowledgeable to stand on end.

Cecil J Allen, The Railway Magazine, *quoted in* Trains Magazine, *November 1949.*

627.38 Festiniog lente.

Bill Bellamy, sometime accountant of the Festiniog Railway, when asked about the rate at which some work was to be done.

627.39 Extra 3005 East, now no less than 98 cars between tank and caboose, was bearing down on Shelby with all the implications of destiny of the Book of Revelations, gaining momentum with each revolution of those four pairs of 69-inch drivers, making the legal mile-a-minute with ease and perhaps a notch or two better.

David P Morgan, Trains Magazine, *September 1956.*

627.40 Big steel spiked to good ties sunk in high ballast—4472 ate it up, skimming along past miles of slide-detector fences, drinking in the scenery, reciprocating her Walschaerts in that imperious, impeturbable British manner of hers.

David P Morgan, Trains Magazine, *January 1972.*

627.41 So speed yes, but let there be money in it.

Gerard Fiennes, Fiennes on Rails, *1986.*

627.42 . . . a thump of squashed air against the window and a blur of blue and white livery . . .

Nicholas Whittaker, Platform Souls, *1995.*

627.43 There was a rhythm to it. The B&O planted its lineside poles 60 to the mile, a pole each second. A fine running train literally was like clockwork—whoosh, whoosh, whoosh, whoosh—the pole line keeping time with the second hand.

John P Hankey, Trains Magazine, *November 2000.*

627.44 Outside, rabbits could be seen outrunning the train on a stretch of track south of Indianapolis. The engineer said quite a few rabbits did that.

Richard Saunders Jr, after an article by Daniel Machalaba, Wall Street Journal, *7 April 2000, concerning the Amtrak train* Kentucky Cardinal; *in* Main Lines, *2003.*

628. SPIKES

628.1 The common rail spike was devised in the New World almost at the very beginning because it greatly conserved labor and material. It remains in use a century and a half later because it still has those virtues.

James Vance Jr, of the American variety of dogspike, used for fastening flat-bottom rails. The North American Railroad, *1995.*

629. SPITTING

629.1 The Superintendent begs leave to remind gentlemen passengers who SPIT, that the car floors cannot be washed while the train is in motion.

A rule of the Salem & Lowell Railroad, 1856, reproduced in RLHS Bulletin 50, *October 1939.*

629.2 In order to aid in the prevention of consumption you are earnestly requested to refrain from the dangerous and objectionable habit of spitting

Sign, Midland Railway, typical of those erected by many British railways at their stations, at the time, c. 1900.

629.3 The presence of rich Turkey carpeting underfoot eventually inhibited the almost universal American habit of spitting tobacco juice, and where it was unable to stem the tide, it at least improved the national aim by providing tasteful spittoons located at strategic intervals.

Lucius Beebe, of the furnishings of Pullman Palace Cars in the nineteenth century; Mr Pullman's Elegant Palace Car, *1961.*

630. STANDARDIZATION

630.1 In the constructive department of locomotives engineers have been and are widely at variance, as anyone may judge from the variety of stock now made; . . . probably five distinct classes of locomotive would afford a variety sufficiently accommodating to suit the varied traffic of railways, whereas I suppose the varieties of locomotives in actual operation in this country and elsewhere are very nearly five hundred in number. Everyone cannot be right, and most of them must be wrong, and it would be for the best interests of railways if the proper authorities could be unanimous in the selection of a good number of classes to uniform patterns to be adopted in future practice. It is doubtful if such an arrangement could be worked out unless there were entire amalgamation of railway interests and the influences of government, or otherwise it is extremely desirable and is certainly practicable.

D K Clark, Preface to Railway Machinery, *1855.*

630.2 . . . each railway has to a very large extent standardised the appliances which it requires for its own use, with very satisfactory results as regards economy, and it seems quite likely that this process has already been carried far enough, and that the introduction of a more rigid standardisation of railway appliances would act as a most undesirable check upon invention.

Lord Monkswell, The Railways of Great Britain, *1913.*

630.3 The mere reduction in the number of railway companies from about 120 to four should in itself pave the way to standardisation as it would be easier for four engineers to agree than 120.

Editorial comment, Railway Gazette, on an address given on 1 June 1921 by Eric Geddes on the 1921 Railways Act to the Junior Institution of Engineers, 9 June 1921.

630.4 The most dramatic thing about railroads in America is not the flashing streamlined train, nor even the mighty freight locomotive, but the fact that any freight car of any railroad can be, and is, moved in the trains of any one of nearly 700 operating railroads.

Robert S Henry, Trains Magazine, January 1945.

630.5 N&W dieselised without diesels.

David P Morgan, of Norfolk & Western's efficient use of steam locomotives through the concentration on four standard designs and the use of modern servicing facilities. Trains Magazine, September 1959.

630.6 It was another example of the tyranny of standardisation.

Oliver Bulleid, of Railway Clearing House standard wagon design, quoted by F A S Brown in Nigel Gresley, Railway Engineer, 1961.

630.7 You tell us what color you want it painted and we'll be responsible for everything else.

Attributed to H L Hamilton of Electro-Motive by David P Morgan. Of the standardization of the American diesel locomotive, Trains Magazine, January 1964.

631. STATION ANNOUNCEMENTS

631.1 In 1861 at almost every station on the Edinburgh & Glasgow Railway a parrot or starling was kept, which called out its name "most distinctly" on the trains' arrival. Elsewhere, porters' voices were used for this purpose. They were untrained, and many of their announcements were inaudible or unintelligible.

Jack Simmons, Oxford Companion to British Railway History, 1997.

631.2 The train hands generally call out the names of stations, but in so disguised and mispronounced a manner that they are unrecognizable.

Edward Bates Dorsey, of stations in England, English and American Railroads Compared, 1887.

631.3 Passing through Newton Abbot we once overheard a discussion on porter's cries. It seems they are fairly divisible into three groups. There are the mere grunts, as at Havant and Yeovil; the abbreviations, as at 'Snks' (otherwise Sevenoaks) and 'Drm', 'S'num', 'Strum', and 'Bl'um' (as at Durham, Sydenham, Streatham, and Balham); and the jubilantly rhythmical, as the 'Woodleywoodleywoodley Junc.' of the M.S. & L., and that glorious crescendo of the Midland, 'Mangots*field*! Mangots*field*! Change here—for Glaster and Chaltenham, Woster and Barmingham, Darby and the Narth!'

W J Gordon, Everyday Life on the Railroad, 1898.

631.4 "Why the deuce don't you sing out the names of the stations clearly?" said an irate railway passenger to a London and North Western Railway porter the other day, who had just delivered himself of a string of unintelligible gibberish.

"Golly!", exclaimed that individual, "'ere's a cove as expects hopera-singers for a porter's wages!"

Anon., quoted in The Railway Magazine, vol. IV, 1899.

631.5 Every now and then the announcer comes to the head of the stairway and calls out something about a train to Harrisburg, Altoona, Pittsburgh and Chicago. There is a note of sadness in his long-drawn wail, as though it would break his heart if no one should take this train, which is a favorite of his.

Christopher Morley, Broad Street Station, within Travels in Philadelphia, 1920.

631.6 For example, on running into Reading, you may hear a measured and somewhat menacing feminine voice declaim very deliberately: "Reading—Reading—Reading", and at times it would seem quite an appropriate sequence if there were to follow: "All hope abandon, ye who enter here."

Cecil J Allen, Trains Illustrated, October 1950.

631.7 By contrast, extreme contrast, there is the jolly prologue of the lady announcers at Kings Cross, who invariably begin their announcements with "Hullo, passengers" so cheerfully enunciated that you almost look round to see if a game of hide-and-seek is about to begin—a thing that you would never *dare* do at Reading.

Ibid.

631.8 . . . a combination of gravel and southern drawl.

O Winston Link, of Buck Stewart, train announcer at Roanoke, Va., in the 1950s. The Last Steam Railroad in America, 1995.

631.9 If any railway enterprise is to be known as One, it should be the Royal Train. But as one always knows where one is going, one does not really need an announcement, does one?

Alan Hamilton, of the much-ridiculed renaming of Anglia Trains as "One," and the confusion it caused at stations; The Times, 6 May 2004.

632. STATION BUFFETS

632.1 I assure you Mr Player was wrong in supposing that I thought you purchased inferior coffee. I thought I said to him that I was surprised you should buy such bad roasted corn. I did not believe you had such a thing as coffee in the place: I am certain that I never tasted any. I have long ceased to make complaints at Swindon— I avoid taking anything there when I can help it.

I K Brunel, letter to S Y Griffiths, December 1842.

632.2 The real disgrace of England is the railway
sandwich.
   Anthony Trollope, He Knew He Was Right, 1869.

632.3 . . . we forced our way into the buffet, where we
yelled, and stamped, and waved our umbrellas for a
quarter of an hour; and then a young lady came and
asked if we wanted anything.
   *Jerome K Jerome, Three Men in a Boat, Advantages of
cheese as a travelling companion, 1889.*

632.4 Snackerie Anglaise
   *Term used in London Stations: A User's Assessment,
published by the Research Institute for Consumer Affairs,
1963, to describe British Railways' station refreshment
services.*

633. STATION MASTERS

633.1 The house ought not to communicate directly
with the booking offices, as too much facility of passing
from one to the other is apt to lead to inattention to
duty.
   *John Wolfe Barry, of station masters' houses, Railway
Appliances, 1887.*

633.2 The Station Master is pompous and grand,
He settles a thing with a wave of his hand.
His coat is trimmed with the finest gold,
And his porters do whatever they're told.
A Station Master I'd like to be,
With no one to ever say "No" to me!
   E V Lucas, poem, Heroes, 1925.

633.3 Mr Arnott of Waverley station
Has a very strong sense of occasion,
When the train's a non-stopper
His topper is proper,
His bowler's for trains of low station.
   *Paul le Saux, commentary in film Elizabethan Express,
1954.*

634. STATION NAMES

634.1 The name-board of the station was faintly visible,
and with a lighted match I went along it letter by letter. It
seemed as if the whole alphabet were in it, and by the
time I had got to the end I had forgotten the beginning.
   *Edith Oenone Somerville and Martin Ross, "Poisson
d'Avril," Further Experiences of an Irish R M, 1908.*

634.2 At half-past six Clare's train was running westward.
But this time, as she retraced her course, the station
names which earlier in the day had been beaten on her
mind with the finality of nails hammered into a coffin
succeeded each other in a crescendo of hopefulness:
Winsworth, New Bromwich, Astbury, Dingley Regis,
Mawne Road Halt. With each of them in succession she

felt the grime and heaviness of Alvaston rolling away
from her. The woods of Mawne drooped over beside the
track. The little train careered and rattled down the gradient as though a breath of clean air had gone to its head.
Clare heard the familiar sigh of Westinghouse brakes.
Wychbury! At last . . .
   Francis Brett Young, Portrait of Clare, 1927.

634.3 What railway historian has not been stirred, angrily
or amusedly according to his line of sentiment, at that
complete victory of logic over history, "Rugby Midland"!
   *Derek Barrie, referring to the renaming of the LNWR
station in Rugby by British Railways, notwithstanding its
lack of any connection with the old Midland Railway, British Railways: A Survey, 1948–1950, within The Railway
Magazine, vol. 97, 1951.*

STATIONS

*Entries on stations are divided into subheadings, thus:
Character, Commerce, Disdain, In Society.*

635. STATIONS—Character.

See also Architecture, Arrival, Departure, Whistles

635.1 Within, there was a spacious breadth, and an airy
height from floor to roof, now partially filled with smoke
and steam, which eddied voluminously upward, and
formed a mimic cloud-region over their heads.
   *Nathaniel Hawthorne, The House of the Seven Gables,
1851.*

635.2 Station very gritty, as a general characteristic. Station very dark, the gas being frozen. Station very cold, as
any timber cabin suspended in the air with such a wind
making lunges at it would be. Station very dreary, being a
station.
   Charles Dickens, Household Words, 21 January 1854.

635.3 Even the bustle and confusion at the railway terminus, so wearisome at other times, roused me and did
me good.
   *Wilkie Collins, of Euston station, The Woman in
White, 1861.*

635.4 'This is the work of devils!' said the lama, recoiling
from the hollow echoing darkness, the glimmer of rails
between the masonry platforms, and the maze of girders
above. He stood in a gigantic stone hall paved, it seemed,
with the sheeted dead—third-class passengers who had
taken their tickets overnight and were sleeping in the
waiting rooms. All hours of the twenty-four are alike
to Orientals, and their passenger traffic is regulated
accordingly.
   Rudyard Kipling, Kim, 1901.

635.5 The lama, not so well used to trains as he had pre-
tended, started as the 3.25 a.m. south-bound roared in.
The sleepers sprang to life, and the station filled with

clamour and shoutings, cries of water and sweetmeat vendors, shouts of native policemen, and shrill yells of women gathering up their baskets, their families, and their husbands.
Ibid.

635.6 What haste and exquisite confusion at the station!
Arnold Bennett, Anna of the Five Towns, *1902.*

635.7 Along the wind-swept platform, pinched and white,
The travellers stand in pools of wintry light,
Offering themselves to morn's long, slanting arrows,
The train's due; porters trundle laden barrows.
Siegfried Sassoon, poem, Morning Express.

635.8 The officials seem to waken with a shout,
Resolved to hoist and plunder; some to the vans
Leap; others rumble the milk in gleaming cans.
Boys, indolent eyed, from baskets leaning back,
Question each face; a man with a hammer steals
Stooping from coach to coach; with clang and clack
Touches and tests, and listens to the wheels.
Ibid.

635.9 In the excellence of their illumination, the breadth of platform, in ease of movement from one to another, the convenience of the entrances and exits, booking offices and the comforts of the waiting, dining and refreshment rooms, they are as different from the Mugby Junction of fiction, and too often of fact, as the Midland trains are in comfort from the old-time trucks in which railways stored their passengers.
Anon. staff member of The Times, *in* The Complete Railway Traveller, *part I,* The Times, *21 November 1905.*

635.10 . . . you will find in a railway station much of the quietude and consolation of a cathedral. It has many of the characteristics of a great ecclesiastical building; it has vast arches, void spaces, coloured lights, and above all, it has recurrence of ritual.
G K Chesterton, Tremendous Trifles, *1909.*

635.11 A railway station is an admirable place, although Ruskin did not think so. He did not think so because he himself was even more modern than the railway station. He did not think so because he was himself feverish, irritable, and snorting like an engine. He could not value the ancient silence of the railway station.
G K Chesterton, The Prehistoric Railway Station, *Tremendous Trifles, 1909.*

635.12 If you wish to find the past preserved, follow the million feet of the crowd. At worst the uneducated only wear down old things by sheer walking. But the educated kick them down out of sheer culture.
I feel this all profoundly as I wander about the empty railway station, where I have no business of any kind. I

have extracted a vast number of chocolates from automatic machines; I have obtained cigarettes, toffee, scent, and other things that I dislike by the same machinery; I have weighed myself, with sublime results; and this sense not only of the healthiness of popular things, but of their essential antiquity and permanence, is still in possession of my mind.
Ibid.

635.13 They are our gates to the glorious and the unknown. Through them we pass out into adventure and sunshine, to them, alas! we return. In Paddington all Cornwall is latent and the remoter west; down the inclines of Liverpool Street lie fenlands and the illimitable Broads; Scotland is through the pylons of Euston; Wessex behind the poised chaos of Waterloo.
E M Forster, Howards End, *1910.*

635.14 And he is a chilly Londoner who does not endow his stations with some personality, and extend to them, however shyly, the emotions of fear and love.
Ibid.

635.15 The railroad terminal is the city gate. Without, it rises in the superior arrogance of white granite, as an architectural something. It has broad portals, and through these portals a host of folk both come and go. Within, this city gate is a thing of stupendous apartments and monumental dimensions, a thing not to be grasped in a moment. In a single great apartment—a vaulted room so great as to have dimensions run up into distant vistas—are the steam caravans that come and go. It is a busy place, a place of infinite variety of business.
Edward Hungerford, The Modern Railroad, *1911, quoted in* Trains Magazine, *December 1943.*

635.16 Another public clubhouse which the marooned businessman finds delightful and always full of good company is the railroad terminal. A big railroad station is an unfailing source of amusement and interest. From news-stand to lunch counter, it is rich in character study and the humors of humanity in flux.
Christopher Morley, Marooned in Philadelphia, *within* Travels in Philadelphia, *1920.*

635.17 People are rarely at their best when hurried or worried, and many of those who one meets at the terminal are in those moods. But, for any rational student of human affairs, it is as well to ponder our vices as well as our virtues, and the statistician might tabulate valuable data as to the number of tempers lost on the railway station stairs daily or the number of cross words uttered where commuters stand in line to buy their monthly tickets.
Ibid.

635.18 Inside the station the world is divided sharply into two halves. On the trainward side all is bustle and stir;

the bright colors of news-stands and flower stalls, brisk consultation of timetables at the information desk, little telephone booths, where lights wink on and off.

Christopher Morley, Broad Street Station, *within* Travels in Philadelphia, *1920.*

635.19 But abaft the big stairway a quiet solemnity reigns. The long benches of the waiting room seem a kind of Friend's meeting. Momently one expects to see some one rise and begin to speak. But it is not the peace of resignation; it is the peace of exhaustion. These are the wounded who have dragged themselves painfully from the onset, stricken on the great battlefield of Travel.

Ibid.

635.20 The station is a shop in which you receive the customer, and there is a lot to be done with this shop.

Frank Pick, 1923, quoted by John Glover in London Underground, *2004.*

635.21 Then would descend that peculiar silence of a country station that every one is settled, and the guard feels that it is safe to let the train start again.

M Vivien Hughes, A London Child of the Seventies, *1934.*

635.22 There are little stations threaded on the lines like beads on a rosary; they stand in the solitude like places of pilgrimage, far from the profane noises of the world; they are the real chapels dedicated to the silent ceremony of Waiting.

Karel Čapek, Intimate Things, *1935.*

635.23 The vast dingy sweep of the cement concourse outside the train-gates was pungent, as it always had been, with the acrid and powerfully exciting smell of engine smoke, and beyond the gates, upon a dozen tracks, great engines, passive and alert as cats, purred and panted softly, with the couched menace of their tremendous stroke. The engine smoke rose up straight into billowing plumes to widen under vaulting arches, to spread foggily throughout the enormous spaces of the grimy sheds. And beside the locomotives, he could see the burly denimed figures of the engineers, holding flaming torches and an oil-can in their hands as they peered and probed through the shining flanges of terrific pistoned wheels much taller than their heads. And forever, over the enormous cement concourse and down the quays beneath the powerful groomed attentiveness of waiting trains the tides of travellers kept passing, passing, in their everlasting change and weft, of voyage and return—of speed and space and movement, morning, cities and new lands.

Thomas Wolfe, Of Time and the River, *1935.*

635.24 Hail to the New World! Hail to those who'll love
Its antiseptic objects, feel at home.
Lovers will gaze at an electric stove,

Another *poésie de depart* come
Centred round the bus-stops or the aerodrome.
But give me still, to stir imagination
The chiaroscuro of the railway station.

W H Auden, Letter to Lord Byron, II.11, *within* Letters from Iceland, *1937.*

635.25 Here at the wayside station, as many a morning,
I watch smoke torn from the fumy engine
Crawling across the field in serpent sorrow.
Flat in the east, held down by stolid clouds,
The struggling day is born and shines already
On its warm hearth far off. Yet something here
Glimmers along the ground to show the seagulls
White on the furrows' black unturning waves.

Edwin Muir, poem, The Wayside Station, *inspired at Leuchars station, while changing train en route from St Andrews to Dundee, 1941. Published in* Collected Poems, *1943.*

635.26 After World War I Britain had a minor renaissance of station building. In modest stations on limited budgets she showed that she was still capable of distinguished designs. These reflected the functional approach as well as her great 19th-century terminals reflected the picturesque approach. The architects, led by Charles Holden of Adams, Holden and Pearson, were encouraged by the brilliant Frank Pick. From 1925 to 1949 these ingenious, economical buildings restored England's lapsed standing as a contributor to the development of the depot.

Carroll Meeks, The Railroad Station, an Architectural History, *1956.*

635.27 An empty bench, a sky of greyest etching,
A bare, bleak shed in blackest silhouette,
Twelve yards of platform, and beyond them stretching
Twelve miles of prairie glimmering through the wet.

Bret Harte, poem, The Station-Master of Lone Prairie.

635.28 North, South, East, West—the same dull grey persistence,
The tattered vapours of a vanished train,
The narrowing rails that meet to pierce the distance,-
Or break the columns of the far-off rain.

Ibid.

635.29 Britain has no peer for Old Station collectors, for we are immune to dirt, rejoice in age, and cheerfully tolerate the grotesque.

David P Morgan, Trains Magazine, *February 1961.*

635.30 None of the companies had a more lavish provision on paper for station modernisation than the LMS, and none bequeathed a larger backlog of relics to British Railways.

Edward Arkle, The Journal of Transport History, *May 1962.*

635.31 There's always something to see in a railway
station.
John Betjeman, commentary in television film on Somerset & Dorset Railway, 1963.

635.32 ... the stranger at Victoria, Waterloo or Liverpool
Street has to search for an insignificant, Regionally-
coloured sign directing him to buffet, taxi or waiting
room amongst a bewildering collection of neon-lit or
flashing signs extolling beer, daily newspapers, cigarettes,
and so on.
Brian Haresnape, Modern Railways, January 1964.

635.33 I agree that the giant roofs of St. Pancras, Paddington, King's Cross and Liverpool Street have precious
little function now that steam traction has virtually disappeared and that they are costly to maintain; nevertheless to exchange one extreme for another—an artificially-
illuminated cavern sunk beneath giant blocks of flats,
shops and offices—does not seem likely to create a rail
gateway that will appeal to future generations of
travellers.
Ibid.

635.34 Over us stands the broad electric face
With semaphores that flick into the gaps,
Notching the time on sixtieths of space,
Springing the traveller through the folded traps
Downstairs with luggage anywhere to go
While others happily toil upward too;
Well-dressed or stricken, banished or restored,
Hundreds step down and thousands get aboard.
Karl Shapiro, poem, Terminal, 1968.

635.35 The lamps are lit, the gas-lamps and the oil,
And shed their lemon and their yellow light,
Pallid and calm, while overhead the night
Draws up its wings and settles on the soil.
Henry Maxwell, poem, Dusk on a Branch, within A
Railway Rubaiyyat, *1968.*

635.36 But whether seen as a space, a shell, a triumph of
engineering mechanics, or as a solid, a mass, an architectonic design, the railroad station in general came to be
regarded as the symbolic gateway to the city. This kind
of centrally located monument was, in fact, the only
structure capable of serving as such a focus of movement, change, and growth in the evolving megalopolis
with its continually altering borders.
Douglas Richardson, within The Open Gate, *ed.
Richard Bébout, 1972.*

635.37 A microcity that embraces a great multiplicity of
elements divided between those introduced for comfort
and convenience of passengers on one hand and those
essential to the movement of trains and servicing of cars
on the other.
Carl Condit, on great railway termini, The Railroad
and the City, *1977.*

635.38 It is perhaps appropriate that in an age which can
design its high-speed trains to resemble aeroplanes and
its low-speed trains to resemble buses, its stations should
equally have no identifiable association with railways.
*J Richards and J M MacKenzie, The Railway Station,
1986.*

635.39 Where the palace, the temple and the cathedral had inspired their Victorian predecessors, British
station-builders of the inter-war years chose more "democratic" models—the cinema (Surbiton), the bank (Exmouth), and with unintended irony, the garage (West
Monkseaton). They in their turn were to give way after
the Second World War to the models of the bus-shelter,
the airport, and the office-block, the ultimate in bankruptcy of imagination.
Ibid.

635.40 She knew the railway station slightly. It was a place
that she enjoyed visiting, as it reminded her of the old
Africa, the days of uncomfortable companionship on
crowded trains, of the sugar cane you used to eat to while
away the time, and of the pith of the cane you used to spit
out of the wide windows.
*Alexander McCall Smith, The No. 1 Ladies' Detective
Agency, 1998.*

636. STATIONS—Commerce

636.1 Our stations, consequently, are not imitations of
baronial castles, nor are the accommodations furnished
above what the average of the community enjoy. There
may be an *aesthetic* value in the castle, as a beautiful
illustration of art. But it is not the province of railroad
companies to become instructors in architecture. Their
only object should be *money-making.*
*Comment in response to the publication of Herbert
Spencer's* Railway Morals and Railway Policy, *in* American Railroad Journal, *2 December 1854.*

636.2 Even tobacco shops have sprung up on some railway platforms. The platform is to be the free lounge of
the future, and we may, after a while, find a small town
erected upon it. The railway companies will have no
objection, and business could be carried on at a small
outlay, as the trains would deliver the purchases.
*H Stanley Tayler, Some Joys of Railway Travelling,
within* The Railway Magazine, vol. VIII, *1901.*

636.3 There is no money in stations.
Sir Frederick Banbury, noted in The London & North
Eastern Railway, *Cecil J Allen, 1966.*

636.4 A railroad company does not beautify its station
grounds for philanthropic reasons. The basic idea is to
increase traffic, and to do this, surroundings are made as
attractive as possible.
Paul Huebner, Horticulture, 10 March 1906.

636.5 A railway station is one of the essential factors of a railway system as a system of transportation, and may be defined, in very general terms, as one of the fixed or *stated* points at which railway transportation begins and ceases.

B C Burt, Railway Station Service, *1911.*

636.6 It scarcely needs to be stated that the station is the chief instrumentality by which the company is brought into contact with its productive environment, whether this be regarded as to the persons or the property which it includes. The station is the organ through which are rendered effective and manifest the functions constituting the essence of transportation and through which the transportation company derives the means of its sustenance and growth. This being the case, the station agent, it may here be said, is not compelled to regard himself as a mere tool of authority emanating from the company in a purely arbitrary manner but as the organic representative of the company in its dependence upon its environment and also, as well, of the environment in relation to the company. The station, we conclude, is an integrant organic constituent of the railway transportation system as such.

Ibid.

636.7 . . . the station is not a mere resultant, but one of the preconditions of the railway itself. But this must not be taken too one-sidedly. While this is true in a sense, and, as a consequence, it is true, railways are public utilities; they have a justification of their own which must not be entirely forgotten even here. Railways cannot be public utilities if they are recklessly managed, whether in the name of the public or not. There are certain principles according to which alone they may be scientifically, safely, and profitably conducted; and these principles have to be applied at the station as elsewhere on the line.

Ibid.

637. STATIONS—Disdain

637.1 Now, if there be any place in the world in which people are deprived of that portion of temper and discretion which is necessary to the contemplation of beauty, it is there.

John Ruskin, The Seven Lamps of Architecture, *1849.*

637.2 It is the very temple of discomfort, and the only charity that the builder can extend to us is to show us, plainly as may be, how soonest to escape from it.

Ibid.

637.3 The flushpots of Euston and the hanging gardens of Marylebone.

James Joyce, Finnegan's Wake, *quoted by Philip Howard*, The Times, *7 August 1998.*

638. STATIONS—In Society

638.1 It is a lively scene; all the gaiety of the packet-service without the sickness. Indeed, it is better than the packet-service; for while the sea air, salt water, and stuffy cabin deter ladies from expensive dress, so the spacious comfort, and perfect shelter of the railway station, invite a liberal display of clothes.

R S Surtees, Plain or Ringlets? *concerning Roseberry Rocks station, 1860.*

638.2 At the railway stations the loafers carry *both* hands in their breeches pockets; it was observable, heretofore that one hand was sometimes out-of-doors—here, never. This is an important fact in geography.

Mark Twain, quoting from his own journal compiled during a journey from New York to the South, date and place not given, Life on the Mississippi, *1896.*

638.3 The railroad station, being the front door to the neighbourhood, should have the same artistic qualities as the front door of a public building or a private residence.

Frank A Waugh, article, American City, *1905. Quoted by John R Stilgoe,* Metropolitan Corridor, *1983.*

638.4 A railway station speaks of epochs of decision in life, a parting of the ways, cross-roads in conduct. Shall we embark upon this adventure; shall we definitely declare our hand; shall we make a break in habit; or a departure from principle? Are we fleeing from Nineveh and duty, or going where Love and Right beckon us? As we wait at the station are we still counting the cost, and weighing consequences in the balance? Are we making a sacrifice in going away? Have we left a clean record behind? When the train has borne us away and we settle into our corner, have we feelings of remorse or satisfaction? Where will the journey's end be? Ah! where indeed? But have we done the right thing *now* according to all the lights of reason and grace?—then we may in faith leave the end in better Hands.

James Scott, Railway Romance and other Essays, *1913.*

638.5 The men's dressing room is a modern idea and is found in only a few stations. It is for the accommodation of those men who, if they happen to be commuters and do not desire to return home to dress for an evening engagement, may do so at very little expense. It is also available to those men who arrive on long-distance trains and desire to change their linen before attending to business engagements.

John A Droege, Passenger Trains and Terminals, *1916.*

638.6 About fifteen years ago, at the end of the second decade of this century, four people were standing together on the platform of the railway station of a town in the hills of western Catawba. This little station, really just a suburban adjunct of the larger town which, behind the concealing barrier of a rising ground, swept away a mile or two to the west and north, had become in recent years the popular point of arrival and departure for travellers to and from the cities of the east, and now, in fact, accommodated a much larger traffic than did the central station of the town, which was situated two miles westward around the powerful bend of the rails. For this reason a considerable number of people were now assembled here, and from their words and gestures, a quietly suppressed excitement that somehow seemed to infuse the drowsy mid-October afternoon with an electric vitality, it was possible to feel the thrill and menace of the coming train.

Thomas Wolfe, opening paragraph, Of Time and the River, *1935.*

638.7 Admittedly the majority of British railway stations suffer from location in depressing and often squalid surroundings, but the reason for this is not difficult to find. Faced during their birth and early growth by exorbitant prices demanded for land (especially in the towns) and antagonism from short-sighted local authorities and land-owners to "a new-fangled idea", the railways were frequently compelled to build their stations on the outskirts of the centres served. Industrial development in all its Victorian hideousness, quickly took place around many of the stations when once the advantages of the railway were realised.

George Dow, Design for Today, *April 1936.*

638.8 We're citizens of every community we serve, and must keep our houses as nice as those of the most meticulous housewives.

Anon. Seaboard Coast Line official, quoted in Railroad Magazine, *March 1941.*

638.9 A station is an intimate and informal place where even the most conventional people laugh, cry, kiss or curse openly under the perfectly correct assumption that everyone else is too busy to notice or care.

Anne Scott-James, of Paddington, War-time Terminus, Picture Post, *quoted in the* Great Western Railway Magazine, *August 1942.*

638.10 The station, indeed, is the introduction or prologue to the train, its relation to the train is that of the first course in an attractive and well-chosen meal. The relation may also be described as that of the frame of the picture; and many a picture has been spoilt by indifferent framing.

Christian Barman, Next Station, *1947.*

638.11 The railway lover counts no time wasted which he spends sauntering on a good station.

Canon Roger Lloyd, The Fascination of Railways, *1951.*

638.12 Railway termini and hotels are to the nineteenth century what monasteries and cathedrals were to the thirteenth century.

Comment in Building News and Engineering Journal, *1875, quoted by Carroll Meeks in* The Railroad Station, an Architectural History, *1956.*

638.13 The traveller at the depot waiting stands,
Impatient for the coming of the train;
The night is hastening on, the hour demands
That he the shelter of his home shall gain.

Jones Very, poem, The Traveller at the Depot.

638.14 Round and round went my head. It was nothing but trains, depots, crowds,—crowds, depots, trains,—again and again, with no beginning, no end, only a mad dance! Faster and faster we go, faster still, and the noise increases with the speed. Bells, whistles, hammers, locomotives shrieking madly, men's voices, peddlers' cries, horses hoofs, dogs' barkings—all united in doing their best to drown out every other sound but their own, and made such a deafening uproar in the attempt that nothing would keep it out.

Mary Antin, of a journey through Germany at age 13, while emigrating from Russia to America, 1894; The Promised Land, *1969.*

638.15 To understand the real India, the Indians say, you must go to the villages. But that is not strictly true, because the Indians have carried their villages to the railway stations.

Paul Theroux, The Great Railway Bazaar, *1975.*

638.16 ... the place where every invention of the Victorian age could first be seen.

David St John Thomas, The Country Railway, *1976.*

638.17 The Pacific Station looked promising. One man was mopping the floor of the lobby, another washing the windows: such attentions are a good indicator that the trains run on time. And there was an eight-foot statue of Jesus Christ across from the ticket window: Godliness and cleanliness.

Paul Theroux, of one of the stations in San José, Costa Rica; The Old Patagonian Express, *1979.*

638.18 Railway stations are most important in giving places an identity.

John Betjeman, Time with Betjeman, *BBC, 13 March 1983.*

639. STATION STAFF

639.1 North-Western porters and guards do their work with military precision, but with a finished nonchalance which is very appropriate to the oldest and most punctual of our great companies.

E Foxwell and T C Farrer, Express Trains English and Foreign, *1889.*

639.2 Oh Mr Porter, what shall I do?
I want to go to Birmingham
and they're taking me on to Crewe,
Send me back to London as quickly as you can,
Oh! Mr Porter, what a silly girl I am!
   Thomas le Brunn, chorus of song, Oh! Mr Porter, *popular music hall song, 1892.*

639.3 And what is this in blue and gold arrayed,
With linen spotless as th' untrodden snow?
It breathes, it speaks, it moves with stately tread,
And gently tells enquirers where to go.
   Anon., The Platform Inspector, *within* The Railway Magazine, *vol. V, 1899.*

639.4 He should cultivate patience and control his temper. Questions asked may seem, and frequently are, foolish, but civil replies should always be given. On the other hand, the employee should be impressed with the fact that he is not filling his position for the purpose of entertaining the public. He should answer questions carefully and courteously at all times, but it is not part of his duty to enter into needless conversation with passengers.
   John A Droege, Passenger Trains and Terminals, *1916.*

639.5 The most difficult thing to secure, however, is courtesy; and it is the most useful. No quality which a man can possess—save honesty—is as valuable in a ticket office. Patrons of the road will overlook mistakes of all kinds, but if they are discourteously treated, they cannot forget that.
   Ibid.

639.6 America provides a contrivance in a thousand situations where Europe provides a man or perhaps a number of men, and the work of our brass check is here done by porters, directed by the traveler himself. The men lack the memory of the check; the check never forgets its identity. Moreover, the European railways generously furnish the porters at the expense of the traveler.
   Stephen Crane, The Scotch Express, *within* Men, Women and Boats, *1921.*

639.7 "Porter!" Yessir; train to Margit? Yessir, platform number three.
Luggage? Yes, miss. Cab, sir, yessir. Made yer lose yer train, sir—ME?
No, sir, you're mistook, sir. "Porter!" Yes, miss. Lost yer ticket? Oh.
Werry sorry, miss, but—Eh, sir? Where's yer box, sir? I dunno.
Fare to Bournemouth? Couldn't say, mum; there's the booking office. Hi!
Cab there! Seen your missis? No I ain't, sir. You'll report me? WHY?

Change a shillin'? No, I can't, mum. No, sir, mustn't cross the line.
"PORTER!" Yessir. Labels? Right, sir. Scotland: Yes, miss, platform nine.
Yessir. Eight-fifteen just gone, sir. London? Number six, sir. Wot?
Smokers full? Well, 'tain't my fault, sir. Scoundrel, am I? No, I'm not!
Steamboat fare to Jersey? No, sir; couldn't tell yer. OUGHT to know.
Ought I?—Don't you swear at me, sir. Knock my head orf, will yer?—Oh!
"PORTER!" Yessir. Bill, this gent has left his luggage on the rack.
"Porter!" Yes, mum. Mind yer children? Yes, mum; hope you'll soon be back.
Three hours late your train, sir? was it? Traffic's werry thick, yer see.
Write and blame the silly Comp'ny, 'tain't no good to bully me.
"PORTER", yessir. "PORTER!" No, mum! Right, sir! Oh, I'm orf my head.
These Bank Holidays is times when porters wish as they was dead!
   F Raymond Coulson, poem, The Porter's Bank Holiday, *within* The Wonder Book of Railways for Boys & Girls, *c. 1921.*

639.8 No one else alighted at the little junction, and he found himself the solitary occupant of the platform on a soft November morning, with the exception of the station-master, who appeared to be guard, porter, ticket-collector, and everything else connected with the working of the remote country station on the Great Western, and who eyed him suspiciously out of a pair of ferrety green eyes half hidden beneath bushy grizzled eyebrows, that, in conjunction with a hawk-beak of a nose, gave him a vague likeness to some evil bird of prey.
   Ruth Alexander, Ghost Train, *adapted from Arnold Ridley's play of that name, 1927.*

639.9 'A phenomenon I have seldom seen,' he said cheerfully. 'A Wagon Lit conductor himself puts up the luggage! It is unheard of!'
   Agatha Christie; words spoken by Hercule Poirot. Throughout this book, although Agatha Christie refers to the Wagons-Lits company correctly by its full name on one occasion, elsewhere throughout the book she abbreviates it as Wagon Lit; Murder on the Orient Express, *1934.*

639.10 They number about three hundred, the Negro red caps at Grand Central; and if you care about such matters, they do *not* speak the minstrel-show dialect imputed to them by some of our sophisticated magazines. Their grammar is not too bad; for New York, it's better than average. Their diction is good, sound, working-class diction. And years of residence in New York have not robbed them of their deeply melodious voices. All in all, their speech is not unbeautiful; and that's more than can be said for New York as a whole.

David Marshall, Grand Central, *1946.*

639.11 Not enough porters. Such porters as there were seemed to be engaged with mail bags and luggage vans. Passengers nowadays seemed always expected to carry their own cases.

Agatha Christie, 4.50 from Paddington, *1957.*

639.12 . . . I arrived at the station, and by an oversight I happened to go out by the way one is supposed to come in, and as I was going out an employee of the railway company hailed me. "Hey, Jack," he shouted, "where do you think you're going?" That, at any rate was the gist of what he said.

Alan Bennett, Take a Pew, *within* Beyond the Fringe *review, 1961.*

639.13 Many porters are extremely polite and extremely patient. Others are not. The ratio is much the same among the passengers.

Dr Richard Beeching, after dinner, Institute of Directors, 1962.

639.14 Stationmasters as a rule are proper persons. David Worsley was by no means a proper person. He drank hard. His possession of a revolver was a menace to his neighbours.

Sir Sam Fay, quoted by George Dow in Great Central, *vol. III, 1965.*

640. STATISTICS

640.1 I will have no statistics on my railway!

Attributed to George Hudson, in George Hudson, *Alan Bailey, 1995.*

640.2 There is something barbarous about such an approximation, and it is disgraceful that at this late day we should in America be forced to estimate the passenger movement on our railroads in much the same way that we guess at the population of Africa.

Charles Francis Adams, of the lack of statistics upon passenger numbers on American railroads, Notes on Railway Accidents, *1879.*

640.3 Figures may be made to represent facts or become the finest and easiest form of camouflage. There are figures of fact and there are figures of deduction. An average is not always a fact and may easily become a disease.

Sir Sam Fay, general manager of the Great Central Railway, Presidential address to the Institute of Transport, 2 October 1922.

640.4 Operating statistics, like most railroad statistics, published or available, are largely historical. An operat-

ing officer is generally a poor statistician, and a statistician is not, as a rule, a good operating man. At least, this seems to be the opinion each has of the other.

Charles E Lee, Railway Age, *10 May 1924.*

640.5 As a means of checking the shifting of blame from one department to another, and for reducing the friction between departments, statistical work is very valuable. The knowledge that the "head office" is in possession of all the essential facts with which to measure efficiency of the work performed has a most stimulating influence, both upon the sluggard and upon the ambitious.

L F Loree, Railroad Freight Transportation, *1922.*

640.6 The Great Western Railway, though well run, had never been in the *avant-garde* of the pursuit of statistics, as the North Eastern was in the opening years of the twentieth century.

Jack Simmons, The Railway in Town and Country, *1830–1914, 1986.*

641. STEAM

641.1 Soon shall thy arm, Unconquer'd Steam, afar
Drag the slow barge, or drive the rapid car;
Or on the wide-waving wings expanded bear
The flying chariot through the fields of air: . . .

Erasmus Darwin, poem, often quoted in anticipation of railways, 1788.

641.2 Let them meet me on fair grounds, and I will soon convince them of the superiority of the 'pressure steam engine.'

Richard Trevithick, of the opponents of high pressure steam, 1804. Quoted in A Century of Locomotive Building, *J G H Warren, 1923.*

641.3 It's a proof what a strong thing steam is.

William Tayler, Diary, *1837, published 1962, edited by D Wise.*

641.4 Lay down your rails, ye Nations, near and far;
Yoke your full trains to Steam's triumphal car;
Link town to town: and in these iron bands
Unite th' estranged and oft embattled lands.
Peace and improvement round each train shall soar,
And knowledge light the ignorance of yore.

Charles Mackay, in the Railway Examiner, *25 November 1845. This differs somewhat from another version published later in* Voices from the Crowd, *1856, in which it appears as part of his poem* Railways.

641.5 Blessings on Science and her handmaid Steam!
They make Utopia only half a dream.

Ibid., lines 43 and 44 of Railways; *the same words appear in both versions.*

641.6 We speak of the array of a conqueror; where is there a conqueror like steam? Its panoply, too, is of iron; man has made it not less than mortal, as it performs the work of 100,000 of men's hands, and, as it is impatient of delay, it rushes through the bosom of the hills, its white and feathery plume is the ensign of daring, a courage that treads its way through the forest, or climbs the side of a mountain, and a power which, while it may find its comparison in the crest of Henry IV at Ivry, is the precursor of the triumphs, not of war but of peace, as they build up the fame, not of heroes, but of the people. That the fruition of these hopes will disappoint no reasonable expectations, but surpass them all, who of us can doubt?

John H B Latrobe, at banquet to celebrate the opening of the Baltimore & Ohio Railroad's line to Wheeling, 12 January 1853. Quoted by L F Loree, Railroad Freight Transportation, *1922.*

641.7 When I meet the engine with its train of cars moving off with planetary motion,—or, rather like a comet, for the beholder knows not if with that velocity and with that direction it will ever revisit this system, since its orbit does not look like a returning curve,—with its steam cloud like a banner streaming behind in golden and silver wreaths, like many a downy cloud which I have seen, high in the heavens, unfolding its masses to the light,—as if this travelling demigod, this cloud-compeller, would ere long take the sunset sky for the livery of his train; when I hear the iron horse make the hills echo with his snort like thunder, shaking the earth with his feet, and breathing fire and smoke from his nostrils, (what kind of winged horse or fiery dragon they will put into the new Mythology I don't know,) it seems as if the earth had got a race now worthy to inhabit it. If all were as it seems, and men made the elements their servants for noble ends! If the cloud that hangs over the engine were the perspiration of heroic deeds, or as beneficent as that which floats over the farmer's fields, then the elements and Nature herself would cheerfully accompany men on their errands and be their escort.

Henry David Thoreau, Walden, *1854.*

641.8 Steam is almost an Englishman. I do not know but they will send him to Parliament, next, to make laws.

Ralph Waldo Emerson, Essay English Traits—Ability, *1856.*

641.9 For ten decades and without challenge or questioning railroad history, practice and progress were postulated on the inescapable and demonstrated circumstance that water transformed into steam occupies sixteen hundred times its original space.

Lucius Beebe, Trains in Transition, *1941.*

641.10 Steam turned out to be capricious. It proved to be as much a master of what Americans called their Destiny as it was their slave. It carried the individual wherever he would go; and it carried away whole communities who

did not want to go anywhere at all. Either that, or it buried them where they were.

Stewart H Holbrook, The Story of American Railroads, *1947.*

641.11 There was no telling what steam would do, and many a fortune was made or lost because of its perversity. Before men realised what was going on, steam had moved the center of population from near the Atlantic seaboard to a point that existed in the school books of the same generation only as a deep wilderness. More than one pioneer related, in no more than his middle years, how the last Indian whoop and the last sad cadence of the owl had died in the echo of the first locomotive. Wilderness one year, metropolis the next. It constantly astonished those who had been through it, to their last day, as well it might.

Ibid.

641.12 Floreat vapor.
(*May steam flourish*)
Motto of the Bluebell Railway, incorporated in the coat of arms. 1950s.

641.13 With a steam locomotive, you create the power, you maintain the power, and you control the power.

Anon. locomotive man, The Ballad of John Axon, *broadcast 2 July 1958.*

641.14 Diesel is for unbelievers, electricity is wrong, steam has got the power that will pull us along.

Richard Stilgoe, song, Light at the End of the Tunnel, *within musical* Starlight Express, *1993.*

642. STEAM AGE

642.1 It is conceivable that the end is in sight for much that is picturesque and stirring to the imagination in the railroad scene. Electrification, Diesel power and the frequently fatuous devices of streamlining do not quicken the heart. Romance and glory are implicit in outside motion, in side rods, crossheads, eccentrics and the implacable rhythm of counter-balanced driving wheels reeling off the miles . . .

Lucius Beebe, Introduction to High Iron, *1938.*

642.2 English Electric and GE salesmen are doing to the Garratts of Africa what the ivory hunters did to the elephants of the same continent.

David P Morgan, Trains Magazine, *January 1966.*

643. GEORGE STEPHENSON

643.1 He is either ignorant or something else which I will not mention.

Edward Hall Alderson, Counsel in Parliamentary Committee hearings on the Liverpool & Manchester Bill, March 1825.

643.2 Your talents are of a much more valuable nature than that of a witness in the House of Commons.

John Moss, Deputy Chairman of Liverpool & Manchester Railway, letter to George Stephenson, 1825.

643.3 ... a gentleman thoroughly acquainted with practical mechanics, and possessing more experience in the construction and working of Railways than perhaps any other individual.

Henry Booth, An Account of the Liverpool and Manchester Railway, 1830.

643.4 That the directors cannot allow this opportunity to pass, without expressing their strong sense of the great skill and unwearied energy displayed by their engineer, Mr George Stephenson, which have so far brought this great national work to a successful termination, and which promise to be followed by results so beneficial to the country at large, and to the proprietors of this concern.

Resolution of the directors of the Liverpool & Manchester Railway, 14 June 1830.

643.5 Now for a word or two about the master of these marvels, with whom I am horribly in love. He is a man of from fifty to fifty-five years of age; his face is fine though careworn, and bears an expression of deep thoughtfulness; his mode of explaining his ideas is peculiar and very original, striking and forcible; and although his accent indicates strongly his north-country birth, his language has not the slightest touch of vulgarity or coarseness. He has certainly turned my head.

Fanny Kemble, actress, letter, 26 August 1830.

643.6 The Father of the Locomotive.

Dr Dionysius Lardner, speech, Literary & Philosophical Society of Newcastle, December 1836. This claim for Stephenson has not been widely accepted, simply on grounds of historical accuracy. For Stephenson himself on the subject see 209.2.

643.7 George Stephenson died the other day in England, the man who made the locomotive, the father of railroads,—and not an engineer on all our tracks heeded the fact, or perhaps knew his name. There should have been a concert of locomotives, and a dirge performed by the whistles of a thousand engines.

Ralph Waldo Emerson (who had met Stephenson on a visit to England), Journal LM, page 119, 1848.

643.8 Untaught, inarticulate genius.

Anon., quoted by John Francis, A History of the English Railway, 1851.

643.9 Born in a small cottage in Newcastle, and dying owner of the fine estate of Tapton; commencing life on a coal-heap, and ending it in a mansion; mending the peasants' clocks to pay for his son's schooling, and living

to see that son a senator; dining in his youth in the mine of Killingworth, and amusing his age in a horticultural contest with a duke; taught arithmetic at four-pence a week, and planning the most difficult railways in the kingdom; consulted by the premier, receiving honour from kings, a kind son, a faithful friend, and a loving father, the name of George Stephenson is one to which all men delight in doing homage.

John Francis, A History of the English Railway, 1851.

643.10 Unfortunately he was introduced to George Hudson; still more unfortunately, Faust was not inattentive to Mephistopheles.

Joseph Devey, The Life of Joseph Locke, 1862.

643.11 ... that great master of mechanical instinct ...

Walton Evans, in discussion papers within English and American Railroads Compared, 1887.

643.12 Think of him going after the job on the Stockton and Darlington Railway! With Nicholas Wood, he started from Killingworth and rode six miles, then went in a coach 30 miles to Stockton, then walked 12 miles through the fields over the line of the proposed railway, then had an interview with Edward Pease, the projector of the road, then walked on the return journey 18 miles from Darlington to Durham. What the "eight hour clock-watcher" might call a full day's work.

L F Loree, Railroad Freight Transportation, 1922.

643.13 To perceive the giant's power latent in a machine so crude, so hesitant and breathless, and to labour stubbornly in the face of every kind of odds both mechanical and human until that power was loosed upon the world, this was the historic role of George Stephenson.

L T C Rolt, George and Robert Stephenson, 1960.

644. STEPHENSON LOCOMOTIVE SOCIETY

644.1 When I became a member you had to tell them your favourite railway, and you had to give six reasons why it was the London Brighton and South Coast Railway.

Peter Semmens, of his joining in 1952, and of then president, John Maskelyne, who loved the Brighton line above all others. Address at the Annual General Meeting, 27 April 2002.

645. ROBERT STEPHENSON

645.1 ... he is decidedly the only man in the profession whom I feel disposed to meet as my equal or superior.

I K Brunel, quoted without source by Adrian Vaughan in Isambard Kingdom Brunel: Engineering Knight-Errant, 1991.

646. STOPPING

646.1 When necessary to stop a Train from any obstruction on the Line, or other cause, the Engineman shall give a tremulous sound of the Whistle as a Signal for the Guard to apply the Brakes; and should the Guard wish to attract the attention of the Engineman, he must do so by suddenly and repeatedly checking the Train by means of his Brake.

Regulation 20 of Regulations for observance by officers and servants of the London and Brighton Railway Company, *1845.*

646.2 If he'd tried to stop like that with the old Westinghouse he would have ended up going through the buffers.

Richard Hardy, of the air brake on Great Eastern locomotives, and of the cautious approach to Leeds Platform 3 by the driver of a modern electric multiple unit, 27 April 2002.

647. STORES

647.1 He was a little man with a straggling moustache and a humpy back, a pencil behind his ear, a *Park Drive* smouldering in his mouth, and an indignant and frequently outraged voice; indeed he *was* the Stratford stores.

Richard Hardy, of Charlie Lock at Stratford locomotive depot, Steam in the Blood, *1971.*

648. STRATEGIC RAIL AUTHORITY

648.1 The SRA is a slight misnomer: it isn't strategic and it has no authority.

Professor Roderick Smith, Lecture: Two and a half Millennia of Railways, *York, 5 June 2001.*

649. STRATFORD UPON AVON & MIDLAND JUNCTION RAILWAY

649.1 The Shakespeare Route
Advertising slogan 1911.

650. STREAMLINE TRAINS

650.1 Less esoteric but no less deserving of our consideration is the streamliner, that soundproof mix of roller bearings and double bedrooms and tinted dome glass and leg-rest seats and stainless steel that remains the civil way of getting from here to there.

David P Morgan, Trains Magazine, *October 1964.*

651. STRIKE

651.1 Red are the rails with rust to-day,
Red is each standing wheel;
No cheerful clank from the gleaming crank,
Or the kiss of steel on steel.
No whistle shrill awakes the hill
To fling an echo back,
Nor piercing beam of a signal's gleam
To give, or bar the track.

Henry Chappell, porter, Great Western Railway, Bath; poem, The Coal Strike, *within* The Day and other Poems, *1918.*

652. SUBSIDY

652.1 We in Central would rather give good service and be subsidized than to give poor service and be criticized by the passengers.

Robert R Young of the New York Central, quoted in Trains Magazine, *September 1955.*

652.2 We are all in favour of reducing the so-called subsidy to the railways, as long as it is not on a line that serves some people who might vote for us at the next election.

Robert Adley MP, reported in Modern Railways, *April 1983.*

653. SUBURBAN SERVICE
See also Season Tickets and Season Ticket Holders

653.1 Let no one, unless he has stood on the bridge at Liverpool Street, say, at 9 o'clock in the morning or 6 o'clock in the evening, ever suppose he knows what suburban traffic really can be.

W M Acworth, The Railways of England, *1899.*

653.2 Thus the company was in a fair way to the speedy completion of duplicate, and in places triplicate, tracks throughout the whole of its suburban area between Barnet and King's Cross. Nevertheless it was no nearer getting rid of the "suburban incubus" than it had been at any time in within the previous twenty years.

Charles Grinling; it is not known if Grinling was quoting another with his reference to suburban incubus, *or if he had coined this expression himself.* The History of the Great Northern Railway, *1903.*

653.3 "Romance!" the season-tickets mourn,
    "He never ran to catch his train,
But passed with coach and guard and horn—
    And left the local—late again
Confound Romance!"....And all unseen
Romance brought up the nine-fifteen.

Rudyard Kipling, poem, The King.

653.4 So Swiftly Home.
Southern railway poster for electrified services, 1932.

653.5 Hauling urban and suburban passengers is a sick business wherever found. It is sick from undernourishment. It is undernourished, not because of any fault of management, employees or owners—but as the result of joint action on the part of the regulatory authorities who limit rates, of the political authorities who surround the business with inequitable competition, and of the taxing authorities who have taken their pound of flesh regardless of its lethal effect on their victim.

Editorial Comment, Railway Age, *quoted* in extenso *in* Trains Magazine, *February 1951.*

653.6 If you don't like it, walk.

Alfred E Perlman, President, Penn Central Railroad, to his son, Lee. Quoted by Joseph Daughen and Peter Bintzen in The Wreck of the Penn Central, *1971.*

SUBWAY
See Underground

654. SUICIDE
See also Death

654.1 I'm gonna lay my head
On some lonesome railroad line,
And let that midnight freight train
Pacify my mind.

Song, Trouble in Mind, *stated to be by a man called Jones, date not known.*

654.2 But always the wheels of the good old-fashioned railway train provide the merit of being self-operating. All you need to do is to make a four-foot jump.

Ian Fleming, You Only Live Twice, *1964.*

655. SUMMER

655.1 The sun has beaten on the track all day,
The glinting rails have shimmered in the heat,
The cinder path has burned the passing feet,
The weeds have withered in the six-foot way.

The lineside grass is powdery with dust,
The singing wires flash lightly overhead;
A water-butt is dappled black and red
Where blistered paint exposes ancient rust.

Old tar upon the ganger's hut is soft,
The sleepers breathe a scent of creosote,
There is a noise of reaping far remote
And swift and swallows circle high aloft.

Henry Maxwell, poem, Lineside in Summer, *within* A Railway Rubaiyyat, *1968.*

655.2 And once in every while there comes a crack
From heat-expanded metals that contract,
And with each foot placed perfectly exact,
A questing cat walks primly down the track.

Ibid.

656. SUMMIT STATION

656.1 Just a ornery little station at the toppin' of the Pass—
Nary tree nor ary water, an' uncommon short on grass;
Jest a measly little shack a-squattin' down inside the wye,
Safe corralled there by its lonely,—sassy-like and high and dry,
Sand around it, sun above it, purty shy on proper shade,
But a-standing like I tell yo' some four thousand up the grade.

J C Davis, poem God's Country, *verse 1, Summit California, 1907, quoted in full in* Trains Magazine, *September 1974.*

657. SUNDAY

657.1 Persons purchasing tickets for Sunday trains will be required to sign a pledge they will use the tickets for no other purpose than attending church.

Timetable annotation, 1850, attributed to a predecessor company of the Boston & Maine Railroad (probably the Boston & Portland) by Stewart H Holbrook in The Story of American Railroading, *1947.*

657.2 Now, how different might have been the railway system if George Stephenson had resolved "to keep the Sabbath-day holy;" his influence would at least have limited Sunday trains, if it had not prevented them.

Edward Corderoy, Lecture to the Young Men's Christian Association, *Exeter, published 1858.*

657.3 The established Sunday tyranny, which is one of the institutions of this free country, so times the trains as to make it impossible to ask anybody to travel to us from London.

Wilkie Collins, The Moonstone, *1868.*

657.4 Why, Soondays hain't cropt out here yet!

Anon. worker, recorded by D W Barrett in Life and Work among the Navvies, *1880.*

657.5 Whatever the original inferences of the title of the Greek film *Never on Sunday,* there still seems to be on the part of British Rail a resolve to make the phrase one of admonition to potential Sabbath travellers.

Editorial comment, Modern Railways, *September 1975.*

658. SURVEYORS

658.1 Our party consisted, besides myself, of a transitman, head chainman, rear chainman, back flagman, stake marker, stake driver, levelman, rodman, topographer, assistant topographer, draftsman, two line teamsters, a supply teamster, and a cook. We also had to have a dog, for a camp without one was, to us, the same as a home without children.

Edward Gillette, Locating the Iron Trail, *1925, quoted by Richard Reinhardt in* Workin' on the Railroad, *1970.*

659. SWINDON

659.1 It was noted for its spacious and gorgeously decorated refreshment room, the finest in provincial England. Artists had illuminated its lofty walls terminating in gilded cornices, with the designs then in vogue, while winged cherubs, if memory serves me, blew trumpets and waved flowing garlands among clouds upon the high ceilings.

A G Bradley, Exmoor Memories, *1926.*

659.2 They *can't* leave Swindon under ten minutes, no matter how late we are.

Mrs Hughes, (mother of author Vivien Hughes), recorded by the latter in A London Child of the 1870s, *1934.*

660. SWISS RAILWAYS

660.1 The Swiss trains go upwards of three miles an hour, in places, but they are quite safe.

Mark Twain, A Tramp Abroad, *1880.*

SWITCHING
See Shunting

661. TALGO TRAINS

661.1 ... the train that looks like a half-used toothpaste tube.

Anon., caption to photograph, Trains Magazine, *September 1958.*

662. TAUNTON STATION

662.1 Everything is hideous, dirty and disagreeable; and the mind wanders away, to consider why stationmasters do not more frequently commit suicide.

Anthony Trollope, The Belton Estate, *1866.*

662.2 Taunton is not like Crewe, which is wide awake all night.

C Hamilton Ellis, Trains Illustrated, *May 1955.*

663. TAY BRIDGE

663.1 Beautiful Railway Bridge of the Silv'ry Tay!
Alas! I am very sorry to say
That ninety lives have been taken away
On the last Sabbath day of 1879,
Which will be remember'd for a very long time.

William McGonagall, The Tay Bridge Disaster, *1879.*

663.2 The conclusion, then, to which I have come is that this bridge was badly designed, badly constructed, and badly maintained, and that its downfall was due to inherent defects in the structure, which must have, sooner or later, brought it down.

H C Rothery, Wreck Commissioner, quoted in The Railway Magazine, vol. VIII, *1901.*

664. TEA

664.1 Railway tea is usually an abomination, but the "cup that cheers" provided here is really excellent.

J W Walker, of the refreshments rooms at Stoke on Trent, Railway Notes No. 23, November 1910.

664.2 Finally, after much cogitation, some genius, whose name ought to be immortalised with that of George Stephenson, suggested that the tea should be boiled in the tenders of a number of locomotives set apart for the purpose. I believe the original idea was to empty cases of tea into the tenders full of cold water and then turn live steam from the engine loose into the mixture. The drawback was that the resultant tea might have left something to be desired as a cheering beverage ....

E L Ahrons, The Railway Magazine, *July 1916, on the arrangements for the celebration of Queen Victoria's jubilee, 1887, at a fete in Swindon.*

665. TELEGRAPH

665.1 To agent and operator at Goshen: Hold the train for further orders.

Charles Minot, the first train orders transmitted by electric telegraph in America, 22 September 1851, quoted by means of an illustration of a commemorative tablet by L F Loree, Railroad Freight Transportation, *1922. The order (not the tablet) was stated by Stewart H Holbrook to have read: "Hold eastbound train till further orders";* The Story of American Railroading, *1947.*

665.2 Near at hand the wires between the telegraph poles vibrated with a faint humming under the multitudinous fingering of the myriad of falling drops, striking among them and dripping off steadily from one another. The poles themselves were dark and swollen and glistening with wet, while the little cones of glass on the transverse bars reflected the dull grey light of the end of the afternoon.

Frank Norris, The Octopus, *1901.*

665.3 I would rather have a road of a single track with electric telegraph to manage the movement of its trains than a double track without it.

D C McCallum, Superintendent Erie Railroad, quoted by L F Loree, Railroad Freight Transportation, *1922.*

665.4 The night telegraph-operator at the railroad station was the most melodramatic figure in town; awake at three in the morning, alone in a room hectic with the clatter of the telegraph key. All night he "talked" to operators twenty, fifty, a hundred miles away. It was always to be expected that he would be held up by robbers. He never was, but round him was a suggestion of masked faces at the window, revolvers, cords binding him to his chair, his struggle to crawl to the key before he fainted.

Sinclair Lewis, Main Street, *1922.*

665.5 The Mail roars by 'neath a starlit sky
As the dark, wee hour appears,
The desert broods in a thousand moods
As she has for a million years.
The wall clock's crack in my lonely shack
Is time to a nocturne's flow,
And the sounder's spree is the song to me,
From files of long ago.

Charles D Dulin, poem, first verse, Morse Memories, *within* Railroad Magazine, *1945.*

666. TELEPHONING THE RAILWAY

666.1 It is curiously difficult to feel self-assured when ringing up a railway.
Paul Jennings, The Jenguin Pennings, *1963.*

666.2 There follows a long, hollow silence, punctuated by someone making squeaky noises with a toy balloon, or by frantic clickings, like desperate machine-gun fire, as the call is transferred from the slick, sophisticated Post Office system to the heavy, ironclad telephones and tremendous alarm bells of the railway: it is easy to visualise the wires looped along black walls, through tunnels and warehouses, and indeed it would not be surprising if the calls were finally answered from one of those mysterious little sheds, just outside any big station, where men in shirtsleeves are always drinking tea by gaslight.
Ibid.

667. TEMPERANCE
See also Alcohol, Beer

667.1 Happy Camborne, happy Camborne,
Where the railway is so near,
And the engine shows how water
Can accomplish more than beer.
Anon., poem, written for the occasion of a temperance special, West Cornwall Railway, 1852.

667.2 What a glorious thing it would be if the newly developed powers of railways and locomotion could be made subservient to the promotion of temperance.
Thomas Cook, describing the thought which led eventually to his creation of Thomas Cook Ltd., originally appearing in Leisure Hour *but also quoted by Piers Brendon in* Thomas Cook, *1991.*

668. TENDERS

668.1 This is a coal cart, not a hearse.
Nigel Gresley, to the Chief Draughtsman at Darlington Works, describing what he wanted for what became the LNER Group Standard tender. Letter from ex-draughtsman R W Taylor to Geoffrey Hughes, 29 December 1981.

669. THE TEXAS EAGLE

669.1 My Grandaddy was a railroad man
When I was young he took me by the hand
Dragged me to the station at the break of dawn
Said "boy I got to show you somethin' 'fore it's gone"
She was blue and silver—she was right on time
We rode that Texas Eagle on the Mopac line.
Steve Earle, song, The Texas Eagle, *no date found.*

670. MARGARET THATCHER

670.1 ... Downing Street, where Margaret Thatcher can sound like a railway enthusiast as she recalls great steam engines thundering through Grantham; but she also appears to regard BR as a symbol of nationalised profligacy to be damned at all points while her newly knighted adviser Alfred Sherman is dedicated to having every railway line concreted over to create a motorised nightmare.
Ian Waller, The Daily Telegraph, *10 August 1983.*

671. THEATRICAL SPECIALS

671.1 Here was a quiet, unseen, essential service afforded by the railways: part of the assumptions on which the Victorian theatre rested in the small island of Great Britain.
Jack Simmons, The Victorian Railway, *1991.*

672. THEFT

672.1 A man who has never gone to school may steal from a freight car, but if he has a university education he may steal the whole railroad.
Attributed to Theodore Roosevelt by en.thinkexist.com. This quotation appears on many websites. No source for it, nor confirmation of the precise words, has been found.

673. THINKING

673.1 As Father Brown wrote the last and least essential part of his document, he caught himself writing to the rhythm of a recurrent noise outside, just as one sometimes thinks to the tune of a railway train.
G K Chesterton, The Innocence of Father Brown (The Queer Feet), *1911.*

673.2 Nowhere can I think so happily as in a train.
A A Milne, A Train of Thought, *within* If I May, *1920.*

673.3 Train travel animated my imagination and usually gave me the solitude to order and write my thoughts: I travelled easily in two directions, along the level rails while Asia flashed changes at the window, and at the interior rim of a private world of memory and language. I cannot imagine a luckier combination.
Paul Theroux, The Great Railway Bazaar, *1975.*

674. THIRD CLASS
See also Class Distinction, First Class, Second Class

674.1 ... third class facilities are a breach of contract, a premium to the lower orders to go uselessly wandering about the country.
Duke of Wellington, quoted by Clement Edwards, Railway Nationalisation, *1898.*

674.2 ... the very lowest orders of passengers ...

Charles Saunders, Secretary, Great Western Railway, evidence before a Parliamentary Committee, July 1839.

674.3 I do not know what number of passengers are put into the third-class trucks, because there is a great difference in the way of packing them.

Attributed to George Hudson by C F Dendy Marshall, A History of the Southern Railway, *1936.*

674.4 At Yarmouth the passenger finds himself at the station, provides himself with his ticket, and takes his seat in a 1st class carriage; or, if economy be his object, in a 2nd class; or if desirous of plenty of air, much smoke, but a more extensive view, in a 3rd class.

Anon. Commentary upon the earliest carriages of the Yarmouth & Norwich Railway, 1845. Quoted in The First Railway in Norfolk, *George Dow, 1947.*

674.5 Pity the sorrows of a third-class man,
Whose trembling limbs with snow are whitened o'er,
Who for his fare has paid you all he can:
Cover him in, and let him freeze no more!

First verse of poem, The Third-class Traveller's Petition, Punch, *1845.*

674.6 Rock away, passenger, in the third class,
When your train shunts, a faster will pass.
When your train's late your chances are small,
Crushed will be carriages, engines and all.

Anon. poem, Punch, *late 1852, quoted by Jack Simmons in* The Express Train and Other Railway Studies, *1994.*

675. THOMAS THE TANK ENGINE

See also Ivor the Engine

675.1 Thomas was a tank engine who lived at a Big Station. He had six small wheels, a short stumpy funnel, a short stumpy boiler, and a short stumpy dome.

Rev Wilbert Awdry, Thomas the Tank Engine, *1949.*

675.2 Understand this: preservation is very expensive. The more fares we collect, the more we preserve, and no-one has figured a better way to get people to line up at a tourist railroad than Thomas. Our employees call him "Thomas the Bank Engine."

Linn Moedinger, President, Strasburg Railroad, quoted in Trains Magazine, *May 2001.*

676. EDWARD THOMPSON

676.1 He was a most imposing figure, tall, aristocratic looking, immaculately dressed, whose wont at the beginning of a meeting was to set out on the table before him an assortment of gold pencils, chains, watches and other symbols of well-being.

E S Cox, A Locomotive Panorama, *1965.*

677. SIR HENRY THORNTON

677.1 There is no man in England big enough for this job; here is the man who will do it!

Lord Claud Hamilton, of his selection of Thornton, a Canadian, for the General Managership of the Great Eastern Railway, May 1914, quoted by D'Arcy Marsh, The Tragedy of Sir Henry Thornton, *1935.*

678. THREATS

678.1 You may seek it with thimbles—and seek it with care;
You may hunt it with forks and hope;
You may threaten its life with a railway share;
You may charm it with smiles and soap—

Lewis Carroll, The Hunting of the Snark, *1876.*

679. TOM THUMB

679.1 This little engine was so small,
It hardly seemed like one at all.
Ran a race with a horse one day,
Engine broke down—horse ran away!

Attributed to M T S, A Study of Railway Transportation, *vol. 2, Association of American Railroads, 1942.*

TICKET AGENTS

See Booking Clerks

680. TICKET COLLECTORS and COLLECTING

680.1 Some of the railway conductors are particularly coarse-minded fellows and placed as they imagine, in some authority, they occasionally assume the manners of petty tyrants. It is scarcely possible to describe the style of contemptuous insolence with which they sometimes treat the passengers. However in justice to the more refined States in the North-eastern portion of the Union, I am ready to testify that the unpleasant behaviour to which I allude in no instance there occurred to me.

Lt Col Arthur Cunynghame, A Glimpse at the Great Western Republic, *1851.*

680.2 Conductors, when you receive a fare,
Punch, in the presence of the passenjare:
A blue trip slip for an eight-cent fare,
A buff trip slip for a six-cent fare,
A pink trip slip for a three-cent fare;
Punch in the presence of the passenjare.
CHORUS.
Punch, brother! punch with care!
Punch in the presence of the passenjare.

Mark Twain, Punch, Brothers, Punch!, *1878.*

680.3 Ticket-collecting is a slow business in the East, where people secrete their tickets in all sorts of curious places.
Rudyard Kipling, Kim, *1901.*

680.4 All Passengers Will Be Required to Show Their Tickets at the Gates.
Sign in bronze letters over each of the New York Central gates within Grand Central Terminal, New York. The politeness of the wording is said to result from the NYC's desire to counter the bad press it got from William Vanderbilt's use of the expression 'The Public be Damned' (q.v.). The lettering was removed for scrap on the day after the attack on Pearl Harbor, according to David Marshall, in Grand Central, *1946.*

680.5 The North Eastern of those days without its ticket platforms was like a ship without a rudder: at least this would appear to be the case, for wherever there was a nice open space the "land development committee" of the company promptly filled it up with a ticket platform. If one took a journey of any length on this railway there was very little of the original ticket left at the end of it. The North Eastern collectors were the most expert ticket whittlers that the country possessed.
E L Ahrons, The Railway Magazine, *December 1916.*

680.6 No ticket! Dear me, Watson, this is really very singular. According to my experience it is not possible to reach the platform of a Metropolitan train without exhibiting one's ticket.
Sir Arthur Conan Doyle, The Adventure of the Bruce-Partington Plans, His Last Bow, *1917.*

680.7 Every year or so one reads about a railroad conductor on a suburban train making his last run and being feted by the passengers, most of whom know him well. This sort of farewell celebration seems to be peculiar to railroad men. All sorts of other people step out of harness and nobody thinks much about it, but a railroad man finishing his work excites the populace unduly. I think this is probably because commuters see, in conductors and brakemen and engineers, the personification of their own frustrated transit—a man who has ridden far and got nowhere.
E B White, quoted by Caskie Stinnett in Will Not Run, *Feb. 22nd, 1949.*

680.8 It isn't easy to be a conductor on a commuter train despite the general feeling that anyone with an irascible nature and strong misanthropic drive can qualify.
Ibid.

680.9 The principal function of a ticket collector on British trains will no doubt continue to be that of waking up sleeping passengers.
· *George Chowdharay-Best, letter to the* Daily Telegraph, *of changes in railway operation after privatisation; 9 June 1995.*

680.10 Like a ticket inspector recently retrained in the art of customer relations, Gordon Brown was all smiles and small talk.
Matthew Parris, The Times, *17 May 2001.*

681. TICKETS
See also Fares and Charges

681.1 Every now and then a well-dressed man hurries past into the booking-office and takes his ticket with a sheepish air as if he was pawning his watch.
Samuel Sidney, of Euston station, Rides on Railways, *1851.*

681.2 "Don't make excuses," said the Guard: "you should have bought one from the engine-driver."
Lewis Carroll, Through the Looking Glass, *1872.*

681.3 List of booking. You passengers must careful. For have them level money for ticket and to apply at once for asking tickets when will booking window open. No tickets to have after departure of the train.
Sign at booking office window, station in Wales, 1875. Quoted in Railway Adventures and Anecdotes, *Richard Pike, dated by Jack Simmons in* The Victorian Railway, *1991.*

681.4 Tickets once nipped and defaced at the barriers, and the passengers admitted to the platform, will have to be delivered up to the Company, in event of the holders retiring from the platform without travelling, and cannot be recognised for re-admission.
South Eastern Railway sign at Cannon Street and elsewhere, 1880. Quoted by Jack Simmons in The Victorian Railway, *1991.*

681.5 "You have come in by train this morning, I see."
"You know me, then?"
"No, but I observe the second half of a return ticket in the palm of your left glove."
Sir Arthur Conan Doyle, The Adventure of the Speckled Band, *1892.*

681.6 Such is the simple system of passenger booking, designed to give the minimum of trouble to the travelling public, and generally thought well of. There are, however, a few ignorant persons who strive to cripple it by travelling without tickets, or putting them to uses for which they were not intended. But with the spread of education and the increase of idiot asylums, such people become fewer; and every year there is a slight increase in the percentage of tickets returned from the collectors in good condition.
W J Gordon, Everyday Life on the Railroad, *1898.*

681.7 School tickets will not be sold to persons attending musical, dancing, riding, dress-making, type-writing, stenographic or schools of similar character.

Regulations, suburban timetables, Delaware Lack-
awanna & Western Railroad, 1901, quoted in Trains Magazine, *January 1985.*

681.8　Tickets Please! So do Bovril sandwiches.
Advertisement commonly seen on British stations,
1930s. Nearly thirty such eye-catching slogans are recorded
in The Railway Magazine, vol. LXXI, *1932.*

681.9　No other printed contracts on paper of the size has
furnished as many examples of confusion, frayed as
many tempers, and caused as many frustrations between
people.
E E Gordon, quoted from the Ticket Agent *in* Trains
Magazine, *September 1957.*

681.10　Ticketless travel is a social evil.
Sign on Indian railway station, noted by Paul Theroux
in The Great Railway Bazaar, *1975.*

681.11　Delay advised.
Legend stamped upon a railway ticket issued to Prof Jack
Simmons. Neither its meaning nor its intention is known.

681.12　Bath tickets, please!
Conductor on Bristol-Weymouth train, 1 August 2000.

681.13　All valid tickets are available on this train.
Notice on departure board, First Great Western, Pad-
dington, December 2003.

682. TILTING TRAINS

682.1　I just stand there with my legs open.
BR stewardess, interviewed by The Guardian, *when*
asked how she coped with the tilting of the APT train,
December 1982, quoted by Sir Peter Parker, For Starters,
1989.

683. TIME

683.1　By your cursed neglect of that you have wasted
more of my time than your life is worth.
I K Brunel, letter to Fripp (thought to be L C Fripp),
concerning his practice, forbidden by Brunel, of making
drawings on the backs of others. Quoted by Adrian
Vaughan in Isambard Kingdom Brunel: Engineering
Knight-Errant, *1991.*

683.2　When the poor man travels he has not only to pay
his fare but to sink his capital, for his time is his capital;
and if he now consumes only five hours instead of ten in
making a journey, he has saved five hours of time of useful labour, useful to himself, his family and society.
Sir James Allport, speech quoted by Hamilton Ellis in
Nineteenth Century Railway Carriages, *1949.*

683.3　The richest cargo in the world is a cargo of TIME,
and the locomotive was made to draw it.
Benjamin Taylor, The World on Wheels, *1874.*

683.4　No single element enters into transportation so
continuously, so pervasively, or with such vital importance as that of time. Not only are the train movements
related to time but their safety is dependent upon its
exact observance, while in the conduct of transportation
it is essential that all regulate their conduct with reference to the same exact standard.
L F Loree, address at memorial service in honor of Webb
C Ball, leader in the accurate measurement and use of rail-
road time; Washington, 11 May 1922. The selfsame words
also appeared in Loree's book Railroad Freight Transportation *which was published that year.*

683.5　One hears of time standing still; in my case it took
two paces smartly to the rear.
Peter Fleming, of a journey on the Trans-Siberian Rail-
way; One's Company, *1934.*

683.6　Let it suffice to say that the Silver Jubilee has caused
many inhabitants of the ancient city of York to revise
their life-long habit of correcting their time-pieces by the
strokes of Big Peter. They now rely upon the Silver Jubilee's distinctive whistle as she passes through the station
at 8.9 p.m. each day, and Big Peter is outdone.
Anon., LNER Magazine, October 1936.

683.7　These cottagers told the time by the smoke of the
trains which passed in the valley. They got up by the milk
train between four and five; they had breakfast by the
paper train at half past seven: the London express at half-
past twelve was their dinner-bell. If they had a clock, they
did not use it or even wind it up.
Edith Olivier, Without Knowing Mr Walkley, 1939.

683.8　If you will follow the simile, regard history as a train
speeding along through time. Birds and animals are disturbed by the noise and tumult of the train's passage,
they fly away from it or run fearfully or cower, thinking
they hide. I am the hawk that follows the train—you have
no doubt seen them doing this, in Greece, for instance—
ready to pounce on anything that may be flushed by the
train's passage, by the passage of history.
Ian Fleming, Goldfinger, 1959.

684. TIMETABLES
See also Bradshaw's Guide, The Official Guide

684.1　A Double Special Express Service to Paris daily in
11 hours, *via* Boulogne and Calais alternately: changing
the hours of departure, both from Paris and from London, to suit the tide and prevent all delay.
Timetable announcement by the South Eastern Rail-
way, 1851, recorded by G P Neele, Railway Reminiscences,
1904.

684.2　I desired also to abolish timetables; and, further, to
deal with traffic as if it were a constant and not a fluctuating quantity. My train would always consist of the same

number of carriages, and that train would pendulum between point and point, as the trains of the Metropolitan and Greenwich lines do at present. Thus there would be no attaching and detaching.

Sir Edward Watkin, of the services on the then recently opened Cheshire Lines Committee, letter, 18 June 1877, quoted by George Dow in Great Central, *vol. II, 1962.*

684.3　　　　　　　　　　A modern Hotspur
Shrills not his trumpet of 'To Horse! To Horse'
But consults columns in a Railway Guide,
A demigod of figures, an Achilles
Of computation.
A verier Mercury, express come down
To do the world with swift arithmetic.

Attributed to "Dipsychus" in Express Trains English and Foreign, *E Foxwell and T C Farrer, 1889.*

684.4 "This is a sad case," said the attendant at an insane asylum, pausing before a padded cell. "There is no hope for the patient whatsoever."

"What's his trouble?" asked the visitor.

"He thinks he understands an Erie time-table."

Joke collected and published as part of a successful campaign by the Erie Railroad to stop the flow of Music-Hall jokes about the railroad. The Railroad Man's Magazine, *January 1910.*

684.5 On another small table stood Zuleika's library. Both books were in covers of dull gold. On the back of one cover BRADSHAW, in beryls, was encrusted; on the back of the other, A.B.C. GUIDE in amethysts, beryls, chrysoprases, and garnets.

Max Beerbohm, Zuleika Dobson, *1911.*

684.6 When we by train decide to go,
A pleasant trip anticipating,
Alas! how little do we know
What troubles we're accumulating.

For instance, when we're striving hard
To reach some West-bound station lonely,
How oft we find upon the card *

And if on Monday we should try
To catch this train with grim precision,
Upon the bill we're sure to spy †

Another train we struggle then
To find—and are informed with unction
That passengers to Cowslip Glen ‡

Or if instead we think we'll go
To Mudcombe Marsh, ere summer's ended,
This sign the table's sure to show §

Undaunted, still we persevere
And settle on Hawthornley Dene
To find that passengers from here ‖

Through seas of figures then we wade,
And hieroglyphic signs unfeeling
To see, when next a choice is made, ¶

Once more the direful sheet we scan,
Our travel prospects nearly flitted,
To meet another cheerless ban ☞

Enraged, our quest we still pursue,
And from our course brook no deflection
Until this symbol comes in view °

And further down, we note this sign,
Which worst of all that we can quote is *₊*

* This train stops here on Mondays only.
† Runs only on the East Division.
‡ Must change at Muddle-Puddle Junction.
§ The Mudcombe service is suspended.
‖ Alight at Brambleborough Green.
¶ This train now only runs to Ealing.
☞ No third-class passengers permitted.
° This train no longer has connection.
₊ All trains are subject on this line
　　　To alteration without notice.

Entitled Holiday Troubles, *quoted from* London Opinion *within* The Railway Magazine, *vol. XXXI, 1912.*

684.7 People don't like timetables, make it easy for them.

Sir Herbert Walker, quoted by John Glover in Southern Electric, *2001.*

684.8 A lady tubed in a tight skirt totters valiantly down the road towards the station, and the courteous train waits for her. If the director general of railroads were a bachelor perhaps he would insert a new footnote in his timetables: "Sk," will not wait for ladies in hobble skirts.

Christopher Morley, referring to the government's director general of railroads during and after the Great War; The Paoli Local, *within* Travels in Philadelphia, *1920.*

684.9 Mixed Train Daily.

Timetable annotation, possibly apocryphal, that inspired the title of the Lucius Beebe book of that name, and which has been described as the only railroad book to be included in a best seller list. Symbols for frequency and train type were usually shown separately, but the Missouri Pacific Lines did give an annotation Mixed train daily, except Sunday; *seen in* The Official Guide of the Railways, *March 1937.*

684.10 Short train occasionally at 10 a.m.

Entry in early timetables of the Maryport & Carlisle Railway. Quoted by Jack Simmons in The Maryport & Carlisle Railway, *1952.*

684.11 On these blank white sheets, as uncompromising as the Finance Act, there are always three dots, instead of

a time, at the place where one wants to go to; and some trains appear to leave places before they arrive there.
Paul Jennings, The Jenguin Pennings, *1963.*

684.12 Extra LNER 4472 West.
Designation of train, classified as an extra, hauled by Flying Scotsman *in America, 1971, recorded by David P Morgan in* Trains Magazine, *January 1972.*

684.13 The seasoned traveller, or even the sensible one, looked up his train times before going on a journey: in Bradshaw if he could find his way round its mysteries, and had enough fingers to leave in pages where he had to change or had an alternative route; or the ABC if he knew only his destination and did not want to use his imagination.
Jeoffrey Spence, Victorian and Edwardian Railway Travel, *1977.*

684.14 To Brighton on the hour, in the hour, at the hour.
Attributed to Sir Herbert Walker, of London to Brighton services on the Southern Railway; Michael Bonavia, The Four Great Railways, *1980.*

684.15 Railroads, you toy with time, think time your bauble;
You run on time, shrink time, cut time in two.
Do all time all kinds of wrong, bind *time* to *table*,
Brag that tomorrow is today for you.
Your goal: arrive—though heat, hail, heaven's fire
Snatch at the brass and varnish of the Flyer.
John Frederick Nims, poem, Freight, *1989.*

684.16 If you can't timetable then you haven't got a functioning railway. The timetable determines the journey time, service frequency, capacity and thus revenue. It establishes the size of traction and rolling stock fleets. Traction and timetable interact, trading off speed and acceleration against headways, journey times and recovery times. By establishing the intensity of service, the timetable determines the level of maintenance required, by both traction and rolling stock, and infrastructure. From maintenance requirements stem the possessions regime. Investment in enhancements is justified in part by the ability to time table more or faster or longer trains. The timetable can determine the need, scale, and location of enhancements.
Editorial comment, Modern Railways, *January 2002.*

685. TIMING

685.1 "All right now?"
"No, no. You want two bridges and forty-five beats."
Exchange between members of Travelling Post Office crew, upon the precise timing of extending an arm to drop mailbags, film Night Mail, *1936.*

686. TIPS

686.1 We tip the man who brings our hat,
The man who brings our cane,
The waiter in the restaurant,
The porter on the train.
But how about the man who risks
His life and knows no fear,
Did you ever hear of anyone
Who tipped the engineer?
Anon., first verse of poem Who Tips the Engineer?, *quoted from the* Denver Times *in* Railroad Magazine, *April 1943.*

687. TIRE HEATERS

687.1 Now, for the information of those who have never played hide and seek in and about engine stalls, it is here stated that a tire heater is the greatest mental and physical hazard ever conceived by man. It is really a portable furnace constructed of a small pipe full of holes, made of joints and dojiggers and snarled-up sections, and it has the faculty of winding itself into more knots than an octopus with cramps ever dreamed of. When you have no earthly use for it, you will stumble over it and skin up your shins or go sprawling into an engine pit; but when you do want it, and have to have it, you might as well look for a clean face on a coal passer. It is made so as to encircle the tire of a driving wheel and the heat is thrown against the tire, expanding it and allowing it to be sledged off the wheel center.
A W Somerville, short story, Counterbalance, *within* Headlights and Markers, *ed. by Frank P Donovan and Robert Selph Henry, 1946.*

688. TOILETS ON TRAINS AND IN STATIONS

688.1 The lavatory accommodations, even in a first class car, was limited, and as it was for the joint use of both sexes it was a cause of frequent embarrassments. Ablutions had to be performed singly, and for two hours each morning there was a little crowd of unwashed and semi-dressed men and women standing about the corridor, all smoking cigarettes, women as well as men, and each eyeing their neighbour with side glances of distrust lest there was some underhand move to get possession of the lavatory first. . . .
John Foster Fraser, of trains on the Trans-Siberian Railway, quoted by Harmon Tupper in To the Great Ocean, The Real Siberia, *1902.*

688.2 I suppose that roadbeds are always going to be open sewers; nothing so advanced as the chemical toilets used in many brush-country farms can be expected before Utopia comes in—two hours late.
Bernard de Voto, Harpers Magazine, *April 1947.*

688.3 That wonderful little room where there's that marvellous unpunctuated motto over the lavatory saying *Gentlemen Lift the Seat*. What exactly does this mean? Is it a sociological description, a definition of a gentleman which one can either take or leave? Or perhaps it's a loyal toast.

> *Jonathan Miller, sketch* The Heat Death of the Universe, *within* Beyond the Fringe Review, *1961.*

688.4 You'd think going to the lavatory was a sin.

> *Comment in* The Observer *(selected as "saying of the week"), referring to a report on London stations by the Research Institute for Consumer Affairs, 1963.*

688.5 All one could do, if travelling any long distance, was to buy an "India Rubber Urinal for Male and Female Railway Travellers", made by Sparks & Son, and others, and advertised in Bradshaw. Whether an example of this device is hidden away in some museum is not known, but it would have been a godsend. One might visualise the hunted look which must have crept into a passenger's eyes when wearing one of these things in a crowded compartment of mixed company, when he had come to the conclusion that it must serve its purpose.

> *Jeoffrey Spence,* Railway Travel, *1977.*

688.6 From the day that feces flew off a train at some feckless Florida fishermen and stirred up a national controversy, a change in how Amtrak handles waste treatment has been in the offing.

> *Bob Johnston,* Trains Magazine, *December 1993.*

688.7 A 12-minute "toilet talk" video of edited responses told the two interior-design teams that passengers 1) thought shelf-type toilets reminded them of an outhouse, 2) hated wet countertops caused by shallow wash basins, 3) found lighting inadequate and glaring, and 4) had nowhere to put their laptop computers when they visited the bathroom. Designers came up with windows, indirect lighting, and a real, white commode.

> *Bob Johnston, of Amtrak's consultation of passengers when designing* Acela *trains.* Trains Magazine, *May 1999.*

688.8 What's the point of paying to use toilets at the train station, when the toilets are free on the trains. Although you're not supposed to use train toilets when the train is in the station, you should be able to sneak on and do this. Just check the departure time before you do this, otherwise you could end up in Orleans.

> *Anon., advice to American travelers in Europe, found 14 March 2004 on website www.bugeurope.com/ destinations/fr-toilet.html*

689. TOLEDO DELPHOS & BURLINGTON RAILROAD

689.1 . . . that magnificent stretch of narrow-gauge rails and mortgages . . .

> *Comment,* The Boston Advertiser, *10 May 1893.*

690. TOWNS

690.1 Railways have set all the Towns of Britain a-dancing. Reading is coming up to London, Basingstoke is going down to Gosport or Southampton, Dumfries to Liverpool and Glasgow; while at Crewe, and other points, I see new ganglions of human population establishing themselves, and the prophecy of metallurgic cities that were not heard of before. Reading, Basingstoke, and the rest, the unfortunate Towns, subscribed money to get railways; and it proves to be for cutting their own throats.

> *Thomas Carlyle,* Latter-day Pamphlets, *1858.*

690.2 With us railways run to the towns; but in the States the towns run to the railways.

> *Anthony Trollope,* North America, *1862.*

690.3 Every railway takes trade from the little town to the big town, because it enables the customer to buy in the big town.

> *Walter Bagehot,* The English Constitution, *1866.*

691. TRACK

See also Permanent Way Men

691.1 You cannot have a good railroad without good track and good equipment, and good men to maintain and operate that track and equipment.

> *Howard Elliott, address before the Tri-State Grain and Stock Growers' Association, Fargo, N.Dak., 17 January 1912, published in* The Truth about the Railroads, *1913.*

691.2 The essential and unique thing about a railroad is the track. There were tracks long before there were locomotive steam engines, or even stationary steam engines, and no matter what may be the locomotive power of the future, still there will be tracks.

> *Robert Selph Henry,* This Fascinating Railroad Business, *1942.*

691.3 The engineer's seat in the cab was beautifully sprung. I sat in it, and I've never seen anything that nice for engineers before. It's so good that when you're sitting in it and hit a lousy piece of Penn Central track you don't get that terrible *thump*. I'm thinking of suggesting that this seat be made mandatory in New York City taxicabs. Fifth Avenue certainly looks as though it were laid and maintained by Penn Central track crews.

> *Rogers E M Whitaker,* All Aboard with E. M. Frimbo, *1974.*

692. TRAFFIC

692.1 Traffic, like water, will find its own level; it will go by the shortest route.

> *Attributed to Edmund Denison by Charles H Grinling in* The Way of Our Railways, *1905.*

693. TRAINING

693.1 We had taught him too much!

Sir Richard Moon, of W D Phillipps, General Manager of the North Staffordshire Railway, who had spent several years on Moon's railway, and who had proved a worthy adversary in business negotiations, 1868. Recorded by G P Neele, in Railway Reminiscences, *1904.*

693.2 Never feel ashamed to ask for information where it is needed, and do not imagine that a man has reached the limit of mechanical knowledge when he knows how to open and shut the throttle-valve.

Angus Sinclair, Locomotive Engine Running and Management, *1902.*

693.3 To obtain a thorough insight into the working of the locomotive, no detail of its construction is too trifling for attention. The unison of the aggregate machine depends upon the harmonious adjustment of the various parts; and, unless a man understands the connection of the details, he is never likely to become skilful in detecting derangements.

Ibid.

693.4 In this country we spend millions in an endeavor to make our apparatus fool-proof, while in England they spend hundreds to eliminate the fool.

James Latimer, Railway Signaling in Theory and Practice, *1909. Quoted by Steven W Usselman in* Regulating Railroad Innovation, *2002.*

TRAINS

Entries on trains are divided into sub-headings, thus: Appearance, Character, Commuter, Express, Freight, Frightening, Last, Local, In the Night, Sound, Symbols of Freedom.

694. TRAINS—Appearance

694.1 As you walk by that long perspective of windows you are aware that they are not just a string of ten Pullman cars. They are fused by something even subtler than that liaison of airy pressure that holds them safe. They are merged into personality, become a creature loved, honored, and obeyed.

Christopher Morley, A Ride in the Cab of the Twentieth Century Limited, *1928.*

694.2   the varnished teak
That, North or South, was never far to seek,
But had for apogee East Coast Joint Stock
That left King's Cross each morning, ten o'clock—
Though many held this did not equal quite
The West Coast purple-brown and "spilt-milk" white.

Gilbert Thomas, poem, Nostalgia. *Published in* Punch, *7 November 1949, but also appearing in* Double Headed, *co-authored by Gilbert Thomas and his son David, 1963.*

695. TRAINS—Character

695.1 I watch the passage of the morning cars with the same feeling that I do the rising of the sun, which is hardly more regular. Their train of clouds stretching far behind and rising higher and higher, going to heaven while the cars are going to Boston, conceals the sun for a minute and casts my distant field into the shade, a celestial train beside which the pretty train of cars which hugs the earth is but the barb of the spear.

Henry David Thoreau, Walden, *1854.*

695.2 "Harness me down with your iron bands,
Make sure of your curb and rein,"
But I'll snap the pens in your puny hands
That would stop the wheels of the Train!
With logs of wood to upset me, try,
From the skulls of blockheads cleft.
Scrunch! they go—sticks, stones, let fly—
But behind in scorn they're left.

The steam is up and away we go!
On a road that's bright and new;
Old coach proprietors, heavy and slow,
Our course with anguish view.
Meetings they call, and bills they print,
And their feeble voices strain:
"Travel with us, nor your life's term stint,
By that murderous headlong Train."

Anon., poem, first two verses, The Song of the Train, *metaphorically launching the magazine* The Train *in 1856.*

695.3 To remember a train in such a spot would have been rank sacrilege.

Anthony Hope, The Prisoner of Zenda, *1894.*

695.4 A passenger train is like the male teat—neither useful nor ornamental.

Attributed to James J Hill by several writers, including Peter Lyon, To Hell in a Day Coach, *1967.*

695.5 Under the rolling clouds of the prairie a moving mass of steel. An irritable clank and rattle beneath a prolonged roar. The sharp scent of oranges cutting the soggy smell of unbathed people and ancient baggage.

Sinclair Lewis, Main Street, *1922.*

695.6 There is no smug Pullman attached to the train, and the day coaches of the East are replaced by free chair cars, with each seat cut into two adjustable plush chairs, the head-rests covered with doubtful linen towels. Halfway down the car is a semi-partition of carved oak columns, but the aisle is of bare, splintery, grease-blackened wood. There is no porter, no pillows, no provision for beds, but all today and all tonight they will ride in this long steel box—farmers with perpetually tired wives and children who seem all to be of the same age; workmen going to

new jobs; traveling salesmen with derbies and freshly shined shoes.
Ibid.

695.7 Trains!
At the lake cottage she missed the passing of the trains. She realized that in town she had depended upon them for assurance that there remained a world beyond.
Ibid.

695.8 Unlike American trains that were absorbed in an intense destiny of their own, and scornful of people on another world less swift and breathless, this train was part of the country through which it passed.
F Scott Fitzgerald, of a train to Cannes, France, Tender is the Night, *1934.*

695.9 There is a great deal to be said against trains, but it will not be said by me. I like the Trans-Siberian Railway. It is a confession of weakness, I know, but it is sincere.
Peter Fleming, One's Company, *1934.*

695.10 Trains sum up, to my mind, all the fogs and muddled misery of the nineteenth century. They constitute, in fact, so many slums on wheels.
Sir Osbert Sitwell, Penny Foolish, *1935.*

695.11 There is more poetry in the rush of a single railroad train across the continent than in all the gory story of burning Troy.
Joaquim Miller, quoted by Oscar Lewis in The Big Four, *1938.*

695.12 Some trains remind me of the pompous Big Business Man, complete with hard hat, beribboned eyeglasses and pearl gray spats. Others have a highty-tighty air and are, figuratively, wearing a lorgnette.
Roy Clark, Trains Magazine, *November 1944.*

695.13 People in the mass think of trains as things to ride on or things that carry coal and scrap iron and oranges, never as art and fun, drama and tragedy.
David P Morgan, Trains Magazine, *May 1957.*

695.14 Who shot the passenger train?
David P Morgan, title of article, Trains Magazine, *April 1959.*

695.15 It would be simpler if the train were just a means of moving people. But the train is also photogenic and nostalgic and fun, a source of employment and taxes, the pride of every Chamber of Commerce, and something on which you hang mail and express cars at one end and a business car at the other. And because trains stir up so much emotion, they are susceptible to the magnificent myth, if not the big lie.
Ibid.

695.16 It is not so much opinion as fact that the finest passenger trains in the world operate west of Chicago.
Ibid.

695.17 The world judges the railways by their passenger services. If this is the window through which we are viewed, we must wash it and shine it, or else cover it with a dark shade.
John Budd, Great Northern Railway (U.S.), quoted (a second time) in Trains Magazine, *November 1969.*

695.18 My friend, if you have never been on a train west of Chicago, *you* have never *been* on a *train.*
Ernest Marsh, AT&SF, 1964, quoted by Peter Lyon, To Hell in a Day Coach, *1967.*

695.19 I have often wondered if the beauty of trains is artistically legitimate or if it is a specialized thing appealing only to the neuroses of a fortunate few.
John Robinson, Letter, Trains Magazine, *August 1964.*

695.20 Perhaps there were once perfect passenger trains in point of fact as well as in mellow memory.
David P Morgan, Trains Magazine, *May 1966.*

695.21 Less flexible than the automobile, more expensive than the bus, slower than the plane, the passenger train has lost all its advantages, save only an excellent safety record.
Professor George Hilton, at California Public Utilities Commission hearings; reported in Trains Magazine, *July 1966.*

695.22 Trains are for the carriage trade. Planes are strictly for peasants.
Attributed to Lucius Beebe in a letter in Trains Magazine, *August 1966.*

695.23 Daddy, what's a train?
Is it something I can ride?
Does it carry lots of grown-up folks and little kids inside?
Is it bigger than our house?
Well how can I explain
When my little boy asks me,
Daddy, what's a train?
Utah "Bruce" Phillips, song, Daddy, What's a Train?, *1973.*

695.24 When I was just a boy
A-livin by the track,
Us kids'd gather up the coal in a big ol' gunny sack,
Then we heard the warning sound
As the Train pulled into view,
The engineer'd smile and wave
As she went rollin' through.

He blew so loud and clear
We had to cover up our ears,
And we counted cars just as high as we could go,
I can almost hear the steam
And the big ol' drivers scream
With a sound my little boy will never know.
Ibid.

695.25 The train can reassure you in awful places—a far cry from the anxious sweats of doom aeroplanes inspire, or the nauseating gas-sickness of the long-distance bus, or the paralysis that afflicts the car passenger.
  Paul Theroux, The Great Railway Bazaar, *1975.*

695.26 The Jarocho Express was one of those trains—rarer now than they used to be—which you board feeling exhausted and disembark from feeling like a million dollars.
  Paul Theroux, The Old Patagonian Express, *1979.*

695.27 The Train is Your Friend.
  Sign on station in Patagonia, seen by Paul Theroux, The Old Patagonian Express, *1979.*

695.28 I'd rather be on the Train!
  National Association of Railroad Passengers, bumper sticker, 1980s.

695.29 But when I say I love trains, I mean I love them all. Not just the wild and wonderful ones, but the ones that go from Liverpool Street to King's Lynn and Colchester, and the ones from Paddington to Bristol Temple Meads, the night trains and the day trains and even the little shuttles from Waterloo to Kingston-upon-Thames and others of that ilk. Brighton was my first dream to come true, and the train journey there was my first taste of this recurring magic.
  Lisa St Aubin de Terán, Off the Rails, *1989.*

695.30 The trains from Liverpool Street run at two-hourly intervals towards King's Lynn. I used to metamorphose on those trains. By Audley End I would have calmed my manic spirits and the effect of the gin and tonics had begun to wear off. At Cambridge I would remind myself of the heritage and at Ely of my lot.
  Ibid.

695.31 The train is both the antidote to temporary insanity and a straitjacket in itself, and it is a moment of release for those already incarcerated by the stranglehold of themselves, and, mostly, it is a capsule of life held in suspension, a vehicle for thought and emotion and a vehicle in which ordinary people may move across the land.
  Ibid.

695.32 By the turn of the century, however, Americans had come to expect from their passenger trains great regularity of departure and reliability of arrival, and the energetic society that had poured so much of the old century's sweat and treasure into building the railroads had learned to use them as the chief tool of an expansive material civilization.
  Albro Martin, Railroads Triumphant, *1992.*

695.33 Though I could hardly have been more traumatized just then, the trains kept me sane, cocooned in

safety, tied securely to home by those ribbons of steel. Inch by inch they would pull me safely home, through Paris and Boulogne and the endless suburbs of south London, all the way to the Tube station at Belsize Park.
  Nicholas Whittaker, Platform Souls, *1995.*

696. TRAINS—Commuter

696.1 . . . and the smoky Conrail commuter
  that rocked and screeched through chemical air
  to New York, a rolling lurching urinal
  carrying bankers and middlemen
  and secretaries smoothing
  their weekday best and putting new lipstick on
  as we coasted out of the tunnel
  into Penn station and waited—standing
  jammed together with salesmen whose sportcoats
  would never hang right from their tired shoulders,
  and teachers needing new heels, and lawyers—
  waited for the doors to open onto the hot black
      platform; . . .
  Reginald Gibbons, poem, American Trains, *1986.*

696.2 Oh Caltrain! My Caltrain! Our fearful trip is done,
  The engine's clattered every mile, arriving on track one,
  the station's near, the bells I hear, the people all exulting,
  while follow eyes the steady steel, the coach at last
      arriving;
  But oh start! start!
  Oh oily drops of red,
  Where in the back the engine dies,
  Fallen cold and dead.
    Sam Freund, poem, Silent Steel, *(with apologies to Walt Whitman), published on the website of Santa Clara Valley, http://www.metroactive.com/papers/metro/09.17.98./cover/best-verse-9837.html, 1998.*

696.3 Oh Caltrain! My Caltrain! rise up and hear the yells;
  Rise up—for you the bell is rung—for you the woman
      shrills,
  For you the screaming passengers, for you the tracks
      a-crowding,
  For you they call, the swaying mass, their angry voices
      shouting;
  Oh Caltrain! Oh bother!
  When all is done and said,
  We'll still be stalled upon the track,
  Still fallen cold and dead.
    Ibid., second verse.

697. TRAINS—Express

697.1 First, the shrill whistle, then the distant roar,
  The ascending cloud of steam, the gleaming brass,
  The mighty moving arm; and on amain
  The mass comes thundering like an avalanche o'er
  The quaking earth; a thousand faces pass—

A moment, and are gone, like whirlwind sprites,
Scarce seen; so much the roaring speed benights
All sense and recognition for a while;
A little space, a minute, and a mile.
Then look again, how swift it journeys on;
Away, away, along the horizon
Like drifted cloud, to its determined place;
Power, speed, and distance, melting into space.
> *Anon., poem, quoted by Frederick Williams on* Our
> Iron Roads, *1852.*

697.2 An earthquake accompanied with thunder and
lightning, going up express to London.
> *Charles Dickens,* Mugby Junction, *published in* All the
> Year Round, *1866.*

697.3 In the still evening air I hear a sound,
A roaring like a distant cataract,
It dies away, and wholly ceases;
Suddenly and louder still it roars again
Along the valley and the wooded hill. . . .
Again it dies away, yet keeps on muttering,
Till on a sudden bursting from the hills
And thundering like an alpine avalanche,
The railway train dashes along the line—
A fiery centipede, terribly beautiful,
More wonderful than any fabled dragon
Disgorging sulphorous smoke and clouds of steam,
The embodiment of swiftness and power.
> *James Hurnard,* The Setting Sun, *1870.*

697.4 These two, it will be noticed, run through the
quickest-witted portion of the island. The northern part,
having only eleven miles of water to separate it from
Scotland, and boasting Scotchmen for its chief inhabitants, can hardly be expected to show much enthusiasm
on the subject of railway smartness.
> *E Foxwell and T C Farrer, commenting upon trains
> between Dublin and Cork,* Express Trains English and
> Foreign, *1889.*

697.5 Out of the blackness, a globe of fire;
Out of the silence a clank and roar;
Out of the restfulness, dead before,
A mad rush, monstrous in strength and ire:
A spangle of sparks on night's dark dress;
A tempest of cinders, whirlwind tost;
An echo dying—in distance lost
thus comes and passes the Night Express.
> *Anon., poem,* The Express, Great Western Railway
> Magazine, *July 1892.*

697.6 A monster taught
To come to hand
Amain,
As swift as thought
Across the land
The train.

The song it sings
Has an iron sound;
Its iron wings
Like wheels go round.
> *John Davidson, poem,* Song of a Train.

697.7 Crash under bridges,
Flash over ridges,
And vault the downs;
The road is straight—
Nor stile, nor gate;
For milestones—towns!

Voluminous, vanishing, white,
The steam plume trails;
Parallel streaks of light,
The polished rails.
> *Ibid.*

697.8 A green eye—and a red—in the dark.
Thunder—smoke—and a spark.

It is there—it is here—flashed by.
Whither will the wild thing fly?

It is rushing, tearing thro' the night,
Rending the gloom in its flight.

It shatters her silence with shrieks.
What is it the wild thing seeks?

Alas! For it hurries away
Them that are fain to stay.

Hurrah! For it carries home
Lovers and friends that roam.

Where are you, Time and Space?
The world is a little place,

Your reign is over and done,
You are one.
> *Mary Coleridge,* The Train, *1907.*

697.9   An ashen canopy of cloud,
The dense immobled sky, high-pitched above
The wind's terrestrial office, overhung
The city when the morning train drew out.
Leaping along the land from town to town,
Its iron lungs respired its breath of steam,
Its resonant flanges, and its vertebral
Loose-jointed carcase of a centipede
Gigantic, hugged and ground the parallel
Adjusted metals of its destined way
With apathetic fatalism, the mark
Of all machinery.
> *John Davidson, poem,* Rail and Road.

697.10   From Paddington
To Basingstoke the world seemed standing still:
Nothing astir between the firmaments
Except the aimless tumult of the wind,

And clanging travail of the ponderous train
In labour with its journey on the smooth,
The ineludible the shining rails.
> *Ibid.*

697.11 When a stillness reigns in the country lanes
And the wayside station's bare,
Stirs a faint, far hum that seems to come
From the spirits of the air;
And the long rails thrill with a murmur till
There's a bursting shell of sound,
A clattering roar, like the rumble of war,
And a trembling of the ground—
A scudding blast has come and passed
With a shriek as of tortured souls
And along the track is the echoing back
That slowly to silence rolls.
> It is I the proud, the strong,
> I who sway the lives of men,
> Beating out my deathless song
> As I speed through field and glen.
> *R Gorell Barnes, poem, first verse,* The Express, *1913, in* The Railway Library, *1914, 1915.*

697.12 I am weighted down with the spoils of the town
And the harvest of the field:
Gaunt famine shrinks back at my sudden attack
And Plenty stands there revealed.
Though I travel afar as the servant of War,
I am foster-mother of Peace;
I bind the world's charms on her outstretched arms
And bring to her power increase.
In my strength and pride I am deified
As the emblem of mortal command,
For I spread o'er the world with the banner unfurled
On the march of a mighty band
And lead a great train, like a thought through a brain
To illumine the darkest land.
The chimney tall starts up at my call
And the factory whistle screams,
As from slumber I wake the shores of the lake
And shatter the valley's dreams.
I am clad in the dress of stern usefulness
And I build with a tyrannous rage:
In my pride I roll on over all that is gone
And I reck not of Beauty nor Age.
> For I am Progress, I am Power,
> I am the spirit of today:
> I fell the forest, clear the glade,
> I drain the marsh and crowd the earth.
> I roll onward, ever on
> Down my God-appointed way,
> Herald of the breaking morn,
> Calling to a nobler birth
> All the forces yet unborn
> And the greatness still to be.
> *Ibid., fifth (last) verse.*

697.13 Every day, a few minutes after two o'clock in the afternoon, the limited express between two cities passed this spot. At that moment, the great train, having halted for a breathing space at the town near by, was beginning to lengthen evenly into its stroke, but it had not yet reached the full drive of its terrific speed. It swung into view deliberately, swept past with a powerful swaying motion of the engine, a low smooth rumble of its heavy cars upon pressed steel, and then it vanished into the cut. For a moment the progress of the engine could be marked by the heavy bellowing puffs of smoke that burst at spaced intervals above the edges of the meadow grass, and finally nothing could be heard but the solid clacking tempo of the wheels receding into the drowsy stillness of the afternoon.
> *Thomas Wolfe,* The Cottage by the Tracks, *in* Cosmopolitan Magazine, *July 1935.*

697.14 After the first powerful manifesto
The black statement of pistons, without more fuss
But gliding like a queen, she leaves the station.
Without bowing and with restrained unconcern
She passes the houses which humbly crowd outside,
The gasworks and at last the heavy page
Of death, printed by gravestones in the cemetery.
Beyond the town there lies the open country
Where, gathering speed, she acquires mystery,
The luminous self-possession of ships on ocean.
> *Stephen Spender, poem,* The Express.

697.15 Sentimentality, snobbishness, romanticism, call the weakness by what name you please, will always make the passenger prefer to travel by a train with a name rather than by the 9.15, though few of us, except boys small enough to travel on half a ticket, bother our heads much about the engine which draws us.
> *Naomi Royde-Smith,* The Times LMS Centenary Number, *20 September 1938, quoted in* The Coronation Scot, *Edward Talbot, 2002.*

697.16 I'm goin' to send a wire,
Hoppin' on a Flyer,
Leavin' today.
> *Mack Gordon, song,* I've got a gal in Kalamazoo, *1942.*

697.17 Night falls. The dark expresses
Roll back their iron scissors to commence
Precision of the wheels' elision
From whose dark serial jabber sparks
Swing swaying through the mournful capitals.
> *Lawrence Durrell,* Night Express, *in* Collected Poems, *1960.*

698. TRAINS—Freight

698.1 Mysterious goods trains, covered with palls and gliding on like vast weird funerals, conveying themselves guiltily away from the presence of the few lighted lamps, as if their freight had come to a secret and unlawful end. Half miles of coal pursuing in a Detective manner, following when they lead, stopping when they stop, backing when they back.

> Charles Dickens, Mugby Junction, *published in* All the Year Round, *1866.*

698.2 Bond felt the heavy barrel of the gun slip into place along his right shoulder.

The moonstruck silence was broken by a loud iron clang from the signal box behind the hoarding. One of the signal arms dropped. A green pinpoint of light showed among the cluster of reds. There was a soft slow rumble in the distance, away to the left by Seraglio Point. It came closer and sorted itself into the heavy pant of an engine and the grinding clangour of a string of badly coupled goods trucks. A faint yellow glimmer shone along the embankment to the left. The engine came labou[ ]g into view above the hoarding.

The engine slowly clanked by on its hundred-mile journey to the Greek frontier, a broken black silhouette against the silver sea, and the heavy cloud of smoke from its cheap fuel drifted towards them on the still air. As the red light on the brake van glimmered briefly and disappeared, there came the deeper rumble as the engine entered a cutting, and then two harsh, mournful whoops as it whistled its approach to the little station of Buyuk, a mile further down the line.

The rumble of the train died away. Bond felt the gun press deeper into his shoulder.

> *Ian Fleming,* From Russia With Love, *1957.*

698.3 Freight train, freight train, run so fast,
Freight train, freight train, run so fast,
Please don't tell what train I'm on,
An' they won't know where I'm goin'.

> *Elizabeth Cotten, song,* Freight Train, *date not known. The popular version recorded in the 1950s by Nancy Whiskey differs somewhat.*

698.4 It's all science, no fiction.

> *David P Morgan, of 223-car freight trains on the Southern Railway with remote-controlled locomotives in the middle of the consist.* Trains Magazine, *October 1964.*

699. TRAINS—Frightening

699.1 Away, with a shriek, and a roar, and a rattle, plunging down into the earth again, and working on in such a storm of energy and perseverance, that amidst the darkness and whirlwind the motion seems reversed, and to tend furiously backward, until a ray of light upon the wet walls shows its surface flying past like a fierce stream. Away once more into the day, and through the day, with a shrill yell of exultation, roaring, rattling, rearing on, spurning everything with its dark breath, sometimes pausing for a minute where a crowd of faces are, that in a minute more are not: sometimes lapping water greedily,

and before the spout at which it drinks has ceased to drip upon the ground, shrieking, roaring, rattling through the purple distance!

> *Charles Dickens,* Dombey and Son, *1846.*

699.2 Well, Sir, that was a sight to have seen; but one I never care to see again! How much longer shall knowledge be allowed to go on increasing?

> *William Hinton, after five speechless minutes, having seen a train (a broad gauge express) for the first time, recorded in* Memoir of Charles Mayne Young, *J C Young, 1871.*

699.3 . . . the great straddling, bellowing railway, the high, heavy, dominant American train that so reverses the relation of the parties concerned, suggesting somehow that the country exists for the "cars," which overhang it like a conquering army, and not the cars for the country.

> *Henry James,* The American Scene, *1907.*

699.4 The Negro has no dragon in his mythology, but he sees a modern one in an engine and train—a fierce creature stretching across the country, breathing out fiery smoke, ruthless of what comes in its path.

> *Dorothy Scarborough,* On the Trail of Negro Folk-Songs, *1925.*

700. TRAINS—Last

700.1 "Porter, what time does the last train go to Walthamstow?"

"Lor' bless you marm, there ain't no last train to Walthamstow."

> *Anon., an illustration of the ceaseless activity at Liverpool Street station,* The Railway Magazine, *vol. V, 1899.*

700.2 Well, the last train to San Fernando
Well, the last train to San Fernando
And if you miss this one
You never get another one.

> *Randolph Padmore, song,* Last Train to San Fernando. *It is assumed that this refers to the Pacific Electric interurban line to San Fernando, which was closed north of Van Nuys in 1938.*

701. TRAINS—Local

701.1 Alone, in silence, at a certain time of night,
Listening, and looking up from what I'm trying to write,
I hear a local train along the Valley, And "There
Goes the one-fifty", think I to myself; aware
That somehow its habitual travelling comforts me,
Making my world seem safer, homelier, sure to be
The same tomorrow; and the same, one hopes, next year.
"There's peacetime in that train." One hears it disappear
With needless warning whistle and rail-resounding wheels.
"That train's quite like an old familiar friend", one feels.

> *Siegfried Sassoon, poem,* A Local Train of Thought.

701.2 But sure enough, just before dusk a gentle sighing sound came across the distant meadow, and a train swam slowly toward us through the tall grass. It was a beautiful and wonderful train advancing with an elaboration of caution under a cloud of undulant soot, like some aged beldame in a wreath of widow's weeds. Its capped-stack locomotive combined extreme elegance with imponderable age. Its half a dozen high cars swayed and dipped over the almost non-existent iron in an ecstasy of improbability.

    Lucius Beebe, Highball: a Pageant of Trains, *1945.*

701.3 The crack express trains—the bullet trains in Japan, 'The Blue Train' from Paris to Cannes, 'The Flying Scotsman'—these are joyrides, nothing more; the rapidity diminishes the pleasure of the journey. But the Local to Cutuco is a plod through the spectacular.

    Paul Theroux, The Old Patagonian Express, *1979.*

702. TRAINS—In the Night

702.1 There's a light at last in the sable mist, and it hangs
    like a rising star
On the border-line twixt earth and sky, where the rails
    run straight and far,
And deeply sounds from hill to hill, in mighty
    monotone,
A distant voice—a hoarse, wild note with savage warning
    blown.
'Tis the night express, and well 't is named, for behold!
    from out the night
It comes and darkly adown the rails it looms to the
    startled sight—
Larger, nearer, nearer yet—till at last there's a clang and
    roar,
A wave of heat, and a gleam of red from a closing furnace
    door;
Then the crash and shriek of the rushing train—and our
    hearts beat fast and high
When sudden and swift through the shadowy mist the
    night express goes by!

    William Hurd Hillyer, The Night Express, *1901.*

702.2 From out the mist-clad meadows, along the river
    shore,
The night express-train whistles with eye of fire before.
A trail of smoke behind her enclouds the rising moon
That gilds the sighing poplars and floods the wide
    lagoon.
Through yellow fields of harvest and waving fields of
    corn
The night express-train rumbles with whistle low and
    lorn.
The silent village harkens the sound it knows so well,
And boys wait on the siding to hear the engine-bell,
While lads who used to loiter with wistful steps and slow,

Await to-night a comrade who comes, but will not go.
The train that brings to mothers the news of sons who
    roam
Shoots red from out the marshes to bring a rover home.

    Willa Cather, poem, The Night Express, *first verse.*

702.3 'A man come to Arizona,' he said, 'with one of them telescopes to study the heavenly bodies. He was a Yankee, seh, and a right smart one too. And one night we was watchin' for some little old fallin' stars that he said was due, and I saw some lights movin' along across the mesa pretty lively, an' I sang out. But he told me it was just the train. And I told him I didn't know yu' could see the cyars that plain from his place. "Yu' can see them," he said to me, "but it is las' night's cyars you're lookin' at." '

    Owen Wister, The Virginian, *1902.*

702.4 The Flying Fornicator.

    Anon. Applied by university students to the last trains each night from King's Cross (to Cambridge) and Paddington (to Oxford); probably 1930s.

702.5 Who has not had upon some wakeful night,
For company, the passing of the a train;
And heard its coming with a child's delight
And in his bed more comfortably lain,

Conjuring images of fire and steam
And signal lamps that beckon in their green
And enginemen who crane to catch their gleam,
And sleeping cars which ride the night, serene?

And regal Postal Vans that in their nets
Gather the mailbags as they speed along
On rail-joints clacketing like castanets
And rails that elevate a wordless song?

    Henry Maxwell, poem, Good Companions, *in* A Railway Rubaiyyat, *1968.*

703. Trains—Sound

703.1 Even now Bond could hear the quick silver poem of the level-crossing bells, the wail of the big wind-horn out front and the quiet clamour at the stations when they lay and waited for the sensual gallop of the wheels to begin again.

    Ian Fleming, Goldfinger, *1959.*

703.2 Down in the meadow
Meadow so low
Late in the evening
Hear the train blow.

    Anon. Quoted in Hear the Train Blow, *Lucius Beebe and Charles Clegg, 1952.*

703.3 Hear the train blow!

    Chapter title, The Man with the Golden Gun, *Ian Fleming, 1965.*

704. TRAINS—Symbols of Freedom

704.1 I first discovered trains as a means of truancy, and thus they have remained, irrevocably linked in my mind with the idea of escape.

Lisa St Aubin de Terán, Off the Rails, 1989.

705. *TRAINS MAGAZINE*

705.1 ... devotional reading of the faithful ...

Lucius Beebe, Trains Magazine, November 1960.

705.2 This magazine is published by and for laymen, lineside observers of railroading, people moved by an enthusiasm that stems not from paycheck or dividend.

Editorial note, Trains Magazine, January 1965.

706. TRAINSPOTTING

See also Watching Trains

706.1 In one part of the railroad the rails are laid straight for more than a mile together. Here I used to feel much gratification, by seating myself to watch the approach of the several heavy trains of coal-waggons, on their way backwards and forwards, laden and unladen, between Darlington coal-fields, and the staiths at Middleborough or Stockton.

Sir George Head, A Home Tour through the Manufacturing Districts and other parts of England, Scotland and Ireland, 1835.

706.2 Steam train, steam train,
What's your number
What's your name,
Collecting trains, a fine day
I've got more than you.
Waiting for the Arpley train
On the road to Diggle,
Saw the local passing by,
Pop! goes the diesel.

Ewen McColl, chant by schoolboys. The connection between Arpley (in Warrington) and Diggle (north east of Manchester) is not understood. The Ballad of John Axon, broadcast 2 July 1958.

706.3 Trainspotting has always been a democracy, embracing all men, from right scruffs to Right Honourables.

Nicholas Whittaker, Platform Souls: The Trainspotter as Twentieth Century Hero, 1995.

706.4 Beige Marks and Spencer anorak, elasticated cuffs, fully enclosable hood ... I don't believe I've got that one.

Adey Bryant, cartoon showing conversation between train driver and his mate as they see a trainspotter on the platform. Catalogue Ref: abro114 on website www .cartoonstock.com, March 2004.

707. TRAMS and TROLLEYS

707.1 They came, of course, at their master's call,
The witches, the broomsticks, the cats, and all;
He led the hags to a railway train
The horses were trying to drag in vain.
"Now, then," says he, "you've had your fun,
And here are the cars you've got to run.
The driver may just unhitch his team,
We don't want horses, we don't want steam;
You may keep your old black cats to hug,
But the loaded train you've got to lug.

Oliver Wendell Holmes, penultimate verse of poem, The Broomstick Train; or, The Return of the Witches. Complete Poetical Works, 1895.

707.2 Since then on many a car you'll see
A broomstick as plain as plain can be;
On every stick there's a witch astride,—
The string you see to her leg is tied.
She will do mischief if she can,
But the string is held by a careful man,
And whenever the evil-minded witch
Would cut some caper, he gives a twitch.
As for the hag, you can't see her,
But hark! you can hear her black cat's purr,
And now and then, as a car goes by,
You may catch a gleam from her wicked eye.
Often you've looked on a rushing train,
But just what moved it was not so plain.
It couldn't be those wires above,
For they could neither pull nor shove;
Where was the motor that made it go
You couldn't guess, *but now you know.*

Remember my rhymes when you ride again
On the rattling rail by the broomstick train!
Ibid., last verse.

707.3 There once was an old man who said, 'Damn!
It is borne in upon me I am
An engine that moves
In determinate grooves,
I'm not even a bus, I'm a tram.'

Maurice Evan Hare, poem, Limerick, 1905.

707.4 To ride on these cars is always an adventure. Since we are in war-time, the drivers are men unfit for active service: cripples and hunchbacks. So they have the spirit of the devil in them. The ride becomes a steeplechase.

D H Lawrence, Tickets Please, date not found.

707.5 This, the most dangerous tram-service in England, as the authorities themselves declare, with pride, is entirely conducted by girls, and driven by rash young men, a little crippled, or by delicate young men, who creep forward in terror. The girls are fearless young hus-

sies. In their ugly blue uniform, skirts up to their knees, shapeless old peaked caps on their heads, they have all the sang-froid of an old non-commissioned officer. With a tram packed with howling colliers, roaring hymns downstairs and a sort of antiphony of obscenities upstairs, the lasses are perfectly at their ease. They pounce on the youths who try to evade their ticket-machine. They push off the men at the end of their distance. They are not going to be done in the eye—not they. They fear nobody—and everybody fears them.
Ibid.

707.6 They told me to take a streetcar named Desire, and then transfer to one called Cemeteries and ride six blocks and get off at—Elysian Fields!
Tennessee Williams, spoken by Blanche outside a "two-storey corner building on a street in New Orleans which is called Elysian Fields and runs between the L&N tracks and the river." A Streetcar Named Desire, 1947.

707.7 For I want to hire out as the Skipper
(Who dodges life's stresses and its strains)
Of the Trolley, the Toonerville Trolley,
The Trolley that Meets all the Trains.
Donald Marquis, poem, The Toonerville Trolley.

708. TRANSCONTINENTAL RAILROAD

708.1 I see over my own continent the Pacific railroad
surmounting every barrier,
I see continual trains of cars winding along the Platte
carrying freight and passengers,
I hear the locomotives rushing and roaring, and the shrill
steam-whistle,
I hear the echoes reverberate through the grandest
scenery in the world, . . .
Walt Whitman, Passage to India, lines 49–52, 1871.

708.2 Let the railroad alone; it will cost so much money that it will break down any administration that adopts it as a party measure.
Major General William T Sherman, quoted in Building the Pacific Railway, Edwin L Sabin, 1919.

708.3 . . . completed the work of Columbus.
Anon. after dinner speaker at a banquet in celebration of the completion of the road; quoted by Oscar Lewis in The Big Four, 1938.

708.4 There was to be no more loneliness; no more sections. The oceans were joined, and all who dwelt between them might be neighbors and friends in a real sense. Never again could distance or isolation be decisive factors in the life, social conditions, culture or opportunities of the people. All might mingle with one another, get really acquainted, discover mutual needs, and work in better harmony for the common advancement. Such was the realization that swept over multitudes as they lifted

up their rhythmic shouts in answer to the bells. It was as though they were chanting the last, triumphant words in a long epic of human endeavor. And if those of future times should seek for a day on which the country at last became a nation, and for an event by virtue of which its inhabitants became one people, it may be that they will not select the verdict of some political campaign or battle-field but choose, instead, the hour when two engines—one from the East and the other from the West—met at Promontory.
Seymour Dunbar, of the completion of the transcontinental railroad on 10 May 1869; A History of Travel in America, 1915.

708.5 The building of the first transcontinental railroad was to that generation what the moon project was to ours, the planting of a first tentative foot in unknown space.
Maury Klein, The Overland Route: First Impressions, within Unfinished Business, 1994.

709. TRANSPORT

709.1 The public interest in transportation is as keen as it is misinformed.
William Sproule, President, Southern Pacific Company, address to Traffic Club of Chicago, 25 February, 1915, reproduced in The Railway Library, 1915, 1916.

709.2 The public interest in transportation really rests in the one word service. To meet the public interest service must be adequate to the public needs. To be adequate to the public needs it must be safe; it must be frequent and varied enough for the requirements of commerce and travel; flexible enough to fit the varying wants of specialized business; it must be maintained at a standard high enough to be dependable; and in all it must be an energetic auxiliary for business and the pioneer of business by extension of service as needed.
Ibid.

709.3 One of our difficulties is that the subject when thus regarded looms so large as to extend far beyond the horizon of most minds. This produces a confusion and bewilderment that have for their result discursive generalities upon what people are pleased to call broad lines, only because they talk in a loose way upon subjects beyond their comprehension. We throw broad phrases as a blanket over our ignorance.
Ibid.

709.4 Transport, of all industries, needs flexibility and prompt decision. It is fighting against time all the time.
Sir Sam Fay, general manager, Great Central Railway, Presidential address, Institute of Transport, 2 October 1922.

709.5 The British grow up with the impression that trains are an archaic and outmoded means of transport, something to be taken on a Sunday between Keighley and Howarth in Yorkshire as a treat, or used by people who do not have cars to get around. In other places, however, they are still seen as integral and fully-functioning members of transport society.

Kate Matthams, Spotting trains, Europe, *found 14 March 2004 on website www.bootsnall.com/travelstories/oct03trains.html.*

710. TRANSPORT POLICY

710.1 What this country needs as urgently as disarmament is a transport policy. No railwayman has the right or the brains to say what that policy should be. But the transport crisis deepens: and the time has come when the nation should devote to the most important single item in its economy its finest characters and its finest brains. Transport has always been governed—and on the whole governed badly. The nation must at long last start to govern it well and to govern it as a whole.

Gerard Fiennes, paper, The Future of British Railways, *c. 1968.*

710.2 The French ministry runs a transport policy, it doesn't try to manage the railway. Ours is precisely the opposite, they're not interested in transport policy but they want to manage the railways.

Anon. British Rail official, quoted by Nicholas Faith, The Economist, *24 August 1985.*

710.3 Too many passengers, not enough trains,
Not enough passengers, too many planes—
Too many cars, not enough roads,
Not enough buses, bedlam bodes.

Tony Hemmings, poem for the day Rail/Transport Problems, *verse 3, 9 November 2001.*

711. TRANS-SIBERIAN RAILWAY

711.1 With the iron road awakening the echoes of the vast tracts of solemn forest where, three centuries ago, the Tunguz and Buriat might only note the cries of animals scarcely wilder than themselves; and bridging rivers where, till yesterday, the fisherman's birchbark canoe alone glided through the solitary reaches, Siberia will be, indeed, conquered, and, with a steel yoke about her neck, compelled to yield her all: of grain and cattle, furs, fish, and timber; porphyry and gold; coal, lead, and mercury; silver, copper, and iron—all the wealth she has, under guard of eternal snow and ice, so long held in trust for centuries.

Archibald R Colquhoun, Overland to China, *1900.*

711.2 . . . rusty streaks of iron through the vastness of nothing to the extremities of nowhere.

Anon., quoted by Harmon Tupper in To the Great Ocean, *1965.*

TRAVEL

Travel is divided into sub-headings thus: Character, In Clean Surroundings, Desirability and Otherwise, Need, Relaxation, Speed.

712. TRAVEL—Character

712.1 The whirl of the confused darkness, on those steam wings, was one of the strangest things I have experienced—hissing and dashing on, one knew not whither. We saw the gleam of towns in the distance—unknown towns. We went over the tops of houses—one town or village I saw clearly, with its chimney heads vainly stretching up towards us—*under* the stars; not under the clouds but among them. Out of one vehicle into another, snorting, roaring we flew: likest thing to a Faust's flight on the Devil's mantle; or as if some huge steam night-bird had flung you on its back, and was sweeping through unknown space with you, most probably towards London.

Thomas Carlyle, letter to John Carlyle, 13 September 1839, quoted in Carlyle, A History of His Life in London, *1884.*

712.2 I can't make out. I am never sure of time or place upon a Railroad, I can't read, I can't think, I can't sleep—I can only dream. Rattling along in this railway carriage in a state of luxurious confusion, I take it for granted I am coming from somewhere, and going somewhere else. I seek to know no more. Why things come into my head and fly out again, whence they come and why they come, and where they go and why they go, I am incapable of considering. It may be the guard's business, or the railway company's; I only know it is not mine. I know nothing about myself—for anything I know, I may be coming from the Moon.

Charles Dickens, Household Words, *10 May 1856.*

712.3 O luxury of travel! joy refined!
To fly steam-harness'd, in the ponderous train,
And feel the victory of mighty Mind
O'er space and time, for uses not in vain!
Yet ever in this world must loss and gain
Balance each other. Is it speed we prize?
'Tis edged with danger, equipoised by pain,
And aids our business but to cheat our eyes.
Th' unsocial Rail affords no varied pleasure
Like yours, ye coaches of a former day:
Apt for our haste, delightful for our leisure;-
We miss the cantering team, the winding way,
The road-side halt, the post-horn's well-known air,
The inns, the gaping towns, and all the landscape fair.

Charles Mackay, poem, The Stage Coach and the Steam Carriage, *1856.*

712.4 There is a certain etiquette in connection with the retaining of seats which it is considered both rude and unjust to disregard. Thus, the placing of a coat, a book, a newspaper, or any other article, on the seat of a carriage, is intended as a token that such a place is engaged.

Advice, The Railways Traveller's Handy Book, *1862.*

712.5 This system of occupation by proxy refers, however, more especially to the first class. With the majority of travellers by second and third class this delicate intimation does not appear to be understood, or, if understood, not recognised.

Ibid.

712.6 In the third-class seat sat the journeying boy,
And the roof-lamp's oily flame
Played down on his listless form and face,
Be-wrapt past knowing to what he was going,
Or whence he came.

Thomas Hardy, poem, Midnight on the Great Western.

712.7 Perhaps the most beautiful part of America is the West, to reach which, however, involves a journey by rail of six days, racing along tied to an ugly tin-kettle of a steam engine. I found but poor consolation for this journey in the fact that the boys who infest the cars and sell everything that one can eat—or should not eat—were selling editions of my poems vilely printed on a kind of grey blotting paper, for the low price of ten cents.

Oscar Wilde, Impressions of America, *1906.*

712.8 We were never less bored in our life than during that two-hour ride. In the first place, the line of march of the B. and O. gives one quite a different view of the country from the course of the P.R.R., with which we are better acquainted. From the Pennsy, for instance, Wilmington appears as a smoky, shackish and not too comely city. In the eye of the genteel B. and O. it is a quiet suburb, with passive shady lawns about a modest station where a little old lady with a basket of eggs and black finger-gloves got gingerly on board.

Christopher Morley, On the Way to Baltimore, *within* Travels in Philadelphia, *1920.*

712.9 Eight days of solid travel: none of spectacular but unrevealing leaps and bounds which the aeroplane, that agent of superficiality, to-day makes possible. The arrogance of the hard-bitten descends upon you. You recall your friends in England, whom only the prospect of shooting grouse can reconcile to eight hours in the train without complaint. Eight *hours* indeed . . . You smile contemptuously.

Peter Fleming, in anticipation of a journey on the Trans-Siberian Railway, One's Company, *1934.*

712.10 Towards nine o'clock at night there is a pause to switch cars and change engines at a junction town. The traveller, with the same feeling of wild unrest, wonder, and nameless excitement and wordless expectancy leaves the train, walks back and forth upon the platform, rushes into the little station lunch room or out into the

streets to buy cigarettes, a sandwich—really just to feel this moment's contact with another town.

Thomas Wolfe, Of Time and the River, *1935.*

712.11 Some people like to hitch and hike;
They are fond of highway travel;
Their nostrils toil through gas and oil,
They choke on dust and gravel.
Unless they stop for the traffic cop
Their road is fine-or-jail road,
But wise old I go rocketing by;
I'm riding on the railroad.

Ogden Nash, Riding on a Railroad Train, *c. 1937.*

712.12 With a farewell scream of escaping steam
The boiler bows to the Diesel;
The Iron Horse has run its course
And we ride a chromium weasel;
We draw our power from the harnessed shower,
The lightning without the thunder,
But a train is a train and will so remain
While the rails glide glistening under.

Ibid.

712.13 Oh, some like trips in luxury ships,
And some in gasoline wagons,
And others swear by the upper air
And the wings of flying dragons.
Let each make haste to indulge his taste,
Be it beer, champagne or cider;
My private joy, both man and boy,
Is being a railroad rider.

Ibid.

712.14 You need never be bored in a train.

John Betjeman, BBC broadcast, quoted in the Great Western Railway Magazine, *September 1940.*

712.15 Trains were made for meditation.

Ibid.

712.16 Pardon me boy
Is that the Chattanooga Choo-choo,
Track twenty-nine,
Boy you can gimme a shine.

I can afford to board,
a Chattanooga Choo-choo
I've got my fare,
and just a trifle to spare.

You leave the Pennsylvania Station 'bout a quarter to four,
read a magazine and then you're in Baltimore,
Dinner in the diner,
nothing could be finer
than to have your ham 'n eggs
in Carolina.

When you hear the whistle blowin' eight to the bar
Then you know that Tennessee is not very far,
Shovel all the coal in,
Gotta keep it rollin'
Woo, Woo, Chattanooga
There you are.

    Mack Gordon, song, Chattanooga Choo-choo, *1941.
This song includes several instances of poetic license and
should not be regarded as an accurate portrayal of service
between New York and Chattanooga. The* Memphis Special *and later the* Tennessean *of the Pennsylvania, Southern and Norfolk & Western left Pennsylvania Station
rather later than 3.45 p.m., and that station did not have as
many as twenty-nine tracks. It would be churlish to point
out that when this song was written, the steam locomotives
used on it invariably had mechanical stokers—shoveling
coal was not necessary.*

712.17 Going to take a sentimental journey,
   Going to set my heart at ease,
   Going to make a sentimental journey,
   To renew old memories.

Got my bag,
I got my reservation,
Spent each dime I could afford,
Like a child in wild anticipation
Long to hear that All Aboard!

Seven, that's the time we leave, at seven,
I'll be waiting up for heaven,
Counting every mile
Of railroad track
That takes me back

Never thought my heart could be so yearny
Why did I decide to roam,
Going to take a sentimental journey,
Sentimental journey home.

    Green, Brown, Homer, song, Sentimental Journey,
*1945. In a version of the song recorded by Dinah Shore, she
adds at the end "We've got sixteen tons of number nine
coal, the Rock Island Line is going to rock and roll . . .
roll . . . roll . . . roll." taking a line from the song* Sixteen
Tons *(about coal mining) and referring to the song* Rock
Island Line *(see Chicago, Rock Island and Pacific
Railroad).*

712.18 That we halted at every station goes without saying. Few sidings—however inconsiderable or, as might
seem, fortuitous, escaped the flattery of our prolonged
sojourn. We ambled, we paused, almost we dallied with
the butterflies lazily afloat over the meadow-sweet and
cow-parsley beside the line; we exchanged gossip with
station-masters, and received the congratulations of sig-
nalmen on the extraordinary spell of fine weather. It did
not matter. Three market-women, a pedlar, and a local

policeman made up with me the train's complement of
passengers. I gathered that their business could wait; and
as for mine—well, a Norman porch is by this time
accustomed to waiting.

    Sir Arthur Quiller-Couch, en route to an old church,
Pipes in Arcady, *1944.*

712.19 A hog can cross America without changing
trains—But you can't.

    *Advertisement, headline, C&O and Nickel Plate roads,
inspired by company president Robert Young, 1946.*

712.20 Why should travel be less convenient for people
than it is for pigs?

    Ibid., within text.

712.21 The train came to a halt with an escape of steam,
then silence. I stuck my head out of the window and saw
the gleaming railway lines, the immaculate tidiness of the
permanent way as it stretched and pointed to a distance
of green and wooded futurity. I was journeying not the
mere fifty miles or so that divide Manchester from
Shrewsbury; I was shedding a skin.

    Neville Cardus, Autobiography, *1947.*

712.22 To one who has never heard the special sound
made by the whistle of this particular train there may be
some minor satisfaction in tuning in to the radio of the
car. To one who never had luncheon in a diner, Mr Johnson may be able to provide just as many flavors of ice
cream as he says he can. One who has never set out to
explore a train, from baggage car on back, may feel that
US Route 86 may lead ultimately to the rainbow, but
rainbows definitely are train borne. Driving into its
motel early in the evening, the lost generation does not
know the supreme pleasure of climbing into its berth
and watching the lights of the country pass by. The lost
generation thinks in its ignorance that the movies are
pretty fair, that television is pretty fair, but it does not
know true art. The true art is not 10,000 extras, jumping
screaming from technicolor cliffs, but is one small light
in a lonely farmhouse, seen at night—from a train. Not
even Raphael could improve on that.

    Anon., The Lost Generation, *in the* New York Times,
reprinted in Trains Magazine, *October 1949.*

712.23 When I *plan* a trip, I'm a staunch Republican. I
chart my own course and I'd rather not have any help,
thank you. But in traveling,—particularly on secondary
lines—it is expedient to mix a little religion with politics. I turn to Taoism. Branch-line travel is arduous: so I
resign myself to fate, and take connections as they come,
and get up and leave anytime around the clock that the
timecard dictates. In short, I have a little Asiatic submission to things as they are—not to things as I wish them
to be.

    Frank P Donovan Jr, Trains Magazine, *May 1952.*

712.24 The emptying train, wind in the ventilators,
puffs out of Egloskerry to Tresméer
Through minty meadows, under bearded trees
And hills upon whose sides the clinging farms
Hold Bible Christians. Can it really be
That this same carriage came from Waterloo?
John Betjeman, Summoned by Bells, *1960.*

712.25 The best way to see a country is from the footplate
of a locomotive.
*George Dow, conversation with his son, the compiler,
c. 1961.*

712.26 Modern civilisation is moronic in many ways, but
none more than its devaluation of travel from a pleasur-
able end to an inconvenient means.
Geoffrey Freeman Allen, review of book, Modern Railways, *December 1964.*

712.27 Good journeys by rail have no history. It is the
exceptions which are memorable.
Gerard Fiennes, I Tried to Run a Railway, *1967.*

712.28 Faraway is near at hand in images of elsewhere.
*David Hall and Geoffrey Hall; graffiti on railway retain-
ing wall on the approach to Paddington station, applied 24
December 1974, and claimed to have been placed to be visible
from trains bound for Oxford. The wall has since been
demolished. The first five words come from the first line of
Robert Graves' poem* Song of Contrariety *(1923) and the last
four adapted from the title of Ruth Padel's poem* Imagery of
the Elsewhere: Two Choral Odes of Euripides, *published in*
Classical Quarterly, *December 1974.*

712.29 One of the virtues of train travel is that you know
where you are by looking out of the window.
Paul Theroux, The Old Patagonian Express, *1979.*

712.30 The journey through the suburbs was like a slow
escape route taking me away from all that I disliked and
feared to be contagious towards all that I loved and
hoped to be fired by. The litany of grey places remains
stamped on my mind, from Clapham Junction with its
seething jumble of tracks strewn over a dead zone of gasworks and scrap metal and blackened bricks to the first
glimpse of the sea itself: Wandsworth Common, back
gardens and tacked-on bathrooms; Balham, slipped
slates huddled in grime; Streatham Common, spiked
railings edged in blackberries; Norbury, tidy patios in
place of the more southerly junk. Each side of the track is
lined with black spikes, iron railings to impale the trai-
tors on. So on and on to Thornton Heath, Selhurst, and
East Croydon and South Croydon, appearing like a suburban stutter anxiously repeating all the stations before.
Lisa St Aubin de Terán, Off the Rails, *1989.*

712.31 The next five months of my life were to be lived on
trains, criss-crossing Italy from Sestri Levante to Brin-

disi, and back and forth, covering thousands and thousands of miles of track along which I shed thousands of
delusions and reinvented hundreds of illusions. My life
was a mish-mash of the insides of compartments and
corridors and platforms, my friends were the transitory
friendships of long journeys and my loves were the vol-
atile loves where passion is quickened by the hourglass it
can run through.
Ibid.

712.32 The circus is in training again, which is to say the
old leather cases are being sorted and allotted ready for
the night train to Guadalajara. It steams into my dreams
with such regularity that I feel my new address must be
care of its guard.
Ibid.

712.33 We pulled into Manchester two minutes early, but
this was at a time when railway companies, terrified of
being fined and mad on bumping up those percentages
in their passenger charters, produced timetables with
delays built-in. So even if you're a little early, you're actually a little late; and if you're truly and expensively late
according to the timetable, there is a modern ritual to
perform: a wait of uncertain duration is announced by a
uniformed soul sounding unnaturally positive, contrac-
tually apologetic and inevitably nasal; weary passengers
take this as a cue to list the best excuses they've ever
heard—'unexpected fish stock on the line'; and then
everyone on board speed-dials home with details of their
precise location and a stoic determination to make it
home against unenviable odds.
Simon Garfield, The Last Journey of William Huskisson, *2002.*

713. TRAVEL—In Clean Surroundings

713.1 The roads are dusty and the use of wood for fuel
sends a quantity of charcoal into your carriage, mixed
with the dust, so that when you have travelled all day you
are as black as a sweep. People are also constantly passing
through the carriages selling papers, books &c, and when
the front door is opened the rush of dust and dirt that
comes in is very disagreeable.
Sir Daniel Gooch, of travel in America in 1860;
Memoirs and Diary, *1972.*

713.2 Says Phoebe Snow
About to go
Upon a trip
To Buffalo
My gown stays white
From morn till night
Upon the road of anthracite.
*Earnest Elmo Calkins. The best-known of over 60 rhym-
ing jingles written to feature Phoebe Snow and used for
advertising by the Delaware Lackawanna & Western Rail-*

road between 1900 and 1917. The road served anthracite areas of Pennsylvania and burned that almost smokeless fuel in its steam locomotives.

713.3 The occasional freshening-up of a closely occupied passenger-car may be associated with the germ theory of diseases, and with the attention that should consequently be given to the sanitation of all travelling conveyances.

Henry S Haines, Efficient Railway Operation, *1919.*

713.4 And outside there was the raw and desolate-looking country, there were the great steel coaches, the terrific locomotives, the shining rails, the sweep of the tracks, the vast indifferent dinginess and rust of colors, the powerful mechanical expertness, and the huge indifference to suave finish. And inside there were the opulent green and luxury of the Pullman cars, the soft glow of the lights, and people fixed there for an instant in incomparably rich and vivid little pictures of their life and destiny, as they were all hurled onward, a thousand atoms, to their journey's end somewhere upon the mighty continent, across the immense and lonely visage of the everlasting earth.

Thomas Wolfe, Of Time and the River, *1935.*

714. TRAVEL—Desirability and Otherwise

714.1 And when the road is made,
With the pick and the spade,
In the locomotive engine, they will put a little fire,
And while the kettle boils,
We may ride three hundred miles
Or go to bed in Baltimore and breakfast in Ohio.
Where they're all waiting, hoping, praying
For a quick way to come to Baltimore!

Arthur Clifton, song, The Carrollton March, *thought to be the first railroad song written in America. 1828.*

714.2 . . . and all this for the sake of doing very uncomfortably in two days what could be done delightfully in eight or ten.

Anon. passenger, Boston & Providence Railroad, 1835, quoted by Charlton Ogburn, Railroads, *1977.*

714.3 Singing through the forest,
Rattling over ridges,
Shooting over arches,
Dashing under bridges,
Whizzing through the mountain,
Buzzing o'er the vale,
Bless me ain't it pleasant,
Riding on a rail.

John G Saxe, chorus of song, Rhyme of the Rail, *(one of three known titles), c. 1840.*

714.4 I'm going by the rail, my dears, where the engines puff and hiss; and ten to one the chances are that something goes amiss;

And in an instant, quick as thought—before you could cry "Ah!"

An accident occurs, and—say good-by to poor Papa!

Anon., poem, The Railway Traveller's Farewell to his Family, *c. 1850.*

714.5 But railway travelling in America is wretched; their republican notions of having only one class makes your company very mixed, and the carriages being all large open saloons with a door at each end and passage down the middle, prevents your having the slightest privacy even if you were a good large party of your own.

Sir Daniel Gooch, of journeys in 1860; Memoirs and Diary, *1972.*

714.6 From the time the avant courier of the omnibus bellowed up the steps of our hotel, "Passengers for the Pacific railroad." we have been wild with pleasure. The courtesy and extreme politeness of the conductor, the brakesmen, and even the train boy, take away our breath; and the fidelity and the courage of the engineer commend our admiration.

Henry Morton Stanley, 22 July 1867, republished in My Early Travels and Adventures in America and Asia, vol. I, *1895.*

714.7 Women both old and young regarded travelling by steam as presumptuous and dangerous, and argued against it by saying that nothing should induce them to get into a railway carriage; . . .

George Eliot, Middlemarch, *1871.*

714.8 At Third Avenue they took the elevated, for which she now confessed an infatuation. She declared it the most ideal way of getting about in the world, and was not ashamed when he reminded her of how she used to say that nothing under the sun could induce her to travel on it. She now said that the night transit was even more interesting than the day, and that the fleeting intimacy you formed with people in second and third floor interiors, while all the usual street life went on underneath, had a domestic intensity mixed with perfect repose that was the last effect of good society with all its security and exclusiveness.

William Dean Howells, A Hazard of New Fortunes, *1889.*

714.9 He said it was better than the theatre, of which it reminded him, to see those people through their windows: a family party of work-folk at a late tea, some of the men in their shirt sleeves; a woman sewing by a lamp; a mother laying her child in its cradle; a man with his head fallen on his hands upon a table; a girl with her lover leaning over the window-sill together. What suggestion! what drama! what infinite interest!

Ibid.

714.10 Stand not upon the order of your going but go to Stratford upon Avon, by the Great Central Railway.
Anon., text of poster, borrowing William Shakespeare's words from Macbeth, *c. 1905.*

714.11 God! I will pack, and take a train,
And get me to England once again!
Rupert Brooke, poem, The Old Vicarage, Grantchester.

714.12 Some people travel by aeroplane,
and others travel by car,
but those who desire comfort with speed,
Travel by L. & Y. R.
Poster, Lancashire & Yorkshire Railway, against illustration of a man falling from an aircraft, a car crashing into a ditch and spilling its passengers, and a speeding LYR express train, 1914.

714.13 The midnight train is slow and old,
But of it let this thing be told,
To its high honor be it said,
It carries people home to bed.
My cottage lamp shines white and clear.
God bless the train that brought me here.
Joyce Kilmer, The Twelve Forty-Five, *1918.*

714.14 I would much sooner go by *wagonlit* from Calais to Monte Carlo in twenty hours, than by magic carpet in twenty seconds.
A A Milne, A Train of Thought, *within* If I May, *1920.*

714.15 The railroad track is miles away,
And the day is loud with voices speaking,
Yet there isn't a train goes by all day
But I hear its whistle shrieking.
Edna St Vincent Millay, Travel, *1921.*

714.16 My heart is warm with friends I make,
And better friends I'll not be knowing;
Yet there isn't a train that I wouldn't take,
No matter where it's going.
Ibid.

714.17 Down here there were no trains. The stillness was very great. The prairie encircled the lake, lay round her, raw, dusty, thick. Only the train could cut it. Some day she would take a train; and that would be a great taking.
Sinclair Lewis, Main Street, *1922.*

714.18 There was an old dame of Dunbar
Who took the 4.4 to Forfar
But went on to Dundee,
So she travelled, you see,
Too far by 4.4 from Forfar.
Anon., The Complete Limerick Book, *edited by Langford Reed, 1924.*

714.19 A violent jerk to the right, followed by a lurch into the waistcoat of the man on my left, with a full head butt at the boards in front, told me that we were swerving through Clapham Junction.
Anon. writer in the Daily Express, *of a journey inside the indicating shelter on Southern Railway locomotive* Sir Lamorak, *quoted in* Southern Railway Magazine, *January 1926.*

714.20 George had never travelled under the seat of a railway-carriage; and, though he belonged to the younger generation, which is supposed to be so avid of new experiences, he had no desire to do so now.
P G Wodehouse, Meet Mr Mulliner, *1927.*

714.21 Ride the Big Red Cars
Advertising slogan used by the Pacific Electric Railway Company, 1929.

714.22 George sat by the window and saw the smouldering dumps, the bogs, the blackened factories slide past, and felt that one of the most wonderful things in the world is the experience of being on a train. It is so different from watching a train go by. To anyone outside, a speeding train is a thunderbolt of driving rods, a hot hiss of steam, a blurred flash of coaches, a wall of movement and of noise, a shriek, a wail, and then just emptiness and absence, with a feeling of "There goes everybody!" without knowing who anybody is. And all of a sudden the watcher feels the vastness and loneliness of America, and the nothingness of all those little lives hurls past upon the immensity of the continent. But if one is *inside* the train, everything is different. The train itself is a miracle of man's handiwork, and everything about it is eloquent of human purpose and direction. One feels the brakes go on when the train is coming to a river, and one knows that the old gloved hand of cunning is at the throttle. One's own sense of manhood and mastery is heightened by being on a train.
Thomas Wolfe, You Can't Go Home Again, *1934.*

714.23 In America, the train gives one a feeling of wild and lonely joy, a sense of the savage, unfenced, and illimitable wilderness of the country through which the train is rushing, a wordless and unutterable hope, as one thinks of the enchanted city toward which he is speeding; the unbroken and fabulous promise of the life he is to find there.
Thomas Wolfe, Dark in the Forest, Strange as Time, *1935.*

714.24 In Europe, the feeling of joy and pleasure is more actual, ever-present. The luxurious trains, the rich furnishings, the deep maroons, dark blues, the fresh, well-groomed look of the travelers—all of this fills one with a powerful sensual joy, a sense of expectancy about to be realized. In a few hours' time one goes from country to

country, through centuries of history, a world of crowded culture, and whole nations swarming with people, from one famous pleasure-city to another.
Ibid.

714.25　Taking the Century.
A figure of speech for traveling to New York from Chicago or vice versa, recorded by A C Kalmbach in Railroad Panorama. *The reference is to travel by the New York Central's* Twentieth Century Limited, *a luxury express train.*

714.26　As it had always done, the movement and experience of the train filled him with a sense of triumph, joy and luxury. The crack express, with its gleaming cars, its richly furnished compartments, its luxurious restaurant, warm with wine and food and opulence and suave service, together with the appearance of the passengers, who had the look of ease and wealth and cosmopolitan assurance that one finds among people who travel on such trains, awoke in him again the feeling of a nameless and impending joy, the fulfilment of some impossible happiness, the feeling of wealth and success which a train had always given him, even when he had only a few dollars in his pocket, and that now, in the groomed luxury of this European express, was immensely enhanced.
Thomas Wolfe, of a journey by PLM in France, Of Time and the River, *1935.*

714.27　You are not going to break records today. Our drivers are confident they could do 120 m.p.h. with these engines, but we do not encourage them to try. We want to show you that you can travel at a cruising speed of 90 m.p.h. in perfect comfort and that it will feel like 40. If we can achieve that with our new train we shall please our customers and we shall have carried rail travel one stage forward on the road of progress.
Sir Ralph Wedgwood, before the inaugural run of the Coronation, *30 June 1937.*

714.28　I must go back to the friendly train,
I'm tired of the service car;
And I long for a carriage seat again,
Where safety and comforts are.
And all I ask is the porter's call,
And the cheerful station bell,
The guard's congenial "tickets all,"
And I know I will travel well.

I must go back to the friendly train,
For I love the gleaming rails,
And the travelling folk I used to know
With their smoking-carriage tales.
And all I ask is a whistle clear,
And the signal falling,
The engine's hiss, and a through express,
And a trip enthralling.

O Wadham, poem, Train Fever, *with apologies to John Masefield;* New Zealand Railways Magazine, *reprinted in* Railroad Magazine, *October 1943.*

714.29　There she blows! Do you know what the three most exciting sounds in the world are? Anchor chains, plane motors, and train whistles.
Frank Capra, Frances Goodrich, Albert Hackett, and others, scriptwriters, spoken by Peter Bailey (played by Jimmy Stewart) at the station, in film It's a Wonderful Life, *1946.*

714.30　Surely it was always summer when we made our first railway journeys!
C Hamilton Ellis, The Trains We Loved, *1947*

714.31　The priceless advantage of railroad travel is that the eye-level scenery it affords is neither obscured by cloud formations nor blanketed by billboards.
Caption to photograph, Trains Magazine, *February 1952.*

714.32　I'm glad I took the train. This is a fine train and a nice route.
Nikita Krushchev, after a journey on the Southern Pacific between Los Angeles and San Francisco, 20 September 1959; recorded in Those Pullman Blues, *David D Perata, 1996.*

714.33　The chief wonder was one of America's first operating railroads, a gravity affair, which switchbacked down from the mines to the river at the mad pace of sixteen miles per hour, carrying alternative loads of coal, shrieking ladies and mules. The mules rode down, calmly chomping hay, and then pulled the cars back up again. It was said that a decent Mauch Chunk mule could not be got to walk downhill after growing used to rail travel.
Nathaniel Burt, of Mauch Chunk, Pennsylvania; The Perennial Philadelphians, *1963.*

714.34　... downtown to downtown in one comfortable seat.
Anon., stated in quotation marks by David P Morgan, but without source, concerning the new bullet trains in Japan, Trains Magazine, *June 1964.*

714.35　The view of the world from the windows of a train is like no other. You are in a superior position, slightly above and removed from ordinary life. Cars wait for you to pass in your lighted mystery. But you are seeing a real world of people and places. It is neither the child's game city with its toy cars that you see from planes, nor the irritating confrontation at car level. We could wonder, really curious, about the lives of people in West Covina, instead of wishing that they would get out of the way.
Lois Dwan, Calendar, within Los Angeles Times, *29 March 1970.*

714.36 Children find trains more exciting than planes, and I think it is because a train is ingenious and remarkable in a way they can understand. Children (and I) do not know why a plane stays up, or why a computer is so smart, or what makes radar blip. But we know that whoever thought of putting a single yellow rose in a bud vase in a spiral holder that allows it to sway with the motion of the train was a remarkable man.

> *Ibid.*

714.37 The only X-rated railroad ride in the United States, if not the world.

> *Description of the* Coast Daylight *train, according to Amtrak employees, because of its passing a beach frequented by nudists, south of Santa Barbara, California. Amtrak travel news brief quoted in* Trains Magazine, *December 1973.*

714.38 Ever since childhood, when I lived within earshot of the Boston and Maine, I have seldom heard a train go by and not wished I was on it.

> *Paul Theroux,* The Great Railway Bazaar, *1975.*

714.39 There is an English dream of a warm summer evening on a branch-line train. Just that sentence can make an English person over forty fall silent with the memory of what has now become a golden fantasy of an idealized England: the comfortable dusty coaches rolling through the low woods, the sun gilding the green leaves and striking through the carriage windows; the breeze tickling the hot flowers in the fields, birdsong and the thump of the powerful locomotive; the pleasant creak of the wood-panelling on the coach; the mingled smells of fresh grass and coal smoke; and the expectation of being met by someone very dear on the platform of a country station.

> *Paul Theroux,* The Kingdom by the Sea, *1983.*

714.40 It was like a ghastly parody of hard times. In what had been the greatest railway country in the world, and the easiest and cheapest to traverse, the traveller was now told with perfect seriousness, "You can't get there from here."

> *Ibid.*

714.41 Most people have that fantasy of catching the train that whistles in the night.

> *Attributed to Willie Nelson,* Travel Notes and Quotes *(a note book for travelers to record their thoughts), 1997.*

714.42 All of us, I imagine, will go to our graves lamenting all those engines and trains and railroads which we never experienced because of personal constraints of age, location, finance, whatever. I have my Dad's word for it that a first-class compartment in a Midland Railway carriage departing St. Pancras station behind a Deeley compound 4-4-0 was accommodation worth all the half-crowns in one's purse.

> *David P Morgan,* Trains Magazine, *November 1986.*

714.43 I like trains about as much as I don't like airplanes. I don't like the congestion in and around airports, or the parking fees. I don't like the lines. I especially don't like being searched and X-rayed before boarding. The only other place they do that is in prison. Trains are different. It's possible to board at Hudson, park for free, go to New York and then on to wherever for a minimum of fuss. And the fares are reasonable.

> *Annabar Jensis, comment in* Daily Mail *of Greene County (Catskill, N.Y.), 2 October 1987.*

714.44 The present InterCity service, combined with a laptop computer, is one of the most congenial working environments available to a Member of Parliament.

> *Tim Devlin (Member for Stockton South), Commons debate, second reading, Railways Bill, Hansard, 2 February 1993.*

715. TRAVEL—Need

715.1 De railroad bridge's
A sad song in de air.
Ever' time de trains pass
I wants to go somewhere.
> *Langston Hughes, poem,* Homesick Blues.

715.2 Oh, a-standing in the railroad yards
A-waiting for a train;
Waiting for a westbound freight,
But I think it's all in vain.

Going east they're loaded,
Going West sealed tight;
I think we'll have to get aboard
A fast express tonight.
> *Anon., song,* The Hobo's Hornbook, *1930.*

715.3 Perhaps this is our strange and haunting paradox here in America—that we are fixed and certain only when we are in movement. At any rate, that is how it seemed to young George Webber, who was never so assured of his purpose as when he was going somewhere on a train. And he never had the sense of home so much as when he felt that he was going there. It was only when he got there that his homelessness began.
> *Thomas Wolfe,* You Can't Go Home Again, *1934.*

715.4 Is your journey really necessary?
> *Slogan used by British Government to discourage railway travel during the Second World War, believed to have been coined by Ministry of Information, 1942, but stated to have been "copied from the Germans" according to Sir John Elliot,* On and Off the Rails, *1982. Described in* Oxford Dictionary of Quotations *(p. 554:7) as 1939, coined to dissuade Civil Servants traveling home over Christmas. Attributed by Alan Middleton to artist Bert Thomas, who devised posters using the slogan, in* It's Quicker by Rail, *2002.*

715.5 Americans never had it so good as in the decades directly after Appomattox, when the national economy was one of limitless abundance, a continental dimension beckoned and there was gold in the hills and buffalo on the prairie. A restless people rode to aureate destinies, and most of them rode the palace cars that were the showcases of the railroads of a land that could never have too many railroads. Amidst fringes, ferns and tassels, on plush and shrouded in velour, it beheld its own image in a thousand bevel-edged French mirrors set on mahogany bulkheads and between the wide picture windows of Pullmans and Wagners beyond counting. It drank champagne on the diners for breakfast and had supper off eight varieties of game on the dollar dinner and felt that it didn't need to tip its hat in the direction of either J. P. Morgan or the Medici. It was the most sumptuously upholstered landfaring in the history of human movement.

Lucius Beebe, Mr Pullman's Elegant Palace Car, *1961.*

716. TRAVEL—Relaxation

716.1 It is as well to have a newspaper—or, say, this book—in your hand to resort to in case tiresome people will talk—a purpose for which railway travelling was never intended.

R S Surtees, Hints to Railway Travellers and Country Visitors to London, *in the* New Monthly Magazine, *1851.*

716.2 I had arrived at one of the transpontine stations of the Pennsylvania Railroad; the question was of proceeding to Boston, for the occasion, without pushing through the terrible town—why "terrible," to my sense, in many ways, I shall presently explain—and the easy and agreeable attainment of this great advantage was to embark on one of the mightiest (as appeared to me) of train-bearing barges and, descending the western waters, pass round the bottom of the city and remount the other current to Harlem; all without "losing touch" of the Pullman that had brought me from Washington. This absence of the need of losing touch, this breadth of effect, as to the whole process, involved in the prompt floating of the huge concatenated cars not only without arrest or confusion, but as for positive prodigal beguilement of the artless traveller, had doubtless much to say to the ensuing state of mind, the happily-excited and amused view of the great face of New York.

Henry James, of passing through New York City before the tunnels under the Hudson River had offered uninterrupted railway travel to Manhattan and onwards. The American Scene, *1907.*

716.3 The general American theory is that railway-travel within the confines of the Republic is a matter of majestic simplicity and facility—qualified at the worst by inordinate luxury . . .

Ibid.

716.4 Where was the charm of boundless immensity as overlooked from a car-window?—with the general pretension to charm, the general conquest of nature and space, affirmed, immediately around you, by the general pretension of the Pullman, the great monotonous rumble of which seems forever to say to you: "See what I'm making of all this—see what I'm making, what I'm making!" I was to become later on still more intimately aware of the spirit of one's possible reply to that, but even then my consciousness served, and the eloquence of my exasperation seems, in its rude accents, to come back to me.

Ibid.

716.5 Slow travel by train is almost the only restful experience that is left to us. I love it.

A P Herbert, Slow Train, *reproduced from* Punch *in the* Great Western Railway Magazine, *August 1938.*

716.6 It is all soothing. I am almost at sea. For six hours nobody can call me on the telephone. It is better than the sea, for no one can send me a wireless. Nobody can knock at the door and ask difficult questions. Nobody can expect me to answer letters.

Ibid. Herbert was a Member of Parliament at this time.

716.7 My husband doesn't like airplanes. He says they're nothing but buses with wings. He likes to get in an air-conditioned train, get single occupancy of a drawing room, stretch out, relax and do his thinking. He says he does some of his best thinking and nearly all of his best sleeping on a train.

Erle Stanley Gardner, The Case of the Lucky Loser, *1956.*

716.8 Any time you choose,
Kick off your shoes,
Rest your weary eyes,
Or catch up with the news.
A favourite book and you
Are perfect company.
Relax. . . .

Forget about the blues,
And you will find,
Leave your cares and worries
Far behind,
Loosen up your tie,
Watch the world speed by.
Relax. . . .

Anon., song in Inter-City television advertisements, 1988.

716.9 To those critics, it was a restful, tranquil, relaxing experience, an adventure in tranquility and nostalgia whose enjoyment required painstaking attention to passenger comfort. The critics didn't simply want to be transported from point A to point B; they expected to be cosseted, comforted, nurtured and stroked by a succession of happy, smiling, ever-attentive railroad employees. SP's perception was different; we felt that we were not in the business of creating transcendental experiences, but rather in the business of transporting people.

J M Smith, Southern Pacific, letter to Don Hofsommer, 29 March 1985; quoted by H Roger Grant, We Took the Train, *1990.*

717. TRAVEL—Speed

717.1　The whole system of railroad travel is addressed to people who, being in a hurry, are therefore, for the time being, miserable.

John Ruskin, The Seven Lamps of Architecture, *1849.*

717.2　And so they dropped the dry sands and moonstruck rocks of Arizona behind them, and grilled on till the crash of couplings and the wheeze of the brake-hose told them they were at Coolidge by the Continental Divide.

Rudyard Kipling, Captains Courageous, *1897.*

717.3　At night the bunched electrics lit up that distressful palace of all the luxuries, and they fared sumptuously, swinging on through the emptiness of abject desolation. Now they heard the swish of a water-tank, and the guttural voice of a Chinaman, the clink-clink of hammers that tested the Krupp-steel wheels, and the oath of a tramp chased off the rear platform; now the solid crash of coal shot into the tender; and now a beating back of noises as they flew past a waiting train. Now they looked out into great abysses, a trestle purring beneath their tread, or up to rocks that barred out half the stars. Now a scaur and ravine changed and rolled back to ragged mountains on the horizon's edge, and now broke into hills lower and lower, till at last came the true plains.

Ibid.

717.4　No one loves it but the passengers.

Anon. headline covering article on Amtrak Acela train, Trains Magazine, *December 2002.*

718. TRENT STATION

718.1　Trent station, for passenger traffic purpose, is a central ganglion of the Midland system, . . .

Frederick Williams, The Midland Railway, *1878.*

719. TRESPASS

719.1　Notice is hereby given, that all Persons trespassing on the Railway, or the Works thereof, are liable *to a considerable Penalty* for each Offence. And that the Punishment doing any Injury or Damage to the said Railway is Transportation for 7 Years.

Poster, Newcastle & Carlisle Railway, 1835.

719.2　Neither walk nor trespass upon the bridge.

Standard sign, Norfolk & Western Railway.

719.3　No rationalizations apply to the "itinerant workman" who pillows his alcohol-clouded noggin on a No 14 switch frog and gets his hair parted down to his umbilicus by an impersonally insistent leading truck. He is held to the natural and probable consequences of his voluntary acts, and we do not inquire whether he erred more in getting stewed to the gills or in assuming a position of peril.

Louis F Meyer Jr, Trains Magazine, *November 1944.*

719.4　Nor do legal benefits accrue to the gas-buggy gaucho who departs this life in the tinny tatters of the joy-wagon he so confidently intruded on a grade crossing. The mere presence of rails is in itself a most eloquent warning of danger, and if the motorist appears on the tracks too late for the train to halt, his heirs, executors and administrators mourn him with prospects as empty as pockets.

Ibid.

719.5　He who pre-empts the rails, be he man or beast, is pushing Providence a smidgen farther than will the law.

Ibid.

719.6　Do not trespass on railway property nor hinder railway servants in the execution of their duty and endeavour to stop others from doing so.

Ian Allan, the creed of the Locospotter's Club, advertised in Trains Illustrated, *March 1952.*

720. RICHARD TREVITHICK

720.1　The publick untill now call'd mee a schemeing fellow but now their tone is much alter'd.

Richard Trevithick, letter to Davies Giddy, 22 February 1804, eight days after the first steaming of his locomotive.

720.2　Trevithick's experiments in steam locomotion were mere incidents in a career which comprised incursions into many varied branches of engineering. He was indeed an adventurer, always anticipating success but never realising it.

Robert Young, Timothy Hackworth and the Locomotive, *1923.*

721. TUCSON, Arizona

721.1　Thirty-two years ago, we welcomed the Southern Pacific, a transcontinental line. It made Tucson what it is today; raised it from an insignificant adobe village into a thriving city. Bob Leatherwood, who was mayor at that time, announced by cablegram to the Pope that Tucson was now connected with the outside world by railroad. Our worthy mayor should now inform that Pope that we have another railroad.

John B Wright, former Attorney General of Arizona, November 1912, at a civic luncheon to mark the arrival of the El Paso & South Western Railroad in Tucson.

722. TUNNELS

722.1 We cannot well conceive anyone so destitute of curiosity, to use no higher term, as not to desire to behold a work so unique in its kind.
Liverpool Mercury, of the public inspection of a tunnel on the Liverpool & Manchester Railway, July 1829. Quoted by Simon Garfield in The Last Journey of William Huskisson, *2002.*

722.2 I wonder why it is that when I shut my eyes in a tunnel I begin to feel as if I were going at an Express pace the other way. I am clearly going back to London, now.
Charles Dickens, Household Words, *30 August 1851.*

722.3 Whenever I pass through a tunnel I meditate upon these things, and wish heartily that I were a poet, that I might tune my heart to sing the poetry of railway tunnels.
George A Sala, Poetry on the Railway, *within* Household Words, *2 June 1855.*

722.4 Male passengers have sometimes been assaulted and robbed, and females insulted, in passing through tunnels. And this has been most frequently the case when there have been only two occupants in the carriage. In going through a tunnel, therefore, it is always as well to have the hands and arms ready disposed for defence, so that in the event of an attack, the assailant may be instantly beaten back or restrained.
Comment, before the age of carriage lighting, The Railway Traveller's Handy Book, *1862.*

722.5 We left for Turin at 10 the next morning by a railway which was profusely decorated with tunnels. We forgot to take a lantern along, consequently we missed all the scenery.
Mark Twain, A Tramp Abroad, *1880.*

722.6 Somerset looked down on the mouth of the tunnel. The popular commonplace that science, steam and travel must be always unromantic and hideous was not proven at this spot.
Thomas Hardy, A Laodicean, *1881.*

722.7   So we passed
The liberal open country and the close,
And shot through tunnels, like a lightning wedge
By great Thor-hammers driven through the rock,
Which, quivering through the intestine blackness, splits,
And lets it in at once; the train swept in
Athrob with effort, trembling with resolve,
The fierce denouncing whistle wailing on
And dying off smothered in the shuddering dark,

While we, self-awed, drew troubled breath, oppressed
As other Titans underneath the pile
And nightmare of the mountains. Out, at last,
To catch the dawn afloat upon the land!
Elizabeth Barrett Browning, verse Aurora Leigh: A Poem in Nine Books, *1856.*

722.8 This bottle neck was a marvel of traffic working, and an unceasing wellspring of torrid language, in which both the railwaymen and the general public took a hearty share.
E L Ahrons, The Railway Magazine, *September 1917, writing of the tunnel at the approach to Bradford Exchange station.*

722.9 If we see light at the end of the tunnel,
It's the light of the oncoming train.
Robert Lowell, Since 1939, *1977.*

722.10 The Channel Tunnel starts at Edinburgh.
Headline of advertisement, in anticipation of the completion of the Channel Tunnel, and particularly of through services from cities such as Edinburgh. Such services have yet to materialise. British Rail, late 1980s.

722.11 The dark at the end of the Chunnel
Anon., headline over article about the Channel Tunnel's continuing financial problems, The Daily Telegraph, *11 April 1995.*

722.12 The Mobile Telephone User's Guide to British Railway Tunnels.
Anon., humorously suggested title of book, necessary to stop irritation to other passengers, late 1990s.

723. TWENTIETH CENTURY LIMITED

723.1 The Greatest Train in The World.
George Henry Daniels, general passenger agent, New York Central; of the Twentieth Century Limited, 1902. According to Al Kalmbach, Daniels coined the name of the train; Railroad Panorama, *1944.*

723.2 This train makes Chicago a suburb of New York.
John W Gates, on the occasion of the inaugural run of the train, June 1902, to reporters in Manhattan. Quoted by Lucius Beebe, Mr Pullman's Elegant Palace Car, *1961.*

723.3 This train makes New York a suburb of Chicago.
John W Gates, on the occasion of the inaugural run of the train, June 1902, to reporters in Chicago. Quoted by Lucius Beebe, Mr Pullman's Elegant Palace Car, *1961.*

723.4 The conservative blue tail sign—almost like a coat of arms—proclaims it to be the 20th Century. It is a symbol and colophon of all that is proper, permanent and reliable in luxury travel.
Frank P Donovan Jr, Trains Magazine, *August 1948.*

723.5 I take *The Century* because I want very little in this world, only the best and there's so little of that.
Michael Arlen, quoted by Lucius Beebe in 20th Century, *1962.*

723.6 For more than a generation the *Twentieth Century Limited* had been to railroading what Tiffany was to silver and Cartier was to diamonds.
Robert C Reed, The Streamline Era, *1975.*

724. TWOPENNY TUBE
See also Underground

724.1 "Hi, guv'nor, there ain't no station named on this ticket!"
"No; all our tickets are alike."
"Then, 'ow do I know where I'm going?"
Cartoon caption in Mr Punch's Railway Book, *5 June 1901.*

725. TYNESIDE

725.1 The native land of railways.
Thomas Sopwith, Guide to Newcastle, *1838.*

726. ULLSWATER

726.1 The English Lucerne.
Advertisement, Great Northern Railway, 1901.

727. UNDERGROUND

See also Elevated Railways and Twopenny Tube

727.1 The Greyhound of Travel
London Underground poster, 1927.

727.2 Our trains are run by lightning
Our Tubes like thunder sound
But you can dodge the thunderbolt
By going underground.
Anon. Design for poster illustrated in Johnston's
Underground Type, *2000.*

727.3 The moment you enter the London Underground
you feel, though you may not be able to explain exactly
how you feel it, that you are moving in an environment
of order, of culture.
Steen Eiler Rasmussen, quoted by Christian Barman in
The Man who Built London Transport, *1979.*

727.4 Ooh! Here comes another one!
George Axelrod, scriptwriter, film, The Seven Year Itch,
*spoken by Marilyn Monroe, on hearing the approach of a
subway train, the updraft of which lifts her skirts in a memorable scene, 1955.*

728. UNION PACIFIC RAILROAD

728.1 The Pacific Railroad is already a power in the land,
and is destined to be a power vastly greater than it is now.
It already numbers its retainers in both houses of Congress, and is building up great communities in the heart
of the continent. It will some day be the richest and most
powerful corporation in the world; it will probably also
be the most corrupt.
Charles Francis Adams, 1869, quoted by Dee Brown in
Hear that Lonesome Whistle Blow, *1977.*

728.2 To undertake the construction of a railroad, at any
price, for a distance of nearly seven hundred miles in a
desert and unexplored country, its line crossing three
mountain ranges at the highest elevations yet attempted
on this continent, extending through a country swarming with hostile Indians, by whom locating engineers
and conductors of construction trains were repeatedly
killed and scalped at their work; upon a route destitute of
water, except as supplied by water trains, hauled from
one to one hundred and fifty miles, to thousands of men
and animals engaged in construction; the immense mass
of material, iron, ties, lumber, provisions and supplies
necessary to be transported from five hundred to fifteen
hundred miles—I admit might well in the light of subsequent history and the mutations of opinion, be regarded
as the freak of a madman if it did not challenge the recognition of a higher motive.
*Oakes Ames, who took responsibility for the task,
quoted by Edwin L Sabin in* Building the Pacific Railway,
1919.

728.3 I've seen every railroad movie ever made, and all
were full of discrepancies. This one is going to be
correct.
*William Jeffers, president, Union Pacific, of the film of
that name, to which he sent advisers to ensure its accuracy.
Quoted in* Railroad Magazine, *March 1939.*

728.4 The fat old lady with a bag of candy.
*Anon. A reference to the cash reserves of the apparently
old fashioned UP, immediately after World War Two.
Quoted by Maury Klein in* Union Pacific, vol. 2, *1990.*

728.5 Be Specific—Say Union Pacific
Advertising slogan.

728.6 Mother of Western Main Lines and a railroad
whose name itself is forever part of the lexicon of the Old
West, Union Pacific has been at various times the pioneer
transcontinental, a synonym for political corruption, a
component of the ambitious transport empire of Jay
Gould, a gilt edge investment under Edward H Harriman, and at all times the strong bond and connecting
link that joined America west of the Missouri River with
The States in post-Civil War times.
Lucius Beebe, The Trains We Rode, *1965.*

728.7 It has been a name at whose mention the American
imagination has always stood at attention.
Ibid.

728.8 Union Pacific is the only major railroad in the west
that has an appreciation for its heritage and an appreciation of the value of positive public relations. The UP has
decided that people need to know their name and think
well of them. Not shippers, not the industry, but ordinary people; and they have succeeded remarkably well.
If you say CSX to a thousand people only six of them
will know what you are talking about. If you say Union

Pacific to a thousand people, 994 will know what you are talking about. And that is worth something.

> *Walter Gray*, Locomotive & Railway Preservation, *Sept/Oct, 1991.*

728.9 . . . Union Pacific, the Railroad that Calibrated the Gauges that Established the Standards for the Standard Railroad of the World.

> *Ed Ellis. The Standard Railroad of the World was the title adopted by the Pennsylvania Railroad,* Trains Magazine, *June 2001.*

729. UNIONS

729.1 For fear that you may think the other speakers have the word efficiency copyrighted, I desire to inform you that the Brotherhood of Locomotive Engineers has always stood for efficiency and always expects to.

> *Warren S Stone, concerning the popular discussions of railroad efficiency at that time; Eleventh Annual Meeting of Grand Chief Brotherhood of Locomotive Engineers, quoted in* The Railway Library, 1911, *1912.*

729.2 Associated Society of Leprechauns, Elves and Fairies.

> *Anon., alternative meaning for ASLEF (Associated Society of Locomotive Engineers and Firemen), recalled by Peter Coster.*

730. URANIUM

730.1 At least, uranium oxide was valueless until Albert Einstein wrote a letter to President Roosevelt. Suddenly an entire region of which the faltering Rio Grande Southern Railroad was the nerve center, became the most jealously guarded mineral deposit in the world and Federal Agents were riding the tops of ore cars carrying cargoes that only yesterday vanadium mill owners were throwing out of the window. Strangers around Placerville and Telluride had to explain their business; a loan from the Reconstruction Finance Corporation bolstered the railroad's faltering economy, and the leased Rio Grande locomotives hauling machine-gun-guarded high cars over Dallas Divide had an ultimate destination, although nobody at the moment knew it, at Hiroshima.

> *Lucius Beebe,* Narrow Gauge in the Rockies, *1958.*

731. URINATION

731.1 That's dandy! I've always wanted to pump ship into the Atlantic and Pacific at the same time. Now I can say I've done it. Let's go!

> *Rear Admiral Robley D Evans, USN, after a visit in 1908 to the apex of the Continental Divide in Galera Tunnel, Central Railway of Peru, whence water is drained in opposite directions to the two oceans. Told by Brian Fawcett in* Railways of the Andes, *1963.*

732. VALVE GEAR

732.1 If it answers, it will be worth a Jew's eye and the contriver of it should be rewarded.

Robert Stephenson, letter, 31 August 1842. The inventors are generally accepted to have been William Williams and William Howe, both employed at Robert Stephenson's locomotive works, although some believe that William T James of New York first produced what was also then known as link motion.

732.2 No innovation, in respect of locomotive engines, ever divided practical and professional opinion so completely; and none has at last so firmly established itself in general favour.

Zerah Colburn, of Stephenson's valve gear, Locomotive Engineering, *1872.*

732.3 Walschaerts was easy to understand but Baker always seemed to have a lot of mysterious little parts performing devious little unnatural acts.

Ed King, caption to photograph, Trains Magazine, *February 1985.*

733. VANDALS

733.1 The idiot who, in railway carriages,
Scribbles on window panes,
We only suffer
To ride on a buffer
In Parliamentary trains.

W S Gilbert, song, To let the punishment fit the crime, The Mikado, *1885.*

733.2 ... a crime usually associated with steam-locomotive builder plates: fandalism?

Nathan S Clark Jr, letter to Trains Magazine, *October 1976.*

734. SAMUEL M VAUCLAIN

734.1 He has a powerful personality that seems to find the most perplexing problems of engineering, trifles worthy to beguile a leisure hour.

Angus Sinclair, Development of the Locomotive Engine, *1907.*

735. VICTORIA STATION (London)

735.1 Jack: The late Mr Thomas Cardew, an old gentleman of a very charitable and kindly disposition, found me, and gave me the name of Worthing, because he happened to have a first-class ticket for Worthing in his pocket at the time. Worthing is a place in Sussex. It is a seaside resort.

Lady Bracknell: Where did this charitable gentleman who had a first class ticket for this seaside resort find you?

Jack: In a hand-bag.

Lady Bracknell: A hand-bag?

Jack: Yes, Lady Bracknell. I was in a hand-bag—a somewhat large, black leather hand-bag, with handles to it—an ordinary hand-bag, in fact.

Lady Bracknell: In what locality did this James, or Thomas, Cardew come across this ordinary hand-bag?

Jack: In the cloak-room at Victoria Station. It was given to him in mistake for his own.

Lady Bracknell: The cloak-room at Victoria Station?

Jack: Yes. The Brighton line.

Lady Bracknell: The line is immaterial. Mr Worthing, I confess I feel somewhat bewildered by what you have just told me. To be born, or at any rate bred, in a hand-bag, whether it had handles or not, seems to me to display a contempt for the ordinary decencies of family life that reminds one of the worst excesses of the French Revolution. And I presume that you know what that unfortunate movement led to? As for the particular locality in which the hand-bag was found, a cloak-room at a railway station might serve to conceal a social indiscretion—has probably, indeed, been used for that purpose before now—but it could hardly be regarded as an assured basis for a recognised position in good society.

Oscar Wilde, The Importance of Being Earnest, *1894.*

735.2 It was like entering another world. The tiled and brilliantly-lit departure platforms, with the special liveried trains waiting, recall artist's impressions in the 1960s of a railway of the future. In comparison, the neo-brutalism of Euston is pure 1984-Orwell's, not Bob Reid's.

Roger Ford, Travellers' Tales, *in* Modern Railways, *January 1985.*

736. VIRGINIA & TRUCKEE RAILROAD

736.1 Sing, therefore, O Muse of Tractive Force and Valve Gear, of the Virginia and Truckee, a railroad of such superlatives that, like the Comstock it served and the San Francisco it enriched, its name will be forever currency in the language of the trans-Mississippi.

Lucius Beebe, Virginia and Truckee, *1949.*

737. VIRGINIAN RAILWAY

737.1 The rectifier locomotives had the beauty of mis-shapen bricks.

H Reid, of class EL-2 electric locomotives, The Virginian Railway, *1961.*

738. VOCABULARY

738.1 Two of the finest ways in which to enlarge a man's vocabulary are joining the army or taking railway photographs.

James I C Boyd, Railway World, vol. XIV, *May 1953.*

739. W1 CLASS LOCOMOTIVE

739.1 A giant in the noble regiment of blue streamliners.
O S Nock, The Locomotives of Sir Nigel Gresley, *1945.*

740. WABASH CANNONBALL

740.1 From the waves of the Atlantic Ocean to the wild
Pacific shore,
From the coast of California to ice-bound Labrador,
There's a train of doozy layout that's well-known to us all,
Its the boes' accommodation called the Wabash
Cannonball.
Traditional song, Wabash Cannonball. *This is one of
many recorded versions. Others may be found in* Long
Steel Rail, *by Norm Cohen, 1990.*

740.2 Now listen to her rumble, now listen to her roar,
As she echoes down the valley and tears along the shore.
Now hear the engine's whistle and her mighty hoboes'
call,
As we ride the rods and brakebeams on the Wabash
Cannonball.
*Ibid. The Wabash Cannonball is a mythical train; a ver-
sion of this song recorded by Roy Acuff, refers to a combina-
tion (i.e., a baggage/coach combine) on the Wabash Can-
nonball. It runs everywhere that a hobo may wish to travel.*

740.3 I've headed for Tolono
On the Wabash Cannonball,
Norfolk & Western all the way.
St Louis down the line,
Detroit somewhere behind,
I thought I heard that old conductor say
"No round trip ticket, you're on the final run,
This Cannonball is never coming back,
Tomorrow she'll just be
Another memory,
And an echo down a rusty railroad track."
Bruce "Utah" Phillips, song, No Round Trip Ticket,
sung to the same tune as, and as a coda to, Wabash Can-
nonball, *1973.*

740.4 I got off at Tolono,
Just below Champaign,
A flag stop on the edge of yesterday.
The whistle blew a song
I whispered "so long",
Waved my hand and slowly walked away.
Ibid.

741. WAGGONWAYS

741.1 Lo! here a lifeless waggon at thy will,
True to its ambit circles round a hill.
Down the descent, so vast the power of art,
The waggonway retains the flying cart.
So far from being cumbered with its freight,
Like virtue—see! it grows beneath a weight.
Down planes inclined the self moved engines fly
To lad the ships which near the Hurries lie.
Through wooden spouts descends the sooty ores,
An export grateful to Hibernia's shores.
Anon. verse, Poetical Prospect of Whitehaven, *quoted
in* The Cumberland Pacquet, *10 August 1802.*

742. WAGONS

742.1 The rapidly increasing size of cars is a growing evil
because it increases costs and risk without increasing
revenue.
*John Stubbs, Southern Pacific Railroad, February
1899. Quoted by Maury Klein in* Union Pacific, vol. 2,
1990.

742.2 I'm a bumped and battered freight car on a
sidetrack in the yard;
I am resting—resting gladly, for my life is cruel hard,
And I seldom find an hour when I'm soberly at home,
For I'm usually loaded and am out upon the roam.
I've been shunted in Seattle, I've been switched in Boston
town;
I've been stranded in St. Louis, where I saw the train crew
drown.
I've been snowed in up by Denver, I was wrecked at
Council Bluffs,
When the strike was in Chicago, I was stoned by thugs
and toughs.
Strickland W Gillilan, poem, Song of the Freight Car,
first verse, 1908.

742.3 I have often lost an axle when the train was
wrecked, and stood
For a week until the workmen found the time to make
it good.
I've been everywhere, seen all things, been in sunshine,
rain and snow.
I've been idle for a fortnight, then for months upon
the go.
I'm a bumped and battered freight car on a sidetrack in
the yard;

There are chalk marks on my body—these my only
  calling card.
But I see the pony engine coming for me on the fly—
No idea where I'm going or what for, but—bump—
  good by!
  Ibid., fourth verse.

742.4 Box cars run by a mile long.
And I wonder what they say to each other
When they stop a mile long on a sidetrack.
    Maybe their chatter goes:
I came from Fargo with a load of wheat up to the danger
  line.
I came from Omaha with a load of shorthorns and they
  splintered my boards
I came up from Detroit heavy with a load of flivvers.
I carried apples from the Hood River last year and this
  year bunches of bananas from Florida; they look for
  me with watermelons from Mississippi next year.
  Carl Sandburg, Work Gangs, *published in* Smoke and
Steel, *1920.*

742.5 The ultimate ownership of all wagons by the railway companies, added to the introduction of larger wagons, will eliminate a very large proportion of shunting
expenditure. The constant sorting and arrangement of
the traders' wagons which have daily to be sent back to
destinations ordered by the 6,450 separate people who
own them is like the perpetual piecing together of a jigsaw puzzle with all the disadvantages of wasted work.
  *Sir John Aspinall, of private-owner wagons in the
United Kingdom;* Proceedings of the Institute of Civil
Engineers, *vol. CCXIV.*

742.6 STOP LOOK LISTEN
  as gate stripes swing down,
  count the cars hauling distance
  upgrade through the town:
  warning whistle, bellclang,
  engine eating steam,
  engineer waving,
  a fast-freight dream:
  B&M boxcar,
  boxcar again,
  Frisco gondola,
  eight-nine-ten,
  Erie and Wabash,
  Seaboard, U.P.,
  Pennsy tankcar,
  twenty-two, three,
  Phoebe Snow, B&O,
  thirty-four, five,
  Santa Fe cattle,
  shipped alive,
  red cars, yellow cars,
  orange cars, black,

Youngstown steel
down to Mobile
on Rock Island track,
fifty-nine, sixty,
hoppers of coke,
Anaconda copper,
hotbox smoke,
eighty-eight,
redball freight,
Rio Grande,
Nickel Plate,
Hiawatha
Lackawanna,
rolling fast,
and loose,
ninety-seven,
coal car,
boxcar,
CABOOSE!
  Philip Booth, poem, Crossing, *1953.*

742.7 I can't even remember the name of the one who fell
Flat on his ass, on the cinders, between the boxcars.
I can't even remember whether he got off his yell
Before what happened had happened between the
  boxcars.
  Robert Penn Warren, verse one of Ballad: Between the
Boxcars, *1960.*

742.8 And there's one sure thing you had better
    remember well,
You go for the grip at the front, not the back, of the
  boxcars.
Miss the front—you're knocked off—miss the back, you
  never can tell
But you're flat on your ass, on the cinders, between the
  boxcars.
  Ibid., verse three.

742.9 I asked one eastern railroad president the other day
why he didn't get himself some big cars. The answer that
I got was that they wear out the rails. It seems to me that
what the railroad industry needs more than anything else
is more wearing out of their rails.
  *William Brosnan Jr, Southern Railway, of the intro-
duction of high capacity cars such as the "Big John" covered
hopper on the Southern; interview in* Business Week,
8 June 1963.

742.10 By curtaining the door with faded laundry, and
adding a chicken coop and children, and turning up the
volume on his radio, the Mexican makes a bungalow of
his boxcar and pretends it is home.
  Paul Theroux, of a boxcar village in San Luis Potosi;
The Old Patagonian Express, *1979.*

742.11 We worship the locomotive, countenance the passenger car, and abide fixed plant. But the common freight car? Bless me, no.

David P Morgan, of the fickle nature of interest in railroads. Caption to photograph of coal hopper car, Trains Magazine, *July 1987.*

742.12 I have not polled enough youngsters of the 1970s to determine whether among the sophistications of contemporary space-age life is a vaccine against the compulsion to count the cars of a passing freight train: but in my own youth I know that the ailment was incurable.

Norm Cohen, Long Steel Rail, *1990.*

743. WAITING ROOMS

743.1 Passengers do not like waiting when once a train is due, except in waiting rooms or waiting sheds, on the platform. The want of waiting rooms on the platforms is in winter much felt on the Metropolitan lines.

John Wolfe Barry, Railway Appliances, *1887.*

743.2 On a morning sick as the day of doom
With the drizzling gray
Of an English May,
There were few in the railway waiting room.
About its walls were framed and varnished
Pictures of liners, fly-blown, tarnished.
The table bore a Testament
For travellers' reading, if suchwise bent.

Thomas Hardy, poem, In a Waiting Room.

743.3 The atmosphere of the waiting-room set at naught at a single glance the theory that there can be no smoke without fire. The station-master, when remonstrated with, stated, as an incontrovertible fact, that any chimney in the world would smoke in a south-easterly wind, and, further, said that there wasn't a poker, and that if you poked the fire the grate would fall out. He was, however, sympathetic, and went on his knees before the smouldering mound of slack, endeavouring to charm it to a smile by subtle proddings with the handle of the ticket-punch.

Edith Oenone Somerville and Martin Ross, "Poisson d'Avril," Further Experiences of an Irish R M, *1908.*

743.4 The place had an air of meanness which no fine words about modernization could disguise.

Nicholas Whittaker, of the rebuilt station at Burton on Trent, Platform Souls, *1995.*

744. WALES

744.1 Cymru a fu a Chymru a fydd.
(Wales hath been and Wales shall be)

Motto of the Taff Vale Railway, which appeared on its coat of arms, quoted by Derek Barrie in The Taff Vale Railway, *1939.*

745. WANTAGE TRAMWAY

745.1 A curious race has come to pass
Between an engine and an ass,
The Wantage Tram, all steam and smoke,
Was beat by Arthur Hitchcock's moke.

Anon., quoted in The Railway Magazine, *vol. LXIII, 1928.*

746. WAR

746.1 The effect of rapid communication by railway is, that it enables you to do with a small army the work of a large one.

Sir Willoughby Gordon, Quartermaster General, quoted by Thornton Hunt in Unity of the Iron Network, *1846.*

746.2 It is strange that the first use—perhaps the only use—the Crim-Tartar will ever witness for centuries of the great invention of recent days should be to facilitate the operations of war and destroy life.

W H Russell, of the natives of the Crimean peninsular; quoted by R M Robbins in The Balaklava Railway, *within* Journal of Transport History, *May 1953.*

746.3 We have hitherto regarded the rail merely as a vehicle of transport, to carry materials which are not to be set in work till off the rails. If we look at the rail as part of an instrument of warfare, we shall be startled at the enormous means we have at hand, instantly available, from mercantile purposes, to convert to engines of war.

William Bridges Adams, article, English Railway Artillery, *within* Once a Week, *13 August 1859.*

746.4 Send your companies here by railroad.

Simon Cameron, Secretary of War, recognising vital importance of railroads in the coming civil war; telegram to Gov Andrew of Massachusetts, 15 April 1861. Quoted in Victory Rode the Rails, *George Edgar Turner, 1953.*

746.5 It cannot be ignored that the construction of railroads has introduced a new and very important element into war by the great facilities thus given for concentrating at particular positions large masses of troops from remote sections and by creating new strategic points and lines of operations.

George B McClellan, to President Lincoln, August 1861. Quoted in Victory Rode the Rails, *George Edgar Turner, 1953.*

746.6 Railroads are at one and the same time the *legs* and the *stomach* of an army.

Brigadier General J H Trapier, 26 December 1861, quoted by Robert C Black, Railroads of the Confederacy, *1952.*

746.7 The idea of a few disguised men suddenly seizing a train far within the enemy's lines, cutting the telegraph

wires, burning bridges, and leaving the foe in helpless rage behind, was the very sublimity and romance of war.
William Pittenger, Daring and Suffering, *1887.*

746.8 ... from a military point of view every railway is welcome, and two are still more welcome than one ...
Attributed to Prussian General Von Moltke by Edwin A Pratt in British Railways and the Great War, *1921.*

746.9 The railway companies in the all-important matter of the transport facilities have more than justified the complete confidence reposed in them by the War Office, all grades of railway services having laboured with untiring energy and patience.
Earl Kitchener, of the mobilization of the British Army at the outbreak of the Great War, House of Lords, Hansard *for 25 August 1914.*

746.10 Apart from any considerations as to the nature of the motives that led Prussia to adopt a policy in which she was followed by other German States and other countries, it must be admitted, as a general principle, that, without an organisation more or less perfected in peace time, railways may not only be deprived of most of their usefulness in time of war, but even may become a source of danger to the army or to the country depending on them.
Edwin A Pratt, British Railways and the Great War, *1921.*

746.11 Timetables must be drawn up in advance for the running of every troop train that will be required for the initial purposes of mobilisation and concentration so that not a minute may be lost on the declaration of war, when, indeed, even minutes may be of vital importance. For each separate unit the mobilisation time-tables must show, among other things, what station it will go from to a given destination; at what time the train will start on the particular day of mobilisation assigned to it, and, what is of no less importance, the time it will, or should, arrive.
Ibid.

746.12 All blinds down, please.
Call from railway porters to passengers of departing trains, during the period of air raids over Great Britain in the First World War, noted by Edwin A Pratt in British Railways and the Great War, *1921.*

746.13 The First World War had begun—imposed on the statesmen of Europe by railway timetables. It was an unexpected climax to the railway age.
A J P Taylor, The First World War, *1963.*

746.14 This is a great railway war. The battle of the Marne was won by the railways of France.
General Joffre, quoted by Samuel Vauclain in Steaming Up!, *1930.*

746.15 The extent to which railways are being used in the present War of the Nations has taken quite by surprise a world whose military historians, in their accounts of what armies have done or have failed to do on the battle-field in the past, have too often disregarded such matters of detail as to how the armies got there and the possible effect of good or defective transport conditions, including the maintenance of supplies and communications, on the whole course of a campaign.
Edwin A Pratt, The Rise of Rail Power in War and Conquest, *1915.*

746.16 Resolved, That the railroads of the United States, acting through their chief executive officers here and now assembled and stirred by a high sense of their opportunity to be of the greatest service to their country in the present crisis do hereby pledge themselves, with the government of the United States, with the governments of the several States, and with one another, that during the present war they will co-ordinate their operations in a continental railway system, merging during such period all their merely individual and competitive activities in the effort to produce a maximum of national transportation efficiency.
Fairfax Harrison, President, Southern Railway, to create the Railway War Board, 11 April 1917, quoted by Edward Hungerford, The Story of the Baltimore & Ohio Railroad, *vol. 2, 1928.*

746.17 We must not take Medina. The Turk was harmless there. In prison in Egypt he would cost us food and guards. We wanted him to stay at Medina, and every other distant place, in the largest numbers. Our ideal was to keep his railway just working, but only just, with the maximum of loss and discomfort. The factor of food would confine him to the railways, but he was welcome to the Hejaz Railway, and the Trans-Jordan railway, and the Palestine and Syrian railways for the duration of the war, so long as he gave us the other nine hundred and ninety-nine thousandths of the Arab world.
T E Lawrence, The Seven Pillars of Wisdom, *1926.*

746.18 The railroads are best suited for mass transportation due to their large and efficient transport units. Requiring little service and providing maximum of speed and safety, they are almost independent of season and weather. They operate according to schedule and can be reached at any moment by military orders. Nothing, therefore, can replace the railroads.
Ernst Marquardt, German Ministry of Transportation, during the Polish Campaign, 1939. Quoted by Paul Wohl in Railroad Magazine, *translator unknown, January 1942.*

746.19 The Lines behind the Lines
Attributed to Cecil Dandridge, Advertising Manager, LNER by Allan Middleton, It's Quicker by Rail, *2002. Thematic slogan used on posters issued by Railway Exe-*

cutive to inform the public of the importance of railway traffic, 1940.

746.20 Vibration due to heavy gunfire or other causes will be felt much less if you do not lie with your head against the wall.

Anon., advice to those sleeping in the tube station at Swiss Cottage to shelter from air-raids; Bulletin No 2, The Swiss Cottager, de profundis, September 1940. Quoted in Lines on the Underground, compiled by Dorothy Meade and Tatiana Woolf, 1994.

746.21 Next time you are on the train, *remember the kid in Upper 4.*
If you have to stand enroute—*it is so he may have a seat.*
If there is no berth for you—*it is so that he may sleep.*
If you have to wait for a seat in the diner—*it is so he . . . and thousands like him . . . may have a meal they won't forget in the days to come.*
For to treat him as our most honoured guest is the least we can do to pay a mighty debt of gratitude.
Advertisement, New York, New Haven, & Hartford Railroad. The "kid in Upper 4" was the archetypal soldier, off to war, shown in an upper berth in a Pullman Sleeper. Christmas 1942.

746.22 The Canadian National is Canada's greatest single war industry.
Anon., wartime slogan on advertisements.

746.23 Has anybody an octopus that isn't working?
Anon., slogan in wartime advertisements for Boston & Maine Railroad, indicating the need for many pairs of hands for all the work that needed to be done.

746.24 A pig has no place in a coach seat.
Anon., Boston & Maine wartime advertisement drawing attention to selfish passengers who take up much space and prevent others sitting.

746.25 Railroaders make good soldiers, for there is no industry in the world that respects authority as much as theirs.
Brig Gen Carl R Gray Jr, quoted in Railroad Magazine, *February 1943.*

746.26 Railway travel now places a considerable strain on one's physical powers of endurance, not to speak of one's powers of resistance to irritation.
Lord Royden, Chairman, LMS, at 1944 Company general meeting.

746.27 The war is practically over as far as the railroad passenger is concerned. Quite suddenly the heavy load of troop return has lifted and empty seats, even double seats, have begun to appear in the coaches. Sleeping car reservations can now be had, in many cases, on evening of departure, and the traveller no longer asks hopefully for "anything". He asks for a lower or a roomette or a

bedroom and usually gets it. Dining car stewards again have time to ask if "everything's all right" and in a few instances the extinct finger bowl has been resurrected.
Editorial comment, Trains Magazine, May 1946.

746.28 Trains lead to ships and ships to death or trains
And trains lead to death or trucks and trucks to death
Or trucks lead to the march, the march to death
Or that survival which is all our hope;
And death leads back to trucks and trains and ships.
Karl Shapiro, Poems, 1940–1953, 1953.

746.29 The cold logic that Allied conquest or defeat in certain stages of World War II hinged upon the traffic density of Pennsy's electrified Washington-New York railroad renders any other example of its value superfluous.
David P Morgan, Trains Magazine, February 1951.

746.30 The fact is that there is no fully effective way of putting a railroad out of service without disproportionate outlay of time and resources. This has been demonstrated over and over again, in every war since railroads became an important element in warfare. Railway durability and recuperability have become thoroughly established as principles of military doctrine.
General James A Van Fleet, Rail Transport and the Winning of Wars, 1956.

746.31 Sharp instituted a system of rail removal which might be described as a disassembly line.
Edward B Burns, of the means of railway tack removal adopted by Thomas Sharp during the American Civil War. Bulletin No 104, RLHS, April 1961.

746.32 Virgil Caine is my name and I drove on the Danville train
'til so much cavalry came and tore up the tracks again
In the winter of '65, we were hungry, just barely alive
I took the train to Richmond that fell
It was a time I remember, oh, so well.
Song, The Night They Drove Old Dixie Down, as sung by Joan Baez, but differing slightly from the original lyrics by J Robbie Robertson, 1970. The Danville referred to is assumed to be the Virginia town just north of the North Carolina border, at the junction of the Southern and Atlantic & Danville railroads.

746.33 Of course, the GI does not equate troop trains with the statistics they produced. WWII vets recall instead cinders on the window sill, one to an upper and two to a lower, duffle bags in the women's room, slack action, the breadth of Kansas and the altitude of Colorado, USO canteens, waving girls, tedium, kitchen cars and paper plates and fruit cup, rumors, and—if you were bound for basic or a port of embarkation—the lonesomest whistles in the world.
David P Morgan, Trains Magazine, 1979.

746.34 in 1940, in the dazed center of my childhood and
 the last year
before the war—*our* war—we watched my suited,
 suitcased father
descend the tiled concourse to board "the james
 whitcomb riley"
at cincinnati's brand-new, brassy, domed and cavernous,
 art-
decoed union terminal with its floor-to-ceiling mosaics
 of labor
and industry. almost empty even then, it echoed with the
 hard
departures of newsstands, shoeshine boys, our own
 hurried heels
on the marble floors. then it was troop trains, troop
 trains,
all troop trains: the long, thrilling, khaki freight of the
 war.
  Alvin Greenberg, second verse of poem Freight Train,
Freight Train, *1989.*

746.35 When World War II came, millions of boys and
 quite a few girls found out what lay beyond the vanishing
 points of the tracks that ran through their little towns.
 Most of them could not get back home fast enough when
 their tours of duty were over. It was the trains that took
 them away, and the trains that brought them back, in one
 shape or another. It would never be like that again.
  Albro Martin, Railroads Triumphant, *1992.*

747. WASHINGTON UNION STATION

747.1 About half a mile to the north of the Capitol, there
 had existed, since there had been any Washington at all,
 an ungainly and unhealthy swamp tract. Delaware Avenue led from the Senate Wing direct to it, and then, in
 its bog, disappeared. When it was announced that this
 swamp—twenty feet below tidewater level—was to be the
 site for the monumental Union Station, Washington
 lifted its aristocratic eyebrows.
  Edward Hungerford, The Story of the Baltimore &
Ohio Railroad, vol. 2, *1928.*

748. WATCHING TRAINS
See also Trainspotting

748.1 The circus came to town yesterday and every boy
 was at the railroad crossing at dawn to see the show
 unload. Some of the boys were under 40.
  *Attributed to a small city newspaper by Eugene Whit-
more,* Trains Magazine, *August 1945.*

748.2 Deliberate train-watching, on either an amateur
 basis or professional, is a sometimes thankless pastime,
 of course. Unlike stamp collecting or golf, it has never
 received Presidential endorsement.
  David P Morgan, Trains Magazine, *May 1957.*

748.3 By its very nature, railroading does not lend itself to
 armchair enjoyment. Somebody has called it the industry without a roof.
  Ibid.

748.4 And once I even turned and broke away,
 Walking a little distance, then came back,
 Held by the broken promise of that track,
 And fascination I could not gainsay;

And so still lingered while the darkness closed,
 Unsatisfied, while dying hope was fed
 upon the breath of hope already dead,
 And stars above the twilight were disclosed.

But nothing touched the stillness of that scene
 And time gathered, when I must be gone.
 Yet one last look to where the signals shone
 Far in the dusk: one had changed to Green!
  Henry Maxwell, poem, Vigil, *at the end of a long wait
beside the track, in* A Railway Rubaiyyat, *1968.*

749. WATER

749.1 It was with water of zero hardness in its boiler.
  T Henry Turner, of the quality of water used by Mallard
on her record run, quoted by Geoff Hughes in Sir Nigel
Gresley: the Engineer and his Family, *2001.*

750. WATERLOO STATION (London)

750.1 We got to Waterloo at eleven, and asked where
 the eleven-five started from. Of course nobody knew;
 nobody at Waterloo ever does know where a train is
 going to start from, or where a train when it does start
 is going to, or anything about it.
  Jerome K Jerome, Three Men in a Boat, *1889.*

750.2 Waterloo was a wooden, brown and fascinating
 place, alive with the constant flutter of the indicator
 boards and the scurry of pigeons. It hadn't yet been ster-
 ilized and Dixonized and turned into the bland place it is
 today.
  Nicholas Whittaker, Platform Souls, *1995.*

750.3 I've nothing against the Eurostar terminus, but it
 looks nothing like a station. Marketing wisdom insists
 that if railways are to compete they've got to look like airports, which is a sad comment on society.
  Ibid.

751. SIR EDWARD WATKIN

751.1 All hail Edward Watkin,
 Pioneer of the world!
 Who unites great nations
 His Banners unfurled

All hail to our hero,
Great man of the day,
A grand Old Veteran
So youthful and gay!

R D Roberts, ode written for the opening of the Chester & Connahs Quay Railway, 16 August 1887.

751.2 Sir Edward has been the Grand Old Man of the railway world. Masterful and capricious, talented and vain, sanguine and impetuous, he has, at least, shown himself endowed with the courage of his convictions.

The Railway Times, *comparing Watkin with his close friend William Ewart Gladstone, 2 June 1894.*

751.3 He might, in fact, be compared with Hudson in many respects, in his power of persuasion, in his grasp of finance, his intimacy with railway problems, his organising skill and appreciation of able officers, in his far-seeing estimates of future development—above all, in the great ambition that distracted him and led to wasteful struggles and unnecessary promotions. But, unlike Hudson, he was honest.

Edward Cleveland-Stevens, English Railways and Their Development and Relation to the State, *1915.*

751.4 Watkin's ambitions were considerable, but they were general in nature. He was an opportunist without a detailed blueprint.

David Hodgkins, The Second Railway King, *2002.*

752. WAVERLEY STATION (Edinburgh)

752.1 On the platforms of the Waverley station in Edinburgh may be witnessed every evening in summer a scene of confusion so chaotic that a sober description of it is incredible to those who have not themselves survived it. Trains of caravan length come in portentously late from Perth, so that each is mistaken for its successor; these have to be broken up and remade on insufficient sidings, while bewildered crowds of tourists sway up and down amongst equally bewildered porters on the narrow village platform reserved for these most important expresses; the higher officials stand lost in subtle thought, returning now and then to repeated enquiries some masterpiece of reply couched in the cautious conditional, while the hands of the clock with a humorous air survey the abandoned sight, till at length, without any obvious reason and with sudden stealth, the shame-stricken driver hurries his packed passengers off into the dark.

E Foxwell and T C Farrer, Express Trains English and Foreign, *1889.*

752.2 ... Edinburgh Waverley as a railway station belongs to its city, and is a part of its city, in a way that cannot be seen comparably in any other city of the United Kingdom.

C Hamilton Ellis, The North British Railway, *1955.*

753. WEATHER

753.1 While the sudden stop that sends out a flagman with protecting signals may give him only a pleasant walk along a meadow-lined track on a rare June day, it may cause him to crawl in the piercing wind of a dark winter night over sleety ties on an ice-bound trestle. He may be required to ride the decks of the cars with the mercury down to 20 or 30 below zero, running over the tops of the cars because it is so much easier to keep the footing running than walking, and safer to jump from car to car than to step deliberately across the space between them, the black smoke and steam from the engine meanwhile blinding him as it rolls back over the train in dense volume.

L F Loree, Railroad Freight Transportation, *1922.*

753.2 During blizzards everything about the railroad was melodramatic. There were days when the town was completely shut off, when they had no mail, no express, no fresh meat, no newspapers. At last the rotary snow-plow came through, bucking the drifts, sending up a geyser, and the way to the Outside was open again. The brakemen, in mufflers and fur caps, running along the tops of ice-coated freight-cars; the engineers scratching frost from the cab windows and looking out, inscrutable, self-contained, pilots of the prairie sea—they were heroism, they were to Carol the daring of the quest in a world of groceries and sermons.

Sinclair Lewis, Main Street, *1922.*

753.3 It's far too lovely a day to go in by train.

Francis Brett Young, of a couple who decide instead to drive (by horse and carriage) to North Bromwich for shopping; Portrait of Clare, *1927.*

753.4 Meet the Sun on the East Coast.

Cecil Dandridge; advertising slogan for LNER, in praise of East Coast Resorts, 1939.

753.5 The whole continent seemed to be gasping for its breath. In the hot green depths of the train a powder of fine cinders beat in through the meshes of the screens, and during the pauses at stations the little fans at both ends of the car hummed monotonously, with a sound that seemed to be the voice of the heat itself. During these intervals when the train stood still, enormous engines steamed slowly by on adjacent tracks, or stood panting, passive as great cats, and their engineers wiped wads of blackened waste across their grimy faces, while the passengers fanned feebly with sheaves of languid paper or sat soaked and sweltering in dejection.

Thomas Wolfe, You Can't Go Home Again, *1934.*

753.6 When the days grow colder, and the darkness noticeably longer and deeper, and the roads are slippery and headlights dazzling, there will be pleasure in the thought of a nice, warm, well-lighted, roomy, comfortable railway carriage in which a man may travel fast and free of responsibility at a penny a mile.

> The Times, *1935, quoted by William B Stocks,* The Railway Magazine, *vol. 108, 1962.*

753.7 Once, an engine attached to a train
Was afraid of a few drops of rain—
—It went into a tunnel
And squeaked through its funnel
And never came out again.

> *Rev Wilbert Awdry,* The Sad Story of Henry, *in* The Three Railway Engines, *1945.*

753.8 T'was in the year of '89 on that old Great Western line,
When the winter wind was blowin' shrill,
The rails were froze, the wheels were cold, then the air brakes wouldn't hold,
And Number 9 came roaring down the hill—oh!

The runaway train came down the track and she blew,
The runaway train came down the track and she blew,
The runaway train came down the track, her whistle wide and her throttle back,
And she blew, blew, blew, blew, blew.

> *Robert E Massey, song,* The Runaway Train, *first two verses.*

753.9 The runaway train went over the hill and she blew,
The runaway train went over the hill and she blew,
The runaway train went over the hill and the last we heard she was going still,
And she blew, blew, blew, blew, blew.

> *Ibid., last verse.*

753.10 I agree with you. This service is bad. The railroad shouldn't let a little thing like a snowstorm keep it from being on time. I think if the railroad can't do better than this, it should get some sense and do like the airlines, close up business whenever the weather is bad.

> *Anonymous railroad executive, on a westbound train from New York, to a group of stranded airline passengers complaining about the train; recorded in* Trains Magazine, *April 1948.*

753.11 Grey weather goes with the railways.

> *Peter Handford, sound recordist, quoted in* The Observer, *5 February 1961.*

753.12 The wrong kind of snow.

> *Nigel Griffiths, Production Editor,* Evening Standard, *11 February 1991. Often attributed incorrectly to Terry Worrall, Director of Operations, British Rail, after he had explained, during a radio interview on the* BBC's Today *earlier that day, what had caused the failure of electric trains with the words "It is not the volume of snow that has caused us problems but the type of snow, which is very dry*

and powdery." The phenomenon caused trouble on other European railways also.

753.13 Rolling popsicles.

> *Anon. headline, concerning ice-laden trains in severe weather;* Trains Magazine, *April 2004.*

754. FRANCIS WEBB

754.1 Some forms of four-cylinder compound locomotives are exceedingly popular on some European railways, but I never heard of any lavish praise being expended on the Webb compounds outside of the designer's immediate friends.

> *Angus Sinclair,* Development of the Locomotive Engine, *1907.*

754.2 . . . a cold, harsh, lonely man considering no ideas but his own, fecund in his original inventiveness, instant as cordite to take offence, intolerant of his subordinates and icily munificent in the great causes of Victorian charity.

> *C Hamilton Ellis,* The Trains We Loved, *1947.*

754.3 There has never been anyone quite like Francis Webb, but much has been written about him, so suffice it to say that he was an autocrat among CMEs, brooking no interference or criticism from anyone.

> *James Lowe,* British Steam Locomotive Builders, *1975.*

755. DANIEL WEBSTER

755.1 Daniel Webster struck me much like a steam-engine in trousers.

> *Sydney Smith,* Lady Holland's Memoir.

756. SIR RALPH WEDGWOOD

756.1 His letters on major subjects, and his policy directives, were couched in language which had all the force and authority of Papal encyclicals.

> *Cecil J Allen,* The London & North Eastern Railway, *1966.*

756.2 Through his railway life it was a long, rather lonely and very hard furrow that Sir Ralph Wedgwood was destined to plough, but plough it he did, and looking back we can see that the furrow was straight.

> *Ibid.*

757. WEEDON

757.1 There was a young lady of Sweden,
Who went by the slow train to Weedon;
When they cried "Weedon Station!"
She made no observation,
But she thought she should go back to Sweden.

> *Edward Lear,* Nonsense Rhymes.

758. WELWYN GARDEN CITY

758.1 You lovely station of a lovely town,
Your breadth and loveliness combine to make
A beauty that can cause the heart to ache
Sweetly, before the eye can travel down
The glorious gate of Howard, with its crown
Of green, for very pleasant is the brake
That closes it, and forces it to take,
For right reward, the symbol of renown.

Time marches on. You win to a fair May.
Lovers from Royston, Hitchin and King's Cross
Delight to dawdle with you night and day.
The Flying Scotsman *may* treat you as dross.
Anxiety is his. A mighty way,
And to be covered quickly, said the Boss.
Terence Greenidge, verse For the Twenty-First birthday
of Welwyn Garden City Station, *in* Girls and Stations,
1952.

759. WESTERN CLASS DIESEL LOCOMOTIVES

759.1 Even the names were a last fling of defiance. Every
one was prefixed with the word that had given the class
its name—*Western Queen, Western Pathfinder, Western
Stalwart*—and which also seemed a last two-fingers up at
the corporate ethos of British Railways.
Nicholas Whittaker, Platform Souls, *1995.*

760. WESTERN PACIFIC RAILROAD

760.1 Western Pacific speaks loudly and carries a small
stick.
*Anon., of Alfred Perlman's period as CEO of the rail-
road. Quoted in letter from G R Green to* Trains Magazine,
June 2002.

761. WEYMOUTH, England

761.1 The Naples of England.
Advertisement, Great Western Railway, *1907.*

762. WHEEL ARRANGEMENTS

762.1 No offense to the General, but he's no substitute for
Mike.
*David P Morgan, of the ill-fated wartime effort to
rename the Mikado type MacArthur;* Trains Magazine,
January 1957.

762.2 This continent didn't contain one handsome 4-8-0
and very few attractive Decapods, but it took a lot of
work to mess up a Pacific.
Ed King Jr, Trains Magazine, *February 1985.*

763. WHISTLES

See also Arrival, Departure, Sounds, Stations

763.1 Railroad whistle like the whir of a gigantic musqitoe
close to your ear at night
And sometimes like an Aeolian harp.
Ralph Waldo Emerson, note in Journal CD, *page 90,
1847.*

763.2 . . . the whistles on the railway . . . . which recalled to
us the familiar sounds of Wolverhampton or of Swindon,
and made us believe for the moment that we were in a
civilized country.
W H Russell, of the Balaklava Railway, The Times,
November 1855.

763.3 The switchmen knew by the engine's moans
That the man at the throttle was Casey Jones.
Wallace Jones, The Ballad of Casey Jones, *1908.*

763.4 . . . soft as the woodwinds of a symphony orchestra.
*Of a new whistle introduced on the Chicago Milwaukee
& St Paul Railway in response to public complaints about
the existing whistles,* Railway Mechanical Engineer, *October 1926.*

763.5 There was an accent of King Tamerlane in 5217's
whistle as she shouted a blasphemy in steam.
*Christopher Morley, of the locomotive whistling when
delayed by a preceding train;* A Ride in the Cab of the
Twentieth Century Limited, *1928.*

763.6 Why, we demand, this convention of hooting and
whistling, when the noise alone of an approaching train
is sufficient to herald its arrival? Perhaps this sound is
due, again, to that prevalent sin of our time, an "inferior-
ity complex". Certainly the smaller the engine, the louder
the screech; and I suspect that, as trains become more
and more supplanted by other forms of locomotion,
they will yell more loudly yet.
Sir Osbert Sitwell, Penny Foolish, *1935.*

763.7 Do you hear that whistle down the line
I figure that it's engine number forty-nine
She's the only one that'll sound that way
On the Atchison, Topeka and the Santa Fe
Johnny Mercer, song from film The Harvey Girls, *1946.*

763.8 The Grand Trunk engines possessed whistles built
by a master who was a combination of Thor himself and
some brilliant esthete, for he perfected an art form fitted
wonderfully to that northern clime and splintery air
where an obbligato was supported by the cracking of
Aurora Borealis.
Stewart H Holbrook, The Story of American Railroading, *1947.*

763.9 It was during this transitional period that a typical
railway function took place at Horwich, namely the

choosing of standard whistles for locomotives. On this occasion all the chief officials turned up, and it was only lacking H M Bateman, or Fougasse, to make the party historical.

T Lovatt Williams, The Railway Magazine, *vol. 95, 1949.*

763.10 SERENA: Do your beloved American locomotives whistle so hysterically?

AXEL: No, ma'am, they sound deeper and more mournful. They have vaster distances to travel, wilder territories to cross, and heavier, more virile responsibilities: they have no time for shrill, Gallic petulance.

Noel Coward; exchange after a French locomotive whistle is heard in the Buffet de la Gare, Boulogne; Quadrille, *1952. For a description of a typical French steam locomotive whistle, see 763.12.*

763.11 The whistle of the night train, with all its terrible urgency and demands upon the heart, seems not mechanical and man-made but as natural as the quiet of fresh fallen snow, the rumble of the river.

Anon., caption to photograph by O Winston Link, Trains Magazine, *November 1957.*

763.12 ... and French steam locomotives ... are tremendous except for their whistles, which sounded like a girl with a mouse in her dressing gown.

C Hamilton Ellis, Rapidly Round the Bend, *1959.*

763.13 Then the shriller freight whistle let go, crackling my eardrums and sending shivers through my timbers—agreeable shivers, like the puckers you get when you bite into a crisp, rosy stalk of rhubarb freshly picked from the garden and lightly sprinkled with salt.

Rosemary Entringer, in the cab of Reading Railroad 2124; Trains Magazine, *June 1962.*

763.14 Long moments later the sound reached my ears—the haunting, romantic, deep-toned whistle-blast of an American locomotive. I felt my skin crawl. I trembled with excitement. This is my welcome to America, I told myself, this powerful, vigorous, masculine voice of a giant steam engine, a sound so completely unlike the thin screeching of European engine whistles.

Otto Kuhler, My Iron Journey, *1967.*

763.15 The majesty of that engine I did not see. But after I had gone back to bed I heard, and heard, and heard it, moaning its herald for each grade crossing (many there are) on down the track for miles and miles in the still clear air. Moaning, crying, wailing—the call of an extinct dinosaur.

Sally Jayne, letter to Joseph Y Jeanes Jr, following the night time passage of Southern Pacific 4449 on the American Freedom Train, in California. Quoted in Trains Magazine, *July 1976.*

763.16 Never has man produced a more lonely sound than the whistle of a steam locomotive. It was a sad sound that seemed to say to each of us who heard it: 'Come with me and I'll show you America. Follow me all the days of your life, and as you lie down to die, you'll pray with your last breath to follow me once again.'

Hood River Blackie (Ralph Goodings), Quest Magazine, *August/September 1978.*

763.17 Sir, we do not blow whistles at people from Newbury.

Guard of train, upon being encouraged to get his train started by General Manager Gerard Fiennes, 1960s, recorded in Modern Railways, *October 1986.*

764. WILLIAM WHITELAW

764.1 He represents the bygone feudal system of railway management, when railway directors regarded the general manager much as they would their bailiff or gamekeeper.

Attributed to Sir Eric Geddes, in The London & North Eastern Railway, *Cecil J Allen, 1966.*

765. WINCHESTER

765.1 The city of Winchester is famed for a cathedral, a bishop—he was unfortunately killed some years ago while riding—a public school, a considerable assortment of the military, and the deliberate passage of the trains on the London and South Western line.

Robert Louis Stevenson, The Wrong Box, *1889.*

766. WINE

766.1 Wine is good for the production of words as well as steam ...

Richard Hardy, Railways in the Blood, *1985.*

767. WOLVERHAMPTON (England)

767.1 Snow, wind, ice, and Wolverhampton—all together.

Charles Dickens, Household Words, *21 January 1854.*

768. WOMEN

768.1 It is not customary, it is not polite, it is not right or just for a lady to occupy one whole seat with her flounces and herself, and another with her satchel, parasol, big box, little box, and bundle.

Anon., Comment in a guide book for passengers on the transcontinental railroad, noted by Edwin L Sabin in Building the Pacific Railroad, *1919.*

768.2 First, the 'emancipation of Woman.' Without claiming that this movement arose directly from the opening of railways, we may firmly maintain that it has been greatly strengthened by the mere fact of railway travel. Women are so tightly moored by Nature that if locomotion is also for them an impossibility they must indeed feel slaves. Compare the portentousness of a hundred miles journey for a girl last century with the ridiculous ease of travel now, and we see how they cannot bask in the new freedom long without tingling to assert their own individuality.

E Foxwell and T C Farrer, Express Trains, English and Foreign, *1889.*

768.3   At nine in the morning there passed a church,
At ten there passed by me the sea,
At twelve a town of smoke and smirch,
At two a forest of oak and birch,
     And then, on a platform, she:

A radiant stranger, who saw not me,
I queried, 'Get out to her do I dare?'
But I kept my seat in search for a plea,
And the wheels moved on. O could it but be
     That I had alighted there!

Thomas Hardy, poem, Faintheart in a Railway Train.

768.4   Promontory town, however, contributed a quota of furbelows.

Edwin Sabin, put "delicately," according to Lucius Beebe, who stated elsewhere that this reference to women attending the celebrations of the completion of the Transcontinental Railroad at Promontory, 10 May 1869, was to "a generous group of prostitutes" from the nearby construction camp at Corinne, Utah, without invitation from the management; Building the Pacific Railroad, *1919.*

768.5   Woman has not now got the same desire to appear always graceful; she adopts a manly gait, talks louder, plays hockey, rides horseback astride, and boldly enters hotel smoking rooms and railway smoking compartments without apology.

Joseph Tatlow, Fifty Years of Railway Life, *1920.*

768.6   What do you do besides lure men to their doom on the Twentieth Century Limited?

Ernest Lehman, scriptwriter, North by Northwest, *asked by Cary Grant of Eva Marie Saint, 1959.*

768.7   She sounded like the Book of Revelations read out over a railway station public address system by a headmistress of a certain age wearing calico knickers.

Clive James, of Margaret Thatcher, date not known.

768.8   You've got to hand it to these girls, they've a lovely touch on the brake.

Anon. locomotive driver in conversation with colleagues, in the early days of employment of female drivers, Great Britain; overheard by Peter Coster.

769. WOODEN RAILWAYS

769.1   No man is rich enough to run a wooden railway.

Of the cost of maintenance on wooden railways, Railroad Gazette, *1886, quoted by John H White Jr, in Chapter 6 of* Material Culture of the Wooden Age, *1981.*

770. WORKING ON THE RAILWAY
See also Railwaymen

770.1   When, however, he came to report his proceedings, he did not appear much elated, and remarked that he thought he had seen human nature in all its forms; but he had come to the conclusion that if a man desired to be fully acquainted with human nature, he must be an agent to obtain land for a railway.

John B Jervis, of a man who had sought to buy land on behalf of an American railroad. Railway Property, *1866.*

770.2   I've been railroading since 1889. I've pleased and displeased people ever since. I have been cussed, discussed, lied about, lied to, hung up, held up, and helped up. There's been times when I've stayed in the railroad business only to see what the hell was going to happen next. But no matter what I've thought, felt, or wanted to do, I've had sense enough to remember that I picked my job and the job did not pick me, that the railroad can get along a damn sight better without me than I can without it, and I've never shortchanged my job.

J J Bernet, President, Nickel Plate Railroad, American Magazine, *July 1928.*

770.3   Railroading is the greatest sport on earth. Everyone is a railfan at heart. There is far more enjoyment to be had along a railroad and in its shops and yards than there is on any golf course or at a baseball stadium or a movie. Working on a railroad is really being a professional sportsman, getting paid for playing in the most fascinating game on earth.

John W Barriger, President, Monon Railroad, Trains Magazine, *September 1950.*

770.4   I worked on freight trains, passenger trains and cattle trains. To me the future always lay at the end of the line. The Iron Horse was my Godhead; as cruel and unpredictable as any of the Gods of Antiquity. He dragged me back and forth across the prairies and deserts and mountains of my country. He took my father from me and two of my brothers. He also, more benevolently, disposed of an uncle in Wisconsin who left me eighteen thousand dollars in his will. That was the beginning of fortune. It seemed to me then as it seems to me now that the destiny of America lies in the increasing power and expansion of her railroads—the annihilation of distance—the drawing together of isolated people—the—I am boring you. This is my hobby horse, my iron hobby horse. You must not encourage me.

Noel Coward; spoken by Axel Diensen, Act II scene III,
Quadrille, *1952.*

770.5 So like a railroad: You know your job is in trouble
when they paint the building you're working in.
Chard Walker, Trains Magazine, *February 2004.*

WRECKS

See Accidents, Derailment, Safety

771. WRITING ABOUT RAILWAYS

See also Books on Railways, Dedications, History—
Railway

771.1 At the back of my mind, in writing of railway matters, is the desire that railwaymen may be better understood and sympathised with by a public which is usually
too busy to heed.
James Scott, Railway Romance and Other Essays, *1913.*

771.2 Make the platform speak. It should have some tales
to tell.
Advice from Rudyard Kipling to Henry Chappell, 1924,
quoted in The Railway Station, *Richards and MacKenzie,*
1986.

771.3 It's one thing to sit behind a typewriter in white
shirt and tie and editorialize on what's wrong with
unionized rail labor; it's another to hold down a cupola
seat in blue shirt and overalls and hang on as 110 feet of
slack contends with a razorback grade at 50 mph.
David P Morgan, referring to correspondence he had
received from Brotherhood members. Trains Magazine,
November 1973.

771.4 Starlight Express began life in 1973. I was asked to
compose the music for a series of cartoons for television
based on the Thomas the Tank Engine stories. Two years
later my life was complicated by meeting a soul singer
who had the unusual gift of being able to sing three notes
at once in the exact pitch of an American steam whistle.
Andrew Lloyd Webber, Introductory Notes *to the music*
of Starlight Express, *1993. He named his production com-*
pany The Really Useful Group following the example of
Thomas the Tank Engine, which aspired to be a Really Use-
ful Engine. See 373.7.

771.5 I should also apologize for arranging temporary
breakdown of service to the London Underground and
the rail link between Cambridge and London, but travellers by public transport may feel that this is a fictional
device which imposes no great strain on their credulity.
P D James, Author's Note in novel The Murder Room,
2003.

772. WRITING ON THE TRAIN

772.1 A tune was born in my head last week,
Out of the thump-thump and the shriek-shriek
Of the train, as I came by it, up from Manchester;
And when, next week, I take it back again,
My head will to the engine's clack again,
While it only makes my neighbour's haunches stir,
—Finding no dormant musical sprout
In him, as in me, to be jolted out.
Robert Browning, Christmas-Eve and Easter-Day,
1850.

772.2 Like others I used to read,—though Carlyle has
since told me that a man when travelling should not
read, but 'sit still and label his thoughts'.
Anthony Trollope, An Autobiography, *1883.*

772.3 I made for myself therefore a little tablet, and found
after a few days exercise that I could write as quickly in a
railway carriage as I could at my desk. I worked with a
pencil and what I wrote my wife copied afterwards. In
this way was composed the greater part of Barchester
Towers and of the novel which succeeded it, and much
also of others subsequent to them.
Ibid.

772.4 Long haul streamliners must be excepted from this.
Their accommodations are admirable but they sway so
much that you cannot read for very long and cannot
write at all.
Bernard de Voto, Harpers Magazine, *April 1947.*

772.5 My notes, barely legible, say "I hope they ride better
than this."
George H Drury, of a passing Eurostar while a passenger
in a 1960s Connex South East EMU. Trains Magazine,
December 1997.

Yy

773. YARDS

773.1 The way freight, like a snail, dragged past him, opening, as it were, the panorama of the scene in the yard. The low switch lights, red, green, purple, and white, like myriad and variegated fireflies hovering everywhere over the ground; the bobbing lantern of a yardman here and there; the dancing gleam of a headlight, as the little yard engine shot fussily away from a string of lighted coaches—the Eastern Express—which it had evidently just made up and backed down on the main line beside the station.
> *Frank Packard,* The Wire Devils, *1918.*

773.2 The yardmaster leads an eventful life, one that requires alertness, precision, and accuracy. Thousands of cars, occasionally broken and always breakable, come in irregular flow. Every unusual incident hinders him, but none helps him, for his work is movement, his danger is blockade. Always the cars are to go forward, for his bailiwick is simply a part of the main-line movement, slightly expanded for his work of breaking up and marshalling trains.
> *L F Loree,* Railroad Freight Transportation, *1922.*

773.3 A blockaded yard means a blocked road, an absolutely useless, expensive tool, and this the yardmaster can make in a day, not necessarily by doing the wrong thing but by not doing enough. Nor can he merely watch and wait; this is fatal; he must do something vigorously and keep going without admitting an impossibility. The ideal man for this work should have an aptitude and ingenuity for meeting small and great emergencies, something quite beyond the ability to follow rules.
> *Ibid.*

773.4 The yardmaster leads an eventful life.
> *John Droege,* Freight Terminals and Trains, *1925.*

774. YORK

774.1 Mak all t'railways cum t'York.
> *George Hudson, quoted by Richard Lambert,* The Railway King, *1934.*

774.2 ... three different lines of railway assemble their passenger mobs, from morning to night, under one roof; and leave them to raise a travellers' riot, with all the assistance which the bewildered servants of the company can render to increase the confusion.
> *Wilkie Collins,* No Name, *1862.*

774.3 With its handsome hotel, it is a gentleman among stations.
> *J Pendleton,* Our Railways, *1896.*

774.4 That lovely York railway station, which always gives one the idea of being on pleasure bent.
> *David Joy, quoted in* Extracts from the Diaries of David Joy, *in* The Railway Magazine, vol. XXIII, *1908.*

775. YORK & NORTH MIDLAND RAILWAY

775.1 London will be the head of our railway, Edinburgh the feet and York the heart. I hope that the head will never be afflicted with apoplexy, nor the feet with gout, and that York will continue sound at the heart.
> *James Meek, at dinner celebrating the opening of the railway, 29 May 1839, quoted by Robert Beaumont,* The Railway King, *2002.*

776. ROBERT R YOUNG

776.1 He's teaching railroads to an age group that cut its teeth on the automobile door handle and was weaned on aircraft fuel.
> *Anon., of Young's efforts to interest young men in railroading. Noted in* Trains Magazine, *November 1947.*

776.2 Robert Young is more than a forced draft under the railroad boiler.
> *Stewart H Holbrook,* The Story of American Railroading, *1947.*

776.3 The Gadfly of the Rails.
> Life Magazine, *noted in* Chessie's Road, *Charles W Turner, 1956. Robert Young was one-time chairman of the Chesapeake & Ohio Railroad.*

The Engine Driver's Epitaph

This verse is known to have appeared on the headstones of four graves, and may have been used on others. It has been attributed to Thomas Codling of Wylam. The version shown under "Epitaphs" (221.2) is the earliest of the four that has been identified.

It predates and differs slightly from its second known use, on the better-known headstone of Thomas Scaife, killed at age twenty-eight by a locomotive boiler explosion at Bromsgrove, Worcestershire, on 10 November 1840. The headstone was erected at St. John the Baptist Church, Bromsgrove, 1842. It bears a likeness of a Norris 4-2-0 at the top, but this choice of locomotive was incorrect, it in fact having been an 0-2-2 of British manufacture that killed Scaife. It was called *Surprise*; given the nature of boiler explosions the name was unexpectedly apt.

The precise spelling and punctuation on the headstone was checked on 20 July 2003. It should be noted that the stone has at some time been painted with a protective coating, with the lettering and punctuation picked out in white paint. The coating is now brown, but may be the same as the black applied to it in 1959 when British Railways undertook the restoration of the headstone, together with that of Joseph Rutherford, Scaife's mate on the footplate. Some of the punctuation is no longer as originally rendered: a photograph taken in 1959 shows that all lines ended in full points. The spelling, however is still discernible, including the misspelling of *o'er* in the fifteenth line:

My *engine* now is cold and still.
No water does my *boiler* fill:
My *coke* affords its flame no more.
My days of usefulness are o'er
My *wheels* deny their noted speed:
No more my guiding hands they heed.
My *whistle* too, has lost its tone.
Its shrill and thrilling sounds are gone.
My *valves* are now thrown open wide.
My *flanges* all refuse to guide.
My *clacks*, also, though once so strong,
Refuse to aid the busy throng.
No more I feel each urging breath;
My *steam* is now condens'd in death.
Life's railway's oe'r, each *station's* past.
In death I'm stopp'd and rest at last.
Farewell dear friends, and cease to weep.
In Christ I'm SAFE, in Him I sleep.

There is a fairly clear photograph of the stone, before restoration, in *Railway & Locomotive Historical Society Bulletin*

No 10, 1925, and the punctuation and use of italics given above agrees with that visible in the photograph. On the stone it is stated that the verse was "composed by an unknown friend as a memento of the worthiness of the deceased."

A third use of the verse may be found in St. Peter's Church, Newton-le-Willows, Lancashire, where Peers Naylor, an engineer who died on 10 December 1842 at the age of twenty-nine, is buried. The stone once stood erect but now lies on the ground, and although no longer in good condition, it bears a likeness of a contemporary 2-2-2 locomotive in bas-relief at the top. Some of the punctuation at the ends of lines has disappeared as the top layer of stone, in which the inscription is cut, is crumbling away. The verse on Peers Naylors' headstone was checked for precise spelling and punctuation on 26 June 2003. Like the Scaife memorial, it uses italics for some words, and reads:

My *engine* now is cold and still,
No water does my *boiler* fill;
My *coke* affords its flame no more
My days of usefulness are o'er.
My *wheels* deny their noted speed
No more my guiding hand they heed.
My *whistle*, too, has lost its tone
Its shrill and thrilling sounds are gone.
My *valves* are now thrown open wide
My *flanges* all refuse to guide.
My *clacks*, also, though once so strong
Refuse to aid the busy throng.
No more I feel each urging breath,
My *steam* is now condensed in death
Life's *railway's* o'er each *station* past
In death I'm stopp'd, and rest at last.
Farewell, dear friends and cease to weep,
In Christ I'm safe in Him I sleep.

There is a statement in the London Midland & Scottish Railway staff magazine of May 1924 that the verse appeared in the churchyard at Winwick, near Warrington, but this is incorrect. There is no such gravestone there, and a search through records held at the church suggests that no such grave existed at that church. However, the reference in the LMS magazine is specific that the memorial appeared on twelve separate stones. No such memorial has yet been found.

The website www.perthshirediary.com gives the text of the poem under its entry for 3 May 1965, stating that the verse once appeared on a stone near the station at Dalnaspidal. No other details are given and an enquiry to the website was not answered. A visit to Dalnaspidal yielded no more.

It is stated in *Railway Adventures and Anecdotes*, by Richard Pike, 1888, that the verse appears on a headstone in Alton, Illinois. It is further stated therein that it was "written by an engineer on the old Chicago and Mississippi Railroad, who was fatally injured by an accident on the road; and while he lay awaiting the death which he knew to be inevitable, he wrote the lines which are engraved on his tombstone." Pike does not, however, give the name of the engineer or any other detail. Diligent searches in Alton by Cathie Lamere, a genealogist at The Hayner Public Library in Alton, have revealed that the grave, in the "Old Yard" burial ground, is that of George Senior, whose gravestone gives the verse (though now barely legibly), and gives his date of death as 4 November 1853. The year is consistent with the fact that the Chicago & Mississippi Railroad existed under that name only from 1852 to 1855. A contemporary account in the *Alton Telegraph* records that, contrary to the account in Pike's book, George Senior wrote the verse while in good health and later told his fireman how appropriate it would be for use in the newspaper if he, Senior, were to be killed on the railroad. The account in the *Alton Telegraph* gives the date of the accident as 8 November and states that Senior died on 11 November. These dates are consistent with reports in the *Alton Weekly Courier*, while the gravestone itself gives the date of death as 4 November. This would require the date on the gravestone to be wrong. The *Alton Telegraph* is incorrect in stating that Senior was employed by the Chicago & Alton Railroad, because that name was not used until October 1862.

Clearly, neither the account of the composition of the verse in Pike's book nor that of the *Alton Telegraph* can be true, for the three uses of the verse, in England in 1840 and 1842, all predate Senior's death by several years, and of them,

two have claims of authorship by friends of the deceased. There is no reason to assume that Senior was making a claim for authorship of the verse, if what he had actually said was that he had written it, meaning that he had written it down, rather than meaning that he had composed it. If in fact, however, he made a claim to authorship, we can now see that he was wrong. It should be pointed out that had Senior written this verse, it is highly unlikely that he would have referred to the fuel as coke, for early American locomotives burned wood or, later, coal. Further, he would have referred to a check valve rather than to a clack.

The Alton inscription is now largely illegible; the text shown below is as given by Pike:

> My engine now is cold and still,
> No water does my boiler fill.
> My coke affords its flame no more,
> My days of usefulness are o'er.
> My wheels deny their wonted speed,
> No more my guiding hand they heed;
> My whistle—it has lost its tone,
> Its shrill and thrilling sound is gone;
> My valves are now thrown open wide,
> My flanges all refuse to glide;
> My clacks—alas! though once so strong,
> Refuse their aid in the busy throng;
> No more I feel each urging breath,
> My steam is now condensed in death;
> Life's railway o'er, each station past.
> In death I'm stopped, and rest at last.

The small differences between this and the other versions will be noted.

George Senior was twenty-nine years of age when he died. The other three men were twenty-seven, twenty-eight, and twenty-nine years old respectively. This is a young man's epitaph.

The Public Be Damned

The public be d——d. I don't take any stock in that twaddle
about working for the good of anybody but ourselves.
I continue to run the 'limited express'—well, because I want to.

William Henry Vanderbilt, in response to a question from
a reporter from the *Chicago Herald*; reported in that newspaper
on Monday, 9 October 1882, and in the *Fort Wayne Daily Gazette*
of 10 October 1882 (entry 542.2).

This is one of several versions of this infamous but variously
reported comment. It may not be possible at this remove to
determine if it is an accurate record of what Vanderbilt said,
but it is used as the appropriate entry in this book for the
reasons given below. The comment was reported to have
been made in one of at least two interviews that Vanderbilt
gave during a visit to Chicago; collectively they have been
the source of a variety of stories of what Vanderbilt said,
when, where, and in what circumstances. The key interview,
from which these words come, was described as having
taken place at the Grand Pacific Hotel in Chicago on the
evening of Sunday, 8 October 1882. The name of the
reporter was not given in the article in the *Herald,* but it is
clearly stated that the *Herald* sent their own reporter.

Edward P Ripley, president of the Atchison Topeka &
Santa Fe, said in a public speech in 1909 that "Mr Vanderbilt
subsequently denied having said it, but whether he did or
not and whatever may have been the provocation, the
phrase has for nearly forty years been used as indicative of
the railway man's attitude towards his patrons."

There were reports of other press interviews granted by
Vanderbilt during this trip. His special train left New York
during the evening of Friday, 6 October 1882, and stopped in
Detroit from 4 p.m. on Saturday, 7 October until 9 a.m. on
Sunday, 8 October. The train stopped in Michigan City
(fifty-six miles from Chicago) to change engines, and
arrived in Chicago at 5 p.m. on the same day. On the next
day, Monday, 9 October, the *Chicago Tribune* printed a
detailed account of an interview given by Vanderbilt. This
report thus appeared on the same day as that in the *Herald*.
No reporter's byline is given, but it appears that he was Clarence Dresser, at that time a keen young reporter who specialized in "railway news." The *Tribune* reporter stated that,
upon presenting his card at Michigan City, he was invited to
join the party and to travel to Chicago on the train. The
interview took place during the remainder of the journey to
Chicago, and therefore preceded the interview granted to
the *Chicago Herald*. There was an exchange between Dresser

and Vanderbilt about limited express trains, but without
any indication of damning the public: when asked about the
profitability of the limited express trains, or whether he ran
them for the benefit of the public, Vanderbilt said "Accommodation of the public? Nonsense, and they do not pay
either." He made it clear that he ran them because he had to
meet competition from other roads. The *Tribune* reporter
made no mention of any other reporter being present.

W A Croffutt, in "The Vanderbilts" (1886), refers to Vanderbilt's words, stating that he and the reporter had an
exchange about whether or not the fast mail-train paid. The
reporter asked if Vanderbilt was working for the public or
his stockholders. The reply, as reported by Croffutt, was
"The public be d——d. I am working for my stockholders! If
the public want the train, why don't they support it?" It is
clear that Croffutt obtained this version from Samuel Bar-
ton, a nephew of Vanderbilt, who was in the party that day.
Croffutt did not name the reporter, but referred to his story
as vicious, omitting all context and surrounding
circumstances.

On 25 August 1918 the *New York Times* published a letter
from Ashley Cole, a journalist who had been a friend of
Clarence Dresser and who described Dresser as being in the
employ of the *Tribune*. This letter is important because it
indicates the tone of Vanderbilt's comment. Cole, writing, it
will be noted, some thirty-five years after the event, said that
he had met Dresser in Chicago a few days after the exchange
with Vanderbilt had been reported. It came up in conversation (Cole having noted that Dresser had "chronicled the
remark in The Tribune"), and Dresser was at some pains
to explain that the comment had been made in fun, not
defiance. He said he had asked Vanderbilt if the New York
Central would put on a luxury train to equal that of the
Pennsylvania Railroad (which was running a luxury train
from Philadelphia to Chicago), on the assumption that the
public would "demand it of the New York Central." It was
this idea that amused Vanderbilt, who was reported as looking at Dresser laughingly "in his usual good-tempered way,

his eyes sparkling, and replied, musingly, 'Oh, the public be damned!' " He went on to say that the administrative and operating staff would determine if a new train should be run, not the public. The whole exchange was distinguished by pleasantry and, it seems, smiles all round. This is in contrast to the *Herald* interview, in which it seems clear that Vanderbilt was not in a good mood.

This letter is not entirely helpful because it contains a number of statements and assumptions which are inconsistent with Dresser's original report in the *Tribune*, not least the fact that Dresser's report did not use the emotive word " 'damned.' " Why did Cole even raise the matter with Dresser when it was not Dresser who had reported the use of this word? Even if Cole read the word in the *Herald*, asking Dresser about it would be pointless, as his interview with Vanderbilt was quite separate in time and place. Was Dresser trying to take credit for getting a socially unacceptable word out of Vanderbilt, or for having been at the heart of an infamous conversation? Or was Dresser, seemingly keen to stress that the comment was in good fun, trying to make up for the damage that the report had done in the following years?

Subsequent references to the interview are less helpful still, but they have probably served to add to the mythology of Vanderbilt's original words and intent. In *Fifty Years a Journalist*, 1921, Melville Stone stated that the infamous comment was made when journalist Clarence Dressler [*sic*] boarded Vanderbilt's private car when he was taking dinner, presumably on the Saturday evening in Detroit. Vanderbilt told him to wait for an interview, but Dressler protested, "But it is late and I will not reach the office in time. The public—" Vanderbilt interrupted him, understandably: "The public be damned; you get out of here."

Much later, Alvin F Harlow in *The Road of the Century*, 1947, gave two versions of the story. He stated that the exchange featuring the infamous phrase took place in Vanderbilt's private car on the *morning* of 8 October, in Michigan City. (If the timings of the train noted in the account in the *Chicago Herald* are correct, Michigan City would almost certainly not have been reached until the afternoon of that day.) In his account, Harlow said that two reporters were present, these being John D Sherman of the *Chicago Tribune* and Clarence P. Dresser of *City News*, which was described as "a Chicago news organization." This allocation of names to employers is at variance with the 1918 letter from Cole. If Stone's account is correct, it would appear that Dresser was trying a second time to see Vanderbilt, perhaps on Sherman's coattails. According to Harlow, Sherman said that he asked Vanderbilt, "Do your limited express trains pay, or do you run them for the benefit of the public?" to which Vanderbilt replied, "The public be damned! We run them because we have to. They don't pay." This does not agree with the language in the account printed in the *Chicago Tribune* the following day, whether the *Tribune* reporter was

Sherman (as stated by Harlow), or Dresser (as stated by Cole and Stone). According to Harlow, Dresser said that Vanderbilt's answer was "The public be damned! What does the public care for the railroads, except to get as much out of 'em for as little as possible." Either way, other newspapers took the story, but it was principally the *Chicago Herald*'s version, reordered a little and using some of the *Tribune* article, that appeared in the *New York Times* on 9 October 1882. Harlow, in telling these two versions, did not state his sources.

In *The Robber Barons*, published in 1934, Michael Josephson stated, without citing any particular newspaper account, that Vanderbilt said, "The public be damned. I am working for my stockholders. If the public want the train why don't they pay for it!" These words appear to have come, though not with total accuracy, from Croffutt's 1886 account, and were repeated precisely by Taylor Hampton in *The Nickel Plate Road*, published in 1947.

And so we see here two contemporary newspaper accounts, both of them to be regarded as primary sources of what Vanderbilt said on two separate occasions, and five later accounts, all secondary sources, some referring to the *Tribune*, none of which agree with each other, and which, because they record the use of the word "damned," are all inconsistent with what was reported in the *Tribune* at the time. The Stone and Harlow accounts, however, show wholly different circumstances from that recorded by the *Chicago Herald*, and there can be little doubt that there were two separate interviews. Both were extensively reported— that of the *Herald* runs to about thirteen hundred words— and go into detail on a number of issues. It is not surprising, therefore, that separate reporters asked similar questions. There is little doubt that Vanderbilt damned the public in the *Herald* interview, for this is hardly something that their reporter would have presumed to invent. Although that interview was given later than that to the *Tribune*, it is cited herein as the source of the infamous words because the contemporary *Tribune* account does not attribute that language to Vanderbilt.

Ripley's description of the importance of the quotation and its consequences is not exaggerated, which is why so much prominence is given to it here, even though Ripley's mention of "nearly forty years" is rather adrift with the facts as reported elsewhere: an elapsed forty years would date it at about 1870, six years after Vanderbilt had joined the board of the then New York & Harlem Railroad.

Bartlett's Familiar Quotations refers to the Stone account and to one by Gustavus Myers. The *Oxford Dictionary of Quotations* (various editions) gives the source as the letter from Ashley W Cole to the *New York Times* on 25 August 1918. Its sister publication, the *Oxford Dictionary of Humorous Quotations*, gives no source, cites the words from Hampton, and gives the incorrect date of 2 October. It does not explain why it considers the words humorous.

Index of Writers and Speakers

PUBLICATIONS

Index of Subjects and Key Words

The compiler particularly wishes to acknowledge the kind help of those who have freely allowed him to reproduce material from their publications. Special thanks are due to Hal Carstens, publisher of *Railroad Magazine*, for permission to reproduce a number of poems that appeared in that magazine, and to Jim Wrinn and Kevin O'Keefe of *Trains Magazine* for helping him to find a number of contributors to their magazine in years past. Credits are listed by entry number.

Extracts from *Collected Poems* and other works by John Betjeman (11.5, 16.1, 28.25, 185.1, 190.1, 428.3–4, 429.12, 477.3, 482.1, 512.1, 541.1, 579.2, 615.9, and 712.24) are reproduced by permission of John Murray Publishers.

The extracts from "I Thought About You" (5.1–2) were written by Johnny Mercer, © 1944 Mercer-Morris Inc, USA. Warner/Chappell North America, London W6 8BS. Reproduced by permission of International Music Publications Ltd. All rights reserved.

Extracts from "Slow Train," by Michael Flanders (entries 44.8–10), are © the Estates of Michael Flanders and Donald Swann and are reproduced by permission of International Music Publications Ltd. All rights reserved.

Extracts from "Blackout" (55.1), by A A Milne, "Bradshaw's Guide" (66.10), by Henry S Leigh, and the anonymous "Crewe" (157.3–5), are reproduced with the kind permission of *Punch* Ltd.

The extract from "The Song of the Caboose" (89.7), by A W Munkittrick, is reproduced by kind permission of Tom Palmer of the Association of American Railroads.

The extract from "Rock Island Line" (121.1) is as written by Huddie Ledbetter in 1934 and is used by permission, © 1959 (renewed) Folkways Music Publishers Inc., New York, N.Y.

The extract from "A Visit from St. Click" (125.1), © Ben L. Dibble, is reproduced by permission granted on website.

The extract from the hymn "God of Concrete" (171.3), by Richard G Jones, is reproduced by permission of Stainer & Bell Ltd.

The extract from "Deltics" (174.2), by Chris Rea, is © 1976 Magnet Music Ltd, and is reproduced by permission of International Music Publications Ltd. All rights reserved.

The extract from "Extra 728 east" (184.11), by Don Buchholz, is reproduced by kind permission of Don Buchholz.

The extract from "Brother Can you Spare a Dime?" (187.1), lyrics by E Y "Yip" Harburg and music by Jay Gorney, is published by Glocca Morra Music (ASCAP) and Gorney Music (ASCAP), and is used by permission. Administered by Next Decade Entertainment, Inc. All rights reserved.

The extract from the song "Fuel Form Blues" (216.16), by Philip Larkin, is reproduced by kind permission of the Society of Authors as the Literary Representative of the Estate of Philip Larkin.

The extracts from "Right of Way" (217.2 and 620.14), by Henry Herbert Knibbs were published in *Songs of the Trail* (Boston: Houghton Mifflin, 1920) and are reprinted by permission of the publisher.

The untitled poem by Margaret Brannagan (218.8) is reproduced by kind permission of Ian Allan Publishing Ltd., publishers of *Trains Illustrated*.

The extract from the untitled poem by Lee Avery (257.14) was first published in *The Saturday Evening Post*, September 14, 1957, and is used with permission.

The extracts from "Freight Train, Freight Train" (257.18 and 746.34), by Alvin Greenberg, are published by kind permission of Alvin Greenberg.

The extracts from "Freight" (293.25 and 684.15), by John Frederick Nims, are reprinted by permission of Louisiana State University Press from *The Powers of Heaven and Earth: New and Selected Poems by John Frederick Nims*. © 2002 by Bonnie Larkin Nims, Frank McReynolds Nims, Sarah Hoyt Nims Martin, and Emily Anne Nims.

The extract from "The River" (294.1) is by Hart Crane and comes from *Complete Poems of Hart Crane,* edited by Marc Simon. © 1933, 1958, 1966 by Liveright Publishing Corporation. Copyright © 1986 by Marc Simon. Used by permission of Liveright Publishing Corporation.

The extract from "King of the Road" (294.3), by Roger Miller, is reproduced by permission of Sony/ATV Songs LLC. © 1964 (Renewed). All rights administered by Sony/ATV Music Publishing, 8 Music Square West, Nashville, TN 37203. All rights reserved.

The extracts from "Old Buddy Goodnight" (294.4), "Phoebe Snow" (390.3–4), "Daddy, What's a Train?" (695.23–24), and "No Round Trip Ticket" (740.3–4), by Utah Phillips, are reproduced by kind permission of Utah Phillips and are © Utah Phillips On Strike Music.

Andrew Dow was born in 1943 into a railway family in Hitchin, England, where his parents worked at the wartime headquarters of the London & North Eastern Railway. He spent thirty years in the aviation industry before being appointed Head of the National Railway Museum in York in 1992.

He joined British Rail as privatization started, in 1994. He played a key role in the successful management buyout of one of the civil engineering units. This company expanded quickly and was highly successful; its resulting sale to another company enabled Dow to take early retirement.

He has participated in a wide variety of railway activities. His first book, *Norfolk & Western Coal Cars* (TLC, 1998), was published in the United States and widely acclaimed. His second book, *Telling the Passenger Where to Get Off* (Capital Transport, 2005), about the evolution of the railway diagrammatic map, in which his father was a key participant, was published in England. He is a fellow of the Royal Society of Arts, president of the Stephenson Locomotive Society, and vice chairman of the Gresley Society.

Dow lives with his wife, Dr Stephanie Dow, who teaches at the Open University, in Newton-on-Ouse, near York. They have a son and a daughter.